Applied Demography Series

Volume 9

Series Editor
David A. Swanson

The field of applied demography is largely driven by the quest for the knowledge required by clients, both in public and private sectors, to make good decisions within time and costs constraints. The book series, Applied Demography, provides a forum for illustrating and discussing the use of demographic methods, concepts, and perspectives in a wide range of settings—business, government, education, law, and public policy—as well as the influence of these settings on demographic methods, concepts, and perspectives. The books within the series can be used as resources for practitioners and as materials serving as case studies for pedagogical uses.

More information about this series at http://www.springer.com/series/8838

David A. Swanson
Editor

The Frontiers of Applied Demography

Springer

Editor
David A. Swanson
Department of Sociology
University of California Riverside
Riverside, CA, USA

ISSN 2352-376X ISSN 2352-3778 (electronic)
Applied Demography Series
ISBN 978-3-319-43327-1 ISBN 978-3-319-43329-5 (eBook)
DOI 10.1007/978-3-319-43329-5

Library of Congress Control Number: 2016957789

Printed on acid-free paper

This Springer imprint is published by Springer Nature
The registered company is Springer International Publishing AG Switzerland

Foreword

Applied demography encompasses a diverse set of endeavors that draw on demographers' specialized knowledge, technical skills, and extensive familiarity with relevant data sources. These proficiencies have enabled demographers to address an ever-wider array of business and public sector concerns. The 23 chapters ahead demonstrate the expanding scope and advancing frontiers of endeavor by applied demographers worldwide.

Applied demography is driven by practical problems; it is not a theory-directed body of knowledge. The wide-ranging concerns addressed in the chapters ahead illustrate the field's scope and emphases. Prominent among those emphases are shaping decisions attuned to a stubbornly uncertain future that resists precise prediction, strengthening the understanding of demographic influences on consumer behavior and service demands, and evaluating the quality of the data applied demographers routinely use so as to better understand and skirt their limitations.

"Frontiers" can be thought of as outer limits in a field of endeavor, from which opportunities yet to be realized can be perceived. The varied contributions ahead define three such frontiers: (1) systematic evaluations of how the status quo evolves, (2) alternative postures for accommodating the future, and (3) emerging data needs presented by the new problems applied demographers are now addressing.

This wide-ranging collection of applications will enjoy a broad and varied readership extending well beyond demographers themselves. It will include analysts gauging the size and makeup of consumer markets at different geographic scales or stages of evolution, municipal planners responsible for designing policies to meet local service demands, analysts framing strategic business decisions with reference to future demographic contexts, economic forecasters concerned with alternative future scenarios of how the demographic status quo might unfold, and government officials heading agencies that collect public data through a periodic national census or ongoing surveys.

In what follows, I offer an overview of how these chapters align with the above concerns, to help direct readers to those chapters likely to be of greatest interest.

Measuring the Present

The production of ever-better "official" current estimates of population size and composition addresses the ongoing needs of national, regional, and local governments. Those estimates figure planning decisions and various formulas for transferring resources from higher levels of government to lower ones.

Traditional censuses are among the largest and costliest statistical activities undertaken by governments and national statistics offices. Chapter 14 examines the prospects for future cost savings as Canada and the United States incorporate technological advancements into census methodologies.

In the past decade, all 50 US states and outlying territories have developed longitudinal administrative records databases, which are essentially partial population registers. Chapter 11 describes the emergence of these State Longitudinal Data Systems and provides a road map for future research possibilities.

Chapter 13 explores the utility of lifetime US migration measures, which link an individual's current place of residence with place of birth. These two building blocks show a basic picture: where a person began the musical game of life and where that person now resides, once the music has stopped. The analysis of lifetime migration is a useful complement to more detailed, shorter-term measures. It provides a long-term perspective on how each state acquired its distinctive mix of population from different geographic origins. These insights to a state's population history can provide useful input into policy and program planning.

Chapter 20 introduces a new approach to analyzing population change over small areas. It makes use of a combined polygon overlay and cell smoothing procedure to derive estimates of population counts for a 1km grid to facilitate assessment of change through time. Population surfaces derived in this way make the most of the available information, according to the author, and provide a sound basis from which to explore longer-term change over very small areas.

Chapter 9 examines the meaning of reported race among different Hispanic groups. It highlights the underlying complexity of Americans' racial and ethnic self-identification and advances the understanding of the responses that census and survey questions elicit from persons of Hispanic origin. The US Census Bureau collects information concerning race and Hispanic origin using two separate questions. Hispanics may report any race or combination of races. Most select a specific race, but some Hispanics report a nationality, a group, or a general category as their "race."

Envisioning and Anticipating the Future

No view of the future can be anything more than an "informed guess." Many non-demographers seek precisely that: an informed guess about how the demographic status quo may evolve. Formulating alternative scenarios of the future is one way to incorporate key demographic contingencies shaping the future, and applied demographers are well situated to offer objective inputs.

Chapter 1 illustrates these possibilities. Both China and India have undergone substantial consumer market expansion. Each, however, is following a distinctive demographic pathway forward into the future. This chapter examines the macro demographic trends behind those pathways and spots demographic contingencies on the horizon. Its insights offer a useful point of departure for envisioning alternative scenarios of how productivity, household purchasing power, and discretionary spending may ripen in other developing country contexts.

Population aging in England is covered in Chap. 2. Here, the authors drill down into the components that are presumed to cause burdens on health-care systems due to an aging population. It does this by employing micro-simulation to examine the health impacts of total population change, population aging, changing ethnic mix of the population, and trends in the age-specific incidence of disease. The results are then used to project the future with implications not only for England as a whole but also for local areas.

Alternative scenarios are no less relevant for understanding demographic contingencies at state and local levels. Applied demographers often are called upon to estimate how a proposed change to public policy would affect future demand for a particular benefit or service. Chapter 3 illustrates this problem via a case study of a long-standing program to assist military veterans and their dependents to obtain college degrees. The authors evaluated alternative proposals for modifying eligibility requirements based on length of military service. Their analysis illustrates how alternative scenarios help to uncover disparate impacts imposed by particular proposals.

Chapter 16 offers a further illustration of how scenarios serve planners' needs. Integrated population, housing, and jobs forecasts involve describing not a single future but a variety of scenarios, to reflect both the uncertainty about the direction of recent trends and the impact of planners' and politicians' conscious intent to *alter* a trend. The approach described here uses standard demographic forecasting methods to foresee the local demand for housing and employment embodied in the current population's growth and age structure. It then calculates the impact on that population of alternative plans for supply of housing or employment.

Population projection is narrowly construed as the science of dissecting demographic change into its constituent processes and as the art of making informative assumptions about the future course of those processes. Projections become forecasts only in the eye of the beholder, who adopts their assumptions as being most probable. Several chapters illustrate the central role of population projections in guiding decision-making and how applied demographers are advancing both the science and the art of projection.

Projections of future population play an important role in supporting planning for the future provision of services and infrastructure, for budgetary allocation, and for staffing publicly provided services. Chapter 5 offers a particular illustration:

anticipating the future demand for Australia's criminal court services and facilities and the future need for related staff, such as judges and magistrates. Crime rates in Australia differ widely between gender, age, and geographic area. The author derives court appearance rates by age, sex, and local area based on projected changes in the size, composition and geographical distribution of the population, and the volume and geospatial dimensions of crime-related court service provision.

Several chapters illustrate ongoing advances in evaluating the accuracy of population projections. Chapter 7 reports an effort to improve the accuracy of estimates and forecasts of the very elderly population of Australia—persons aged 85+. It develops probabilistic forecasts to 2051 by sex and single years of age up to age 110+ using extinct cohort and survivor ratio methods. Chapter 15 describes a detailed evaluation of Japan's official prefectural and municipal projections. Japan's aging and declining population poses future concerns about the repercussions of a shrinking working-age population and an increase in the elderly population. The authors identify factors associated with the accuracy of prefectural and municipal population projections based on a multivariate regression analysis. Chapter 21 undertakes an evaluation of the geographically weighted regression (GWR) method to estimate relationships between population change and a variety of driving factors and consider possible spatial variations of the relationships for small-area population forecasting using 1990–2010 data at the minor civil division level in Wisconsin, USA. The results indicate that the GWR method provides an elegant estimation of the relationships between population change and its driving factors, but it underperforms traditional extrapolation projections. Chapter 23 reports a test of the accuracy of the Hamilton-Perry method for forecasting US state populations by age. This method's minimal data input requirements and its capability to produce age and other characteristics in a forecast are attractive features for use at state and county levels of geography. Chapter 22 proposes a new method for small-area estimation and examines its efficacy and applicability to long-term population projection.

Responsibility for preparing demographic projections for local government areas often rests with local government authorities. Australia offers a useful model for systematizing data assembly for preparing local population projections, as detailed in Chap. 17. The author describes how local governments assemble basic demographic information on dwellings and population, supplemented by property tax and property development databases, Google Street View, and regular aerial view updates to validate demographic forecasts. Chapter 18 identifies and addresses the further issues applied demographers confront when forecasting individual ethnic groups within the populations. The authors incorporate the ethnicity dimension into projection models for local authorities in the United Kingdom. They present a useful checklist of design decisions that arise when building a projection model for subnational populations classified by ethnicity.

Accommodating the Future

In the final analysis, forecasting rests on the faculty of human judgment, and it requires that the future be thought about systematically—and with imagination. Several chapters illustrate the specialized roles applied demographers play in formulating responses to forthcoming change.

Chapter 4 identifies the causes of fire injuries and risk factors and potential preventative measures. Chapter 6 is a case study of a small island community that needed to modernize its outdated facility dedicated to providing assisted living and skilled nursing care to local elderly residents. It illustrates how applied demographers can inform important public choices by drawing attention to the long-range implications of demographic change. The study draws upon publicly available data and standard demographic accounting to show what the future holds and the trade-offs it might impose.

Chapter 12 applies newly devised measures of physical access to primary health care to reveal the hidden barriers to US health care. The question it addresses is: Can an individual with health insurance obtain an appointment with a primary health-care practice? The study shows that simply having health insurance does not equate to access to primary care. It further suggests that barriers to access to primary care (as measured here) actually are lower in rural places than in metropolitan places. It appears that lower operating costs in rural areas, plus smaller patient populations, may provide the financial incentives for physicians to accept Medicaid patients.

Evaluating Data Quality

Data are at the heart of applied demography, which is why its practitioners devote considerable effort to understanding the limits imposed by the data they use. Chapter 8 pursues a puzzling anomaly in historical US Census data: errors in the coverage of Black male cohorts since the late 1930s. Here, the US Census Bureau analysts explore the potential causes of this coverage error and its consequences for the Bureau's current population estimates. Chapter 10 addresses known limitations in the Puerto Rico Community Survey (PRCS) due to the questionable quality of the base population and household estimate controls for Puerto Rico. The authors explore ways to work around these limitations and propose ways to incorporate PRCS data in future estimates despite its known biases. Chapter 19 reports on Australia's efforts to use information from its census post-enumeration survey to correct subsequent population estimates. Its results point toward ways to improve the quality of Australia's population estimates from the 2016 census and beyond.

RAND
Nantucket, MA
USA
e-mail: petermorrison@me.com

Peter A. Morrison

Preface

It was a pleasure to work with the chapter authors and the Springer team to bring this book into existence. It not only covers a wide range of topics of interest to applied demographers and others but provides a wide range of perspectives from around the world while doing so. The topics and perspectives are found in the book's 23 chapters, which are organized into three major sections: (I) "Demographic Information for Decision-Making: Case Studies," (II) "Data: Issues and Analyses," and (III) "Projection and Estimation Methods: Evaluations, Examples, and Discussions."

Part I consists of seven chapters. The topics range from a demographic market analysis of China (Chap. 1) to the future of assisted living in the United States (Chap. 6). India, China, and the United States are not the only geographic areas covered. Chapter 2 deals with health trends in the United Kingdom, while Chap. 7 deals with the elderly population in Australia, and Chap. 5 discusses the future need for courtroom services in Australia. At the subnational level, Chap. 3 looks at the use of veterans' benefits in Texas, while Chap. 4 looks at fire injuries in Anchorage, Alaska.

Part II also consists of seven chapters. Two chapters examine the work of the US Census Bureau at a national level (Chaps. 8 and 9), while one chapter does so at the subnational level (Chap. 10). Two chapters in Part II tackle state-level data issues. In Chap. 11, a newly implemented longitudinal database in Washington is described, while in Chap. 12, access to health care in Mississippi is described. Chapter 13 uses US Census data to examine lifetime migration in the United States, while Chap. 14 looks at census costs in Canada and provides suggestions on how they might be reduced.

What has been described as the "core" of applied demography is covered in the nine chapters found in Part III, which deal with population estimation and projection methods. Two of the Chaps. 15 and 22, are focused on Japan with the former conducting a cross-national comparison of accuracy and the latter introducing a new approach to the development of small-area estimates and projections. Chapters 16 and 21 also introduce new perspectives on small-area population projections using data from the United Kingdom. Chapter 17 examines the challenges of developing population projections for the state of Queensland in Australia, while Chap. 19 takes a look at the validity of historical population estimates for Australia as a whole. In Chap. 18, the challenges that face the United Kingdom in developing population projections by ethnicity are discussed along with suggested solutions. Turning back to the United States, Chap. 21 considers the effect of geographically weighted regression models on the accuracy of small-area population forecasts, while Chap. 21 uses decennial census data for a sample of states over a 100-year period to examine the accuracy of the Hamilton-Perry method of forecasting a population by age.

Riverside, CA, USA David A. Swanson

Contents

Part II Data: Issues and Analyses

Part III Projection & Estimation Methods: Evaluations, Examples and Discussions

Part I
Demographic Information for Decision-Making: Case Studies

Chapter 1
Demographics and Market Segmentation: China and India

Jo M. Martins, Farhat Yusuf, Gordon Brooks, and David A. Swanson

Abstract China and India are the two most populous countries in the world but have followed different demographic courses. Both countries have experienced substantial expansion of their markets for a range of commodities. However, dissimilar household composition and socioeconomic paths have affected household preferences in the two countries. The paper reviews macro demographic trends that have led to different demographic structures with significant implications for productivity and household purchasing power and discretionary spending in the two countries. It then conducts an examination of household expenditures based on household surveys undertaken in 2005 and assesses similarities and disparities in household preferences for broad categories of goods and services in rural and urban areas, and also for households with varying levels of income. This provides a basis for hypothesis building concerned with market growth for progressive commodities, in view of current demographic structures in the two countries and projected fertility and population growth.

Keywords Income • Household expenditures • Discretionary spending • Demographic trends

J.M. Martins (✉) • G. Brooks
Macquarie University, Sydney, Australia
e-mail: jmartins@tpg.com.au; gordon.brooks@mq.edu.au

F. Yusuf
University of Sydney, Sydney, Australia

Macquarie University, Sydney, Australia
e-mail: farhat.yusuf@sydney.edu.au

D.A. Swanson
University of California Riverside, Riverside, CA 92521, USA
e-mail: dswanson@ucr.edu

D.A. Swanson (ed.), *The Frontiers of Applied Demography*, Applied Demography Series 9, DOI 10.1007/978-3-319-43329-5_1

1.1 Introduction

China and India have large populations but they have followed different demographic paths and socioeconomic development. The two populations have experienced substantial growth in income over the last two decades. In view of their relative importance to world markets, it is useful to gain insights into the characteristics of the two markets and how their differences in population and socioeconomic development have affected household choices and market penetration for different types of consumer goods and services. Both countries carried out household consumer expenditure surveys in 2005 that can be used to provide a preliminary examination of household preferences and market segmentation for the range of household consumer goods and services.

1.2 China and India: Demographics and Socioeconomic Characteristics

China and India are the two most populous countries and constituted about 38 % of the World's population in 2005. However, they differed substantially in their demographic structure that influence their economic and market characteristics.

In the context of about 2005, India's fertility rate of 3.1 children per woman is considerably higher than China's 1.7 that is well below the replacement level of 2.1. The difference in fertility has influenced a number of factors. It has resulted in a much higher annual rate of population growth in India (2.0 %) than in China (0.75 %). *Ceteris paribus*, this has retarded the growth in income per capita in India in comparison with China and average household purchasing power. China's dependence rate of 0.41 reflects a much lower proportion of dependent children and a larger proportion of its population in the more economic productive ages of 15–64 years. The lower proportion of children to educate could have facilitated greater average investment in the education of individual children and help raise literacy rates in China (91 % of the adult population) in comparison to India (61 %). This raised one dimension of the quality of human capital in China more than in India and its potential productivity. The larger proportion of women participating in the formal economic sector in China (69 %) than India's (34 %) must also have raised overall economic productivity in China (Table 1.1).

In purchasing power parity terms using the United States as a reference, China ranks second in the World in the size of Gross Domestic Product (GDP). India ranks fifth. Together they represent about 15 % of the world's GDP (WB 2008). This is considerably less than their proportion of the World's population (38 %) and reflects a lower GDP per capita than the more developed countries. Both countries have experienced relatively large growth in GDP per capita in recent years, but China's growth rate was twice the size of India's in the period 1990–2005, and its GDP per capita is almost twice that of India's. Inequality in income distribution as

Table 1.1 China and India demographic and socioeconomic indicators 2005

Indicator	China	India
Population (million persons) (2005)	1313.0	1134.4
Yearly population growth rate (%) (1975–2005)	1.2	2.0
Population age distribution (%) (2005)		
0–14	21.6	33.0
15–64	70.7	62.0
65+	7.7	5.0
Total fertility rate (no. child. per women) (2000–2005)	1.7	3.1
Life expectancy at birth (years) (2005)	72.5	63.7
Dependency rate (child & aged pop./pop. aged15–64) (2005)	0.41	0.61
Urban population (% of total) (2005)	40.4	28.7
Adult literacy rate (% of pop. 15 years & older) (1995–2005)	90.9	61.0
Female economic activity participation rate (15 years & older) (%) (2005)	68.8	34.0
Gross domestic product (billions PPPUS$) (2005)	5333.2	2341.0
GDP per capita (PPPUS$) (2005)	4091	2126
GDP p.c. PPPUS$ yearly average growth rate (%) (1990–2005)	8.8	4.2
Gini index of income distribution (2005)	46.9	36.8

Sources: UNDP (2007) and WB (2008)

measured by the Gini coefficient is greater in China than India. All these factors have implications for the size and characteristics of their markets (Table 1.1).

1.3 Literature Review

There is a wealth of references in the literature on China and India individually and comparisons of economic and social development in China and India. However, a search of bibliographic data bases for material related to consumer behaviour for each country is less abundant. Yusuf et al. (2008) in their review for China noted the paucity of published research in this area. The same can be said in the case of India. Further, comparisons between the two countries on these topics are scarce, especially regarding market segmentation, consumer classes and household expenditure preferences for alternative commodities. According to Sun and Wu (2004), much of the research focused on China is about specific groups, regions or the rural/urban dichotomies rather than on the nation as a whole. Consideration of the following review shows a similar pattern for research on India. More generally, the relevant literature tends to be focused on geographic distinctions within countries, rather than on countries as the unit of analysis. These are often framed around an urban versus rural distinction.

With regard to China, Dickson et al. (2004) focused on urban consumers while McEwan et al. (2006) examined urban versus other consumers and the importance of relative affluence in consumer behaviour. Cui and Liu (2000) researched the psychographic characteristics of urban and rural consumers in China. They studied geographical segmentation with differences in seven geographic areas in regard to education level, line of work and attitudes and product ownership. They found that income was a major variable in determining acquisition of communication appliances and computers. However, no distinction was made between urban and rural consumers' preferences. Sun and Wu (2004) studied the relationship between products that indicate a high standard of living and consumers' brand consciousness and attitudes in rural and urban areas in China. Housing rated highest both in rural and urban areas. There were differences in preferences for other products in rural and urban areas. In rural areas, colour televisions, telephones, washing machines and cars were the most valued items, in that order. While in urban areas, the order of preference was: personal computers, air conditioners, colour televisions and cars. Dickson et al. (2004) provided some evidence that there were differences between population segments in urban areas (Beijing, Guangzhou and Shanghai). They found six identifiable segments and that age, income, sex, employment and type of job were important factors in their segmentation of urban consumer markets for U. S. fashion types of clothing.

Yusuf et al. (2008) in their study of urban/rural segmentation in China noted that household expenditure in urban China between 1995 and 2003 grew about twice as much as that in rural China during the same period. The examination of household expenditure on eight broad commodity groups showed that the proportion on food declined both in urban and rural areas as household income grew in the period 1995–2003. However, the proportional reduction in food expenditures in urban households with higher income growth was greater than that in rural households. They reported that in a complementary way household expenditure on items other than clothing and household facilities and services rose. Thus, household expenditures on housing, education and medical services could be said to be progressive commodities whose expenditures increased more than proportionally as household income grew. This study also found that market penetration of consumer durables such as mobile phones, refrigerators and washing machines was substantially higher in urban Chinese households (about 90 %) than in rural households (about one sixth to one third). The rate of ownership of colour television sets of more than 130 % in urban households compared with half of that rate in rural areas. Motor cycles were the only consumer durable whose rate of ownership was greater in rural households than in urban areas. This study also found that car ownership was growing but rather low, at the time, in China. Yusuf et al. (2008) found that there were large differences in household income between urban and rural areas. The lower household income in rural areas led to a higher proportion spent on food and a lower acquisition of progressive commodities than in urban households.

Jussaume Jr. (2001) was concerned with *modernisation* and how it influenced preferences for different types of food in a Chinese urban context (Quingdao, Shandong Province). The analysis took into consideration age, family type, income levels, as well as attitudes, interests and behaviour in shopping for food. Most people never shopped in supermarkets and more consumed pork than other meats on a weekly basis. Almost half consumed vegetables but less than a fifth ate fresh fruit on a weekly basis. Higher incomes were associated with consumption of more meat and fresh produce. Lack of concern for prices led to a greater use of supermarkets and increased consumption of fresh vegetables and fruit. Age was not associated with the consumption of modern foods, as far as meat and produce were concerned. Households' interest in branded and imported foods was not associated with greater consumption of chicken or fruit.

With regard to India, Ray (2001) dealt with household responses to changes in prices and income in urban and rural India related to food and non-food items. He found that household preferences for clothing and non-food items behaved as that for superior (progressive) goods. In other words, household preferences for these items increased more than proportionately as their income rose. While household preferences for food items and domestic fuel and power declined more than proportionately in both rural and urban settings. He also found that changes in prices affected consumption of food and clothing in rural areas and domestic fuel and power in urban areas. Kumar and Sarkar (2008) used a sample of respondents to a questionnaire in five urban centres (Mumbai, Kolkata, Delhi, Chennai and Bangalore) to segment them into six behavioural groups. They found that age, household income and job stability led to differences in consumer behaviour. Considering two large urban centres in India (Kolkata and Mumbai), Goswami (2008) studied the willingness of consumers to pay a premium for pollution-free garments. Three segments were suggested by the study: light, dark and non-green consumers. The majority were considered light green: they were concerned with the environment but their commitment was not great. A rural versus urban distinction was also used by Kumar et al. (2009) to study the consumption of cereals and pulses by poor people in India. They suggested that improvements in cereal deprivation have taken place in rural areas but not much change was observed in urban areas.

Lancaster et al. (2008) examined urban and rural distinctions within three Indian states. Their study examined the issue of sex bias in the allocation of household expenditures. The identification method involved the additional presence of males or females in the household. The analysis controlled for household size and age. The study found that food was the major household expenditure item (more than half in all three states) followed by transport, domestic fuel and power, clothing, education, medical care and tobacco. In the case of the number of male and female children in the household, there was no associated bias in household expenditures for different items. However, bias was evident in the case of adults aged 17–60. The predominance of adult males in the household led to an increase in food expenditure in the state with middle income. It also tended to raise expenditures on tobacco and alcohol in rural areas of the poorest state. In this state, there was also a bias in favour of expenditure on education for males (11–16 years of age). The bias in education tended to diminish when intergenerational factors were considered. It was found that the higher the level of adult literacy, the lower the bias in expenditure on education for males and females. There was an actual bias toward education expenditure for females in the state that has a matrilineal inheritance custom unlike the rest of India.

In their study of food consumption in India, Jha et al. (2009) also used an urban versus rural distinction, examining the nutrient-income elasticities of calories and protein intake of rural households in India. They ascertained that as income rose, the calorie intake diminished and there was a small degree of substitution between wheat and other food.

Jaffrelot and van der Veer (2008) eschewed the urban versus rural distinction, addressing the relatively large and wealthy middle class segments of both China and India. Their review of middle class characteristics and consumption preferences suggested attitudes that combined cultural traits related to attachment to nationalistic roots but also to modernisation. The latter is reflected in preferences for progressive commodities and away from basics. Preferences in terms of leisure activities are viewed as a factor in the changes to land use in urban settings. A comprehensive national study (Subramanian et al. 2004) found that in India there was an association between expenditure on tobacco and education and income levels. Those with lower incomes and levels of education were more likely to consume tobacco than those with higher education and income levels. The study drew on a representative, cross sectional survey of 92,447 households from 26 - Indian states. While these authors recognised the urban versus rural distinction, it was not the focus of their study, but was utilised as a one of a number of control variables.

In summary, consideration of consumer preferences in China and India has tended to focus on the urban versus rural distinction, and demonstrated the utility of this approach for a range of purposes. However, the limited research available does not provide a comprehensive comparison between these countries utilising the urban versus rural distinction or founded on nationally representative data. This research addresses this limitation.

1.4 Data and Methods

The data used in the analysis came from a number of sources. The demographic and socioeconomic indicators in Table 1.1 for both China and India were sourced from the United Nations Development Programme (UNDP 2007) and the World Bank (WB 2008) to enhance consistency in definitions. The data on Purchasing Power Parities was sourced from the International Comparison Program for 2005 published by the World Bank (WB 2008), again, to improve consistency and comparability. Household expenditure for India was obtained from tabulations prepared by the National Sample Survey Organisation of the Ministry of Statistics and Programme Implementation from their quinquennial household survey on the subject (NSS 61st Round – July 2004–2005) (NSSO 2006, 2007a, b). The NSSO tabulations were prepared separately for urban and rural areas. They are presented as per capita estimates for a 30-day period. Household expenditures were estimated using average household size available. National aggregates were estimated using the number of urban and rural households as weights for the whole of India also available. Annual household expenditures for India were estimated using the equivalent of a 365-day period from the 30-day expenditures. Further household expenditure quintile estimates for urban and rural areas were grossed up, in the first instance, by using household size and then the relative weights of each 20th percentile and or decile also available. The data for China household expenditures was sourced from tabulations prepared by the National Bureau of Statistics of China from their Urban Household Income and Expenditure Survey conducted in 2005 and a separate household survey for rural areas in the same year. The annual estimates of household expenditure in both urban and rural areas were tabulated on a per capita basis. Household expenditures were estimated using household sizes available for both rural and urban areas. National household expenditures for China were estimated using, as weights, the number of urban and rural households in China. Household quintile expenditures were estimated using the number of households for given deciles as weights. Although the urban and rural household surveys use similar concepts, there is some lack of clarity regarding the boundaries of the household income levels in the two surveys. The urban estimates used income deciles and quintiles while the rural estimates refer to: low income, lower middle income, middle income, higher middle income, and high income, as "five equal parts". The authors have used these estimates as representing five household income quintiles. There is also some ambiguity regarding the definition of the estimates, the authors have used tabulations relating to "Household Living Expenditure" for both urban and rural areas in China (Table 1.2).

Both the Chinese and Indian surveys are large national probability and stratified samples to ensure comprehensive geographical and social representation and allow for the derivation of robust estimates. The authors have made efforts to make the composition of the broad groups of commodities in the two countries close. However, it is likely that some differences between the compositions of these groups have persisted and some caution should be used in the comparison of the two countries. The substantial difference in the proportional sizes of the residual *other goods and services* points in that direction. Despite the efforts made using the tabulations available, it has not been possible to include *telephone* and *postal services* in the category of *transport and communication, water* in *housing*, and exclude *bedding* from *clothing and footwear* for India's quintile household expenditures in urban and rural areas. Another constraint is the lack of standard errors of estimates. The authors used purchasing power parities in comparing the aggregated annual household expenditures in China and India. This follows the understanding that exchange rates used in international trade do not reflect the prices of most domestic transactions and that PPPs are an improved basis for such comparisons

Table 1.2 Number of households sampled: China and India 2005

	China	India
Area	No. households sampled	
Urban	54,496	45,346
Rural	68,190	79,298

Sources: NBS (2005); NSSO (2006)

(see Notes). The arc elasticity of expenditures used in the analysis of relative proportional rates of change in expenditure as total household expenditures rise follow conventional economic practice (see Notes).

1.5 Household Budget Expenditures and Allocations: China and India 2005

In 2005, China's average household expenditure expressed in international dollars (PPPs US$) of 4995 was 1.75 higher than India's 2850. China spent 39 % of the average household budget on food compared with 48 % in India (Table 1.3). This is in accordance with Engel's Law that states that the proportion spent on food decreases as household expenditure rises. The corollary of Engel's Law is that market opportunities for progressive commodities tend to rise as household expenditures move to higher levels.

Although caution needs to be used in comparisons of the two countries because of possible differences in definitions, the higher household purchasing power in China is reflected in household allocation to the two broad categories of *transport and communication* and *education and culture* that are substantially higher in China than India (Table 1.3).

The higher household purchasing power in China is reflected in the greater market penetration and household ownership of household appliances such as television sets, refrigerators, air conditioners and also of motor cycles (Table 1.4). There is no readily available data on motor cars from the 2005 rural survey in China to estimate motor car ownership for the whole of China.

Table 1.3 Annual household expenditure in China and India 2005

Expenditure category	Allocation of household expenditure	
	China	India
	%	%
Food	39.3	48.0
Housing, domestic fuel, light & power	11.5	12.4
Household appliances & services	5.2	4.0
Clothing & footwear	8.8	6.9
Transport & communication	11.7	7.5
Medical services	7.3	5.8
Education & culture	13.1	5.6
Other	3.1	9.7
All	100.0	100.0
Total household expenditure	**PPP US$ 4955**	**PPP US$ 2850**

Sources: NBS (2005) and NSSO (2007a, b)
Computations of the authors

Table 1.4 Ownership of household appliances and motor cycles in China and India 2005

Item	Number per 1000 households	
	China	India
Television sets	1206	367
Refrigerators & freezers	595	120
Air conditioners & coolers	443	81
Motor cycles	327	127

Sources: NBS (2005) and NSSO (2007a, b)
Computations of the authors

Table 1.5 Household expenditure in urban areas on food and durable goods as a proportion of total expenditure in China and India 2005

Household expenditure group	Food and durables as % total expenditure				Arc elasticity	
	China		India			
	Lowest quintile	Highest quintile	Lowest quintile	Highest quintile	China	India
Food	45.9	30.5	55.9	32.3	0.709	0.571
Durables	1.2	3.3	1.7	6.2	1.468	1.601

Sources: NBS (2005) and NSSO (2007a, b)
Computations of the authors

1.6 Urban Household Expenditure on Food and Durables: Income Segmentation

The analysis of the allocation of household expenditures to *food* and *durable goods* in urban areas in China and India shows the importance of relative levels of discretionary income in market penetration for durable goods. While the allocation of the household budget to food decreases as income rises, the inverse happens as allocation for *durable goods* rises markedly. India with a lower income per capita shows a greater decline in the proportion allocated to food and a steeper rise in the proportion spent on durable goods. This is reflected in the proportional increase allocated to durable goods for every additional unit of total household expenditure of 1.468 in China and 1.601 in India (arc elasticity of household expenditure) (Table 1.5).

The analysis of the ownership of household appliances by income/expenditure quintiles in urban areas both in China and India provide supporting evidence of the income segmentation of the market for durable goods (Table 1.6).

As noted earlier, market penetration for television sets, air conditioners and refrigerators is greater in China than India. Lower incomes in India have led to lesser ownership of these domestic appliances from the lowest to the highest income groups (quintiles). In accordance with the expenditure elasticities for durable goods, previously examined, the progression is greater in India than China. In other words, for every additional unit of household total expenditure in India a higher proportion is spent on durable goods and leads to higher ownership

Table 1.6 Ownership of household appliances and vehicles in urban areas in China and India 2005

Item	Units owned per 1000 households				Ratio Highest quintile/lowest quintile	
	China		India			
	Lowest quintile	Highest quintile	Lowest quintile	Highest quintile	China	India
Television sets	1127	1624	362	795	1.441	2.196
Air conditioners & coolers	271	1525	46	420	5.627	9.130
Refrigerators & Freezers	783	1102	28	647	1.407	23.107
Motor cars	5	109	2	162	21.800	81.000
Motor cycles	181	273	26	501	1.508	19.269

Sources: NBS (2005) and NSSO (2007a, b)
Computations of the authors

by households with higher levels of income. The examination of the ownership of motor vehicles indicates that while market penetration of motor cars is higher in India than China, the situation is closer in the case of motorcycles. However, the progression from the lowest to the highest quintiles is much greater in India than in China for both types of vehicles (Table 1.6).

1.7 Household Consumption: Urban and Rural Segmentation

Average household expenditures in rural areas are about half those in urban areas both in China and India. The larger number of people in rural households, especially in China, further erodes their per capita household purchasing power and their allocation of expenditures on non-food items (Table 1.7).

The Engel index (expenditure on food as a proportion total household expenditure) is substantially higher for rural than urban areas in the two countries. The higher discretionary household expenditure in urban areas allows higher proportional allocations in urban areas to progressive categories such as *household appliances and service, transport and communication and education and culture* in the two countries. The difference in household allocations between rural and urban areas is particularly marked in India for transport and communication and education and culture. The characteristics of the housing markets in China and India must be substantially different and influence different patterns in household allocations. The same applies to clothing and medical services (Table 1.8).

An examination of the household consumption patterns as income rises clearly indicates that the allocation to food declines substantially in rural as well as urban

Table 1.7 Average household size and average annual household expenditure in urban and rural areas in China and India 2005

	China	India
Area	Average household size (persons)	
Urban	2.96	4.36
Rural	4.08	4.88
Rural/urban ratio	1.38	1.12
Rural/urban ratio	**Average annual household expenditure**	
	0.44	0.56
Rural/urban ratio	**Average annual household expenditure per capita**	
	0.32	0.52

Sources: NBS (2005) and NSSO (2007a, b)
Computations of the authors

Table 1.8 Household expenditure allocation in urban and rural areas: China and India 2005

	China		India		Ratio of proportions Rural/urban	
Expenditure category	Urban	Rural	Urban	Rural	China	India
	%	%	%	%		
Food	37	45	41	53	1.23	1.31
Housing, domestic fuel, light & power	10	14	15	10	1.43	0.68
Household appliances & services	6	4	4	4	0.79	0.88
Clothing & footwear	10	6	6	7	0.57	1.17
Transport & communication	12	10	10	6	0.78	0.54
Medical services	8	7	5	6	0.87	1.21
Education & culture	14	12	9	4	0.84	0.44
Other	3	2	10	10	0.63	1.02
All	100	100	100	100		

Sources: NBS (2005) and NSSO (2007a, b)
Computations of the authors

areas. The noted growing allocations to progressive commodities such as household appliances, transport and communication and education and culture are more substantial in rural than urban areas in both countries (Table 1.9).

The arc elasticities are usually larger in rural with lower household discretionary income than in urban areas. However, the pattern of progression for medical services that is substantial in India both in rural and urban areas is not marked in China with arc elasticities of about unit. Although clothing and footwear receive a larger proportion of the household budget in China than India. The rate of progression in urban and rural areas in China is relatively small and regressive in India (Tables 1.9 and 1.10).

Market penetration of *television sets*, *refrigerators* and *air conditioners* is substantially higher in urban than rural areas in both countries. As mentioned

Table 1.9 Expenditure patterns of urban households by income quintiles in China and India 2005

Expenditure category	Quintiles					Arc elasticity
	1st	2nd	3rd	4th	5th	
	Percentage of total household expenditure					
China urban						
Food	46	42	39	36	31	0.71
Housing, domestic fuel, light & power	12	10	10	10	10	0.89
Household appliances & services	4	5	5	6	7	1.30
Clothing & footwear	9	10	11	11	10	1.06
Transport & communication	8	10	11	11	17	1.38
Medical services	7	7	8	8	7	1.00
Education & culture	12	13	14	14	15	1.13
Other	3	3	3	4	4	1.28
India urban						
Food	56	51	47	42	32	0.57
Housing, domestic fuel, light & power	14	14	15	15	15	1.07
Household appliances & services	2	2	3	4	6	1.60
Clothing & footwear	8	8	7	7	6	0.76
Transport	2	3	4	6	9	1.66
Medical services	4	5	5	5	6	1.22
Education & culture	4	5	7	9	10	1.53
Other	11	11	12	13	16	1.23

Note: In the case of India, *housing* does not include *water, transport* does not include *postage* and *telephone services* and *clothing and footwear* include *bedding*
Sources: NBS (2005) and NSSO (2007a, b)
Computations of the authors

previously, household ownership of these appliances is greater in China than India, and the rural/urban ownership ratio is lower in India than China for TV sets and refrigerators and the inverse in the case of air conditioners. Household ownership of *motor cycles* is higher in rural than urban areas in China but the ownership of *motor vehicles* is not available for rural China, while in India the rural/urban ratio for both motor cars and motor cycles is rather low (Table 1.11).

1.8 Discussion

The discussion of findings must be guarded because of the constraints arising from the nature of the data used and should be viewed as preliminary findings. The authors are concerned with the possible inconsistencies in definitions in the two countries. The lack of standard errors of the estimates is another concern in assessing the significance of differences. Nevertheless, the large stratified probability samples used and the consistency of most findings with empirical evidence

Table 1.10 Expenditure patterns of rural households by income quintiles in China and India 2005

Expenditure category	Quintiles					Arc elasticity
	1st	2nd	3rd	4th	5th	
	Percentage of total household expenditure					
China rural						
Food	51	50	48	45	39	0.69
Housing, domestic fuel, light & power	13	13	14	14	17	1.23
Household appliances & services	4	4	4	4	5	1.23
Clothing & footwear	6	6	6	6	6	1.06
Transport & communication	7	8	9	10	12	1.49
Medical services	7	7	6	6	7	0.96
Education & culture	10	11	11	12	12	1.49
Other	2	2	2	2	2	1.39
India rural						
Food	60	59	57	55	46	0.75
Housing, domestic fuel, light & power	12	11	11	10	9	0.78
Household appliances & services	2	2	2	3	6	1.70
Clothing & footwear	9	9	8	8	7	0.74
Transport	2	2	2	3	6	1.84
Medical services	3	4	5	6	9	1.69
Education & culture	2	2	3	3	5	1.76
Other	11	11	11	11	12	1.11

Note: In the case of India, *housing* does not include *water*, *transport* does not include *postage* and *telephone services* and *clothing and footwear* include *bedding*
Sources: NBS (2005) and NSSO (2007a, b)
Computations of the authors

Table 1.11 Ownership of household appliances and vehicles in urban and rural areas China and India 2005

Item	Units owned per 1000 households					
	China		India		Ratio Rural/urban	
	Urban	Rural	Urban	Rural	China	India
Television sets	1348	1058	661	256	0.78	0.39
Refrigerators & freezers	974	201	319	44	0.21	0.14
Air conditioners & coolers	807	64	213	31	0.08	0.15
Motor cars	34	na	46	8	na	0.17
Motor cycles	250	407	260	77	1.63	0.30

Note: (na) not available
Sources: NBS (2005) and NSSO (2007a, b)
Computations of the authors

from other countries and generic theoretical frameworks are indications of the usefulness of these preliminary findings.

China and India are two large markets by any standards, if for no other reason than their large populations. However, their development has taken place against different demographic trends that have influenced their demographic structures. China has been favoured in terms of lower population growth (with implications for growth in income per capita) and an age structure with a lower proportion of dependent child population and a higher proportion in the more economic productive age of 15–64 years. It is also apparent that productivity in China could also have benefited from a higher literacy rate of its adult population and female participation in the formal economic sector.

The comparison of the two countries shows consistent findings that support the tenet of the importance of rising income per capita in the growth of markets for non-food items, especially in relation to more progressive commodities such education services, transport and communication, and consumer durables. Chinese households with a higher income spend proportionally less on food and more on these progressive commodities. Within each of the two countries, urban households also spend a lower proportion of their expenditures on food and a higher proportion on these items. Ownership of household appliances in the two countries supports the notion of considerable segmentation of markets for progressive commodities between urban and rural areas and between different income groups. In China, the large market penetration of television sets both in urban and rural areas might have been affected by government policies that favoured access to these appliances as a means of providing information. Similarly, the high market penetration of motor cycles and low penetration of motor cars might also reflect government priorities. Household preferences in the two countries show substantial similarities regarding progressive commodities but China's propensity to spend more on education is striking. The differences in literacy rates in the two countries could be partly affected by government policies but could also suggest relative household concern with education. The higher proportion of expenditure on clothing and footwear in China could be partly due to the larger proportion of China's population living in colder climates.

These findings point to greater potential for the growth of the market for progressive commodities in India than China. India's potential could be enhanced by the experienced and expected decline in fertility that should slow down population growth and reduce the proportion of dependent child population and increase that of the age groups in the more economic productive ages (UN 2007). However, even China with an Engel Index (expenditure on food as a proportion of all household expenditure) of 0.39 (compared with India's 0.48) shows considerable potential, as income per capita rises. Countries with higher income per capita such as the United States, Australia and Japan (WB 2008) have Engel's indices of respectively about 0.15, 0.16 and 0.23 (BOL 2007) (ABS 2006) (SB 2006) and considerably higher household expenditures on progressive commodities such as household appliances, transport and communication items.

1.9 Conclusion

Different paths of demographic and socioeconomic development have led to greater household purchasing power in China than India. This has affected the nature of their markets for the range of consumer goods and services. These markets reflect household preferences for progressive goods and services as their discretionary income rises and spend a lower proportion of the household budgets on basic commodities such as food. The Engel indices indicates that households in China have greater discretionary purchasing power than India's and households in rural areas in both countries with higher Engel indices also have lower discretionary spending on progressive commodities. Although this is a preliminary examination guarded by the constraints in the data used, it is clear that in both countries the markets are highly segmented in terms of income groups, and there are also substantial differences between urban and rural segments, partly because of differences in household income. In both countries, household discretionary spending on appliances, transport and communications and education and culture (recreation) reflect this segmentation. The segmentation of markets for these progressive commodities is supported by market penetration in terms of ownership of household appliances such as television sets, refrigerators and air conditioners, and also motor cars and motor cycles that is usually greater in urban than rural areas and households in the higher income quintiles. The relatively high Engel indices in the two countries in comparison with those of more developed countries indicate the potential for future growth in the markets for progressive commodities in China and India. Especially in India that is expected to slow down its population growth, reduce the proportion of dependent children in its population, and raise the proportion of people in more economically productive ages and possibly in women participation in the formal productive sector. This should enhance growth in productivity and income per capita and lead to higher household discretionary spending on progressive commodities.

Notes

Purchasing Power Parities (PPPs) Comparisons of market size and purchasing power in different countries are difficult. Some comparisons use currency exchange rates from international trade. However, most goods and services consumed in domestic markets are not involved in international trade. Consequently, exchange rates from international trade do not reflect the prices in most domestic transactions. To overcome these constraints purchasing power parities (PPPs) have been devised to that allow for the conversion of the rate at which one currency would have to be used to buy the same amount of goods and services in another country. For example, how much of the currency in Country A would be required to buy a kilo of rice in comparison with how much of the currency Country B is needed to buy a kilo of the same rice in Country B. The methodology is described in: Kravis, Irving B., Alan Heston and Robert Summers. 1982. *World Product and Income: International comparisons of real gross product.* Baltimore: The Johns Hopkins University Press.

A brief discussion of the issues involved is contained in:

Cullen, Tim.2007. PPP versus the market: which weight matters? *Finance & Development.* March 2007, Vol.44 (1).

Arc Elasticity Elasticity measures the proportional change in a given item (dependent variable) divided by the proportional change in the independent variable, in this case total household expenditure. Arc elasticity measures elasticity between two points: Allen, R.G.D. 1933. The concept of arc elasticity of demand. *Review of Economic Studies.* 1 (3): 226–229.

An equation used to measure arc elasticity is

$$\boldsymbol{ae} = [(y_i - y_0)/((y_i + y_0)]/[(x_i - x_0)/(x_i + x_0)]$$

where x is the independent variable (total household expenditure) and y is the dependent variable (particular expenditure item).

References

Allen, R. G. D. (1933). The concept of arc elasticity of demand. *Review of Economic Studies, 1*(3), 226–229.

Australian Bureau of Statistics. (2006). *Household expenditure survey – Summary of results – 2003–04.* Catalogue No. 6530.0. Canberra.

Bureau of Labor Statistics (BOL). (2007). *Consumer Expenditures in 2005.* Report 998, February 2007. Washington, DC: US Department of Labor.

Cui, G., & Liu, Q. (2000). Regional market segments of China: Opportunities and barriers in a big emerging market. *Journal of Consumer Marketing, 17*(1), 55–72.

Cullen, T. (2007, March). PPP versus the market: Which weight matters? *Finance & Development, 44*(1).

Dickson, M., Lennon, S., Montalto, C., Dong, S., & Li, Z. (2004). Chinese consumer segments for foreign apparel products. *Journal of Consumer Marketing, 21*(5), 301–317.

Goswami, P. (2008). Is the urban Indian consumer ready for clothing with eco-labels? *International Journal of Consumer Studies, 32*(5), 438–446.

Jaffrelot, C., & van der Veer, P. (Eds.). (2008). *Patterns of middle class consumption in India and China.* Sage Publications: New Delhi.

Jha, R., Gaiha, R., & Sharma, A. (2009). Modelling variety in consumption expenditure on food in India. *International Review of Applied Economics, 23*(4), 503–519.

Jussaume, R. A., Jr. (2001). Factors associated with modern urban Chinese food consumption patterns. *Journal of Contemporary China, 10*(27), 219–232.

Kravis, I. B., Heston, A., & Summers, R. (1982). *World product and income: International comparisons of real gross product.* Baltimore: The Johns Hopkins University Press.

Kumar, R. V., & Sarkar, A. (2008). Psychographic segmentation of Indian urban consumers. *Journal of the Asia Pacific Economy, 13*(2), 204–226.

Kumar, T. K., Mallick, S., & Holla, J. (2009, April). Estimating consumption deprivation in India using survey data: State-level rural-urban analysis before and during reform period. *Journal of Development Studies*, *45*(4), 441–470.

Lancaster, G., Maitra, P., & Ray, R. (2008, June). Household expenditure patterns and gender bias: Evidence from selected Indian states. *Oxford Development Studies*, *36*(2), 133–157.

McEwan, W., Fang, X., Zhang, C., & Burkholder, R. (2006). Inside the mind of the Chinese consumer. *Harvard Business Review*, *84*(3), 68–76.

National Bureau of Statistics (NBS). (2005). *Statistical yearbook of China*. Beijing: Statistical Publishing House of China.

National Sample Survey Organisation (NSSO). (2006). *Level and pattern of consumer expenditure, 2004–05*. Delhi: Ministry of Statistics and Programme Implementation.

National Sample Survey Organisation (NSSO). (2007a). *Household consumption of various goods and services in India, 2004–05, Vol. I: Major states and all-India*. Delhi: Ministry of Statistics and Programme Implementation.

National Sample Survey Organisation (NSSO). (2007b). *Household consumption of various goods and services in India, 2004–05, Vol. II: Minor states and Union Territories*. Delhi: Ministry of Statistics and Programme Implementation.

Ray, R. (2001). Analysis of a time series of household expenditure surveys for India. *Review of Economics and Statistics*, *62*(9), 595–602.

Statistics Bureau (SB). (2006). *Family income and expenditure survey 2005*. Tokyo: Ministry of Internal Affairs and Communications. www.stat.go.jp/english/data. Accessed 19 June 2008.

Subramanian, S. V., Nandy S., Kelly M., Gordon D., & Davey Smith G. (2004, April 3). Patterns and distribution of tobacco consumption in India: cross sectional multilevel evidence from the 1998–9 national family health survey. *BMJ*, *328*(7443), 801–806. doi:10.1136/bmj.7443.81.

Sun, T., & Wu, G. (2004). Consumption patterns of Chinese urban and rural consumers. *Journal of Consumer Marketing*, *21*(4), 245–253.

United Nations (UN). (2007). *World population prospects – The 2006 revision*. New York: United Nations.

United Nations Development Program (UNDP). (2007). *Human development report 2007/2008*. New York: Pulgrave Macmillan.

World Bank (WB). (2008). *2005 international comparison program – Tables of final results*. Washington, DC: World Bank.

Yusuf, F., Brooks, G., & Zhao, P. (2008). Household consumption in China: An examination of the utility of urban-rural segmentation. In S. H. Murdock & D. A. Swanson (Eds.), *Applied demography in the 21st century* (pp. 285–298). Dordrecht: Springer.

Chapter 2
The Drivers of Health Trends: A Decomposition of Projected Health for Local Areas in England

Stephen D. Clark and Philip H. Rees

Abstract Population ageing is commonly cited as one of the main drivers of increasing pressures on health care systems, as more people with chronic morbidities live to older ages. This chapter digs deeper into this presumed relationship by estimating the successive health impacts of: total population change, population ageing, changing ethnic mix of the population and trends in the age-specific incidence of disease. The decomposition of a set of health projections is developed using a micro-simulation model. These projections are based on the population of England aged 50 and over, classified by local authority of residence. The model projects forward, for the 20 years beyond 2011, the prevalence of cardiovascular disease, diabetes and respiratory illness. For diabetes the finding is that population increase alone contributes to a 24 % increase in prevalence by 2031, while the changes in gender, ethnicity and age composition together contribute another 24 %. Taking account of all potential contributions, the overall diabetes prevalence count increases by 57 %. For cardio-vascular disease (CVD), population increase contributes a 23 % increase; demographic composition processes a further 30 %; while decreases in CVD prevalence rates reduce prevalence by 60 %, resulting in an overall decrease of 35 % in those with CVD by 2031. For respiratory illness, population increase contributes 23 %; demographic composition changes 13 %; while a decrease in prevalence rates of 29 % means that the burden of the disease reduces by a modest 1 %. These results underline the potential for successful health intervention (as in CVD), the urgent need for prevention (as in diabetes) and the incentive to continue to care about health to very old ages (as in respiratory illness). These headline results refer to England as a whole but we also show how they vary across local authorities by area type, pointing to models of good practice in morbidity control.

Keywords Health trends • Population change • Population ageing • Diabetes • Cardiovascular disease • Respiratory illness • Impact decomposition methods

---

S.D. Clark (✉) • P.H. Rees
School of Geography, University of Leeds, Leeds LS2 9JT, UK
e-mail: s.d.clark@leeds.ac.uk; p.h.rees@leeds.ac.uk

D.A. Swanson (ed.), *The Frontiers of Applied Demography*, Applied Demography Series 9, DOI 10.1007/978-3-319-43329-5_2

2.1 Introduction

Population ageing is seen to be one of the important challenges facing western societies. Knowledge of the existing population stock and the extrapolation of trends in fertility, mortality and migration all point to increasing numbers of elderly in the populations of all western societies (He et al. 2016). As the report from the United Kingdom's House of Lords (House of Lords 2013) concludes, for the UK:

> The aging of the population is inevitable and affects us all.

This demographic phenomenon has a number of potential impacts on society, one of which is the demands that an ageing population will place on the health services. Again, the House of Lords report states that

> The NHS [National Health Service] is facing a major increase in demand and cost consequent on ageing and will have to transform to deal with this." And the fear is that "Because of this rising demand [...], needs will be unmet and cost pressures will rise inexorably.

These conclusions are, however, based on a purely demographic assessment of the likely situation in years to come. There is a need to also take account of likely health trends, and to ask whether these trends are stationary or static. Take, for example, the link between a person's occupation and health. For most of the twentieth century a large proportion of the male population of western societies were employed in heavy industries: manufacturing, textiles, steel making, mining and agriculture. All these industries had the potential to expose the body to stresses and toxic substances which contribute to later life morbidities. However, the occupational mix is now shifting towards non-manual occupations in finance, administration, retail and software development. These occupations imply a different morbidity profile in terms of later health. Thus it would be wrong to assume that the morbidity profile for a current generation of those aged 65 and older will necessarily apply to those aged 65 and older in future years. This argument is supported by a study by McCulloch (2012a) which shows that since 1985 there has been a downward trend in the age standardised death rate for coronary heart disease in England and Wales, caused not just by changes in occupation but other factors such as reduced smoking prevalence and pharmacological innovations (e.g. prescription of statin drugs, which lower blood pressure and improve cholesterol ratios, Law et al. 2003). Thus those prevalence prediction models that apply current morbidity rates to future populations, e.g. those produced using the POPPI model (Institute of Public Care 2011) are likely to be wrong (Mason et al. 2015) since they only account for the demographic changes in society.

So what are these characteristics of the individual that impact on later health? The first are *socio-demographic* attributes. They include age, because, as people age, they accumulate more morbidities and their heath deteriorates. Some studies have, however, argued that it is not age itself which is important but time to death, with individuals requiring the greatest use of heath care resources in the final years and months of life (Zweifel et al. 1999; Sanderson and Scherbov 2010). If this is the case, and coupled with people living to older ages (McCulloch 2012b), the future burden on the health services will not change much, since the costs are simply pushed back to later ages. Gender shows a clear pattern of longer life expectancies for women than men, although their healthy and disability-free life expectancies are not that much different (Office for National Statistics 2014a). Thus women tend to spend more years in ill health towards the end of their lives than men. Another socio-demographic factor that influences heath is race or ethnicity. Some groups are more prone to certain morbidities than the general population (e.g. diabetes in the south Asian population) and some groups are generally healthier (e.g. the Chinese) (Health and Social Care Information Centre 2005a, b).

The second set of characteristics is *socio-economic* in nature. Persons with greater accumulated wealth or higher incomes tend to have better health outcomes (McCulloch 2011) and Marmot (2015) stresses the additional role of relative deprivation (inequality) in determining health outcomes. Those in higher status occupations, people with more secure housing tenure, those with higher qualifications or with greater assets all enjoy greater wealth and income and therefore better health. Lifestyle factors such as smoking and obesity may also be important, with trends in these factors driving later life health status.

The third set of characteristics is related to *place* rather than person. If people live in an area with low density housing, with plenty of nearby green space and a range of healthy food outlets but away from industrial sites or busy roads, then their health outcomes are generally better.

A fourth set of influences relate to the *supply of health care*. There is evidence to suggest that the availability of health services positively impacts on health. This is not just related to access to medical practitioners through general practice services and hospitals, but also to the availability of cutting edge treatment technologies and pharmaceuticals.

Thus, whilst the future socio-demographic structure of the population in terms of age, gender and ethnicity is able to capture some of the changes in health trends, there are other important socio-economic, place and health service influences on health trends.

Having introduced the health challenges associated with ageing societies and the determinants of health in older population, in the following section we review the different methods used for projecting health conditions: micro-simulation, macro-simulation, hazard models and their combination. The demographic, ethnic and health data employed in the study are described in Sect. 2.3. In particular, key features of the 2011 Census and the English Longitudinal Study of Ageing (ELSA) are outlined. In Sect. 2.4 an account is provided of the methods used to project the health conditions of the population of local authority districts (LADs) in England. These methods include hazard models, spatial micro-simulation, ethnic population projections and health micro-simulation. In Sect. 2.5 we describe the method for decomposing the impact of a succession of processes on health conditions. Decomposition trends in England and decomposition trends by local authority traits are presented in Sect. 2.6. The final part of the chapter (Sect. 2.7) summarises what we now know about the relationship between population ageing and health, the implications of this knowledge for health policy and sets an agenda for future research.

2.2 Review of Methods

The reasoning above suggests that in modelling health care demand there is need for a model that incorporates some measure of the dynamic relationship between health and a range of personal and environmental factors, and one that also anticipates future trends in morbidity outcomes. In the modelling of health status the most common approach adopted is statistical modelling (Brailsford et al. 2009) whilst the second most common approach is simulation.

Of relevance to this study are statistical models that predict the occurrence of a disease event or an incidence, variously called hazard models, event based models or survival models (Singer and Willett 2003). These models are usually estimated on panel/repeated measure/longitudinal data where the life course of an individual can be followed and the occurrence of an event linked to changes in their characteristics. The most obvious application is the prediction of death (hence the term survival analysis) but the approach is adaptable to other events e.g. birth of first child, length of stay in employment, first sexual experience and, of particular importance here, the incidence of a morbidity. Such models yield the predicted odds or probability of occurrence of an event for the individual, given their circumstances. It is these probabilities that help to determine the growth or decline in the occurrence of events over time, and in the context of morbidities, the prevalence within the population.

These probabilities that individuals of type i experiences an event (p_i) can be used in a macro sense, i.e. given the size of the population with characteristics i, where the event has not yet occurred (F_i), this can be converted to a stock who have newly experienced the event by a simple multiplication (p_iF_i). Alternatively, they can be used in a micro sense using a Monte-Carlo simulation approach (Mooney 1997), where individual outcomes are determined by a comparison of a random probability estimate (r) with the probability that the outcome occurs (p_i). If $r < p_i$ then the event is deemed to of occurred, otherwise not. Van Imhoff and Post (1998) outline the similarities, differences, advantages and disadvantages of macro and micro approaches.

The remaining issue is then how to define the population of interest. This population can be described in an aggregate manner (for a macro approach) or individually (for micro-simulation). Sources of comprehensive and high quality aggregate data are either population censuses or population registers. The advantage of population censuses are that they are usually multi-faceted, giving a rich and diverse picture of an area down to small population geographies. The disadvantage is that they are expensive to conduct and therefore are only carried out periodically (every 5 or 10 years). Aggregate data based on population registers typically maintained for administrative purposes can be almost continuous in regards to the information they provide on an area but the range of information available can be limited.

Alternative individual data (for micro-simulation) are available through sample surveys conducted by governments. These surveys can be generic in nature (the UK Annual Population Survey) or specific to a topic domain, such as health (the Heath Survey for England, HSfE). They can be cross sectional in nature, interviewing a different set of individuals on each occasion or longitudinal, where an attempt is made to contact the same individuals at each survey wave (e.g. Understanding Society and ELSA). The advantage of these individual surveys is that they are conducted more frequently than censuses and contain a rich level of detail about the domain of interest. However, they only provide a sample of the population and this makes any estimate based on their data unreliable, particularly at some smaller population geographies. Techniques are however available to combine aggregate data and survey data to provide the required level of detail at smaller population geographies.

2.3 Data

In this study use will be made of four data sets. Two are aggregate in nature: the 2011 United Kingdom census in England and the projected ethnic population of English LADs (ETHPOP). The remaining two data sets are individual, the ELSA and the HSfE, which provide rich domain detail concerning health outcomes and contributory risk factors.

The 2011 Census for England was administered by the Office for National Statistics and these data provide here the required detail of contextual information at the geographic scale of English LADs. It collected information on the individual and household characteristics of the resident and visitor UK population on the census date. These characteristics included three questions directly related to health: self-assessed general health, the presence of a limiting long term illness and the amount of care provided to others (Office for National Statistics 2010). They also collected information on socio-demographic (gender, age, ethnicity) and socio-economic (tenure, vehicle ownership and employment) characteristics. The information is released as univariate tables (Key and Quick Statistics) or as multivariate tables (Details and Local Characteristics) and also as microdata (Office for National Statistics 2014c). There is no information in the census on the presence of specific morbidities. The second aggregate data set are projections of the size of the population by ethnicity in each LAD from 2011 to 2031 (University of Leeds 2016). There is considerable evidence that for some morbidities the outcomes are different by ethnic group thus knowledge of the relative seizes of the ethnic populations within each LAD will influence the prevalence estimates for some morbidities.

The ELSA is a longitudinal survey that collects data from volunteer HSfE participatory households where there is a person aged 50 and older in the household. Equivalent surveys are conducted in the US (the Health and Retirement Study) and Europe (Survey of Health, Ageing and Retirement in Europe). The ELSA survey started in spring 2002 and survey waves are repeated every 2 years. There are currently data available for waves 1 to 6. Information is collected on participant's personal circumstances (employment, income, wealth and assets), their activities (holidays, caring and travel) and health status (doctor diagnosed morbidities, activities of daily living and psychological wellbeing). These data are supplemented with bio-marker data collected by a nurse-practitioner (saliva, blood, physical abilities, e.g. grip strength) and a one-off life history survey asking individuals to recall significant life events (e.g. employment, house moves, births and deaths). Since the ELSA covers individuals aged 50 and older, this is the start of the age range adopted as our population of interest. In application however, supplementary information from the HSfE is required to provide information on the characteristics of individuals younger than 50 that may potentially 'replenish' the population of interest over the period 2011 to 2031. This is primarily related to their smoking history.

2.4 Methods

In this study a framework is adopted that incorporates elements of both simulation and statistical estimation. The basic unit of analysis is the individual whose life course over the period 2011 to 2031 is simulated. To carry out this task there is the need for: a simulated base population for 2011; a population projection by ethnicity; a mechanism to update the morbidity status of individuals; and a way to replenish the population so that it consistently covers the required age range.

The base population is estimated using the technique of spatial micro-simulation. This technique requires a constraint population that describes in an aggregate manner the nature of the area, in this case the 324 LADs in England. These constraint tables should capture the characteristics that are thought to be significant in explaining the variation in the outcome of interest, e.g. such attributes as age, gender, ethnicity, general health or vehicle ownership. Commonly they are provided in multivariate cross tabulations which help to capture the interactions between these characteristics. The technique also requires a sample population of individuals (or households) who can be typified using these same characteristics but possess other characteristics or outcomes of interest that are not available in the aggregate data.

For this study the 2011 Census Detailed Characteristics tables provide the constraint tables and the ELSA the sampling population (where the morbidity outcomes in ELSA are the added information of interest). The significant characteristics identified for the modelling of CVD, diabetes and respiratory illness are: gender, age, ethnicity, presence of a limiting long term illness, living arrangements, vehicle ownership and social class. Only the first five of these characteristics are used to estimate the population, the last two are held back for validation purposes.

The populations are estimated using the combinatorial optimisation approach implemented in the Flexible Modelling Framework (Harland 2013). With the existence of this synthetic population for each LAD it is possible to estimate morbidity prevalence counts and rates for 2011.

The Health micro-simulation estimates future biennial prevalence rates: by ageing individuals in the population by two years (the length of an ELSA wave); by re-structuring the population to agree with an external population projection; by updating the morbidity status of individuals; and by introducing "replenishers" aged 50 and 51. Replenishers are new entrants to the study population who have their 50th and 51st birthdays in each biennial time interval (see Clark 2015, for details).

Ageing is a straight forward process in which individuals have their age increased by 2 years at each time step. Then the aged population is re-structured to agree with the population projections that capture the impact of those entering the population at younger ages, those migrating and mortality. Rather than using the *raw* projections, an update is applied based on their predictive performance between the 2001 and the 2011 Censuses and a desire to constrain them to official subnational population projections (Office for National Statistics 2014b, see Rees and Clark 2014 for details of the updating method). The desired population structure by gender, age and ethnicity is achieved by selecting individuals in the existing population to 'clone' (in-migration) or remove (out-migration or death). The selection probability for each individual is estimated using a hazard model to predict the probability of migration or death. The separate morbidity status updating process uses the predicted probability of acquiring a morbidity estimated using a hazard model that incorporates these terms: time, gender, age, ethnicity, smoking status, the presence of a comorbidity and a place effect through the type of LAD they are resident in (Office for National Statistics 2003). The mechanism for updating the morbidity status is implemented using the Monte Carlo approach. The number of individuals needed to replenish the LADs population aged 50 and 51 every 2 years is derived from the population projections and they enter the population with a credible set of characteristics including their smoking status and any pre-existing morbidity.

This process is implemented using a bespoke piece of C code and at each biennial simulation time step it is possible to tally in each LAD those who now have each morbidity (or a comorbidity) and thereby estimate a prevalence count and rate. Additionally it is possible to segment this information by known characteristics, e.g. ethnicity or smoking status.

Once a method of estimating the morbidity status of the population within a LAD has been established, it is possible to return to an investigation of the original question of what is it that is driving the estimated trends in morbidity status? To recap, these trends may be stationary, upward or downward and are driven by changes in such characteristics as lifestyles, medical interventions, longer life expectancies, the demographic momentum of younger populations ageing, a transition to a more diverse population, an ageing of the population, or a convergence between the genders. All but the first two of these are separately identifiable demographic impacts on health outcomes whilst the first two are a contribution from health specific influences. What follows is an attempt to describe how the separate effects of these demographic and health impacts can be measured (in the spirit of Bongaarts and Bulatao 1999). It will involve, at times, the definition of a population which is clearly artificial, but is required in order to carry out this task.

The decomposition of the total health trends can be imagined as a "Russian doll". The largest doll is the full Health Micro-simulation as described here, accounting for all the drivers. The first inner doll represents the scenario where all the demographic drivers are in place but there are no impacts due to health trends. This achieved by using the same piece of C code but not ageing the population and not updating the morbidity status of individuals. (Since there is no ageing, there is also no need for replenishers.) Thus the population is stationary in terms of its characteristics, people stay at the same age and therefore represent those of that age throughout the micro-simulation and retain the same morbidity and smoking status. What does change is the number of people in each gender, age and ethnic group according to the revised ethnic projections but, crucially, the prevalence rates within each group does not change. The estimates obtained under such a scenario are the closest to the estimates obtained by multiplying some demographic by a (static) rate that is assumed to persist. In this and all 'subsequent' micro-simulations the ageing and morbidity dynamics are not enabled.

The next inner doll additionally removes the impact of mortality improvement. The population projections are revised by removing the mortality improvement factors incorporated in the Office for National Statistics sub-national population projections (as captured in the no-mortality improvement national population projection variant, Office for National Statistics 2013). So whilst individuals still age and the population structure changes due to the relative sizes of younger cohorts, their mortality rate stays static over time meaning these younger cohorts do not live any longer than the existing cohorts. This restriction and all subsequent restrictions are implemented by a revision to the re-structured LAD populations used within the bespoke C code. This population incorporates the projected shifts in the population by gender, ageing and ethnicity.

The next restriction is to assume that there are no cohort size effects due to age. At this stage there are no health impacts, no mortality improvements and each age cohort remains the same size as that in the 2011 Census. However, within each age cohort the gender and ethnic split does change over time, in accordance with the splits implied by the projections. The penultimate restriction is to assume that the ethnic split does not change from that in the 2011 Census, only the gender split varies. The final restriction is rather trivial and assumes that the size of the population is fixed over time. This restriction however allows a series of steps to be taken from an original 2011 Census population to the full Health Micro-simulation estimation, as summarised in Table 2.1.

The execution of the C code under these various restrictions produces as output counts of the population ($P_{.}$) and counts of those with morbidity m ($P_{.}^{m}$). The impact of each demographic and health driver can be quantified in terms of an incremental

Table 2.1 Summary of incremental design of decomposition scenarios

Scenario		Feature: Size	Gender	Ethnicity	Age	Health Dynamics
Original	P_O	As 2011 Census	As 2011 Census	As 2011 Census	As 2011 Census	None
Re-size	P_S	Without mortality improvements	As 2011 Census	As 2011 Census	As 2011 Census	None
Gender	P_G	Without mortality improvements	As revised ETHPOP	As 2011 Census	As 2011 Census	None
Gender and ethnicity	P_E	Without mortality improvements	As revised ETHPOP	As revised ETHPOP	As 2011 Census	None
Gender, ethnicity and age	P_N	Without mortality improvements	As revised ETHPOP	As revised ETHPOP	As revised ETHPOP	None
Mortality improvement	P_M	With mortality improvements	As revised ETHPOP	As revised ETHPOP	As revised ETHPOP	None
Health micro-simulation	P_H	With mortality improvements	As revised ETHPOP	As revised ETHPOP	As revised ETHPOP	All

(I^m) and a relative (R^m) change in the size of these populations. The equations for the incremental changes in the number with morbidity m for each year y for each gender g, age a and ethnicity e combination are:

due to changes in population size

$$I_S^m = P_S^m - P_O^m \tag{2.1}$$

due to changes in gender composition

$$I_G^m = P_G^m - P_S^m \tag{2.2}$$

due to changes in ethnicity composition

$$I_E^m = P_E^m - P_G^m \tag{2.3}$$

due to changes in age composition

$$I_N^m = P_N^m - P_E^m \tag{2.4}$$

due to changes in mortality rates

$$I_M^m = P_M^m - P_N^m \tag{2.5}$$

due to changes in health prevalence rates

$$I_H^m = P_H^m - P_M^m \tag{2.6}$$

where P_O^m is the number of individuals with the morbidity based on the 2011 Census population for the LAD; P_S^m is the number of individuals with the morbidity based on a re-sized population without mortality improvement but with 2011 Census gender, age and ethnicity splits; P_G^m is the number of individuals with the morbidity based on a population without mortality improvement adjusted by gender to revised projections; P_E^m is the number of individuals with the morbidity based on a population without mortality improvement adjusted by gender and ethnicity to revised projections; P_N^m is the number of individuals with the morbidity based on a population without mortality improvement adjusted by gender, ethnicity and age splits in the revised projections; P_M^m is the number of individuals with the morbidity based on a population with mortality improvement adjusted by gender, ethnicity and age but no health dynamics; and P_H^m is the number of individuals with the morbidity based on a population with mortality improvement adjusted by gender, ethnicity and age and with health dynamics.

The final estimate of the number of individuals with a morbidity can be reconstructed from the original estimate P_O^m and the incremental estimates in Eqs. (2.1), (2.2), (2.3), (2.4), (2.5), and (2.6) using Eq. (2.7).

$$P_E^m = P_O^m + I_S^m + I_G^m + I_E^m + I_N^m + I_M^m + I_H^m \tag{2.7}$$

Alternatively, the impacts can be derived as ratios:

$$R_S^m = P_S^m / P_O^m \tag{2.8}$$

$$R_G^m = P_G^m / P_S^m \tag{2.9}$$

$$R_E^m = P_E^m / P_G^m \tag{2.10}$$

$$R_N^m = P_N^m / P_E^m \tag{2.11}$$

$$R_M^m = P_M^m / P_N^m \tag{2.12}$$

$$R_H^m = P_H^m / P_M^m \tag{2.13}$$

In which case the final estimate of the number of individuals with a morbidity condition can be re-constructed using Eq. (2.14).

$$P_E^m = P_O^m \times R_S^m \times R_G^m \times R_E^m \times R_N^m \times R_M^m \times R_H^m \tag{2.14}$$

In Eq. (2.14), the ratios from R_S^m to R_N^m are largely given, since these represent demographic pressures already in the population and the impact of their anticipated out-turns. R_M^m which represents the impact ratio due to improvements in mortality can be influenced by policy makers, but only in a negative sense that they can theoretically 'reverse' the trend seen recently for increases in life expectancies (e.g. by failing to fund health care adequately) and thereby reduce the number of elderly people in the population (see Public Health England 2016, and National Center for Health Statistics 2016, for instances of such recent reversals). The final ratio, R_H^m, incorporating the health dynamic trend, is the one most amenable to policy intervention, either through health education or through innovations in health treatment or technology.

2.5 Results

Here we begin to show the results of this exercise. Firstly, just one of the case study morbidities of diabetes is examined at the national level of England. The tables of projected populations that follow are presented in unit detail. Table 2.2 shows the size of the population at risk in each decomposition scenario, for forecast years to 2031 in 4 year steps.

For the Original scenario, the population size remains constant over the estimation period. The size of the population at risk for the next four scenarios then remains roughly constant, subject to slight variations introduced by the Monte-Carlo sampling used to re-structure the population. For the Mortality Improvement scenario, the size of the population again increases to reflect the fact that mortality is improving and stays the same for the Health Micro-simulation scenario. We now inspect the estimates of those with diabetes in each scenario in Table 2.3.

There is a large increase in those with diabetes when the population at risk is Re-sized, although this is to be expected, with a constant prevalence rate a larger population implies a larger number of individuals with the morbidity. For the next four scenarios there are estimated increases in those with diabetes. There are also large jumps in those with diabetes for the Mortality Improvement and Health Micros-simulation scenarios, although this is again partly influenced by the increase in the size of the population at risk. To discount this size effect, Table 2.4 shows the prevalence rates for diabetes.

As expected, the Original scenario shows the same prevalence rates over time. The Re-size scenario should also show the same prevalence rates because the composition of the population is fixed, but does not quite because of the variability introduced by Monte Carlo sampling. The impact of allowing the Gender split to vary causes only a small increase in the prevalence of diabetes. The big change occurs when both the Gender and Ethnic split are varied. There is an almost +1.0 % estimated increase in the prevalence rates. The remaining two decomposition scenarios also estimate modest increases in the prevalence rate. Perhaps the headline finding here is that the full Health Micro-simulation predicts an increase in both the prevalence counts (Table 2.3) and prevalence rates (Table 2.4) for diabetes from 2011 to 2031 (although between 2027 and 2031 the prevalence rate starts to drop off).

Table 2.5 shows the impacts measured as ratios. This table demonstrates that the cumulative impact of demographic changes contributes the most to the rise in the number of individuals diagnosed with diabetes. In 2031 the demographic contribution is $1.001 \times 1.080 \times 1.045 \times 1.091 = 1.233$ and by comparison the health trends increases this number by a modest 1.029. Looking at individual demographic

Table 2.2 Populations at risk for each decomposition scenario

Total	Original	Re-size	Gender	Gender and ethnicity	Gender ethnicity and age	Mortality improvement	Health micro-simulation
	P_O	P_S	P_G	P_E	P_N	P_M	P_H
2011	18,229,893	18,229,893	18,229,893	18,229,893	18,229,893	18,229,893	18,229,893
2015	18,229,893	19,612,338	19,612,286	19,612,399	19,612,348	19,689,593	19,689,593
2019	18,229,893	20,754,507	20,754,524	20,754,526	20,754,480	21,083,429	21,083,429
2023	18,229,893	21,628,697	21,628,633	21,628,713	21,628,634	22,304,919	22,304,919
2027	18,229,893	22,046,432	22,046,369	22,046,350	22,046,344	23,129,460	23,129,460
2031	18,229,893	22,502,369	22,502,421	22,502,405	22,502,352	24,029,232	24,029,232

Table 2.3 Population with diabetes for each decomposition scenario

	Original	Re-size	Gender	Gender and ethnicity	Gender ethnicity and age	Mortality improvement	Health micro-simulation
Diabetes	P_O^m	P_S^m	P_G^m	P_E^m	P_N^m	P_M^m	P_H^m
2011	2,189,938	2,189,938	2,189,938	2,189,938	2,189,938	2,189,938	2,189,938
2015	2,189,938	2,354,167	2,357,184	2,386,839	2,382,834	2,396,726	2,656,947
2019	2,189,938	2,491,301	2,495,686	2,560,065	2,550,691	2,608,150	2,979,780
2023	2,189,938	2,597,227	2,601,630	2,707,846	2,731,366	2,851,741	3,191,523
2027	2,189,938	2,648,832	2,653,463	2,809,809	2,883,514	3,077,822	3,358,974
2031	2,189,938	2,706,720	2,710,272	2,928,029	3,060,763	3,340,327	3,436,486

Table 2.4 Prevalence rates of diabetes for each decomposition scenario

	Original	Re-size	Gender	Gender and ethnicity	Gender ethnicity and age	Mortality improvement	Health micro-simulation
Diabetes	p^m_O/P_O	p^m_S/P_S	p^m_G/P_G	p^m_E/P_E	p^m_N/P_N	p^m_M/P_M	p^m_H/P_H
2011	12.01 %	12.01 %	12.01 %	12.01 %	12.01 %	12.01 %	12.01 %
2015	12.01 %	12.00 %	12.02 %	12.17 %	12.15 %	12.17 %	13.49 %
2019	12.01 %	12.00 %	12.02 %	12.33 %	12.29 %	12.37 %	14.13 %
2023	12.01 %	12.01 %	12.03 %	12.52 %	12.63 %	12.79 %	14.31 %
2027	12.01 %	12.01 %	12.04 %	12.75 %	13.08 %	13.31 %	14.52 %
2031	12.01 %	12.03 %	12.04 %	13.01 %	13.60 %	13.90 %	14.30 %

Table 2.5 Impact ratios for diabetes for each decomposition scenario

	Re-size	Gender	Gender and ethnicity	Gender ethnicity and age	Mortality improvement	Health micro-simulation	
Diabetes	R^m_S	R^m_G	R^m_E	R^m_N	R^m_M	R^m_H	Product
2011	1.000	1.000	1.000	1.000	1.000	1.000	1.000
2015	1.075	1.001	1.013	0.998	1.006	1.109	1.214
2019	1.138	1.002	1.026	0.996	1.023	1.142	1.361
2023	1.186	1.002	1.041	1.009	1.044	1.119	1.458
2027	1.210	1.002	1.059	1.026	1.067	1.091	1.534
2031	1.236	1.001	1.080	1.045	1.091	1.029	1.568

Table 2.6 Impact ratios for CVD for each decomposition scenario

	Re-size	Gender	Gender and ethnicity	Gender ethnicity and age	Mortality improvement	Health micro-simulation	
CVD	R^m_S	R^m_G	R^m_E	R^m_N	R^m_M	R^m_H	Product
2011	1.000	1.000	1.000	1.000	1.000	1.000	1.000
2015	1.077	1.001	1.004	1.009	1.008	0.865	0.953
2019	1.139	1.002	1.009	1.014	1.033	0.731	0.881
2023	1.186	1.002	1.016	1.034	1.064	0.600	0.798
2027	1.208	1.002	1.025	1.073	1.098	0.494	0.723
2031	1.233	1.001	1.035	1.104	1.133	0.405	0.647

components the largest ratios occur when the ethnic composition of the population changes and when mortality improves.

Equivalent impact ratios for CVD shown in Table 2.6 illustrate a much starker contrast in the relative contribution of demographic and dynamic health impacts.

Here, the largest demographic composition impacts are from to a population whose age structure incorporates the demographic momentum of ageing (1.104 in

2031) and which then incorporates improvements in mortality (1.133). However these demographic drivers that increase the population with CVD are mitigated to a large degree by the dynamics incorporated in the Health Micro-simulation. The result is that the number of people with CVD in 2031 is just 64.7 % of those with CVD in 2011 (McCulloch 2012a also reports a historic downward trend for this morbidity)

The ratios for respiratory illness are shown in Table 2.7, which shows a similar pattern as for CVD, the demographic factors increase those with respiratory illness whilst health dynamics reduce the number. The result by 2031 is that the numbers with respiratory illness is very similar to the numbers in 2011 (although the size of the population at risk has increased, see Table 2.2).

The results in Tables 2.2, 2.3, 2.4, 2.5, 2.6, and 2.7 show the decomposition results on a national scale for England as a whole. It is possible to compute similar ratios for individual LADs in England. For diabetes, the distribution of these ratios in 2031 for each scenario for seven area types in the 2011 Office for National Statistics area classification (Office for National Statistics 2015) is shown in Fig. 2.1 using box plots (the width of the box is in proportion to the number of LADs of this area type).

The top left plot shows the distribution of the impact of an increase in the size of the population from that in 2011 to the population without mortality improvement (R_S^m). This impact captures a mechanistic increase in the size of the population at risk. Here the London Cosmopolitan type of LAD shows the largest ratios, their 50 and older population is projected to relatively increase the most, the outlier is Tower Hamlets.

The top right plot shows the distribution when the gender splits are varied in line with the revised gender-specific projections (R_G^m). These ratios are generally close to 1.00 and there is little differentiation by area types.

The middle left plot shows the ratios when both gender and ethnicity change (R_E^m). Here there is a distinct pattern. The London Cosmopolitan and Suburban Traits LADs have higher ratios and are substantially higher than those for Coastal and Heritage and English and Welsh Countryside LADs. This shows that the impact of the changing nature of the ethnic diversity in London and Suburban type authorities is an important driver of increases in diabetes prevalence. Conversely, the anticipated ethnic composition change is more limited in Coastal and Heritage and English and Welsh Countryside LADs and as a result this impact ratio is between 1.00 and 1.05. The remaining area types have a wide range of impact ratios, suggesting they contain a range of areas whose ethnic composition is likely to impact differently on diabetes.

When the demographic split of the LAD is changed in line with gender, ethnic and age projections (R_N^m), the London Cosmopolitan type of authority now has the lowest ratios whilst the other LADs have higher ratios. Thus the additional impact of the demographic ageing potential of the population is predicted to be greatest in non-London LADs. This is in line with the hypothesis that the more urban LADs, and particularly those in London, will retain a 'youthful' older population.

Table 2.7 Impact ratios for respiratory illness for each decomposition scenario

	Re-size	Gender	Gender and ethnicity	Gender ethnicity and age	Mortality improvement	Health microsimulation	
Respiratory illness	R_S^m	R_G^m	R_E^m	R_N^m	R_M^m	R_H^m	Product
2011	1.000	1.000	1.000	1.000	1.000	1.000	1.000
2015	1.077	1.000	1.001	1.002	1.005	0.938	1.018
2019	1.139	0.999	1.003	1.003	1.019	0.877	1.024
2023	1.187	0.999	1.008	1.004	1.038	0.819	1.020
2027	1.209	1.000	1.013	1.013	1.059	0.768	1.009
2031	1.234	1.000	1.020	1.025	1.082	0.710	0.990

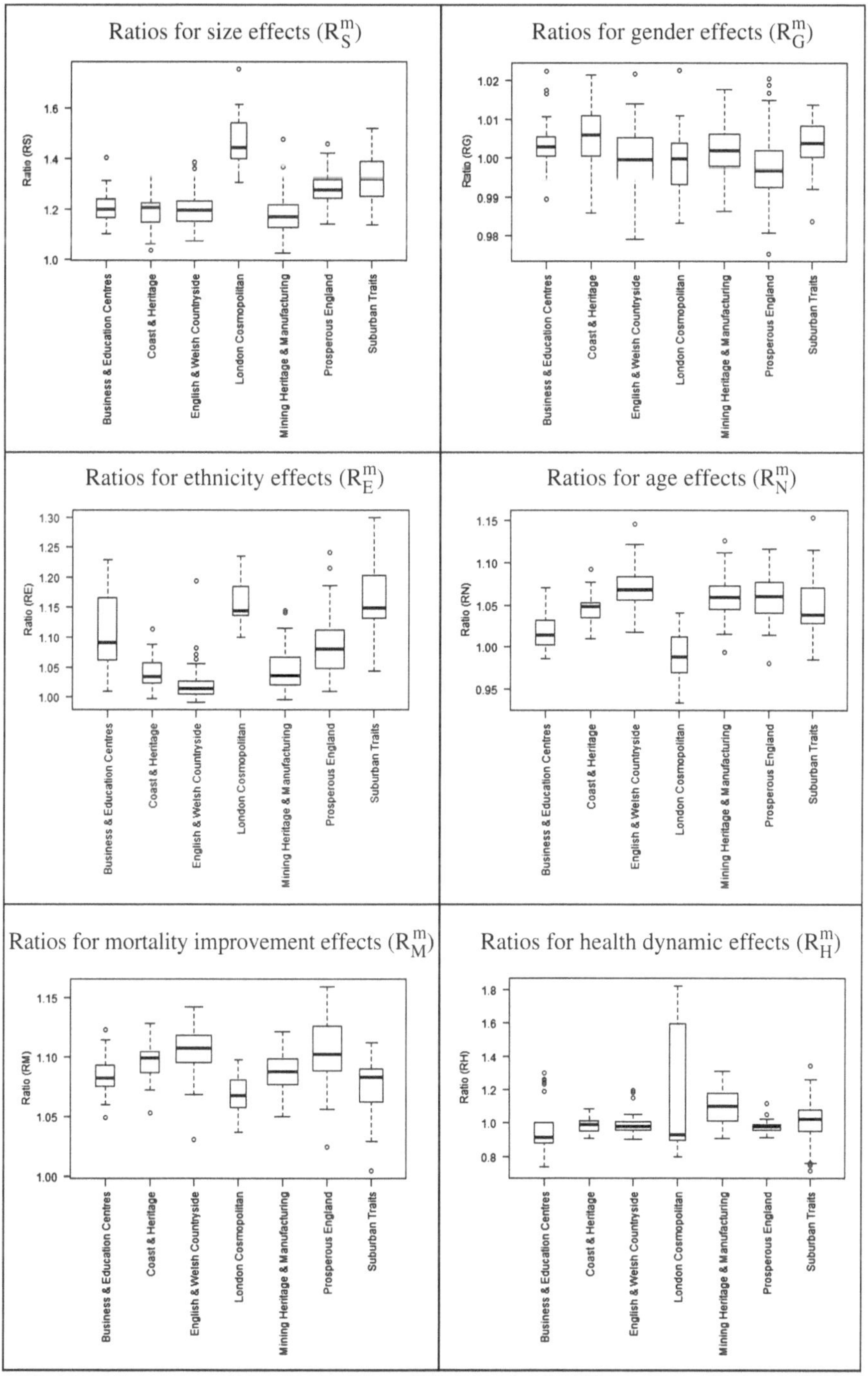

Fig. 2.1 Distribution of decomposition ratios for diabetes in each area type

When the impact of mortality improvements is incorporated (R_M^m) the London Cosmopolitan and Suburban Traits LADs are again those with lower ratios, although the difference with other LAD types is less pronounced.

The final ratios show the impact of the heath dynamics (R_H^m). The result that stands out from this is the wide range and skewed nature of ratios for the London Cosmopolitan LADs. The changes in prevalence rates over time incorporated in the hazard models and driven in part by individual's lifestyles creates a divergence in the population. The reason for this divergence is that the term "cosmopolitan" covers a diverse range of population compositions and only some ethnic minorities (mainly south Asian in origin) experience raised diabetes prevalence rates, whilst other minorities do not.

2.6 Discussion

This chapter has described methods for projecting the future health status of subnational populations in the form of both numbers experiencing particular morbidities and as rates of prevalence in relation to the projected population. The methods combine a micro-simulation of the whole population of the geographical areas, an aggregate projection of those populations, and a hazard model that predicts trends in the morbidities based on population characteristics and recent historical experience. These projections are then decomposed by running a sequence of scenario projections that omit components of the full model in a way that enabled estimation of the impacts of total population change, gender composition shift, ethnic composition shift, age composition shift, mortality improvement and projected changes in prevalence rates.

Conventional wisdom and media discourse emphasises that population ageing is the main driver of increases in particular morbidities. Our analysis shows that this is a gross over-simplification. Population growth and changing ethnic composition also have important roles. But the most important driver is the progress or lack of progress in driving down the prevalence rates by gender, age and ethnicity. For CVD and respiratory illnesses huge progress has been made and we project this to continue. By contrast, our results for diabetes are not encouraging, with adverse trends in prevalence failing to counter-act the effect of a growing and ageing population. Preventing the onset of diabetes and improving treatments should be the focus of health policy in England, and we suspect the whole of the UK and other countries. We also demonstrated that there were very large variations across types of local area. Adoption of best practice in areas experiencing favourable health micro-simulation impacts in all areas will clearly help.

One demographic driver of local variation in morbidity outcomes that we did not identify is that of internal migration, though there were clear signs of its impact in some results. For example, the ratios for age composition (R_N^m) were very low in the London Cosmopolitan type of LAD. This was most probably the result of out-migration of people aged 50 and over from Greater London, so that presentation with diabetes symptoms would occur in their migration destinations such as the English and Welsh Countryside type. Rees et al. (2013a, b) developed their projection model scenarios so that the impact of internal migration on ethnic group population change could be measured. In the future, such effects could be built into the assessment of health trends by extending the micro-simulation to the whole population and integrating it closely with the aggregate population projection model incorporating internal and international migration processes.

One final comment should be made. Our demographic and morbidity analysis only covers the demand side of the story of increasing expenditure on health care needs. It is necessary to look at the supply side as well. Health care costs are increasing much faster than the demographic drivers of health trends that we have examined. Our approach could possibly be extended to incorporate the complex economic, technical and organizational factors at work here. But that is for another day.

References

Bongaarts, J., & Bulatao, R. A. (1999). Completing the demographic transition. *Population and Development Review, 25*(3), 515–529.

Brailsford, S. C., Harper, P. R., Patel, B., & Pitt, M. (2009). An analysis of the academic literature on simulation and modelling in health care. *Journal of Simulation, 3*, 130–140.

Clark, S. D. (2015). Modelling the impacts of demographic ageing on the demand for health care services. PhD dissertation, School of Geography, University of Leeds. http://etheses.whiterose.ac.uk/11676/

Harland, K. (2013). *Microsimulation model user guide: Flexible modelling framework.* National Centre for Research Methods Working Paper 06/13. http://eprints.ncrm.ac.uk/3177/2/microsimulation_model.pdf

He, W., Goodkind, D., & Kowal, P. (2016). *An ageing world.* International Population Reports. United States Census Bureau. http://www.census.gov/content/dam/Census/library/publications/2016/demo/p95-16-1.pdf

Health and Social Care Information Centre. (2005a). *Health survey for England 2004: The health of minority Ethnic Groups – Headline tables.* http://www.ic.nhs.uk/webfiles/publications/hlthsvyeng2004ethnic/HealthSurveyForEngland161205_PDF%20.pdf

Health and Social Care Information Centre. (2005b). The health of minority Ethnic Groups, Vol. 1, K Sproston & J. Mindell. http://www.ic.nhs.uk/pubs/hse04ethnic

House of Lords. (2013). *Ready for ageing?* House of Lords Select Committee Report. http://www.parliament.uk/business/committees/committees-a-z/lords-select/public-services-committee/report-ready-for-ageing/

Institute of Public Care. (2011). Poppi user guide. Projecting Older People Population Information. http://www.poppi.org.uk/POPPI%20User%20Guide.pdf

Law, M. R., Wald, N. J., & Rudnicka, A. R. (2003). Quantifying effect of statins on low density lipoprotein cholesterol, ischaemic heart disease, and stroke: Systematic review and meta-analysis. *British Medical Journal, 326*, 1423–1429.

Marmot, M. (2015). *The health gap: The challenge of unequal world.* London: Bloomsbury.

Mason, T., Sutton, M., Whittaker, W., & Birch, S. (2015). Exploring the limitations of age-based models for health care planning. *Social Science and Medicine, 132*, 11–19.

McCulloch, A. (2011). Health service funding. http://www.significancemagazine.org/details/webexclusive/1385441/Health-Service-funding.html

McCulloch, A. (2012a). *Standardising death*. http://www.significancemagazine.org/details/webexclusive/1501135/Standardising-Death.html

McCulloch, A. (2012b). *Population ageing and mortality*. http://www.significancemagazine.org/details/webexclusive/1664457/Population-ageing-and-mortality.html

Mooney, C. Z. (1997). *Monte Carlo simulation. Quantitative applications in the social sciences*. Sage University Paper, Number 07–116.

National Center for Health Statistics. (2016). *Health, United States, 2015, with special feature on racial and ethnic health disparities*. U.S. Department of Health and Human Services, Centers for Disease Control and Prevention, NCHS, Hyattsville MD. http://www.cdc.gov/nchs/data/hus/hus15.pdf

Office for National Statistics. (2003). *National statistics 2001 area classification*. Office for National Statistics. http://www.ons.gov.uk/ons/guide-method/geography/products/area-classifications/ns-area-classifications/index/index.html

Office for National Statistics. (2010). *Final recommended questions for the 2011 census in England and Wales on Health*. Office for National Statistics. http://www.ons.gov.uk/ons/guide-method/census/2011/the-2011-census/2011-census-questionnaire-content/question-and-content-recommendations/final-recommended-questions-2011---health.pdf.

Office for National Statistics. (2013). *National population projections, 2012-based extra variants report*. Office for National Statistics. http://www.ons.gov.uk/ons/dcp171776_338113.pdf

Office for National Statistics. (2014a). *Health expectancies at birth and at age 65 in the United Kingdom, 2009–2011*. Office for National Statistics. http://www.ons.gov.uk/ons/rel/disability-and-health-measurement/health-expectancies-at-birth-and-age-65-in-the-united-kingdom/2009-11/index.html

Office for National Statistics. (2014b). *Methodology: 2012-based sub national population projection*. Office for National Statistics. http://www.ons.gov.uk/ons/rel/snpp/sub-national-population-projections/2012-based-projections/rpt-snpp-2012-based-methodology-report.html

Office for National Statistics. (2014c). *Microdata*. Office for National Statistics. http://www.ons.gov.uk/ons/guide-method/census/2011/census-data/2011-census-prospectus/release-plans-for-2011-census-statistics/subsequent-releases-of-specialist-products/microdata/index.html

Office for National Statistics. (2015). *Methodology note for the 2011 area classification for local authorities*. Office for National Statistics. http://www.ons.gov.uk/ons/guide-method/geography/products/area-classifications/ns-area-classifications/ns-2011-area-classifications/methodology-and-variables/methodology.pdf

Public Health England. (2016). Recent trends in life expectancy at older ages: Update to 2014. https://www.gov.uk/government/uploads/system/uploads/attachment_data/file/499252/Recent_trends_in_life_expectancy_at_older_ages_2014_update.pdf

Rees, P., & Clark, S. (2014). *The projection of Ethnic Group populations aged 18 and over for Westminster parliamentary constituencies in Great Britain for election years 2015, 2020, 2025, 2030 and 2035*. London: A Report to the Policy Exchange.

Rees, P., Zuo, C., Wohland, P., Jagger, C., Norman, P., Boden, P., & Jasinska, M. (2013a). The implications of ageing and migration for the future population, health, labour force and households of Northern England. *Applied Spatial Analysis and Policy, 6*(2), 93–122.

Rees, P., Wohland, P., & Norman, P. (2013b). The demographic drivers of future ethnic group populations for UK local Areas 2001–2051. *The Geographical Journal, 179*(1), 44–60.

Sanderson, W. C., & Scherbov, S. (2010). Remeasuring aging. *Science, 329*, 1287–1288.

Singer, J. D., & Willett, J. B. (2003). *Applied longitudinal data analysis*. Oxford: Oxford University Press.

University of Leeds. (2016). *ETHPOP database*. http://www.ethpop.org/

Van Imhoff, E., & Post, W. (1998). Microsimulation methods for population projection. *Population-E, 10*(1), 97–136.

Zweifel, P., Felder, S., & Meiers, M. (1999). Ageing of population and health care expenditure: A red herring? *Health Economics, 8*, 485–496.

Chapter 3
Demographic Analyses for Public Policy: Projecting the Use of Veteran Educational Benefits in Texas

Michael E. Cline, Steve H. Murdock, and Mary Zey

Abstract Demographers are often called upon to project change in public programs and to analyze alternative scenarios based upon proposed changes to public policy. In response, methods must be found to adequately prepare research despite limitations in time, data, and information. This chapter describes one such project completed to estimate the number of potential participants in a state-specific program to help veterans and their dependents obtain college degrees or technical credentials. We use the Hamilton-Perry method to prepare projections of veterans and veteran dependents as a basis for projecting the change in the use of a Texas specific educational benefit for veterans and dependents – the Hazlewood Exemption. We then use these projections of Hazlewood participants as well as estimates of the characteristics of veterans and veteran households to analyze the impacts of proposals for managing the future use of the Hazlewood Exemption. Despite recent wars in Iraq and Afghanistan, the size of the United States military is much smaller than it was during the Vietnam and Cold War eras which will lead to a smaller veteran population as older cohorts of veterans age and die. The analysis indicates that the age characteristics of these veterans as well as the characteristics of their households are critical to understanding future levels of educational benefit usage.

Keywords Veterans • Projections • Military

M.E. Cline (✉) • S.H. Murdock
Rice University, Houston, TX 77005, USA
e-mail: mec6@rice.edu; Shm3@rice.edu

M. Zey
University of Texas at San Antonio, San Antonio, TX 78249, USA
e-mail: maryzey@gmail.com

D.A. Swanson (ed.), *The Frontiers of Applied Demography*, Applied Demography Series 9, DOI 10.1007/978-3-319-43329-5_3

3.1 Introduction

The Hazlewood Exemption is an educational benefit established by the Texas Legislature that enables veterans and select veteran family members to obtain education and technical training beyond high school. As of 2015, veterans can use the Hazlewood Exemption to obtain 150 h of college credit at one or more Texas public universities, medical schools, community colleges, or technical schools. In addition, a veteran can transfer any unused credit hour exemptions to a dependent who is younger than 26 years of age. A Hazlewood Exemption can also be obtained for the spouse or child of a military service member killed in action, deceased during military service or who has a 100 % disability rating. This educational benefit originated in post World War I Texas, when the 38th Texas Legislature first instructed public educational institutions to exempt veterans who were not receiving federal rehabilitation act benefits from paying certain fees and charges (Legislative Budget Board 2014: 2–3, 33). Since that time, this benefit has been modified in ways that expanded eligibility in some cases and limited eligibility in others. The first major set of amendments to this program were made in 1943 during World War II by Senators Grady Hazlewood and George Moffett which extended benefits to veteran's children under conditions in which the veteran's service resulted in death during wartime service. These education benefits are known simply as the Hazlewood Exemption or Hazlewood Program today.

For most of the program's history, the Hazlewood Exemption was restricted to those who did not receive federal educational benefits and, beginning in 1953, to those veterans who were legal residents of Texas at the time they entered military service. At the end of the 2000s, several changes to the program expanded program eligibility significantly resulting in precipitous growth in the number of recipients of the Hazlewood Exemption. In 2007 the exemption was extended to dependents of veterans with a 100 % disability rating and veterans were allowed to receive federal educational benefits simultaneously as long as the total amount of federal and state benefits did not exceed the total amount of tuition and fees. In 2009 the exemption was extended to allow a veteran to reassign unused hours of benefits to dependents less than 26 years of age – establishing the Legacy provision of the Hazlewood Exemption. With these changes, public educational institutions saw the number of Hazlewood Exemptions increase from 9,882 in 2009 to 38,946 in 2014 – an increase equating to a change in value of the exemptions from $24.7 million in 2009 to $169.1 million in 2014 (Legislative Budget Board 2014). Much of this increase has been attributed to the growth in the Legacy program, although all of the categories of exemptions have experienced growth. Importantly, most of these costs are borne by the institutions themselves since only a portion of the costs for the foregone tuition and fees are reimbursed by a fund set aside by the Texas Legislature. In addition, a federal court ruling in January of 2015 led to concerns about the state's ability to limit the benefit to only those veterans who were Texas residents at the time of enlistment (referred to as the "fixed-point residency requirement"), thus opening the benefit to even more veteran families (Weirlein, Jr. 2015).[1] As a result of this rapid growth in recipients and the Federal court decision, legislation to slow the growth in the program's use was proposed so that the Hazlewood Exemption could be maintained long-term.

State decision-makers are now faced with the dilemma of ensuring that the services of Texas veterans are adequately recognized and supported by providing financial aid that enables veterans to obtain education and skills while at the same time ensuring that the fiscal effects of the Hazlewood Exemption do not have deleterious impacts on the economic viability of state institutions of higher education. Critical to this analysis was the determination of how many veterans and their dependents are likely to be using the Hazlewood Exemption in the coming years. In this chapter, we make use of the Hamilton-Perry method to project future Hazlewood Exemption use (Hamilton and Perry 1962). We then utilize information from the American Community Survey and other data sources to analyze the effects of various public policy proposals on future exemption use.

The major policy focus related to the continuation of the Hazelwood exemption involved addressing two key issues:

1. How many Texas Veterans and their spouse and children are likely to use Hazlewood Exemptions in the coming years?
2. How might selected policy changes in eligibility criteria alter the growth in the use of the Hazlewood Exemption?

3.1.1 Previous Analyses of the Future Number of Hazlewood Recipients

The Texas Legislative Budget Board (LBB) published an analysis of the Hazlewood Exemption program in December of 2014. It provided an overview of the program and of federal policy and education benefit use, examined the program relative to similar programs in other states and described the demographic and program participation characteristics (e.g., credit hours, degrees and cost of attendance) of recipients. The LBB also provided projections of future impacts of the program from 2015 to 2019. The LBB projections were based on an assumption that the weighted annual average numeric change from 2012 to 2014 would prevail through 2019. As shown in Figure 14 of the LBB report the largest increases were from 2012 to 2013 when veteran recipients increased from 15,732 in 2012 to 17,256 in 2013; dependent recipients increased from 751 to 1,003; spouse recipients increased from 91 to 132, and Legacy recipients from 12,288 in 2012 to 17,153 in 2013. These were increases of 1,524; 252; 41; and 4,865 for veteran, dependent,

[1] As of November 2015, the federal district court decision was on appeal. The court decision found that the fixed-point residency requirement was unconstitutional under the Equal Protection Clause because the policy was not "rationally related to any legitimate state interest."

spouse, and Legacy recipients, respectively or 6,766 overall.[2] Nearly 72 % of the increase was due to Legacy recipients.

From 2013 to 2014 the number of veterans increased from 17,256 to 17,290; the number of dependents from 1,003 to 1,513; the number of spouses from 132 to 183. The number of Legacy recipients from 17,153 in 2012 to 19,715 in 2014. These were increases of 34, 510, 51, and 2,562 respectively, a total increase of 3,177. Assuming a weighted average annual change from 2012 to 2014, the LBB projected that the number of Hazlewood recipients would increase by 19,100 to 47,940 (or by more than 43 %) from 2015 to 2019 with associated costs increasing from $208.6 million in 2015 to $379.1 in 2019. The LBB projections were limited by the fact that the data for use of the Hazlewood Exemption was not previously maintained by a central depository – with each educational institution keeping records of the exemption use and associated costs. The state only recently began to standardize the record keeping and maintain a central depository of data on Hazlewood use. In addition, the Legacy program was established in 2009 and saw tremendous growth over the last few years as families became aware of the program. As a result, the LBB was limited to projecting the use of the Hazlewood Exemption over a very short historical period. Although the method used was reasonable given these limitations, using the weighted average use ignored the fact that there was a declining level of growth over this timeframe. More importantly, the projection method failed to account for the changing size and aging of the veteran population that affects the number of potential Legacy and dependent recipients. It thus does not allow the underlying patterns of change in the veteran population to be adequately examined.

3.1.2 The Veteran Population

The size of the veteran population in Texas grows or declines as a result of three main factors: (1) additions to the veteran population as people are honorably discharged from active duty; (2) subtractions from the veteran population as veterans die; and (3) additions or subtractions to/from the veteran population as veterans move into or out of Texas. The number of people becoming veterans is influenced by the size of the United States military at any given point in time. The largest cohort of veterans served during World War II when the total size of the United States military reached 12.2 million people (National WWII Museum 2015). Many of the World War II and Korean War veterans have died or are in older ages today. The next largest cohort of veterans served during the Vietnam War, a war that ended 40 years ago. There were 3.5 million military personnel in 1968 during the peak of the Vietnam War (see Fig. 3.1). Many of those veterans have now

[2]There were 147 exemptions in 2012, 225 exemptions in 2013, and 245 exemptions in 2014 that were not classified by type.

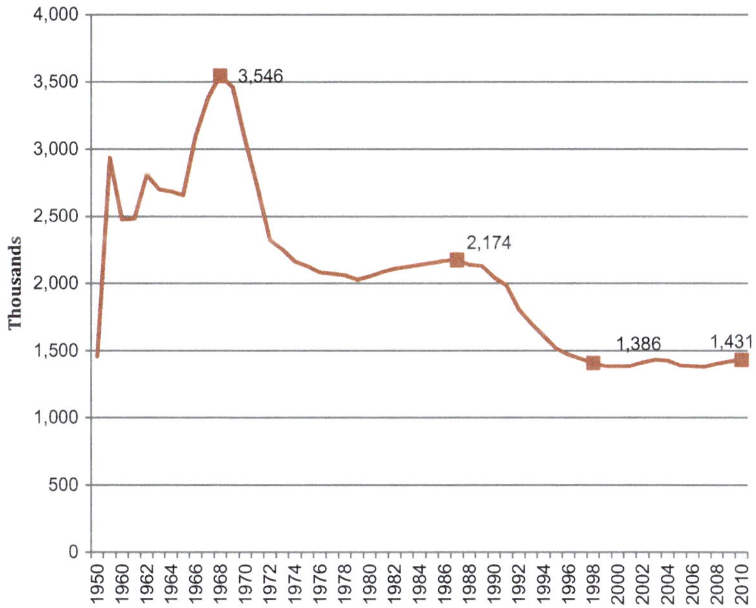

Fig. 3.1 Department of Defense Personnel: 1950 to 2010 (*Source:* U.S. Department of Defense, Selected Manpower Statistics as reported in the *2012 Statistcal Abstract of the United States (U.S. Census Bureau 2011)*)

reached retirement age or will be reaching retirement age in the next few years. From the Vietnam War peak, the military declined to about 2.0 million by 1979 before rising again to 2.2 million in 1987 during the Cold War build-up of the 1980s. As the Cold War ended, the size of the military decreased to about 1.4 million by 1998 – a size that has remained relatively unchanged despite the wars in Iraq and Afghanistan. The effect of these differences in the size of the military over the years can be seen in the age structure of the Texas veteran population in 2000 and 2010 (Fig. 3.2). Thus any projection model must take into account the differences in cohort sizes of veterans over time.

Data from the National Center for Veterans Analysis and Statistics in the United States Department of Veterans Affairs for 2014 (see Table 3.1) provide information on the age of Texas Veterans. These data show that many of the 1.7 million veterans are likely to be veterans from WWII and Vietnam and unlikely to still be seeking college degrees or to have children who are doing so (National Center for Veterans Analysis and Statistics 2014). Assuming that those involved in the Vietnam War (the more recent of the two wars) that ended in 1975 were 21 years of age or older when the war ended, they would be at least 60 in 2014. As shown in Table 3.1, 826,073 veterans (or 49.2 % of all veterans) are 60 years of age or older. It is highly unlikely that they are now seeking a college education. In addition, if one assumes that most of the veterans who are seeking college degrees for themselves are likely to be less than 35 years of age then the number of all veterans using Hazlewood benefits directly in the near-term is likely to be a maximum of 200,049 (see

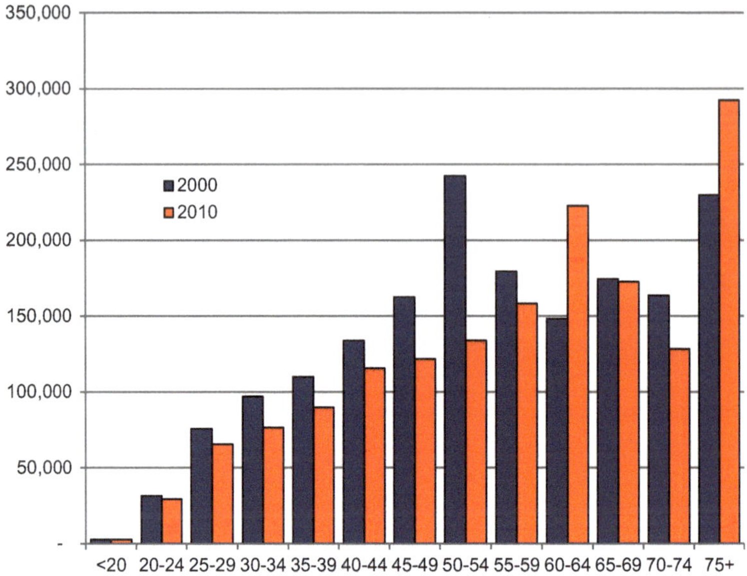

Fig. 3.2 Veteran Population in Texas by Age, 2000 and 2010 (Source: Integrated Public Use Microdata Series, 2000 Census and 2008–2012 American Community Survey)

Table 3.1 Estimated number of Veterans in Texas by age, 2014

Age	Veterans	Percent of total
<20	505	0.0
20–24	23,188	1.4
25–29	70,234	4.2
30–34	106,122	6.3
35–39	108,274	6.4
40–44	124,009	7.4
45–49	132,991	7.9
50–54	136,401	8.1
55–59	152,660	9.1
60–64	162,763	9.7
65–69	227,447	13.5
70–74	146,341	8.7
75–79	106,046	6.3
80–84	94,887	5.6
85+	88,549	5.3
Total	1,680,418	100.0

Source: VetPop2014, Office of the Actuary, Office of Policy and Planning, U.S. Dept. of Veterans Affairs

Table 3.1). In sum, although there were 1.7 million veterans in Texas in 2014 (according to the Department of Veterans Affairs) the number currently seeking degrees and funds of any type to support them seem unlikely to continue to increase over time.

3.1.3 *Educational and Household Characteristics of Texas Veterans*

Many jobs today require a high school diploma and some require at least some education beyond high school. Today's United States military enlistment requirements are more stringent than they were historically. As a result, the majority of veterans, and especially the youngest veterans, have completed at least some college. In 2010, 68.0 % of all veterans had completed at least some college (compared to 52.2 % of nonveterans) and the percentage of Texas veterans with at least a bachelor's degree in 2010 was, at 27.7 %, larger than that of nonveterans [22.6 % for nonveterans (see Table 3.2)]. Although veterans overall are more educated than nonveterans, a large percentage of veterans have not yet obtained an Associate's degree (63.3 %) and would benefit from the educational benefits provided by the Hazlewood Exemption.

The Texas Legislature, recognizing the sacrifices that military families make in order to support their family's military service member, extended the Hazlewood Exemption to children of veterans by allowing veterans to assign unused Hazlewood Exemption benefits to a child. The use of the Legacy exemption will be affected by the number of children in veteran households which is, in turn, dependent upon the age structure of the veteran population. For our purposes, we assume that a veteran household is any household in which the householder or householder's spouse was a veteran. Table 3.3 summarizes Texas veteran households by age and presence of children in the household. In 2010, 31.0 % of all veteran households had at least one child living in the household. As would be expected, most veteran households with a child present are headed by veterans who are aged 25–44. Of those households headed by a veteran age 25–44, 65.9 % had one or more children present, whereas a majority of all veteran households had no children present in 2010.

Table 3.2 Veteran educational attainment by age, 2010

Age	<High school	High school	Some college	Associate	Bachelor	Prof./ Grad.	Total
18–24	3.8	38.8	48.7	4.0	4.5	0.2	32,386
25–44	1.6	22.3	40.0	12.8	16.5	6.8	347,250
45–64	5.1	24.8	31.3	10.7	17.2	10.9	637,100
65+	13.1	25.3	24.9	5.1	18.7	12.9	593,200
25+	7.3	24.5	30.8	9.1	17.6	10.7	1,577,550
All	**7.3**	**24.8**	**31.2**	**9.0**	**17.4**	**10.3**	**1,609,936**
Nonvet.	**22.0**	**25.8**	**24.0**	**5.6**	**15.5**	**7.1**	**17,124,495**

Source: Integrated Public Use Microdata Series, American Community Survey, 2008–2012

Table 3.3 Veteran households in Texas by age of parent and presence of children, 2010

Age	No children present	1 child present	2 children present	3 or more children present	Total
18–24	62.8	25.0	10.0	2.2	12,746
25–44	34.1	20.3	27.4	18.2	273,727
45–64	66.8	19.7	9.5	4.0	560,208
65+	88.9	9.5	1.3	0.3	543,218
All	**69.0**	**15.9**	**9.8**	**5.3**	**1,389,899**

Source: Integrated Public Use Microdata Series, American Community Survey, 2008–2012

3.2 Data and Methods

3.2.1 Data Sources

In addition to current data on Texas veterans and related dependents, the primary data source used in this work was the Integrated Pubic Use Microdata Series (IPUMS) from the Minnesota Population Center (Ruggles et al. 2010). The IPUMS includes a sample of individual's (but without identification) data from the U.S. Census Bureau's decennial census and American Community Survey. Because of change in collection methods, question wording, and survey form layouts, variables describing the characteristics of the population at one point in time may not be identical to the variables at another period of time. The IPUMS, when possible, harmonizes variables so that population characteristics can be compared across time. The IPUMS allowed estimates to be made of the number and characteristics of veterans and their households for historical periods back to 1980. Data were obtained on veterans and the characteristics of veterans from the data in IPUMS for the 1980, 1990, and 2000 Census sample survey as well as the American Community Survey for 2008–2012 (which we used to represent 2010). The IPUMS allowed estimates to be made of the characteristics of veterans and veteran households. These were used as inputs to the projection model used in our analysis.

In addition to the census data, data on Hazlewood Exemption use were obtained from the Texas Higher Education Coordinating Board (THECB). The Hazlewood database records the number of exemptions by year, type of exemption (veteran, dependent, spouse, or Legacy), institution, and characteristics of the recipients. Although these data have been collected for many years, there was no consistency in the reporting of the data until the THECB standardized the data reporting requirements beginning with FY2012.[3] Therefore, we used Hazlewood Exemption data only from 2012, 2013, and 2014.

[3]For additional discussion on this issue see Legislative Budget Board. 2014. *Report on Hazlewood Exemption*. Austin, Texas.

3.2.2 Methodology for Projecting Base Populations

In order to project Hazlewood Exemption use a series of projections were completed for the populations that are likely to be eligible for the exemption. This is primarily the veteran population and their dependents in Texas. After projecting the veteran population we prepared projections of veteran children as well as the disabled veteran population. In this section, we first describe our methods for projecting the veteran population and follow that discussion by outlining our methods for deriving the childhood population and the disabled veteran population.

Veterans are "born" when military members are honorably discharged from military service. However, unlike a real birth, veteran births do not occur at a single given age so that in a way, veteran "births" operate much like population in-migration because individual veterans are entering the veteran population at different ages (with higher probabilities of "birth" occurring at certain ages). In addition, to project veteran population, estimates of the two other components of population change: migration and mortality are required. Due to a limited time frame in which to prepare the projections of veteran populations and their dependents used to analyze alternate public policy proposals, data on military discharges and veteran deaths could not be obtained in order to estimate these components of change and prepare a modified cohort-component projection model. Because it requires fewer data inputs, the Hamilton-Perry method is well suited to projecting special populations such as the veteran population in Texas (Hamilton and Perry 1962; Swanson et al. 2010). In order to project the veteran population (and veteran dependents), estimates of the veteran population by age were obtained from the decennial censuses of 1980, 1990, 2000 and the American Community Survey for 2008–2012 (which we used to represent the 2010 period). From these data, cohort change ratios (CCRs) were calculated for each period (1980–1990, 1990–2000, 2000–2010) for veterans by race/ethnicity (nonHispanic White, nonHispanic Black, Hispanic and nonHispanic Asian and Other groups) and age. After evaluating these different ratios (and average of ratios), final CCRs were selected and used to project the veteran population in Texas for 2020 and 2030. CCRs are calculated in the following way:

$$ {}_{n}CCR_{x} = {}_{n}P_{x,t}/{}_{n}P_{x-k,t-k} $$

where

${}_{n}P_{x,t}$ = the population aged x to x + n at the last of two different points in time (t),
${}_{n}P_{x-k,\ t-k}$ = the population age x−k to x−k + n at the first of two different points in time (t−k),
k = the number of years between the two points in time for which population data are available

These ratios were obtained for all 5-year age groups from 30 to 34 through 75+ years of age using 2000 to 2010 data. These change ratios were then applied to the population by age in 2010 to project the population by age in 2020 and in 2030. For example, the CCR calculated for the veterans obtained from intercensal change for those 30–34 years of age in 2000 to those 40–44 years of age in 2010 was applied to the 30–34 year olds in 2010 and 2020 to project 40–44 year olds in 2020 and 2030. Change ratios were calculated for each period (1980–1990, 1990–2000, and 2000–2010). After carefully reviewing the results of these models as well as the data on the changes in the size of the military, it was determined that the best model for future projections was that based on the change ratios for the most recent period (2000–2010). Despite the wars in Iraq and Afghanistan, the size of the military has remained relatively stable and is much smaller than the military present during the first Gulf War and prior to the fall of the Berlin Wall. During the 1990s, there were additions to the veteran population as a result of the end of the Cold War and Gulf War I. Between 2000 and 2010, there was only a slight increase in the size of the military and no large increases in the veteran population as a result of military downsizing relative to other periods of time. Although the wars in Afghanistan and Iraq had officially ended, the size of the military did not increase significantly so that the rate at which active duty service members leave the military and become new veterans are likely to be similar to those between 2000 and 2010.

Cohort change ratios were not computed for the age groups under 30 years of age for this analysis because there were few veterans in these cohorts 10 years earlier (when the veterans were younger than 20) and thus CCR ratios were not applicable for use in the analysis of veterans 10 years later. We estimated the number of veterans at these youngest ages by assuming that the ratio of the number of veterans to the number of people serving in the military for each of these age groups in 2010 were applicable for each decade through 2030.

The U.S. Department of Defense has indicated that the military will be reduced by 5 % by 2018 (Congressional Budget Office 2013). We assumed that this reduction would be applicable to the military in Texas so a reduction of 5 % (proportionally by age) was completed for 2017–2018 and the new values so obtained used in our analysis for periods after 2018. After establishing the projections for 2020 and 2030, the projections for the interim years were derived by interpolating the projected values for the change between 2010 and 2020 and 2020 and 2030. The final projections were then reviewed for reasonableness by comparing the projections to those produced by the U.S. Department of Veteran Affairs as well as reviewing the change in age structure of the projected veterans over time.

The projections of the base veteran population were used to project the potential recipients of Legacy exemptions (children). First, veteran households were estimated for 2010 by using 2008–2012 estimates of the veterans who were either a householder or the spouse of a householder. The estimated number of veteran households for each age group was used to calculate veteran householder rates that were then applied to the veteran population projections to derive projections of veteran households. An average of the proportions of veteran's households with children from the 2000 Census and the 2008–2012 American Community Survey by age was applied to the veteran household projections. The results of this analysis provided the projected number of veteran households in which at least one child would be present.

The Hazlewood Legacy program can only be passed on to one child at a time and children must use the exemption before age 26. As a result an additional step was taken to estimate and project the number of children potentially eligible for the Legacy program. This number was estimated based on variables in the Integrated Pubic Use Microdata Series (IPUMS) from the Minnesota Population Center (Ruggles et al. 2010) that indicate the presence of children in the household and the age of the youngest child in the household. This estimate of the youngest child in the household was applied to the household projections to estimate the number of veteran children by age. The age of the youngest child was used to determine the maximum length of time that a Legacy exemption could be used by his or her children (a veteran can have more than one child and theoretically could give a portion to each child until the exemptions is used up). The ages of children were grouped into 5-year age groups except for the youngest age group (<1, 1–5, 6–10, 11–15, 26–20, and 21–25). The number of children for each age group was then combined with the age of veteran householders to calculate child to parent ratios for each decade. The ratios were calculated as follows:

Children < 1–5 / Veteran Householders < age 44
Children 6–10 / Veteran Householders < age 49
Children 11–15 / Veteran Householders < age 54

After review of the historical data, the average of 2000 and 2010 child to veteran ratios were selected and applied to the veteran household projections to produce projections of children by age. Some children may leave home for college or work but may still be dependents of their veteran parents (in terms of education financing). Thus, projections for the oldest ages (16–20 and 21–25) were calculated using cohort change ratios similar to the ratios used for the veteran population. These ratios were calculated for the overall population for those age groups (i.e. population 21–25 in 2010/population 11–15 in 2000 and population 26–30 in 2010/population 16–20 in 2000). The resulting CCRs were then applied to the veteran child population for the respective ages to project the potential Legacy eligible population in 2020 and 2030. The average CCRs for 1990–2000 and 2000–2010 were used. The projections for the interim years were derived by interpolation and exponentially smoothed to derive reasonable estimates by year and age group.

3.2.3 *Methodology for Projecting Rates of Exemption Use*

The veteran population projections and the potential Legacy eligible population projections were then used to calculate Hazlewood Exemption rates of use for 2012, 2013, and 2014 by dividing Hazlewood Exemption use by age of the recipient and type of exemption for the specified exemption type (i.e. veteran exemptions by age divided by veteran population projections by age and Legacy exemptions by age divided by veteran children by age). The Legacy rates were limited to the ages 16–20 and 21–25. In addition, the spouse and dependent rates were calculated by dividing the total number of exemptions for these two categories by 5 % of the projected veteran-present households with children where the parent was less than age 60. This was based on the fact that in 2010, 5 % of these households included veterans with a disability rating of 70 % or greater (the highest age category captured in the American Community Survey). The Hazlewood Exemption is only available to spouses and dependents of those with a 100 % disability rating and those whose parent or spouse died in action. Unfortunately, we did not have estimates of those with 100 % disability and can only use the 70 % or greater disability rating, thus our projections of disabled veterans may be biased upwards.

The rates were calculated for each group as described above and for 2012, 2013, and 2014. The change in these rates by year was then trended to 2025 using annualized rates of change determined by using average rates for 2012–2014. The rates for each year were then applied to the projected veterans population by age, eligible children by age, and the disabled veteran household population under age 60 by year. An alternative projection was prepared for comparison. This projection assumed 2014 rates of Hazlewood Exemption use (i.e. that there were not changes in rates of use). After projecting the number of recipients through 2025, we discuss the effects of proposed eligibility changes on these projected number of Hazlewood recipients.

3.3 Projections of Hazlewood Exemption Recipients

The changing age structure of the veteran population in Texas and the United States will lead to fewer veterans at ages most likely to benefit from educational programs. The projected growth in the Hazlewood program is primarily driven by the increase in the age specific rates of use and not by the growth in the underlying populations. The number of veterans under the age of 50 is projected to decline by 68,000 from 2014 to 2025 while the numbers of veteran's children in the ages 18–25 are projected to decline by almost 40,000 people during the same period.

Using historical data on Hazlewood Exemptions from the Texas Higher Education Coordinating Board and historical demographic data about Texas veterans and their households, we projected the number of recipients of the Hazlewood Exemption by category of exemption through the year 2025. The results of these projections are shown in Fig. 3.3 and Table 3.4. From 2012 to 2014, the number of individuals using the Hazlewood Exemption increased by 9,608 people (a 35.4 % increase – see Table 3.4). Most of that growth occurred among those who used the

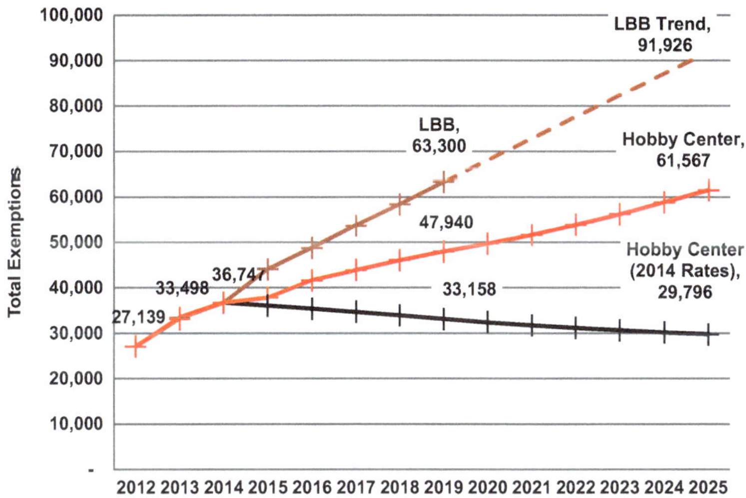

Fig. 3.3 Comparison of Hazlewood Exemption Recipients by Year 2012–2014 and Projected for 2015 to 2025 by the Texas Legislative Budget Board (LBB) and the Hobby Center for the Study of Texas. *Note:* The LBB projected Hazlewood use through 2019. LBB projections to 2025 shown here by a *dashed line* assume that the 2014–2019 trends continue through 2025

Table 3.4 Hazlewood exemptions by type and year, 2012–2014 and projected from 2015 to 2025

Type	2012[a]	2013[a]	2014[a]	2015	2016	2017	2018	2019
Veteran	14,706	16,057	16,297	15,757	15,805	15,615	15,551	15,376
Legacy	11,477	16,037	18,599	19,816	23,192	25,219	27,064	28,729
Spouse & Dependent	956	1,404	1,851	2,276	2,690	3,086	3,469	3,835
Total	**27,139**	**33,498**	**36,747**	**37,849**	**41,687**	**43,920**	**46,084**	**47,940**
Type			**2020**	**2021**	**2022**	**2023**	**2024**	**2025**
Veteran			15,265	15,118	15,008	14,879	14,767	14,644
Legacy			30,216	31,984	33,993	36,205	38,590	41,123
Spouse & Dependent			4,186	4,527	4,860	5,183	5,497	5,800
Total			**49,667**	**51,629**	**53,861**	**56,267**	**58,854**	**61,567**

Source: Texas Higher Education Coordinating Board (2012–2014); Hobby Center for the Study of Texas (2015–2025)
[a]Historical estimates as of January 28, 2015

Legacy exemption (74.1 % of the total growth). The Legacy program began in 2009 and, as with most new programs, the number of users grew quickly as people learned about the program but over time the rate of growth in new recipients has slowed. Our projections incorporate age specific rates of use for each exemption category. The use of such age specific rates is preferable because of the changing age characteristics and size of the veteran population as noted above.

If the historic change in the age specific rates of use continue through the projection period, given the underlying population of veterans and the related potential eligible children and spouses, then the number of people using the Hazlewood Exemption will increase to 47,940 by 2019 and to 61,567 by 2025. This growth, however, will be bounded by the size of the eligible population so that the growth in the Hazlewood Exemption is not likely to be as great as that suggested by the Legislative Budget Board (which merely extrapolated the weighted average change in the number of exemptions from year to year into the future). As the data in Fig. 3.3 indicate, the projected growth in the total number of exemptions is likely to be 16,000 less than the LBB's 2019 projection and 30,000 fewer in 2025 (assuming that the LBBs projection were extended to 2025).

Our projections indicate the Legacy and spouse and dependent exemptions will account for all of the growth in the Hazlewood Exemption program from 2014 to 2025. If current change in the rates of use continue, the number of people using the Legacy exemption will increase to 28,729 by 2019 (an increase of 10,130 people from 2014 to 2019) and to 41,123 by 2025 (an increase of 22,524 – more than doubling the number of people using the Legacy exemption).[4] In addition, given current change in the rates of use, the number of people using the spouse and dependent exemption will increase to 3,835 by 2019 (adding 1,984 from 2014 to 2019) and 5,800 recipients by 2025 (an increase of 3,949 or 213.3 %). Our analysis thus shows that the number of veterans at ages most likely to use the veteran exemption (coupled with the fact that the rates of use of the veteran exemption have been relatively flat over the last few years) results in a projected decline in the use of the Veteran exemption (a decrease of 921 users or −5.7 % by 2019 and a decline of 1,653 users or −10.1 % by 2025). Thus, our analysis projects that by 2019, there will be a projected 15,376 veteran recipients and by 2025, there will be 14,644 veteran recipients (down from 16,297 in 2014).

In order to examine the effects of the change in the age characteristics of the veteran population, an alternative projection was completed and is shown in Fig. 3.3. In this projection, age-specific rates were held constant at the rate that they were in 2014. Under this assumption, the number of people using the Hazlewood Exemption would **decline** by about 7,000 people from 2014 to 2025 (changing from 36,747 recipients in 2014 to 29,796 recipients in 2025). Therefore, if the age specific rates of use remain the same as they were in 2014, then the number of recipients of Hazlewood Exemptions would be less than half the number projected under our baseline projections that assume that age specific rates of use continue to increase (29,796 exemptions in 2025 at a constant rate compared to 61,567 exemptions under our baseline projection).

Even without changes in the eligibility requirements of the Hazlewood Exemption, the number of recipients of the Hazlewood Exemption will increase over the coming years. But this increase is unlikely to be as large as that projected by the

[4]Under these projections, 34 % of eligible veteran children age 18 to 25 will use the Legacy Exemption (up from 13 % in 2014).

LBB because of the change in the age structure of the veteran population and associated change in dependents or descendants. The growth in the Legacy program since it began in 2009 is an indication of its popularity and of the demand for means of financing a college education. Several alternative proposals to change the eligibility criteria in order to slow the growth in the use of the Hazlewood Exemption and associated costs have been presented in proposed legislation. The next section analyzes the potential impacts of these proposals on change in the number of recipients of the Hazlewood Exemption.

3.4 Effect of Proposed Changes to Eligibility Criteria for the Hazlewood Exemption

The previous section reported the results of projections of the Hazlewood Exemption recipients. These projections assumed that there would be no change in the eligibility criteria from the criteria present in 2015. During the 2015 legislative session, several proposals to change eligibility criteria were discussed and legislation was proposed (Birdwell 2012; Zerwas et al. 2015). After the legislative session, the Hazlewood Exemption Program remained as it was before and no change was made to the eligibility criteria. However, concern about growth in the number of Hazlewood Exemptions and associated costs continue and will likely produce proposals to limit the program. Current proposals to limit the exemption include:

1. Limiting eligibility to those veterans who served at least 6 years on active duty;
2. Limiting Legacy eligible children to those whose veteran parent served at least 6 years on active duty;
3. Limiting Hazlewood recipients to those veterans (and their children) who were born in Texas or to those who have lived in Texas for at least 8 years;
4. Limiting the exemption by adding an age limit to recipients that are 60 years of age or younger;
5. Limiting the exemption by adding an expiration date for using the exemption of 10 years (or 15 years) after the veteran's discharge; and
6. Setting additional academic standards to continue eligibility.

Unfortunately, data to fully evaluate the projected impacts of all of these changes are not available. However, several of these alternatives can be evaluated with existing data and the results of such analyses are included in the following sections. For each proposed change, current statistics are provided about veterans that are critical in evaluating the impact of the proposed criteria. Where possible, conclusions are drawn about the potential impact of each criteria change on the projected number of Hazlewood beneficiaries.

3.4.1 Age Restriction and Expiration Date

Two proposals to slow the growth in the Hazlewood Program included restricting the exemption to only those veterans and their families who were younger than 61 and to set a timeframe (since military discharge) in which a veteran or Legacy dependent can use the exemption. The age limitation would have limited impact on the use of the Hazlewood Exemption because few veterans and veteran dependents utilize the exemption at older ages. The expiration date restriction is more difficult to evaluate because the historical data available for the Hazlewood Exemption did not include information on the veteran's date of discharge.

Restricting Hazlewood eligibility to veterans aged 60 years of age or less will not significantly impact the number of people using the Hazlewood Exemption. The majority of Hazlewood Exemption recipients in 2014 were younger than age 30 and the majority of veterans who used the Hazlewood Exemption were under age 40 (Texas Higher Education Coordinating Board 2015). Veteran and spouse exemptions are taken at older ages (average age for veterans was 38 while the average age for spouses was 43). The Legacy and dependent programs limit children of veterans so that they cannot take the benefit beyond age 26. So, as would be expected, most *Dependent* and *Legacy Exemptions* are taken during the late teens and early twenties with the average age being 21 (for Legacy) and 22 (for Dependents). Legacy Data are not available to evaluate proposals that would limit the Legacy and dependent exemption based upon a parental age limit of 60. However, it is highly likely that this limitation would have only a small impact on the number of recipients because most parents of college age children are in their late 40s and 50s.

Although the change in the numbers of exemptions is small, the age limitation would likely impact the spousal beneficiaries the most. The average age for spousal beneficiaries was 43 years of age in 2014 and 12 % of spousal beneficiaries were 60 years of age or older. In order to understand the impact of this limitation, an alternative projection for the veteran and spouse and dependent beneficiaries was prepared. For these alternative projections, the same proportional distribution of Hazlewood Exemption users by type as were present in 2014 was assumed. This age restriction decreases the total number of exemptions to veterans by 5 % and spouse and dependent beneficiaries by 3 % from what they would have been without the restriction (see Table 3.5). The number of veteran beneficiaries would decline to 13,839 in 2025 (a decrease of 2,458 people or 15.1 %). Without this restriction veteran exemptions would only decline by 10.1 % for the same period (or 1,653 people). Spouse and dependent exemptions would increase by 3,793 from 2014 to 2025 (to 5,644). Without this exemption, spouse and dependent exemptions would increase by 3,949 (an increase of 213.3 %). This age restriction would have no impact on the Legacy program because Legacy beneficiaries are younger than age 60 (and most are younger than age 25).

The effects of an expiration date for Hazlewood eligibility are more difficult to estimate. Today's military recruitment standards are more stringent than in

Table 3.5 Veteran and spouse & dependent exemptions in 2014 and projected to 2025 for the Baseline Projection and assuming an age restriction of 60 years or younger

Year	Base projection	Age <60 restriction	Year	Base projection	Age <60 Restriction
	Veteran			Spouse & dependent	
2014	16,297	16,297	2014	1,851	1,851
2019	15,376	14,530	2019	3,835	3,732
2025	14,644	13,839	2025	5,800	5,644
2014–2025			**2014–2025**		
Change in veteran exemptions			**Change in spouse & dependent exemptions**		
Numeric	−1,653	−2,458	Numeric	3,949	3,793
Percent	−10.1	−15.1	Percent	213.3	204.9

previous decades so that those serving in the military are more likely to have at least a high school diploma and some college education relative to previous generations of recruits (United States Department of Defense 2013). Figure 3.4 shows, for alternative periods since military service, the percentage of veterans who would be most likely to benefit from the Hazlewood Exemption: those veterans with no college degree (includes those with no high school diploma, those with a high school diploma, and those with some college but no degree or vocational certification). Evident in Fig. 3.4, the percentage of veterans who would most likely benefit declines as the number of years since military service increases through the period of 10–14 years. The percentage of veterans without college degrees or vocational certification increases for the next two durations; however, this is most likely a function of older generations of veterans who were more likely to have lower educational attainment. The patterns suggest that most veterans go back to school soon after their discharge, typically when they are younger and when federal educational benefits are available. Most federal education benefits are limited to a maximum of from 10 to 20 years after discharge or death of the service member depending on the particular program (U.S. Department of Veterans Affairs 2012).

Limiting the education benefit to no more than 15 years after discharge would likely have the largest impact on the Legacy and dependent programs. In order to use the Hazlewood Exemption, a child must be no younger than age 4 at the veteran parent's discharge or death (assuming that the child entered college and used the exemption at age 18). According to the 2012 Current Population Survey Veteran Supplement, the median age of the youngest child of U.S. veterans who left military service within the previous 5 years was 4 years of age while the median age for the oldest child was 10 years of age.[5] Unfortunately, the Hazlewood database does not have information linking the Legacy and dependent recipients to their veteran parents or information about the time since discharge of the veterans. Thus, we are not able to fully analyze the impact of this criteria change but these data on

[5]We used the Integrated Public Use Microdata Series – Current Population Survey to analyze veteran time since discharge and age of children in households.

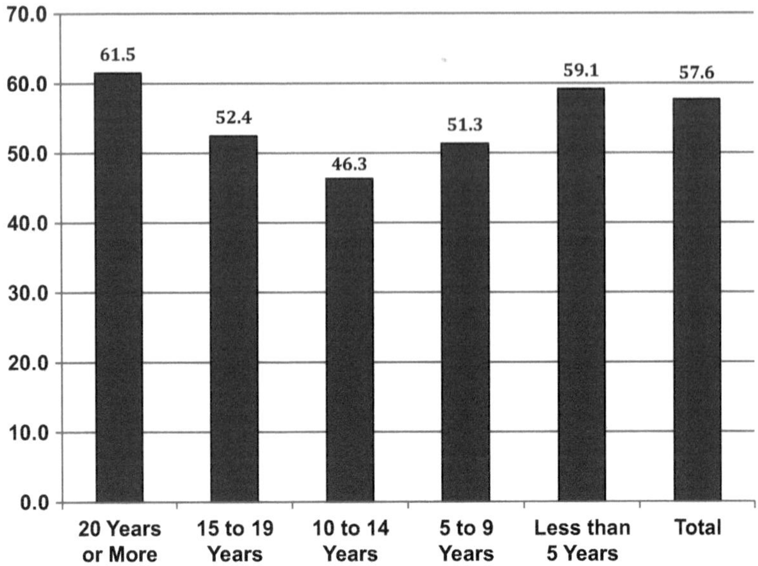

Fig. 3.4 United States Veterans Age Less Than 65 Without College Degrees or Certificates by Length of Time Since Last Military Service

veteran children suggest that at least a part of the veteran dependents may benefit from the Hazlewood Exemption. However, this change would restrict the benefit to only those children who were born prior to or during the veteran's military service (and only those children that were at least 4 years old at the time of the parent's discharge or death).

3.4.2 *Length of Service*

One proposal for changing the eligibility requirements of the Hazlewood Exemption limits the exemption to only those veterans who served at least 6 years of active duty. This change would result in requirements that are more restrictive than the requirements for federal educational benefits and significantly impact the number of veterans who would be eligible for receipt of the Hazlewood Exemption.[6] In the United States in 2012, only 31.2 % of veterans who were less than age 65 served 6 years or more of active duty [King et al. 2010; U.S. Census Bureau and U.S. Bureau of Labor Statistics 2012 (see Table 3.6)]. These veterans served an average of 4.3 years (with 68.8 % of these veterans serving less than 6 years).

[6]For information on federal education benefits see U.S. Department of Veterans Affairs, Veterans Benefits Administration (2012). Summary of VA Education Benefits. http://benefits.va.gov/BENEFITS/benefits-summary/SummaryofVAEducationBenefits.pdf.

Table 3.6 Veterans in the United States less than age 65 by length of military service

Years served	Number	Percent	Cumulative percent
<4	3,963,435	40.9	40.9
4 through 5	2,700,385	27.9	68.8
6 through 9	1,406,864	14.5	83.3
10 through 14	577,251	6.0	89.3
15 or More	1,035,954	10.7	100.0
Total	9,683,889	100.0	

Source: IPUMS-CPS (2012 Current Population Survey, Veteran Supplement)

Table 3.7 Veteran and legacy exemptions in 2014 and projected to 2025 under alternative assumptions for length of service requirements

		Length of service of at least:		
Year	Base projection	4 years	6 years	10 years
	Veteran			
2014	16,297	16,297	16,297	16,297
2019	15,376	9,087	4,797	2,568
2025	14,644	8,655	4,569	2,446
Change in veteran exemptions 2014–2025				
Numeric	−1,653	−7,642	−11,728	−13,851
Percent	−10.1	−46.9	−72.0	−85.0
	Legacy			
2014	18,599	18,599	18,599	18,599
2019	28,729	16,979	8,963	4,798
2025	41,123	24,304	12,830	6,868
Change in legacy exemptions 2014–2025				
Numeric	22,524	5,705	−5,769	−11,731
Percent	121.1	30.7	−31.0	−63.1

We do not have information on the length of service for veterans that were past recipients of the Hazlewood Exemption (or the veteran parents for Legacy exemptions). However, in order to estimate the potential impact of this change we examined the results of alternative projections. Using our original projections reported in the previous section, we provide alternative scenarios of veteran and Legacy exemptions by assuming that the length of service characteristics for Texas veterans are the same as those for veterans nationally. These alternative scenarios show the potential impacts on the program due to the use of the length of service requirement of 4 years or more, 6 years or more, and 10 years or more. The results of this analysis are shown in Table 3.7. The number of 2014 recipients is shown for comparison. The effect of these restrictions would be to decrease the number of recipients by more than 70 % if the veterans were required to have served at least 6 years (compared to a 46.9 % cut in recipients at a required four years or more and a 83 % cut for a period of service of ten years or more). Given the 6 year length of service requirement, the number of veteran exemptions would decline by 12,000 (a 72 % decrease) while the number of Legacy exemptions would decline by 5,769 (a 31 % decrease).

In addition to limiting the number of Texas veterans eligible for Hazlewood, the length of service eligibility criteria will effectively limit the program to veterans (or their children) who served as commissioned officers and career enlisted men and women. The length of service varies between the enlisted and officer ranks. For those who were serving active duty in 2013, officers had served an average of 11 years or more while those in the enlisted ranks had served 6.7 years (United States Department of Defense 2013). In 2013, 46.2 % of active duty enlisted members had served for 3 years or less. On the other hand, commissioned officers served a longer period of time, with 45.1 % of commissioned officers serving 11 years or more (Office of the Under Secretary of Defense 2013).

Importantly, of all of the proposed changes to the Hazlewood Exemption, the length of service requirement would have the largest impact on the number of recipients. This change in criteria would slow the growth in the program, and for the longest service requirements, would cut the number of exemptions below the number in 2014. In addition, veterans who are most likely to need this exemption (those who served in the enlisted ranks) will be the ones least likely to meet this requirement and be eligible for the Hazlewood Exemption.

3.4.3 Residency Requirements

The Hazlewood Exemption was originally intended for Texas veterans – those who were born in Texas or who had lived in Texas for a specific number of years and enlisted or were commissioned in Texas. Data on the state of enlistment for current veterans is not available. However, because of its relative population size and because Texans have a higher propensity to serve in the United States military relative to citizens of other states, Texans account for a significant percentage of military recruits (United States Department of Defense 2013: 19–22). In 2013, of those who enlisted into the active duty military, 9.8 % (or 16,078) enlisted in Texas. These individuals, upon honorable discharge and return will add to the ranks of Texas veterans.

Due to the nature of military service, military service members are more mobile than the population as a whole (Bailey 2011). Service in the military may lead individuals, upon discharge or retirement, to settle far away from the state in which they were born or enlisted. According to the American Community Survey, an estimated 50.1 % of the veterans living in Texas (or 806,598) in 2010 were born in Texas and 46.3 % (or 294,020) of veterans who were 55 years of age or younger in 2010 were born in Texas (see Table 3.8). In addition, today's veterans are more mobile than those in previous decades. Over time, the percentage born in Texas has declined for all veterans as well as veterans who were less than age 55. Limiting the program to only veterans born in Texas would severely curtail the number of

Table 3.8 Number and percent of veterans living in Texas who were born in Texas, 1980, 1990, 2000 and 2010

	All veterans		Veterans less than age 55	
Year	Number	Percent	Number	Percent
1980	1,004,480	58.8	657,100	57.7
1990	1,062,001	55.7	572,153	52.9
2000	925,849	52.9	425,073	49.7
2010	806,598	50.1	294,040	46.3

Source: Integrated Public Use Microdata Series, Census 1980, 1990, 2000 and American Community Survey, 2008–2012

veterans eligible to receive Hazlewood benefits to half of the veteran population living in Texas.

The number and percentage of recruits enlisting in Texas and the number and percentage of veterans living in Texas who were born in Texas provide an indication of the potential beneficiaries of educational benefits but the data do not provide sufficient information to fully analyze the impact of limiting Hazlewood beneficiaries to those who had lived in Texas for at least 6 or 8 years prior (the residency requirements included in recent legislative proposals). The last large scale survey to collect information on the mobility of veterans for a length of time longer than 1 year was the long form sample survey of the 2000 Census. According to the 2000 Census, 91.0 % of veterans living in Texas in 2000 lived in Texas in 1995 as well (Table 3.9). Thus, only 9.0 % of all veterans living in Texas in 2000 lived outside of Texas 5 years prior. For the purposes of the Hazlewood Exemption, the percentage may be even smaller because these statistics do not account for those veterans who lived elsewhere 5 years prior to the census but were on active duty and claimed Texas as their home of record.

The census data on migration mask important differences in migration histories for veterans of various ages. In general, people are more likely to move at younger ages but over time settle in one place. This pattern can be seen in Fig. 3.5. At the youngest ages (a time when most veterans attend college) a large share of the veteran population moved to Texas between 1995 and 2000 (38.5 % of those age 25–29 in 2000). Of those veterans age 55 years and younger, 16.1 % lived in another state or abroad in 1995.

Assuming that the age specific mobility characteristics of veterans in 2000 exist today and continue into the future (and not accounting for those veterans who migrated but who had claimed Texas as their home of record), residency requirements of at least 5 years would limit the number of veteran Hazlewood beneficiaries. At the same time, this change would likely have limited impact to the number of Legacy recipients because the parents of these individuals are older and more likely to have lived in Texas for at least 5 years. Table 3.10 shows the results of alternative projections for a residency requirement of at least 5 years for the veteran and Legacy programs. These alternative projections show that this limitation would reduce the veteran projection to 79 % of what they would be without this limitation (11,500 recipients compared to 14,644 projected recipients in 2025) and the Legacy

Table 3.9 Residence in 2000 of All Texas Veteran Householders and Veteran Households with a Householder Less Than Age 55 (in 2000) Relative to Residence in 1995

Migration status	Number	Percent	Cumulative percent
	All veteran households		
Same House	909,378	64.0	64.0
Elsewhere within Texas	383,234	27.0	91.0
Another State	113,110	8.0	99.0
Abroad	14,350	1.0	100.0
	Householder less than age 55		
Same House	381,143	44.5	44.5
Elsewhere within Texas	337,492	39.4	83.9
Another State	117,806	13.8	97.7
Abroad	19,619	2.3	100.0

Source: Integrated Public Use Microdata Series, Census 2000

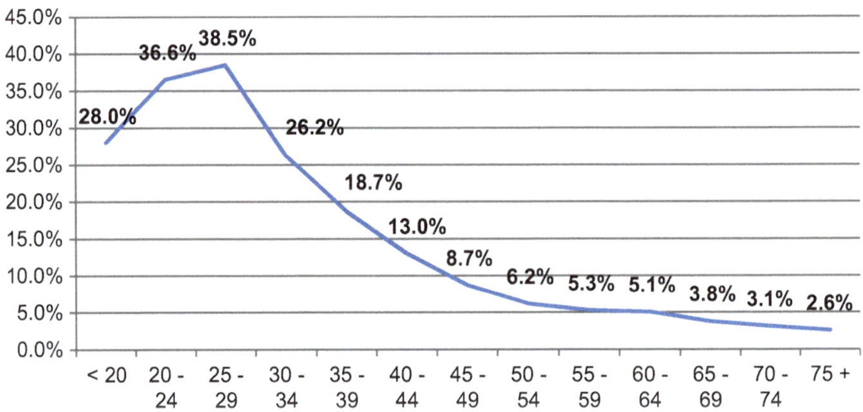

Fig. 3.5 Percent of Texas Veteran Householders by Age Who Moved to Texas from Another State or From Abroad Between 1995 and 2000

Table 3.10 Veteran and legacy exemptions in 2014 and projected to 2025 for the baseline projection and assuming a residency requirement of at least 5 years

Year	Base projection	5 year residency	Year	Base projection	5 year residency
	Veteran			Legacy	
2014	16,297	16,297	2014	18,599	18,599
2019	15,376	12,119	2019	28,729	25,856
2025	14,644	11,500	2025	41,123	37,011
2014–2025			**2014–2025**		
Change in Veteran Exemptions			**Change in Legacy Exemptions**		
Numeric	−1,653	−4,797	Numeric	22,524	18,412
Percent	−10.1	−29.4	Percent	121.1	99.0

recipients to 90 % of what they would be without this limitation (37,011 recipients compared to our 41,123 baseline projection).

Although sufficient information is unavailable to fully evaluate proposals to limit eligibility requirements for the Hazlewood program to a length of residency of 6 or 8 years, an examination of a residency of at least 5 years shows that this criteria change would have a larger impact on veteran exemptions leading to a projected decline of 29.4 % while having a much smaller impact on the Legacy program (which would still grow by about 99 % from 2014 to 2025).

3.5 Conclusion

Texas has long supported men and women who have served in the military. Recognizing the importance of a college education, the Texas Legislature began providing an exemption of tuition and fees to Texas veterans as far back as 1923 and has expanded those benefits over the years. Extensive growth in the use and associated costs of the Hazlewood Exemption, primarily as a result of the growth in the Legacy program, have led to concerns about the exemption's long term viability. According to our projections, the number of recipients will continue to grow over the next several years from 37,000 in 2014 to a projected 48,000 by 2019 and 62,000 by 2025. Although this growth is significant, it is not as large as that previously projected because our projections account for the change in the age structure and size of the veteran population. Despite wars in Iraq and Afghanistan, the U.S. military is smaller than what it was during the Cold War and the first Gulf War which will lead to a smaller veteran population in the future.

Because of the change in the size and age structure of the veteran population, there will be fewer veterans using the Hazlewood Exemption and our projections show that most of the growth in the Hazlewood Exemption will be as a result of the growth in the Legacy program. These projections also show that the number of Legacy recipients will increase to 29,000 by 2019 and 41,000 by 2025 (more than double the number of Legacy recipients in 2014). At the same time, the spouse and dependent exemptions will increase to 3,825 by 2019 and 5,800 by 2025 (up from 1,851 in 2014).

The alternative proposals for modifying the eligibility requirements would have disparate impacts on the growth in the Hazlewood Exemption. Of all proposed changes, the length of service would have the most substantial impact as it would limit the number of exemptions, assuming a 6 year length of service requirement, to a third of what they would be without the restriction. In addition, this requirement would limit the Hazlewood Exemption to those who are least likely to need the exemption (veterans who served as commissioned officers or spent a career in the enlisted ranks) since the average length of service for all veterans in Texas in 2010 was 4.3 years and active duty officers in 2013 had served an average of 11 years compared to an average of only 7 years for the enlisted ranks.

Requiring veterans to live in Texas for at least 8 years would likely impact veteran exemptions the most and do little to slow the growth in the Legacy program. The veteran parents of children attending college are more likely to have been living in Texas at least 5 years (the longest period for which we have data), while young veterans are more likely to have moved to Texas within the last 5 years. Similarly, although the majority of the users of the Legacy and Dependent Programs are in the ages of a typical college student (18–22), restricting the Hazlewood Exemption to veterans less than age 60 would only impact the veteran and spouse program. Finally, although data are not available to fully analyze the effect of adding an expiration date on the use of the Hazlewood Exemption, the information provided in this report suggests that this change would have the largest impact on Legacy exemptions because a child would need to be at least 4 years old at the time of the parent's military service discharge or death to be able to use any of this benefit.

The Hazlewood Exemption has played a critical role in improving the education of veterans, and in turn, improving the economic competitiveness of the Texas labor force. This program, unique among other states, is just one program among many others that shows the state's commitment to the United States military and its veterans. It is this commitment that have led some, who are concerned about its long term viability, to propose legislation to change the eligibility requirements of the program. This analysis was provided in order to help guide decision makers in determining the best paths forward for modifying and maintaining the Hazlewood Exemption.

References

Bailey, A. K. (2011). Race, place, and veteran status: Migration among Black and White Men, 1940–2000. *Population Research and Policy Review, 30*(5), 701–728.

Birdwell, B. (2012). *Relating to tuition and fee exemptions at Public Institutions of Higher Education for Certain Millitary Personnel and their dependents.* SB 1735. Texas Senate, 84th Legislature of Texas.

Congressional Budget Office. (2013). *Long-term implications of the 2014 future years defense program.* Washington, DC: Congressional Budget Office. Retrieved (http://www.cbo.gov/sites/default/files/44683-FYDP.pdf).

Hamilton, C. H., & Perry, J. (1962). A short method for projecting population by age from one decennial census to another. *Social Forces, 41*(2), 163–170.

King, M., Ruggles, S., Alexander, J. T., Flood, S., Genadek, K., Schroeder, M. B., Trampe, B., & Vick, R. (2010). *Integrated public use microdata series, current population survey: Version 3.0.* [Machine-readable database]. Minneapolis: Minnesota Population Center [producer and distributor].

Legislative Budget Board. (2014). *Report on Hazlewood exemption.* Austin: Legislative Budget Board.

National WWII Museum. (2015). The National WWII Museum | New Orleans: Learn: For students: WWII by the numbers: US Military. Retrieved April 30, 2015 (http://www.nationalww2museum.org/learn/education/for-students/ww2-history/ww2-by-the-numbers/us-military.html).

Office of the Actuary. (2014). *Veteran population projections model (VetPop2014)*. Washington, DC: National Center for Veterans Analysis and Statistics, United States Department of Veterans Affairs.

Office of the Under Secretary of Defense, Personnel and Readiness. (2013). *Population representation in the military services*. Washington, DC: United States Department of Defense. Retrieved May 7, 2015(http://poprep.cna.org/2013/).

Ruggles, S., et al. (2010). *Integrated public use microdata series: Version 5.0 [Machine readable database]*. Minneapolis: University of Minnesota.

Swanson, D. A., Schlottmann, A., & Schmidt, B. (2010). Forecasting the population of census tracts by age and sex: An example of the Hamilton-Perry method in action. *Population Research and Policy Review, 29*(1), 47–63.

Texas Higher Education Coordinating Board. (2015). *Hazlewood exemption database*.

U.S. Census Bureau. (2011). *The 2012 statistical abstract – U.S. Census Bureau*. Retrieved May 1, 2015(http://www.census.gov/compendia/statab/).

U.S. Department of Veterans Affairs, Veterans Benefits Administration. (2012). Summary of VA education benefits. http://benefits.va.gov/BENEFITS/benefits-summary/SummaryofVAEducationBenefits.pdf. Retrieved on July 30, 2015.

Weirlein, Jr., E. (2015). Keith Harris vs. Eliseo "Al" Cantu, Jr. U.S. District Court, Southern District of Texas.

Zerwas, J., Miller, R., & Otto, J. (2015). *Relating to tuition and fee exemptions at Public Institutions of Higher Education for Certain Military Personnel and their dependents*. HB 3572. Texas House of Representatives, 84th Legislature of Texas.

Chapter 4
Residential Fire-Related Injuries and Deaths in Anchorage, Alaska, 2007–2012: Causes and Prevention

Donna Shai

Abstract The purpose of the study was threefold: (1) to identify the main causes of fire injuries to the civilian population of Anchorage; (2) to identify risk factors through a data set drawn from fire department records and newspaper surveillance, and (3) to suggest preventative measures. Methods used included an analysis of cases in ANFRS, newspaper surveillance, and cross-tabulations. The analysis finds that Anchorage census tract analysis did not show a relationship between income and fire injuries, however behavior such as entering the fire area before it has been cleared, inadequate supervision of children, widespread use of space heaters, as well as above average smoking and drinking behaviors were likely to affect fires. The addition of sprinklers, repair of equipment such as hydrants, a public health campaign on children and fire play, smoking cessation efforts and increased warnings on holidays could be helpful.

Keywords Income • Smoking • Drinking • Fire play by children • Holiday dangers

4.1 Introduction

In 2010 fire and burn injuries were responsible for 3,194 deaths in the U.S. as a whole (Murphy et al., 2013). A long-established pattern of fire injuries by age group in the United States is that children under the age of 5 and the population over the age of 54 are at the highest risk of death in fire, and that the risk of fire injury is greatest in the 20–44 age group and in those 85 and over (US. Department of Homeland Security 2004, Ehrlich et al. 2008). When compared to other U.S. states, Alaska's population has been assigned a relative risk of 1.4 times more likely to die in a fire than the population as a whole and with a fire death rate of 15.4 (1) to the U.S. general population (U.S. Fire Administration 2013). In Alaska, the highest civilian fire injury rates from 2006 to 2012 included ages 40–49. The highest rate

D. Shai (✉)
Department of Sociology and Criminology, Villanova University, Villanova, PA, USA
e-mail: donna.shai@villanova.edu

D.A. Swanson (ed.), *The Frontiers of Applied Demography*, Applied Demography Series 9, DOI 10.1007/978-3-319-43329-5_4

for fire deaths was more variable, affecting the age group 40–49 only in 2012. The majority of victims in both causes are men (Fire in Alaska 2006–2012). In this paper I will examine fire deaths and injuries in Anchorage from 2007 to 2012 looking at the factors of climate, economic status, population and health behaviors to explain the vulnerability for fire injury in Alaskans in that age group. I then turn to a data set obtained from the Anchorage Fire Department comprised of injurious fires in Anchorage from 2007 to 2012. After making some suggestions for prevention, I also suggest a new way to keep records of fires that may occur after electricity cutoffs.

Anchorage has almost one-half of the Alaskan population and a well-equipped fire department, an advantage not shared with many of the more remote rural communities. It also has detailed data available to analyze the risk of injury and death to civilians within the county. While there are 116 fire departments in Alaska, according to 2011 reports (Fire in Alaska 2007), there are many areas, such as villages in southwest Alaska, where a fire may cause fatalities despite neighbors trying to fight it with buckets and hoses (Theriault Boots 2013). Since Alaska has the longest and coldest winter temperatures in the United States (Osburn 2013), it is not surprising that attempts to heat buildings, including residences, take on great importance. However, the need for heat can also create fire hazards in itself. For example, according to Mitchell and Seifert (2009), if fuel supply lines freeze, electric heat tapes have sometimes burned down the house. They also caution that items such as gas lanterns kerosene heaters when used to warm frozen fuel supply lines can ignite a house fire. Burning wood stoves can be dangerous if not well maintained. If electricity is cut off, the only alternatives, other than a standby generator, are wood- oil-or gas-fired space heaters, but these also can start fires if they are unattended (Mitchell and Seifert 2009). Residents in low income areas are the most vulnerable to electricity cut-offs due to nonpayment of bills, and may innovate by using space heaters and even cooking stoves for heat. Furthermore, firefighting in extremely cold conditions often is challenging and dangerous. In a newspaper article in the *Fairbanks Daily News-Miner* in January 2014 about a blaze in a Chena Ridge home, Richardson reports that given the temperature of 30 below, a sustained effort over 3 h from a number of Fire Departments was needed to extinguish the blaze (2).

4.2 Methods

The data used in this article come from two main sources: ANFRS, the Alaskan version of NFRS, for fires in Anchorage involving injuries and deaths, and newspaper surveillance of *the Anchorage Daily News* to obtain additional information about a particular fire where published, and fires that were not included in ANFRS. In addition, the publications of the Anchorage Fire Marshal's Office were used for yearly information of Anchorage fires. A total of 57 cases were included in the ANFRS data, with 21 additional cases from the newspaper surveillance, for a total of 78. The incidents were analyzed and mapped. In addition, I spoke with representatives of the main electricity suppliers in Anchorage, and a spokesperson from the regulatory council of Alaska about shut-offs of electricity in Alaska when customers fail to pay their electricity bill that might lead to fire injuries.

Another method was a cross-tab of fires involving injuries by census tract income, in order to test for the relationship between income and the risk of fire injuries. While Anchorage as a whole is a city with relatively high income (a median income of $55,546), it is easy to forget that there is a poor population amidst the affluence. There is a fairly extensive literature on the relationship between low income households and fire risk (e.g., Gielan et al. 2012, Runyan et al. 2005). Causes can include living in substandard housing, heating and cooking, and defective smoke detectors (Gielan et al. 2012). The causes extend from the nature of the housing, to the heating and electrical systems, etc., the likelihood of utility shut-offs for nonpayment. To investigate the extent to which injurious fires in Anchorage take place more often in low income households, all structural fires for 2007 to 2012 were mapped by census tract. Census tracts are often used to represent neighborhoods because an attempt is to keep the populations comparable and the boundaries used to demarcate them are highways or rivers or other natural barriers. Anchorage has 59 census tracts. Since social situations can be influenced by conditions that preceded them but not by conditions reached after the year of study we looked at the Census data for 2000 Summary File 3 median household income in 1999 (dollars) for data by tract. The median household income for the Anchorage Municipality was $55,546 in 2000. The chi–square was calculated for census tracts above and below the median income. In addition, census population data was used to determine if Anchorage had an unusual age structure that would explain fire injuries and deaths by age group.

4.3 Results

4.3.1 *Cross-Tabulation*

Census tracts were divided into tracts above the median and tracts below the median income of 73,004 in 2010 (2006–2010 American Community Survey, American Factfinder). Of the 55 tracts, 28 were below and 27 were above. Drawing from the data provided by the Anchorage Fire Department, the tracts were then divided into those that had experienced an injurious fire between 2007 and 2012 (6 years) indicated by "yes" and those that did not indicated by "no". The results were, as displayed in a cross tabulation (Table 4.1).

The numbers of fires per tract ranged from 0 to 10 over the study years. Looking at the 10 tracts with the highest numbers of fires (four and above), it was found that four were above the median income and six were below, suggesting that while income plays a role it is not a major role in predicting fire injuries. A chi-square analysis was then performed to see if there was a relationship between census tract

Table 4.1 Anchorage census tracts, 2007–2012: number of fires by household income[a]

Injurious fires?	Above median household income	Below median household income	Total
Yes	**18**	**15**	**23**
	(64 %)	(56 %)	
No	**10**	**12**	
	(36 %)	(44 %)	
Total	**28**	**27**	**22**
	(100 %)	(100 %)	

[a]See text for data definitions and sources

income and the likelihood of an injurious house fire in the last 6 years in those tracts. After completing the analysis I am 95 % confident that there is no statistically significant differences within these results, suggesting that income is not the crucial factor that it often is in studies of residential fires elsewhere.

4.3.2 Analysis of ANFRS and Newspaper Surveillance

The direct and indirect causes of injury from ANFRS and the newspaper surveillance can be divided into six groups:

1. Self-help before fire department arrives, such as trying to put the fire out with a garden hose, thawing out frozen pipes with heating tape, rescue attempts, returning to the site before the fire is contained, and jumping from a window when the fire company is on the way,
2. Children not given sufficiently close supervision, as when they are playing with matches or a cigarette lighter,
3. Smoking and drinking, cooking not properly attended, persons unable to act either through alcohol or disability, or too young to escape without help,
4. Nonworking equipment such as nonfunctional smoke detectors, nonworking fire hydrants or blocked hydrants, misuse of cooking equipment such as using a stove for heat, flames for light, candle for lighting, and space heaters,
5. Risks associated with holidays: Thanksgiving and stove fires, Christmas decorations, candles, Christmas trees,
6. Inappropriate storage of large amounts of ammunition or blocking space so that firefighters cannot move easily through the building or apartment.

4.4 Discussion

Is the unusual vulnerability of males 40–51 a reflection of an unusually large cohort? Using the range 40–49 from the 2010 Census, a comparison of the percentage of this cohort in Anchorage, Alaska and the U.S., found that males aged

40–49 were 7.2 % of the Anchorage population, 7.5 % of the Alaskan population and 7.0 % of the US population in 2010. Therefore, the cause of this group's involvement in fire injuries is not due to an unusually large cohort. However, compared to other age groups, the age group from 40 to 51, which was cited by the Alaska Fire Marshal as a unique and vulnerable age group for both fire injuries and fire deaths, is relatively large compared to other age groups in Anchorage and in Alaska. For example in Alaska, those aged 40—51, both sexes, (using fire department age designations) numbered 124,765 as opposed to 0–9, at 104, 883 and 70 + at 32,736. In Anchorage the age cohort of 40–51 was 50,922, while the 0–9 was 42,579 and the 70+ was 12,792. Therefore the vulnerability of the 40–51 years cohort may be due to their larger numbers. In addition, In Anchorage in particular, there is likely to be a migration factor of elderly residents leaving the state due to the harsh climate.

A fourth possibility is that the high risk age group of 40–51, mostly males, results from differences in health behavior which may predict fire injury, known as the "Scottish effect" (Popham and Boyle 2011). Health behaviors include rates of smoking and alcohol use. Alcohol plays a role in many fires (see Ahrens 2009). According to annual reports by the Anchorage Fire Marshal, the percentage of victims possibly impaired by alcohol was 27 % in 2008 (Fire in Alaska 2007–2012). Heating and cooking are the major causes of residential fires in Anchorage. In a public information notice in May of 2011, the State Fire Marshal issued a press release concerning a fire in Wasilla (outside of Anchorage). A male victim who died had been intoxicated, and apparently the alcohol abuse prevented him from escaping. He was found in the kitchen asleep while the kitchen cabinets were burning. The Fire Marshal commented "All evidence points to the fact that people who abuse alcohol and other drugs are unable to think clearly and unable to react rationally in an emergency situation. Many times they try to cook food and pass out, leaving their cooking unattended." He also warns the public against using alcohol while smoking (Tyler, D. "Another Tragic Fire Death Related to Alcohol Abuse" DPSPR# 11–027). Alaska State Troopers 2011 Annual Drug Report indicates that the primary drugs abused in Alaska are: alcohol, cocaine, heroin, marijuana, methamphetamine, and prescription drugs (Alaska State Troopers 2011 Annual Drug Report) and contribute to violent crimes and accidental deaths including fire deaths.

While the sale of cigarettes has declined since 1996, until 2008, the percent of adults who smoked in Alaska exceeded the average for the U.S. The percentage of smokers is higher among men than women (Alaska Tobacco Facts, April, 2012). States with high smoking rates generally have high mortality from residential fires (Diekman et al. 2008). When alcohol is combined with tobacco use, there are often deadly consequences.

4.5 Implications for Prevention

Research suggests that, in addition to smoke detectors, sprinklers should be used in the event of a fire. Smoking and drinking in the home should be reduced as much as possible. If possible, stoves should be able to turn off when food begins to burn (Ahrens 2013a, b). Television ads should target main home fire hazards such as candles and holiday decorations. Vulnerable groups should be reached such as men aged 40–51 (Popham and Boyle 2011), and children who require supervision while at play. Finally, electric turn-offs caused by weather or nonpayment should be monitored to prevent subsequent fires.

Acknowledgements I would like to thank the Anchorage Fire Department for its helpful cooperation.

Notes

1. The relative risk of fire deaths in Alaska was calculated by FEMA by dividing the total number of fire deaths in Alaska by the total population in the state, multiplied by 1,000,000. The crude rate is not age-adjusted. (See U.S. Fire Administration, 2004).
2. "Cause of Christmas Eve house fire still unknown" Jeff Richardson, Newsminer.com, Jan. 3, 2014.

References

Ahrens, M. (2009). *Possible impairment by alcohol or drugs as a contributing factor in home fire deaths*. Quincy: National Fire Protection Association, Fire Analysis and Research Division.

Ahrens, M. (2013a). *Home fires involving cooking equipment*. Quincy: National Fire Protection Association: Fire Analysis and Research Division.

Ahrens, M. (2013b). *Home candle fires*. Quincy: National Fire Protection Association: Fire Analysis and Research Division.

Alaska State Troopers, Alaska Bureau of Investigation, Statewide Drug Enforcement Unit (2011) *2011 Annual Drug Report*.

Alaska Tobacco Facts. (2012) The impact of tobacco on the lives of Alaska's people, April 2012 update. Available at: http://www.hss.state.ak.us/dph/chronic/tobacco/Alaska_tobacco_facts.pdf

Diekman, S. T., Ballesteros, M. F., Berger, L. R., Caraballo, R. S., & Kegler, S. R. (2008). Ecological level analysis of the relationship between smoking and residential-fire mortality. *Injury Prevention, 14*(4), 228–31.

Ehrlich, A. R., Bak, R. Y., Wald-Cagan, P., & Greenberg, D. F. (2008). Risk factors for fires and burns in homebound, urban elderly. *Journal of Burn Care & Research, 29*(6), 985–7.

Fire in Alaska. (2007). Department of public Safety; Division of Fire and Life Safety. Available at: www.dps.state.ak.us/fire

Fire in Alaska Issues. (2007–2012). Department of Public Safety, Division of Fire and Life Safety, Kelly Nicolello, State Fire Marshal, Alaska.
Gielan, A. C., Shields, W., McDonald, E., Frattaroli, S., Bishai, D., & Ma, X. (2012). Home safety and low-income urban housing quality. *Pediatrics, 130*(6), 1053–9.
Mitchell, F., & Seifert, R. (2009). *Keeping your house operating during a cold Alaskan winter*. Fairbanks: University of Alaska Fairbanks Cooperative Extension Service.
Murphy, S. L., Xu, J. Q., & Kochanek, K. D. (2013) Deaths: Final data for 2010. National vital statistics reports (Vol. 61, no. 4). Hyattsville: National Center for health Statistics.
Osburn, L. (2013). Coldest states in America. Current results: Research news and science facts. Available at: http://www.currantresults.com-Weather-Extremes/US/coldest-states.php
Popham, F., & Boyle, P. J. (2011). Is there a "Scottish effect" for mortality? Prospective observational study of census linkage studies. *Journal of Public Health* (Oxf), *33*(93): 453–458
Runyan, C. W., Casteel, C., Perkis, D., Black, C., Marshall, S. W., Johnson, R. M., Coyne-Beasley, T., Waller, A. E., & Viswanathan, S. (2005). Unintentional injuries in the home in the United States Part 1: Mortality. *American Journal of Preventive Medicine, 28*(1), 73–9.
Theriault Boots, M. (2013). Victims of fire in Southwest Alaska identified as 8-year-old boy, two men. ADN July 9, 2013, ADN.com. Available at: http://www.adn.com/2013/07/09/2969210/victims-of-fire-in-southwest-alaska.htm
U.S. Fire Administration. (2004). Department of Homeland Security, FEMA. Fire risk. Topical Fire Research Series, volume 4, issue 7, available at: http://www.usfa.fema.gov/statistics/reports/pubs/tfrs.shtm
U.S. Fire Administration/FEMA State fire death rates and relative risk. (2013). Available at http://www.usfa.fema.gov/staitstics/estimates/states.shtm

Chapter 5
Demographic Projections of Demand for Criminal Court Services Across New South Wales, Australia

Imogen Halstead, Brian Opeskin, and Nick Parr

Abstract Demographic projections can provide an evidential basis for the assessment of the future demand for criminal court services and facilities, and the future need for related staff, such as judges and magistrates. This chapter assesses the implications of projected changes in the size, composition and geographical distribution of the population of New South Wales (Australia) for the volume and geospatial dimensions of crime-related court service provision. The method involves application of court appearance rates by age, sex and local area to projections of the population. The results illustrate the pivotal importance of uncertainty relating to whether recent general declines in court appearance rates will continue. Under the assumption of constant court appearance rates, total court appearances are projected to increase by 18 % between 2012 and 2031. This is less than the 27 % projected growth of the population because of the concomitant increase in the share of older people, who have a lower propensity to appear before the courts. The projected percentage increases in court appearances are generally greater in Sydney, especially in Western and South-Western Sydney, whilst reductions are projected in regional New South Wales. In contrast, under the assumption of a linear extrapolation of trends in court appearance rates since 2002, a 10 %

The authors would like to thank Nicole Mahoney for her invaluable assistance in preparing the maps included in this chapter and Alison Taylor, the former Chief Demographer at the New South Wales Department of Planning and Environment, for providing the population projections for 2011–2031 by age and sex for NSW Local Government Areas. We also express our gratitude to Dr. Don Weatherburn, Director of the New South Wales Bureau of Crime Statistics and Research, for guidance in the formative stages of the study.

I. Halstead
New South Wales Bureau of Crime Statistics and Research, Sydney, Australia
e-mail: imogen.halstead@gmail.com

B. Opeskin
Faculty of Law, University of Technology, Sydney, Australia
e-mail: brian.opeskin@uts.edu.au

N. Parr (✉)
Faculty of Business and Economics, Macquarie University, Sydney, Australia
e-mail: nick.parr@mq.edu.au

D.A. Swanson (ed.), *The Frontiers of Applied Demography*, Applied Demography Series 9, DOI 10.1007/978-3-319-43329-5_5

reduction in total court appearances between 2012 and 2031 is projected. The implications of the results for policies relating to court demand mitigation, judicial recruitment and retirement, judicial productivity, the provision of court infrastructure, and the geospatial dimensions of court service provision are discussed.

Keywords Population projections • Courts • Service provision • Age-crime curve • Crime trends • Workforce planning • Applied demography • Australia

5.1 Introduction

Projections of future population play an important role in supporting planning for the future provision of services, infrastructure, budgetary allocation, and staffing requirements of a wide range of publicly-provided services (NSW Department of Education 2015, NSW Treasury 2012; Opeskin and Parr 2014; Parr et al. 2000). Since a cabinet decision in 2013 in the Australian state of New South Wales (NSW), all government policy decisions have been required to refer to the population evidence and population implications (Parr 2014). Only those projections prepared by the NSW State Government Department of Planning and Environment may be used by state government departments as a basis for planning decisions and policy development (NSW Department of Planning and Environment 2015).

Demographic considerations are especially relevant to the future demand for criminal court services and facilities, and the future need for related staff, such as judges and magistrates. Crime rates differ widely between the sexes, between age groups, and between geographical locations with differing socioeconomic status (Opeskin and Parr 2014; Rosevear 2007, 2012). Therefore population projections provide a useful anchor for thinking about the long-term needs of courts. This chapter assesses the implications of projected changes in the size, composition and geographical distribution of the NSW population for the volume and spatial dimensions of crime-related court service provision.

5.1.1 Population Context

Australia has a population of around 24 million people, and NSW, with a population of 7.6 million (32 %), is its most populous state (Australian Bureau of Statistics 2015). The NSW population is highly urbanised. In 2014, 64 % of state's population (4.8 million) lived in the Greater Sydney region. Within Sydney, the populations of suburbs in the West and South-West tend to have higher levels of socioeconomic disadvantage, while the populations of suburbs in the North and East are more prosperous. Apart from Sydney, the only other cities in NSW with populations of over 100,000 people are Newcastle—located to the north of Sydney in the Lower Hunter and Central Coast region—and Wollongong, in the Illawarra region along

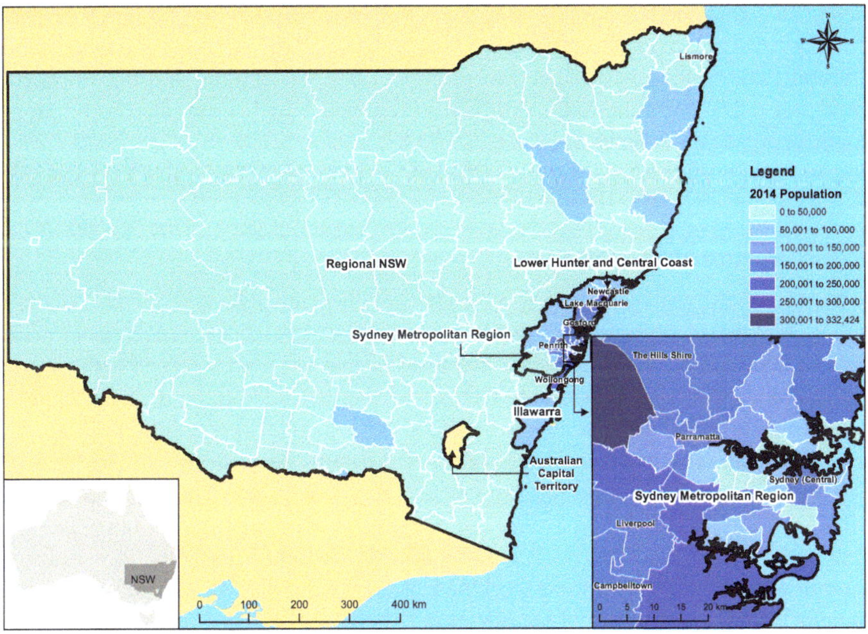

Fig. 5.1 NSW population by Local Government Area, 2014

the coast to the south of Sydney. The inland area of the state is sparsely populated, especially the more arid regions to the West (see Fig. 5.1).

Since 2010, NSW's annual population growth rates have ranged between 1.1 and 1.4 %, and have generally been below Australia's national growth rates. NSW's recent population growth has been the product of considerable gains from net international migration and slightly lower rates of natural increase, which have been partially offset by net out-movement to the other Australian states and territories (Australian Bureau of Statistics 2015; Opeskin and Parr 2014). As with Australia as a whole, the NSW population has been ageing, with the most rapid growth occurring in the oldest age groups and in those filled by the 1940s birth cohorts.

Changes in the level, demographic composition, and geographic distribution of the NSW resident population over the next 20 years are likely to have a direct impact on future demand for court services across the state. The NSW Department of Planning and Environment projections suggest that the NSW population will grow by 28 % between 2011 and 2031—an annualised growth rate of 1.2 %. Population ageing is expected to continue, with the projections implying a 77 % increase in the number of people aged 65 and over by 2031, but only an 11 % increase in the number of 15-29 year-olds. Population growth is expected to be concentrated in the Greater Sydney region, with average growth of 1.6 % per annum projected in this region between 2011 and 2031. All local areas across Sydney are expected to grow in population, with the largest increases anticipated in 'greenfield' designated growth areas, and in suburbs to the West and South-West of the central business district. Outside of Sydney, the population of the Lower Hunter and Central Coast region is projected to grow at 1.0 % per annum, and the Illawarra region at 0.8 % per annum. Growth across regional NSW is expected to be more modest, at 0.5 % per annum.

5.1.2 Criminal Context

As in all Western democracies, the Australian criminal justice system plays a vital role in maintaining the rule of law by ensuring that those who offend against key social norms are held to account. The protections afforded to an accused require that conviction and punishment is preceded by a judicial determination in courts established for that purpose. Like the United States, Australia is a federation in which power to enact criminal laws is shared by state and federal legislatures. However, unlike the United States, the different sources of criminal law—federal and state—are almost wholly administered in state courts (Australian Law Reform Commission 2006). The future workload of criminal courts in Australia's largest state is thus of great practical significance.

The criminal court system of NSW comprises three tiers: a lower court (Local Court); an intermediate court (District Court), and a superior court (Supreme Court), which also includes an appellate division (Crawford and Opeskin 2004). A further appeal is possible, with special leave, to the High Court of Australia, but this is rare in criminal cases. The Local Court is staffed by magistrates, and the District Court and Supreme Court by judges.

The relative workloads of the three courts are reflected in their staffing levels and the annual number of matters finalised. In respect of criminal matters alone, in 2013–2014 the Local Court engaged 100.8 full-time equivalent magistrates, the District Court 38.0 judges, and the Supreme Court 11.6 judges (Productivity Commission 2015, Table7A.27). In the same year, these courts finalised 94 %, 6 %, and less than 1 %, respectively, of the 180,908 criminal matters concluded in NSW courts (Productivity Commission 2015, Table 7A.6). Each of the courts has a civil caseload as well, but this is not analysed in this chapter.[1]

The criminal workload of the courts varies by subject matter. In the Local Court, the largest categories of offence in 2014 were traffic and vehicle regulatory offences (37 %), acts intended to cause injury (23 %) and illicit drug offences (13 %) (NSW Bureau of Crime Statistics and Research 2015, Table 1.1). However, for the higher courts, which hear more serious cases, the workload is skewed toward illicit drug offences (27 %), burglary and unlawful entry (19 %) and acts intended to cause injury (19 %) (NSW Bureau of Crime Statistics and Research 2015, Table 3.4).

[1] For Australia as a whole in 2012, criminal work accounted for 70 % of judicial time in magistrates courts, 68 % in District Courts, and 32 % in Supreme Courts (Opeskin 2013).

The delivery of court services in NSW has a spatial dimension, which reflects the courts' differing functions, staffing and caseloads. In 2014, the Local Court sat regularly in 150 locations throughout NSW, the District Court in 30 locations, and the Supreme Court in just 11. (District Court of New South Wales 2015; Local Court of New South Wales 2015; Supreme Court of New South Wales 2015). One consequence of these considerations is that projections of the size and spatial distribution of future criminal court workload in NSW are driven overwhelmingly by the work of the Local Court.

5.2 Method

5.2.1 Data

Annual population data disaggregated by age, sex and local area were sourced from Australia's national statistical agency, the Australian Bureau of Statistics (ABS). Geographical areas were defined by NSW Local Government Area (LGA) boundaries, which divide the state into 152 areas. The NSW Department of Planning and Environment provided customised data that are not publicly available, namely, population projections disaggregated at the LGA-level, by sex and 5-year age group.[2] Starting from 2011 census-based population estimates, the population projections are available at 5-year intervals from 2016 to 2031.

It is challenging to estimate demand for crime-related court services because there are no readily available time series data that comprehensively capture the hours of court time required to dispose of incoming criminal work. This study employs a proxy measure of demand that counts 'court appearances'. Importantly, for the purpose of linking demand to demographics, this measure can be disaggregated by age, sex and LGA, based on the demographic characteristics and place of residence of the defendant in each court appearance.[3]

More precisely, our measure of annual demand for crime-related court services counts the number of finalised criminal court appearances in NSW courts, plus the number of granted Apprehended Violence Orders (AVOs). AVOs are not criminal proceedings; they are legal orders applying to specific individuals designed to offer protection against violence for persons in need, particularly in the context of family

[2] At the local area level, the most recent publicly available NSW projections are not disaggregated by sex, and a higher level of age group aggregation is reported for small population areas.

[3] Where age or sex data, or both, were missing, court appearances were distributed among age-sex cohorts in the relevant LGA with weights based on the demographic composition of the local population. However, court appearances involving defendants with no fixed or known address, or defendants residing in institutions, interstate or overseas were excluded. As a result, the count of court appearances aggregated over LGAs is lower than the total count of court appearances in NSW. However, there is no marked difference in the evolution of these two series over time, and the lower count thus serves as a satisfactory proxy measure for analysing trends for NSW overall.

violence. They were included in the measure of demand because they generate a significant share of court workloads,[4] and they are qualitatively different from the courts' other civil work. This count of court appearances captures the overwhelming majority of crime-related cases in NSW. However, the study excludes appearances before the Children's Court, and focuses solely on adult jurisdictions. In addition, due to data availability, the measure also captures only AVOs granted, rather than all AVO applications made before the courts. All data used in the court appearances measure were sourced from the NSW Bureau of Crime Statistics and Research—a long-established statistical and research agency within the NSW Department of Justice—and are available from 1995 to 2014.

One issue worth noting is that in focusing on finalised court appearances and AVOs granted, the demand measure captures the output of the courts rather than latent demand for their services, since court output is constrained by capacity (e.g. the number of judicial officers appointed to hear and determine cases). This is less of a concern if supply capacity evolves over time to meet growing demand, so that trends in underlying demand are evident in historically realised data. Data on the growing number of Australian judicial officers over time underpin the reasonableness of this assumption (Opeskin 2013; Opeskin and Parr 2014). Moreover, there is pressure to avoid excessive constraints on supply because the price of not doing so is a mounting backlog of cases, an increase in the number of persons held on remand awaiting trial, and diminished access to justice.

Another limitation of the proxy measure is that it captures only *finalised* court appearances, whereas many cases involve multiple visits to court.[5] Moreover, the time taken in court varies according to the type of criminal matter and the level of the court hierarchy (matters in higher courts tend to be more complex). Thus, court appearances can best be interpreted as an index of demand, in that *changes* in criminal court appearances across time, space and demographic cohorts might be more useful from a policy perspective than the absolute count itself.

Although it is not intrinsic to our projection model, for expository purposes we also refer to ABS data on the level of socioeconomic disadvantage evident in each LGA. Given the well-documented link between socioeconomic disadvantage and crime,[6] this can be a useful lens through which to interpret geographical variation in court appearance rates across the state. The specific measure used is the ABS's Index of Relative Socioeconomic Disadvantage (IRSD). The index reflects a range of variables, including English language ability, educational status, occupational category, employment status, marital status, disability, and income levels (Australian Bureau of Statistics 2011).

[4] A 1999 survey of 68 magistrates revealed that 'two-thirds of all the respondents estimated that they spent between 10 and 20 % of their work time dealing with AVOs' (Potas 1999).

[5] On average there were 2.6 attendances per finalisation in the NSW Magistrates Court and 3.1 in the District Court in NSW in 2013–2014 (Productivity Commission 2015).

[6] For evidence specific to NSW see Devery (1991).

5.2.2 Analysis

The projections were derived in three steps. First, historical data on population levels and court appearance levels for each age-sex cohort in each local area were combined to estimate court appearance *rates*. The rates were calculated as the number of court appearances for each age-sex group in each local area, divided by the population in each age-sex group in each area. These were calculated for each year from 2001 to 2013, which is the period over which the population data were available, thus giving historical time-series.

These data can be volatile because counts of court appearances are very low in some age-sex cohorts (e.g., older females), particularly in geographical areas with small populations. To mitigate the impact of this volatility in estimating court appearance rates over time, a 3-year centred moving average was calculated to smooth the data, yielding a time-series spanning the period 2002–2012; 2012 is therefore used as the base year for the projections. These smoothed data on court appearance rates are referred to hereafter. In addition, 5-year age groups were aggregated into four broader age bands: 15–29 year olds; 30–44 year olds, 45–64 year olds, and 65 years and over.

The historical linear trends in each age, sex and area-specific court appearance rate were estimated by calculating the average annual change in the smoothed court appearance rate over the 10-year period from 2002 to 2012 (that is, one-tenth of the difference between the smoothed court appearance rate in 2012 and the smoothed court appearance rate in 2002).[7] The differencing approach was employed in order to anchor the projections at current levels of court appearances. In some cases, a downward linear trend might imply negative court appearance rates over the projection period for a particular age-sex group in a particular local area. In these instances, the projected court appearance rates were bounded below by zero.

The second step involved projecting future court appearance rates under two alternative assumptions. Under Assumption 1, age, sex and area-specific court appearance rates were assumed to remain unchanged over the projection horizon; i.e., they were held constant at the moving-average rate as at 2012. Assumption 2 supposed that age, sex and area-specific court appearance rates continued to change over the projection period according to their respective historical trends, i.e., those evident in the smoothed time-series over the decade 2002–2012, but subject to a lower limit of zero as described above.

Thirdly, the projected age, sex and area-specific court appearance rates were applied to the NSW Department of Planning and Environment population projections to estimate levels of demand for crime-related court services under the alternative scenarios. Thus, under Assumption 1, age-sex specific court appearance

[7] Least squares methods could alternatively be employed to estimate historical linear trends in court appearance rates. These methods would reduce the influence of the first and most recent observation on the projections. However, the predicted values implied by least squares models can be out of step with observed court appearance rates at any one time.

rates currently evident in each local area were applied to the population projections for each age-sex group in each area at 5-year intervals from 2016 to 2031. Under Assumption 2, projected age, sex and area-specific court appearance rates, which were calculated by extending the historical linear trend over the projection horizon, were applied to the population projections for each age-sex cohort in each local area at 5-year intervals from 2016 to 2031.

The two alternative assumptions regarding court appearance rates are considered in order to acknowledge, and investigate the impact of, the uncertainty surrounding the future trajectory of court appearance rates in NSW. Assumption 1 is the benchmark scenario, since the projections of court demand under this assumption in our model are driven only by changes in demographic variables, namely, the size, age-sex composition, and geographic distribution of the population. Assumption 2 introduces an additional dynamic factor: changes in the tendency for individuals in particular age-sex groups in particular areas to appear before the courts.

In reality, changes in demographic variables might not be entirely separable from changes in underlying court appearance rates. By way of example, strong growth in the resident population of the Sydney (Central) local area over the past 10 years has occurred at the same time as sharp falls in court appearance rates for young males in this area. This correlation could reflect the relatively high socio-economic status of international and internal migrants to the Sydney (Central) area—a change that might be expected to continue, leading court appearance rates ever lower into the future as the population of Sydney (Central) continues its strong growth. In this case, Assumption 2 offers internal consistency to the model; if demographic trends are expected to continue, then it is also reasonable to expect some continuation of historical trends in court appearance rates. However, many factors unrelated to demographics affect court appearance rates, including policing levels and criminal justice policy, and these too are implicit in the trends intrinsic to Assumption 2. We return to these issues in the discussion.

5.3 Results

5.3.1 Baseline Data

Figure 5.2 displays the resident population, court appearance rates, and level of court appearances in each NSW local area in 2012, in Panels a, b and c respectively. Panel a also illustrates the projected population for 2031. The local areas are grouped by region, according to NSW Department of Planning and Environment definitions. The first cluster covers the areas within the Sydney metropolitan region (Kuringgai, Mosman, Woollahra, Lane Cove, North Sydney, The Hills Shire, Manly, Pittwater, Hunters Hill, Hornsby, Willoughby, Waverley, Leichhardt, Warringah, Sutherland Shire, Canada Bay, Ryde, Camden, Randwick, Blue Mountains, Kogarah, Wollondilly, Strathfield, Marrickville, Hawkesbury, Sydney, Ashfield, Hurstville, Penrith, Burwood, Rockdale, Parramatta, Botany Bay,

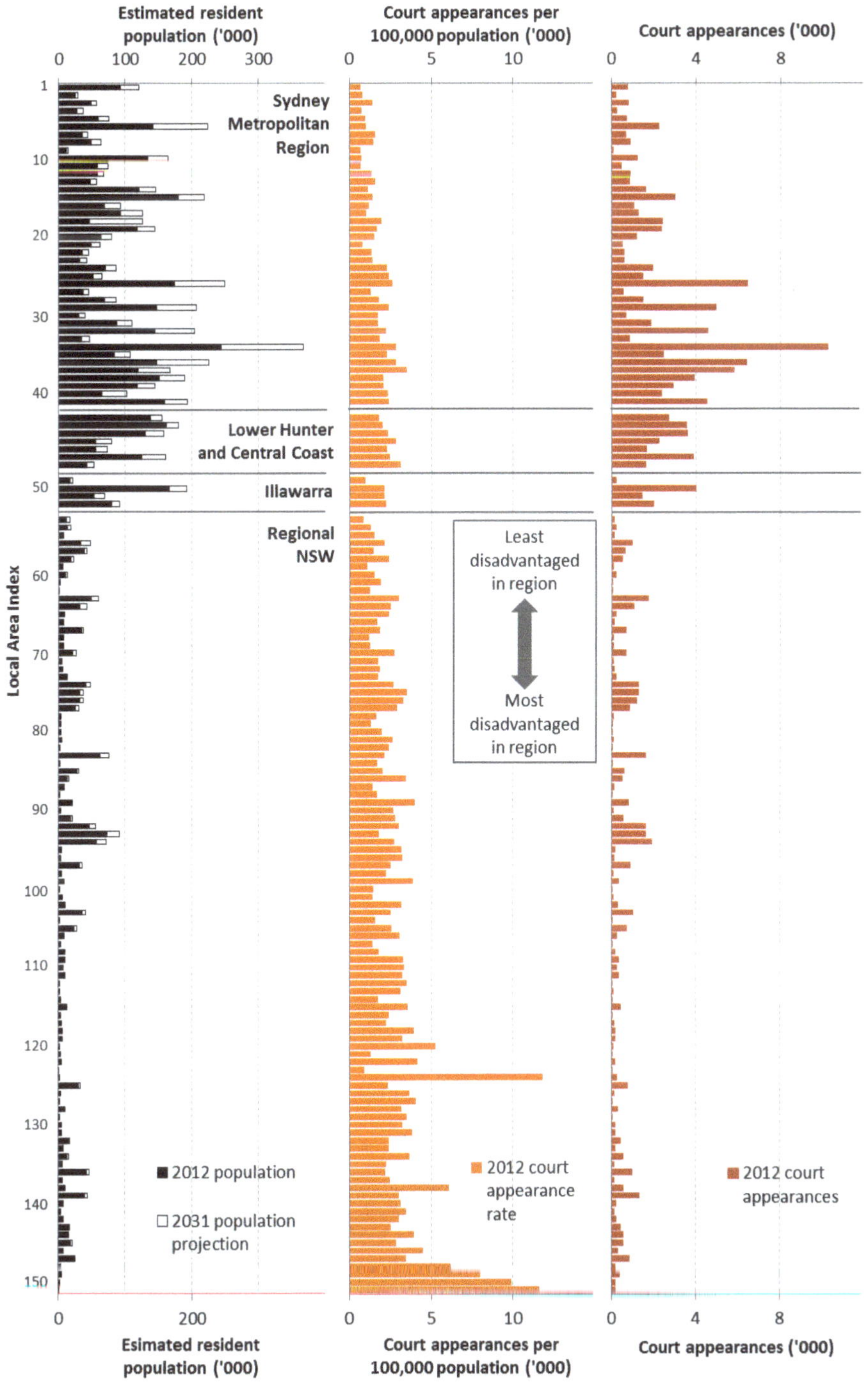

Fig. 5.2 Baseline data: population, court appearance rates and court appearances, by NSW local area, 2012

Blacktown, Holroyd, Liverpool, Campbelltown, Bankstown, Canterbury, Auburn, Fairfield); the second covers the Lower Hunter and Central Coast region (Gosford, Lake Macquarie, Newcastle, Maitland, Port Stephens, Wyong, Cessnock), the third covers the Illawarra region (Kiama, Wollongong, Shellharbour, Shoalhaven), and the fourth covers areas elsewhere in regional NSW (Palerang, Yass Valley, Snowy River, Queanbeyan, Wingecarribee, Singleton, Upper Lachlan Shire, Cabonne, Lockhart, Jerilderie, Wagga Wagga, Bathurst Regional, Cooma Monaro, Greater Hume Shire, Ballina, Dungog, Murray, Armidale Dumaresq, Uralla, Blayney, Upper Hunter Shire, Albury, Dubbo, Orange, Byron, Oberon, Wakool, Coolamon, Bland, Walcha, Port Macquarie Hastings, Carrathool, Bega Valley, Muswellbrook, Corowa Shire, Boorowa, Griffith, Gundagai, Mid Western Regional, Tamworth Regional, Tweed, Coffs Harbour, Wentworth, Cobar, Eurobodalla, Temora, Leeton, Tumbarumba, Berrigan, Narrabri, Lismore, Bombala, Goulburn Mulwaree, Tumut Shire, Gloucester, Bellingen, Young, Forbes, Gunnedah, Balranald, Bogan, Weddin, Parkes, Harden, Junee, Deniliquin, Cootamundra, Warren, Gwydir, Lachlan, Urana, Bourke, Great Lakes, Guyra, Murrumbidgee, Cowra, Hay, Narromine, Narrandera, Lithgow, Glen Innes Severn, Inverell, Liverpool Plains, Clarence Valley, Tenterfield, Moree Plains, Greater Taree, Warrumbungle Shire, Gilgandra, Kyogle, Nambucca, Broken Hill, Richmond Valley, Wellington, Kempsey, Coonamble, Walgett, Central Darling, Brewarrina). Within these clusters, local areas are ordered (and indexed along the vertical axis) in increasing order of socioeconomic disadvantage.

The figure clearly highlights the fact that the vast majority of the NSW population live in the Sydney region, although there are also some densely populated centres in other coastal areas (Fig. 5.2a). Consistent with the link between crime and disadvantage, court appearance rates generally increase with the socioeconomic disadvantage of an area (Fig. 5.2b). Lower court appearance rates are more common among areas in Sydney, while higher court appearance rates are more common in regional NSW.

The numbers of court appearances attributable to residents of each local area reflects the product of population and court appearance rates. Residents of the most disadvantaged areas of the Sydney metropolitan region account for the greatest number of court appearances (Fig. 5.2c). The other coastal population centres also contribute relatively large shares of aggregate demand for court services. Consistent with the sparse population of regional NSW, court appearance levels are relatively low in regional areas, but generally higher than the population alone would imply, due to their relatively high court appearance rates.

Court appearance rates vary markedly across different demographic groups. Figure 5.3 illustrates aggregated court appearance rates (using NSW court appearances aggregated across local areas in the numerator, and the total NSW population in the denominator) for the age-sex groups for the years 2002–2012. In a given year, court appearance rates are much higher for males than females. They are also higher for younger age groups than older age groups for both sexes. This is consistent with the widely-documented 'age-crime' curve, which shows that the probability of criminal activity peaks in the late teenage years and declines thereafter (Farrington 1986; Moffitt 1993; Piquero et al. 2003). In NSW, however, the strength of this

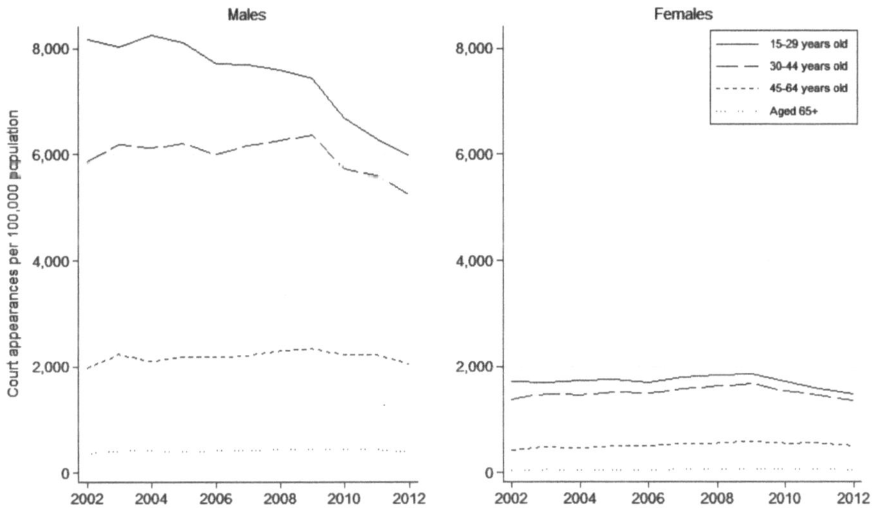

Fig. 5.3 Court appearance rates by age and sex, NSW, 2002–2012

age-crime relationship may have diminished in recent years. In particular, in 2002 court appearance rates for 15–29 year old males were higher than for 30–44 year old males by a third, but recently these age-sex specific rates have become more comparable as court appearance rates for 15–29 year old males have declined much more dramatically than other cohorts.

These aggregate trends in court appearance rates for NSW mask substantial variation at the local level. Figure 5.4 shows the annual change in the court appearance rate between 2002 and 2012 in each NSW local area (indexed as per Fig. 5.2). Marked declines are evident in some of the more remote areas (although in the most extreme cases, from a relatively high base in 2002).[8] Very few local areas demonstrated increasing court appearance rates over this period. To some extent, the geographical variations in court appearance rates reflect variations in the pace of population ageing. However, they also reflect variations in recorded crime trends. For example, Weatherburn and Holmes (2013) found that the drop in property crime in NSW between 2000 and 2012 (attributable in part to reduced drug use, more severe criminal justice sanctions, and improved economic conditions) was much larger in Sydney and other urban areas than elsewhere in NSW. Similar results have been reported elsewhere (Moffatt et al. 2005; Wan et al. 2012).

[8] Local areas with small populations occupy the extremes of this distribution. This partly reflects volatility in measures of court appearance rates, and consequent volatility in trends.

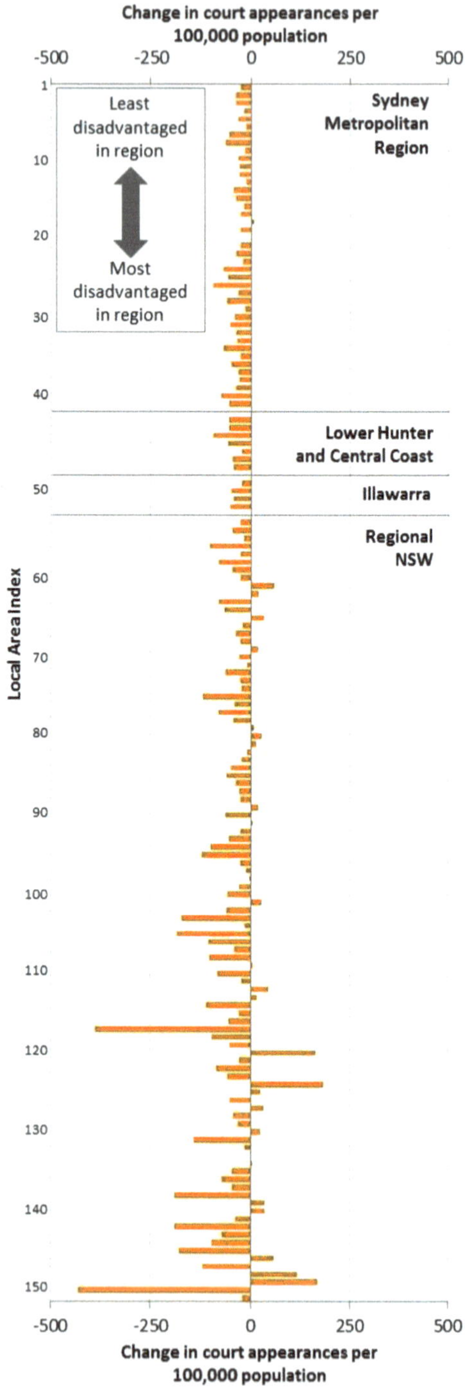

Fig. 5.4 Annual average change in court appearance rates by NSW local area, 2002–2012

Table 5.1 Court appearances by NSW region, 2012–2031

	Actual	Projections							
		Assumption 1: Constant court appearance rates				Assumption 2: 10-year trend court appearance rates			
			Increase: 2012–2031				Increase: 2012–2031		
Region	2012	2031	Level	%	% of total	2031	Level	%	% of total
Sydney Metropolitan Area	67,935	88,338	20,403	30	88	64,827	−3,108	−5	23
Lower Hunter & Central Coast	16,862	19,319	2,457	15	11	13,237	−3,625	−21	27
Illawarra	7,029	7,638	610	9	3	5,905	−1,124	−16	8
Regional NSW	39,780	39,480	−300	−1	−1	34,123	−5,657	−14	42
Total	131,605	154,776	23,171	18	100	118,092	−13,513	−10	100

5.3.2 *Projection Results*

Table 5.1 summarises the projected number of court appearances in 2031 for each of the four regions of NSW under the two alternative assumptions about the trajectory of age, sex and area-specific court appearance rates. The complete results for each local area are presented in Table 5.2 at the end of the chapter. Figure 5.5 illustrates the increase in the level of court appearances across NSW local areas under Assumptions 1 and 2, in Panels a and b, respectively.

Under Assumption 1, total court appearances across NSW are projected to increase by 18 % between 2012 and 2031 (Table 5.1). The vast majority of the increase is expected to occur in the Sydney metropolitan area (88 % of the total), reflecting its more rapid population growth and slower population ageing. In fact, more than one half of the total increase in court appearances is projected to be concentrated in just seven local areas: Blacktown (13 % of the total), Sydney (Central) (8 %), Liverpool (8 %), Camden (7 %), Campbelltown (6 %), Penrith (5 %) and Parramatta (5 %) (Table 5.2). With the exception of the Sydney (Central) area, these local areas are situated in greater West and South-West Sydney (Fig. 5.5). These areas vary in their socioeconomic status (disadvantage is relatively common in Blacktown, Campbelltown and Liverpool, but much less so in Camden), but they are all areas expected to undergo substantial population growth over

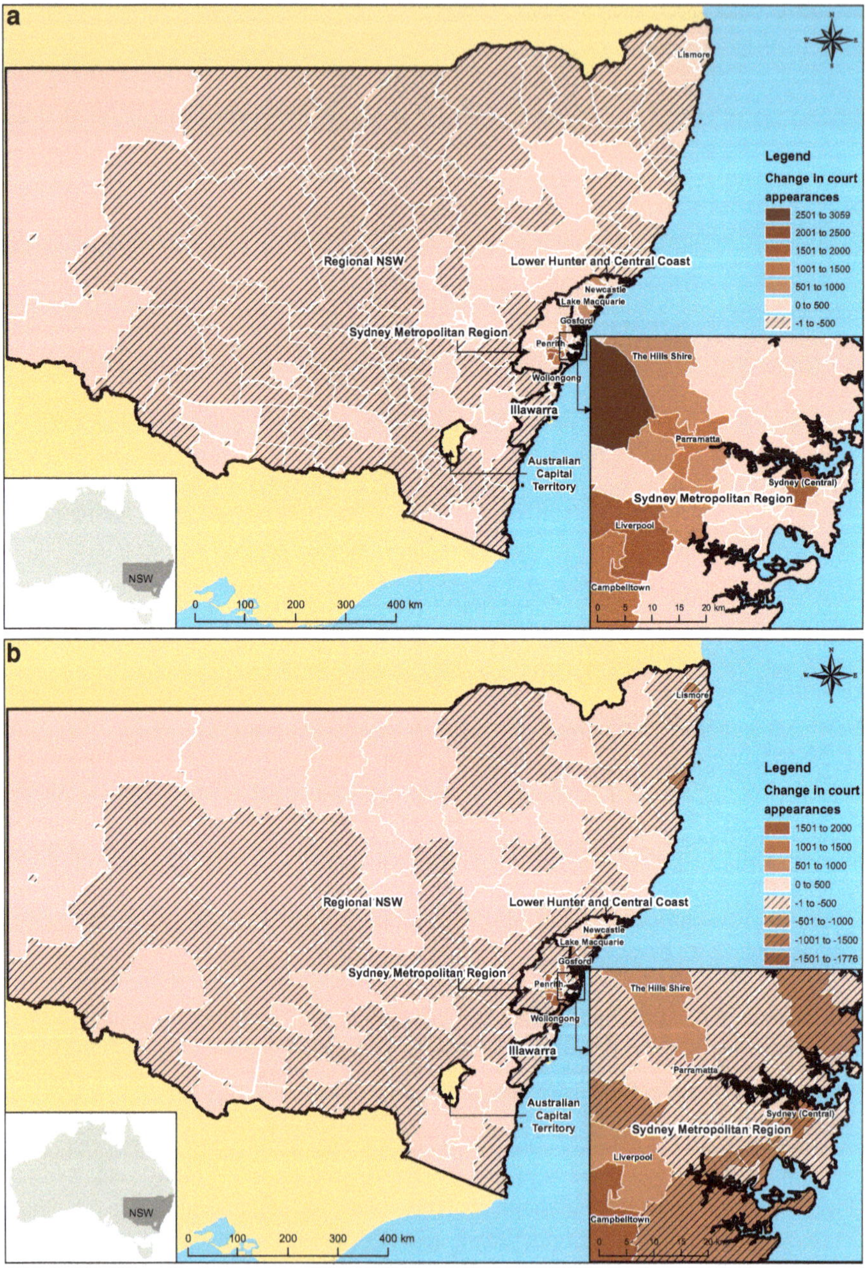

Fig. 5.5 Change in court appearances by NSW local area, 2012–2031. (**a**) Assumption 1 (constant criminal court appearance rates); (**b**) Assumption 2 (10-year trend criminal court appearance rates)

the projection period (Table 5.2). The Lower Hunter and Central Coast region is projected to contribute 11 % to the increase in court appearances across NSW under Assumption 1, and the Illawarra 3 %. Meanwhile, a projected fall in court appearances in regional NSW (due to population ageing and limited population growth) reduces the aggregate increase expected for NSW by 1 % (Table 5.1). Whilst the areas in which reductions in court appearances are projected under Assumption 1 cover a large proportion of the area of NSW, it should be noted that much of this region is relatively sparsely populated (Fig. 5.5a).

The outlook for court appearance levels under Assumption 2 differs considerably. Overall, the level of court appearances is projected to fall by 10 % under this assumption, rather than rise by 18 % as projected under Assumption 1 (Table 5.1). A decline in court appearances is projected in each of the four major regions under this scenario, reflecting the assumed continued decline in court appearance rates—particularly among cohorts of younger males—across most of NSW. In the Sydney region, court appearances are expected to fall by 5 % under Assumption 2, rather than rise by 30 %, as projected under Assumption 1. Most notably, the large increases in court appearances forecast for the areas of Blacktown, Sydney (Central), and Parramatta are reversed (Table 5.2). In fact, three out of every four local areas in Sydney show a decline in court appearances under Assumption 2. In contrast, court appearances were projected to increase across the board for Sydney under Assumption 1 (compare the inset in Figs. 5.5a, b). Positive growth in court appearances is still anticipated in some parts of Sydney under Assumption 2, and indeed larger increases are projected under Assumption 2 in Camden, Penrith and Campbelltown—areas on Sydney's Western or South-Western fringe.

Under Assumption 2, particularly marked falls in the number of court appearances are expected outside Sydney, in coastal and regional NSW. In fact, with the exception of Sydney (Central) area, locations outside Sydney contribute the most to the aggregate decline in court appearances across the state. The Newcastle area in the Lower Hunter and Central Coast region accounts for 13 % of the total fall, with the nearby centres of Lake Macquarie and Gosford contributing 8 % and 7 % of the total, respectively (Table 5.2). Wollongong in the Illawarra, and Lismore in regional NSW, also contribute 6 % each to the total decline under Assumption 2.

In some areas in regional NSW, court appearances are projected to increase under Assumption 2, although declines were forecast under Assumption 1. For Greater Taree (on the Mid-North Coast), for example, the number of court appearances was projected to decrease by 4 % between 2012 and 2031 under Assumption 1, but increase by 35 % under Assumption 2. These increases outside the metropolitan areas are modest in magnitude, and insufficient to offset the overall decline.

5.4 Discussion

The 18 % growth in court appearances projected under Assumption 1 over the period 2012–2031 is broadly consistent with other studies. Opeskin and Parr (2014), for instance, estimated a 16 % increase in demand for a single court (the Local Court), using a different proxy variable (criminal convictions), and a slightly longer projection period (to 2036)—due to demographic change alone. In either case, the projected growth in demand is more modest than the 27 % growth anticipated in the population itself over the same period. This divergence reflects the ageing of the NSW population, with its concomitant increase in the share of older people, who have a lower propensity to commit crime and appear before the courts. In general, however, the projected increase in court appearances is concentrated in those parts of NSW with the largest population increases; in particular, an increase in court appearances between 2012 and 2031 is projected for every local area in the Sydney metropolitan region. In percentage terms, growth in court appearances is highest in 'greenfield' areas of Sydney—Camden and the Hills Shire—which are designated as priority growth sites, and are expected to undergo particularly rapid population growth over the projection period. Only a few areas in regional NSW show a decline in court appearances between 2012 and 2031, in line with more modest population growth and more marked population ageing. The projected geographical distribution of the population also affects the aggregate outcome for NSW because, all else being equal, areas with high current court appearance rates are projected to experience the largest increases in court appearances.

These factors are also at play under Assumption 2. However, introducing trend changes in court appearance rates over the projection horizon leads to a markedly different outlook by 2031: a 10 % decline in court appearances across NSW (rather than 18 % growth). Males aged 15–29 years—who historically have contributed the most significant share of total court appearances—generally showed a trend decline in court appearance rates over the period from 2002 to 2012 (Fig. 5.3). In some areas, the trend decline for this age group was so steep that its continuation implied negative rates by 2031. However, as described above, for the purposes of the model a zero rate was assumed in these areas, and hence zero court appearances were projected for the relevant cohort in 2031. An overall fall in court appearances was often projected under Assumption 2 in areas where the decline in court appearance rates for 15–29 year old males was particularly marked, often reversing increases projected under Assumption 1. However, this shift was not universal across NSW. In some areas, an offsetting rise in court appearance rates for 30–44 year old males led to larger increases in court appearances under Assumption 2 (e.g. in Camden, Penrith and Campbelltown). In many areas, the projections under Assumption

2 imply a shift in the peak of the 'age-crime' curve, with court appearance rates for the 30–44 year old male age group in 2031 exceeding rates for 15–29 year old males. Other scenarios were also evident; in some parts of regional NSW with generally increasing court appearance rates (such as Greater Taree), projected declines in court appearances due to population ageing under Assumption 1 were reversed to show increases in court appearances under Assumption 2.

The divergent results under Assumptions 1 and 2 are consistent with the fact that demographic factors only partially explain variations in crime (Levitt 1999; Rosevear 2007, 2012). At face value, these divergent projections present something of a dilemma for policy making. It is not immediately clear whether court appearance rates are more likely to stay constant over the forecast horizon (Assumption 1) or to continue to follow their decade-long trends (Assumption 2). However, there are several reasons to believe that the projections under Assumption 2 are improbable, and likely to underestimate future court appearances. To some extent this is apparent from the mechanical output of the model. In particular, negative court appearance rates—implied by a continuation of recent trends for some cohorts—are impossible. It follows that there can *only* be upside risk to these particular projections, that is, the projections for these cohorts necessarily underestimate the true number of court appearances likely to occur in 2031. Moreover, the implied shift in the peak of the 'age-crime' curve is also questionable. Although such shifts are not unknown, the age-crime curve is an idée fixe of criminological literature, even if its causes are contested (Sweeten et al. 2013).

From a substantive perspective, the fall in court appearance rates for young males over the past decade mirrors the fall in reported crime. Existing Australian research suggests that the fall in crime is linked to (a) the heroin shortage, which started in 2001; (b) the growth in real average weekly earnings; and (c) the rise in the probability of arrest after offending, and imprisonment after conviction (Moffatt et al. 2005; Wan et al. 2012). While each of these factors may continue to exert downward pressure on court appearance rates, it is easy to envisage softer declines in future if, for example, the availability of alternative drugs increases, economic growth softens, or natural limits to arrest and imprisonment rates draw nearer.

Even if the decline in court appearance rates for young males does not continue in the long term, it is not known when it will end. If crime rates were the only factor at work, one could examine the areas where crime is growing (e.g., drug trafficking, fraud) and estimate the impact of the growth. However, the influences on court appearance rates are numerous. In mid-2002, NSW had 13,716 police in pursuit of vast numbers of offenders (NSW Police 2003, 48); by mid-2012 there were 16,371 police (a 19 % increase) and far fewer offenders (NSW Police Force 2013, 29). This means police have a great deal more flexibility in their resource allocation decisions. The scope for increasing arrest rates in some categories of crime (e.g., illegal drug use, drink driving, drug trafficking, internet pornography) is substantial. Changing practice around criminal diversion (e.g., cannabis cautioning) have also had an impact. Future court workloads thus depend significantly on policing policy, not merely underlying criminality.

In light of this uncertainty, Assumption 2 is one way to consider the implications of continuing decline in court appearance rates for young males. Overall, however, the results of the modelling under Assumption 2 are suggestive of risks surrounding the benchmark Assumption 1, rather than reflecting a balanced projection of future court demand.

5.5 Policy Implications

The significance of this study lies in its capacity to provide an evidential basis for government decision making in the allocation of scare resources. While the size, age-sex structure, and spatial distribution of a population are just some of many factors influencing future demand for criminal court services, the inexorability of these long-term demographic processes provides an obvious starting point for assessing future societal needs.

This study has a number of salient policy implications, which variously address the demand or supply side of the problem. On the demand side, projections of higher aggregate demand for criminal court services may prompt policies that mitigate demand by attacking the root causes of crime, or diverting offenders into non-curial processes through cautioning, treatment or educational programs.

There are also policy implications on the supply side. The first relates to the *retention* of judicial officers. In NSW, all judicial officers face mandatory retirement at 72 years of age, although some exit before that age voluntarily by resignation or involuntarily by reason of incapacity or misbehaviour. Early resignation can be addressed by policies that ameliorate the often stressful working conditions faced by judges and magistrates (Mack et al. 2012). Consideration could also be given to extending the retirement age, especially since it is more cost effective to extend the tenure of existing appointees than to make new appointments (Opeskin 2011). Secondly, there are policy implications for the size and frequency of judicial *recruitment*. Data on the age-sex structure of an existing judicial workforce, combined with statistics on the attrition rate in each court, allow estimates to be made of the number and timing of appointments needed to maintain or enhance supply. This knowledge can aid governments and court administrators in budgeting and planning. Thirdly, improving judicial *productivity* (e.g. through more effective use of technology) may allow rising demand to be met by a static population of judicial officers because it effectively increases supply. Fourthly, projected growth in demand impacts not only court personnel but also *court infrastructure*. In the United States, for example, a facilities planning exercise in the 1990s led to the construction of a large number of new federal courthouses (Fentress 2000), and similar assessments should be made in high-growth areas of outer Sydney. Fifthly, there are *geospatial* implications for future supply because a changing population distribution shifts demand in particular localities, even if aggregate demand is constant. This affects where judicial officers are assigned to sit, the nature of judicial circuits, and the appropriate location of court infrastructure.

It is apparent, however, that policy responses must confront the risks that inhere in any projection. If the model's assumptions are not met, actual and projected values will diverge. In the present study we saw this in the outcomes that flowed from two different, but seemingly reasonable, assumptions about future trends in court appearance rates—Assumption 1 led to a projected 18 % rise in demand by 2031, while Assumption 2 led to a 10 % fall. The policy challenge is acute in the provision of court services because rigidities in the system make it difficult to reverse policy decisions if the model's assumptions are later controverted. Judicial independence generally demands that Australian judges, once appointed, hold office until they reach the mandatory retirement age (Opeskin 2015). A decision to make a new appointment to meet anticipated increases in demand thus has a ratchet effect and cannot be unwound. At the same time, there are costly consequences of failing to address excess demand. For example, a spike in the number of people on remand awaiting trial in NSW in early 2015 contributed to acute prison overcrowding (Ralston 2015).

In NSW, the government has been alive to these challenges and has adopted a range of responses intended to give it greater flexibility on the supply side. These include the appointment of acting judges (usually recent retirees) on short term commissions, part-time judges, and quasi-judicial officers to carry the ancillary procedural burden that usually attends criminal proceedings. There are concerns that some of these initiatives undermine judicial independence (Kirby 1998) but, whatever their merits, they are policy choices made by democratically accountable governments. The methods of applied demography can support those choices by providing policy makers with a sound evidential basis for their decisions.

Table 5.2 Court appearances by NSW local area

Local area	Actual	Projections								Index of Relative Socio-economic Disadvantage (ABS IRSD[a])
		Assumption 1: Court appearance rates held constant				Assumption 2: Court appearance rates follow 10-year trend				
			Increase: 2012–2031				Increase: 2012–2031			
	2012	2031	Level	%	% of total	2031	Level	%	% of total	
Albury	1,111	1,264	153	14	1	1,251	141	13	−1	978.6
Armidale Dumaresq	628	698	70	11	0	682	54	9	−0	986.9
Ashfield	456	569	113	25	0	336	−120	−26	1	1,015.4
Auburn	1,692	2,367	675	40	3	1,238	−454	−27	3	916.7
Ballina	689	679	−10	−2	−0	600	−89	−13	1	988.7
Balranald	78	55	−23	−29	−0	84	6	8	−0	946.2
Bankstown	3,361	3,918	557	17	2	3,107	−254	−8	2	931.7
Bathurst Regional	823	1,045	222	27	1	727	−96	−12	1	991.0
Bega Valley	651	611	−40	−6	−0	524	−127	−20	1	968.7
Bellingen	229	188	−41	−18	−0	94	−135	−59	1	950.1
Berrigan	124	94	−30	−24	−0	146	22	18	−0	954.1
Blacktown	7,181	10,239	3,058	43	13	6,865	−316	−4	2	968.5
Bland	120	101	−19	−15	−0	130	10	8	−0	974.5
Blayney	91	110	19	21	0	69	−22	−24	0	982.2
Blue Mountains	987	1,176	189	19	1	1,421	434	44	−3	1,038.6
Bogan	90	69	−21	−23	−0	85	−5	−6	0	946.1
Bombala	25	28	3	13	0	32	7	30	−0	952.5
Boorowa	31	33	2	6	0	34	3	11	−0	963.8
Botany Bay	653	837	184	28	1	605	−48	−7	0	975.7
Bourke	343	265	−78	−23	−0	372	29	9	−0	932.6
Brewarrina	210	161	−49	−23	−0	212	2	1	−0	788.4

Broken Hill	689	537	−152	−22	−1	417	−272	−39	2	899.6
Burwood	487	677	190	39	1	478	−9	−2	0	996.1
Byron	787	860	73	9	0	653	−134	−17	1	976.6
Cabonne	171	194	23	13	0	170	−1	−1	0	1,000.3
Camden	922	2,443	1,521	165	7	2,907	1,985	215	−15	1,047.1
Campbelltown	4,400	5,795	1,395	32	6	5,938	1,538	35	−11	944.8
Canada Bay	812	1,055	243	30	1	742	−70	−9	1	1,067.0
Canterbury	2,551	2,934	384	15	2	2202	−348	−14	3	922.0
Carrathool	45	30	−15	−33	−0	31	−14	−31	0	968.8
Central Darling	211	156	−55	−26	−0	111	−100	−47	1	824.4
Cessnock	1,300	1,614	314	24	1	1,353	53	4	−0	936.4
Clarence Valley	1,019	998	−21	−2	−0	743	−276	−27	2	919.4
Cobar	121	110	−11	−9	−0	106	−15	−12	0	956.7
Coffs Harbour	1,741	1,934	193	11	1	1,044	−697	−40	5	958.4
Coolamon	71	62	−9	−12	−0	98	27	38	−0	975.1
Cooma Monaro	230	209	−21	−9	−0	290	60	26	−0	990.6
Coonamble	249	179	−70	−28	−0	251	2	1	−0	879.6
Cootamundra	216	174	−42	−19	−0	184	−32	−15	0	941.5
Corowa Shire	160	135	−25	−16	−0	129	−31	−19	0	967.8
Cowra	394	297	−96	−24	−0	308	−86	−22	1	928.0
Deniliquin	302	184	−118	−39	−1	159	−143	−47	1	941.6
Dubbo	1,186	1,287	101	9	0	761	−425	−36	3	977.0
Dungog	89	84	−5	−5	−0	80	−9	−10	0	988.5
Eurobodalla	964	881	−83	−9	−0	1,079	115	12	−1	955.8
Fairfield	4,177	4,529	352	8	2	3,514	−663	−16	5	854.0
Forbes	274	233	−41	−15	−0	174	−100	−36	1	946.8
Gilgandra	166	112	−54	−32	−0	131	−35	−21	0	910.6

(continued)

Table 5.2 (continued)

Local area	Actual	Projections									Index of Relative Socio-economic Disadvantage (ABS IRSD[a])
		Assumption 1: Court appearance rates held constant				Assumption 2: Court appearance rates follow 10-year trend					
			Increase: 2012–2031				Increase: 2012–2031				
	2012	2031	Level	%	% of total	2031	Level	%	% of total		
Glen Innes Severn	220	158	−62	−28	−0	194	−26	−12	0		921.8
Gloucester	79	58	−21	−27	−0	56	−23	−29	0		951.0
Gosford	2,593	2,721	128	5	1	1,635	−958	−37	7		1,006.3
Goulburn Mulwaree	693	707	14	2	0	202	−491	−71	4		951.4
Great Lakes	817	769	−48	−6	−0	1,106	289	35	−2		932.3
Greater Hume Shire	153	133	−20	−13	−0	135	−18	−12	0		989.3
Greater Taree	1,349	1,299	−50	−4	−0	1,817	468	35	−3		913.7
Griffith	805	814	9	1	0	986	181	22	−1		963.7
Gundagai	89	74	−15	−16	−0	65	−24	−27	0		961.6
Gunnedah	326	336	10	3	0	336	10	3	−0		946.7
Guyra	139	138	−1	−1	−0	127	−12	−9	0		931.0
Gwydir	64	45	−19	−30	−0	44	−20	−31	0		939.9
Harden	76	62	−14	−18	−0	63	−13	−18	0		941.9
Hawkesbury	1,331	1,508	177	13	1	1,128	−203	−15	1		1,020.3
Hay	112	62	−50	−44	−0	76	−36	−33	0		927.0
Holroyd	1,872	2,456	584	31	3	2,012	140	7	−1		965.6
Hornsby	1,014	1,224	210	21	1	518	−496	−49	4		1,085.2
Hunters Hill	78	100	22	29	0	82	4	5	−0		1,092.2
Hurstville	1,342	1,504	162	12	1	653	−689	−51	5		1,006.9
Inverell	553	547	−6	−1	−0	628	75	14	−1		921.4
Jerilderie	15	12	−3	−19	−0	17	2	15	−0		997.6
Junee	118	107	−11	−9	−0	6	−112	−95	1		941.7

Kempsey	894	851	−43	−5	−0	561	−333	−37	2	879.7
Kiama	167	196	29	17	0	177	10	6	−0	1,054.6
Kogarah	383	499	116	30	1	241	−142	−37	1	1,036.2
Kuringgai	595	783	188	32	1	270	−325	−55	2	1,120.7
Kyogle	258	230	−28	−11	−0	97	−161	−63	1	907.1
Lachlan	250	183	−67	−27	−0	160	−90	−36	1	938.1
Lake Macquarie	3,548	3,557	9	0	0	2,425	−1,123	−32	8	994.8
Lane Cove	186	258	72	39	0	172	−14	−8	0	1,106.9
Leeton	387	350	−37	−10	−0	389	3	1	−0	954.5
Leichhardt	735	864	129	18	1	842	107	15	−1	1,078.9
Lismore	1,001	1,024	23	2	0	256	−745	−74	6	952.7
Lithgow	516	412	−104	−20	−0	473	−43	−8	0	924.2
Liverpool	4,458	6,372	1,914	43	8	5,363	905	20	−7	951.0
Liverpool Plains	161	143	−18	−11	−0	120	−41	−26	0	921.3
Lockhart	43	39	−4	−10	−0	70	26	60	−0	999.3
Maitland	1,640	2,266	626	38	3	1,731	91	6	−1	992.8
Manly	595	677	82	14	0	293	−302	−51	2	1,099.4
Marrickville	1,681	1,953	272	16	1	1,156	−525	−31	4	1,021.6
Mid Western Regional	557	561	4	1	0	631	74	13	−1	961.5
Moree Plains	721	566	−155	−22	−1	431	−290	−40	2	915.1
Mosman	170	221	51	30	0	93	−77	−45	1	1,110.7
Murray	86	93	7	8	0	139	53	62	−0	987.7
Murrumbidgee	93	60	−33	−35	−0	84	−9	−10	0	928.3
Muswellbrook	450	517	67	15	0	450	0	0	−0	968.2
Nambucca	505	436	−69	−14	−0	376	−129	−26	1	900.0
Narrabri	414	313	−101	−24	−0	284	−130	−31	1	953.4
Narrandera	202	151	−50	−25	−0	99	−103	−51	1	925.2

(continued)

Table 5.2 (continued)

Local area	Actual	Projections								Index of Relative Socio-economic Disadvantage (ABS IRSD[a])
		Assumption 1: Court appearance rates held constant				Assumption 2: Court appearance rates follow 10-year trend				
			Increase: 2012–2031				Increase: 2012–2031			
	2012	2031	Level	%	% of total	2031	Level	%	% of total	
Narromine	197	161	−36	−18	−0	223	26	13	−0	926.6
Newcastle	3,203	3,635	432	13	2	1,427	−1,776	−55	13	993.9
North Sydney	573	734	161	28	1	325	−248	−43	2	1,104.8
Oberon	74	67	−7	−9	−0	66	−8	−11	0	975.9
Orange	1,122	1,203	81	7	0	1,046	−76	−7	1	977.0
Palerang	103	136	33	32	0	118	15	15	−0	1,081.7
Parkes	432	431	−1	−0	−0	435	3	1	−0	943.5
Parramatta	3,396	4,556	1,160	34	5	3,152	−244	−7	2	983.7
Penrith	3,784	4,952	1,168	31	5	5,327	1,543	41	−11	996.3
Pittwater	781	897	116	15	0	424	−357	−46	3	1,094.4
Port Macquarie Hastings	1,500	1,597	97	6	0	1,551	51	3	−0	968.9
Port Stephens	1,358	1,662	304	22	1	1,754	396	29	−3	979.9
Queanbeyan	685	988	303	44	1	305	−380	−56	3	1,045.7
Randwick	1,920	2,382	462	24	2	1,784	−136	−7	1	1,042.7
Richmond Valley	560	570	10	2	0	227	−333	−59	2	899.5
Rockdale	1,542	1,888	346	22	1	812	−730	−47	5	991.2
Ryde	950	1,292	342	36	1	757	−193	−20	1	1,050.4
Shellharbour	1,248	1,437	189	15	1	1,274	26	2	−0	968.6
Shoalhaven	2,053	2,018	−35	−2	−0	1,661	−392	−19	3	954.6
Singleton	486	508	22	5	0	278	−208	−43	2	1,013.0
Snowy River	114	110	−4	−3	−0	116	2	2	−0	1,050.0
Strathfield	478	581	103	22	0	366	−112	−23	1	1,022.1
Sutherland Shire	2,703	3,028	325	12	1	2,184	−519	−19	4	1,074.6

Sydney (Central[b])	4,503	6,431	1,928	43	8	3,056	−1,447	−32	11	1,019.9
Tamworth Regional	1,428	1,636	208	15	1	1,598	170	12	−1	959.9
Temora	128	96	−32	−25	−0	127	−1	−1	0	955.6
Tenterfield	155	145	−10	−7	−0	160	5	3	−0	915.4
The Hills Shire	1,382	2,240	858	62	4	2,140	758	55	−6	1,101.1
Tumbarumba	40	38	−2	−5	−0	30	−10	−26	0	954.3
Tumut Shire	333	262	−71	−21	−0	185	−148	−44	1	951.3
Tweed	1,283	1,608	325	25	1	963	−320	−25	2	958.5
Upper Hunter Shire	175	221	46	26	0	184	9	5	−0	981.5
Upper Lachlan Shire	84	65	−19	−23	−0	41	−43	−51	0	1,006.3
Uralla	89	91	2	3	0	103	14	16	−0	984.7
Urana	9	6	−3	−38	−0	8	−1	−17	0	937.1
Wagga Wagga	1,508	1,764	256	17	1	1,114	−394	−26	3	997.6
Wakool	53	35	−18	−33	−0	51	−2	−3	0	975.7
Walcha	64	53	−11	−17	−0	62	−2	−4	0	973.9
Walgett	505	366	−139	−27	−1	574	69	14	−1	856.2
Warren	149	101	−48	−33	−0	177	28	19	−0	941.2
Warringah	1,426	1,618	192	13	1	713	−713	−50	5	1,077.3
Warrumbungle Shire	255	219	−36	−14	−0	309	54	21	−0	911.3
Waverley	784	904	120	15	1	545	−239	−30	2	1,079.6
Weddin	57	45	−11	−20	−0	36	−20	−36	0	945.0
Wellington	385	298	−87	−23	−0	332	−53	−14	0	893.2
Wentworth	158	165	7	4	0	111	−47	−30	0	957.4
Willoughby	389	487	98	25	0	187	−202	−52	1	1,083.5
Wingecarribee	631	619	−12	−2	−0	636	5	1	−0	1,023.8
Wollondilly	531	604	73	14	0	442	−89	−17	1	1,033.6
Wollongong	3,561	3,987	426	12	2	2,793	−768	−22	6	979.6

(continued)

Table 5.2 (continued)

Local area	Actual	Projections									Index of Relative Socio-economic Disadvantage (ABS IRSD[a])
		Assumption 1: Court appearance rates held constant					Assumption 2: Court appearance rates follow 10-year trend				
			Increase: 2012–2031					Increase: 2012–2031			
	2012	2031	Level	%	% of total		2031	Level	%	% of total	
Woollahra	674	786	112	17	0		439	−235	−35	2	1,107.0
Wyong	3,220	3,865	645	20	3		2,911	−309	−10	2	951.7
Yass Valley	176	228	52	29	0		129	−47	−27	0	1,060.6
Young	370	342	−28	−8	−0		392	22	6	−0	947.3
Total	131,605	154,776	23,171	18	100		118,092	−13,513	−10	100	–

Note. Court appearances capture count of finalised criminal court appearances (finalised local court appearances and finalised trial and sentence cases in the District and High Courts) by defendant's place of residence, plus the number of Apprehended Violence Orders (AVO) granted against persons of interest by that person's place of residence

[a]A lower value on the Index of Relative Socio-economic Disadvantage indicates a higher level of disadvantage. The index is standardised to a mean of 1000 and a standard deviation of 100 across local areas Australia-wide

[b]This is the 'Sydney' Local Government Area; the 'Central' qualifier is added to differentiate this local area from the wider Sydney metropolitan region

References

Australian Bureau of Statistics. (2011). *Socio-economic Indexes for Areas (SEIFA)*, Technical paper, Cat. No. 2033.0. 55.001. Canberra: Australian Bureau of Statistics.

Australian Bureau of Statistics. (2015). *Australian demographic statistics: March 2015*. Cat. No. 3101.0. Canberra: Australian Bureau of Statistics.

Australian Law Reform Commission. (2006). *Same crime, same time: Sentencing of federal offenders*. ALRC No. 103. Sydney: Australian Law Reform Commission.

Crawford, J., & Opeskin, B. (2004). *Australian courts of law*. Melbourne: Oxford University Press.

Devery, C. (1991). *Disadvantage and crime in New South Wales*. Sydney: NSW Bureau of Crime Statistics and Research.

District Court of New South Wales. (2015). *Annual review 2014*. Sydney: District Court.

Farrington, D. P. (1986). Age and crime. *Crime and Justice, 7*, 189–250.

Fentress, K. (2000). Long-range facility planning for federal courts. *Journal of Judicial Administration, 9*, 134–152.

Kirby, M. (1998). Acting judges: A non-theoretical danger. *Journal of Judicial Administration, 8*, 69–76.

Levitt, S. D. (1999). The limited role of changing age structure in explaining aggregate crime rates. *Criminology, 37*, 581–598.

Local Court of New South Wales. (2015). *Annual review 2014*. Sydney: Local Court.

Mack, K., Wallace, A., & Roach Anleu, S. (2012). *Judicial workload: Time, tasks and work organisation*. Melbourne: Australasian Institute of Judicial Administration.

Moffatt, S., Weatherburn, D., & Donnelly, N. (2005). What caused the recent drop in property crime? *Crime and Justice Bulletin, 85*, 1–15.

Moffitt, T. E. (1993). Adolescence-limited and life-course-persistent antisocial behavior: A developmental taxonomy. *Psychological Review, 100*, 674–701.

NSW Bureau of Crime Statistics and Research. (2015). *New South Wales criminal court statistics 2014*. Sydney: NSW Bureau of Crime Statistics and Research.

NSW Department of Education. (2015). *Teaching workforce supply and demand*. Sydney: NSW Department of Education.

NSW Department of Planning and Environment. (2015). *Population projections [Online]*. Sydney: NSW Department of Planning and Environment. Available: www.planning.nsw.gov.au/Research-and-Demography/Demography/Population-Projections. Accessed 19 Nov 2015.

NSW Police. (2003). *Annual report 2002–2003*. Sydney: NSW Police.

NSW Police Force. (2013). *Annual report 2012–13*. Sydney: NSW Police Force.

NSW Treasury. (2012). *NSW long-term fiscal pressures report: NSW intergenerational report 2011–12* (Budget paper No. 6). Sydney: NSW Treasury.

Opeskin, B. (2011). The high cost of judges: Reconsidering judicial pensions and retirement in an ageing population. *Federal Law Review, 39*, 33–70.

Opeskin, B. (2013). The state of the judicature: A statistical profile of Australian courts and judges. *Sydney Law Review, 35*, 489–517.

Opeskin, B. (2015). Models of judicial tenure: Reconsidering life limits, age limits and term limits for judges. *Oxford Journal of Legal Studies, 35*, 627–663.

Opeskin, B., & Parr, N. (2014). Population, crime and courts: Demographic projections of the future workload of the New South Wales Magistracy. *Journal of Judicial Administration, 23*, 233–252.

Parr, N. (2014). *Fertility levels and intentions in New South Wales*. Report commissioned by the centre for demography, policy and research. Sydney: NSW Department of Planning and Environment.

Parr, N., Boyages, J., Taylor, R., Anura, P., & Callaghan, T. (2000). Projecting mammographic screens. *Journal of Medical Screening, 7*, 146–151.

Piquero, A. R., Farrington, D. P., & Blumstein, A. (2003). The criminal career paradigm. *Crime and Justice, 30*, 359–506.

Potas, I. (1999). *Executive summary: Apprehended violence orders: A survey of magistrates [Online]*. Sydney: Judicial Commission of New South Wales. Available: www.judcom.nsw.gov.au/publications/research-monographs-1/monograph20/Executive%20summary.html. Accessed 18 Nov 2015.

Productivity Commission. (2015). *Report on government services* Vol. C. Justice. Canberra: Productivity Commission.

Ralston, N. (2015). NSW prison population hits record high forcing police to 'Babysit' inmates. *Sydney Morning Herald*, 3 February.

Rosevear, L. (2007). Australian crime trends and population ageing: A quantified perspective. *Flinders Journal of Law Reform, 10*, 831–847.

Rosevear, L. (2012). The impact of structural ageing on crime trends: A South Australian case study. *Trends and Issues in Crime and Criminal Justice, 431*, 1–6.

Supreme Court of New South Wales. (2015). *Annual review 2014*. Sydney: Supreme Court.

Sweeten, G., Piquero, A. R., & Steinberg, L. (2013). Age and the explanation of crime, revisited. *Journal of Youth and Adolescence, 42*, 921–938.

Wan, W.-Y., Moffatt, S., Jones, C., & Weatherburn, D. (2012). The effect of arrest and imprisonment on crime. *Crime and Justice Bulletin, 158*, 1–20.

Weatherburn, D., & Holmes, J. (2013). The great property crime drop: A regional analysis. *Crime and Justice Statistics Bureau Brief, 66*, 1.

Chapter 6
Projecting Future Demand for Assisted Living: A Case Study

Peter A. Morrison

Abstract Nantucket is one of only a few Massachusetts communities that offers its elderly residents local assisted living and skilled nursing care partially subsidized by local taxpayers. When contemporary standards of care required modernizing Our Island Home to meet those standards, the question arose: Would a 45-bed facility meet the future needs of this isolated island community? A demographic analysis predicted a widening gap between availability and need within 15 years, disrupting local families' capacity to supplement local institutional care with frequent personal visits to loved ones and obliging Nantucket residents to seek other distant arrangements. The analysis recommended sizing the future facility to accommodate a projected 63 % increase in the future needs of Nantucket *families* with elderly residents. This case study illustrates two distinct contributions of applied demographic analysis: identifying immediate necessary decisions and clarifying distant long-term tradeoffs for public choice.

Keywords Assisted living • Skilled nursing • Institutionalized elderly • Family ties

6.1 Introduction

How does an isolated community anticipate and respond to the future needs of its aging resident population? This case study illustrates how applied demographers can inform important public choices and heighten awareness of their potential long-range implications. It draws upon publicly available data and standard demographic methods to discern what the future could hold and the tradeoffs that future could impose.

The focus is a small island community that needed to modernize its outdated facility dedicated to providing assisted living and skilled nursing care to local elderly residents. In the course of bringing Our Island Home (OIH) up to contemporary standards, its fit with tomorrow's demographic landscape posed an obvious concern: sizing the facility in light of projected future need stemming from the

P.A. Morrison (✉)
RAND, Nantucket, MA, USA
e-mail: petermorrison@me.com

D.A. Swanson (ed.), *The Frontiers of Applied Demography*, Applied Demography Series 9, DOI 10.1007/978-3-319-43329-5_6

community's future elderly resident population. Depicting that future landscape drew attention to an immediate concern–preserving the option to expand OIH to meet potential further growth in local need. It also unveiled potential long-term tradeoffs ahead. Applied demographic analysis, as we shall see, can play two critical roles: identifying immediate issues for decision and clarifying distant long-term tradeoffs for public choice.

The following technical analysis draws upon standard demographic methods and publicly available data to describe the evolving demographic context shaping the community's future need for assisted living and skilled nursing care among its elderly residents.[1]

6.2 Background and Context

The Town of Nantucket is one of only a few Massachusetts communities that offers its elderly residents local assisted living care and skilled nursing care partially subsidized by local taxpayers. Contemporary standards of care required modernizing Our Island Home to meet those standards. That led to a decision to build an entirely new facility scaled to local need. That decision posed the question of how local need might change in the future—and more broadly what the future might hold for OIH's traditional mission.

Questions Posed OIH existed in 2015 as a 45-bed skilled nursing facility whose central location offers the family members of its occupants easy daily access to their loved ones. The facility was fully occupied, almost exclusively by persons 60 years of age or older. How can an applied demographer foresee and quantify what the future holds, to inform immediate and long-term decisions for meeting local needs in the future? This question raises three considerations:

- *Future capacity*: Are 45 beds the upper limit of what the community will need and want?
- *Standards for admission*: Who will be entitled to an available OIH bed–and the attendant peace of mind among the occupant's family members?
- *Length of stay*. Would admission to OIH guarantee the occupant lifelong residence regardless of cognitive or health status, irrespective of other prospective occupants whose local family support networks may be strained to the breaking point?

[1] The analysis set forth here may offer a useful instructional case for inclusion in an advanced undergraduate or graduate course in applied demography. The case can be replicated as a student project using data for any particular area that has a significant presence of elderly residents with local family ties. It trains students to integrate disparate sources of publicly available data to foresee future needs and inform policymaking.

6.3 Analysis

The first step is to project Nantucket's future elderly population by age. We need to foresee the future number of persons 65+ who will call Nantucket home. We can derive a plausible forecast from the official population projection for Nantucket, summarized in Table 6.1.[2]

According to these data, Nantucket's total (all ages) resident population is projected to increase only 13 % over the next 20 years (2015–2035), whereas persons over 65 would increase 56 % and persons 80+ by 80 %. Whatever the margins of uncertainty here, there are sharp increases in store for the 65+ and 85 + elderly population.

How will this aging of Nantucket's population affect the size of OIH's target population–elderly residents who need assistance with activities of daily living (ADLs)? A person's ability to perform certain routine everyday activities without needing assistance often declines markedly after about age 80.[3] Needing the help of other persons becomes imperative: with bathing or showering, dressing, eating, getting in or out of bed or chairs, using the toilet, including getting to the toilet, and getting around inside the home. Such needs rise sharply with advancing age (see Fig. 6.1 and Table 6.2 below).

The specific question here is: How many of Nantucket's future elderly will have multiple limitations in ADLs? We can quantify this outlook using a straightforward demographic projection framework that incorporates publicly-available data. Table 6.2 furnishes the most current national data on the age-specific prevalence of limitations in ADLs for the period 2003–2007.[4] These data provide a sufficiently current and detailed measure of prevalence for the civilian noninstitutionalized population of interest here. We multiply the projected age-specific population shown in the top panel of Table 6.3 by the corresponding "2+ ADL" prevalence rate in Table 6.2 to derive the projected number of persons with limitations in 2+ ADLs (shown in the bottom panel of Table 6.3). For example, there are projected to be 817 persons 75–84 in 2035 (Table 6.3, top panel). The prevalence of multiple limitations in ADLs among persons this age is 4.7 % (Table 6.2). Accordingly, we estimate that 38 of these 817 (4.7 %) will have such limitations.

The national prevalence rates in Table 6.2 can be defended as reasonably accurate prevalence measures for contemporary noninstitutionalized Nantucket residents of comparable age. But what about future elderly cohorts? Our

---

[2] Massachusetts' State Data Center, the UMass Donahue Institute (UMDI), publishes population projections for all Massachusetts municipalities at 5-year intervals to 2035. Accessed at: http://pep.donahue-institute.org

[3] The six basic "activities of daily living" (ADLs) are: eating, bathing, dressing, toileting, transferring (walking) and continence. An individual's ability to perform ADLs is important for determining what type of long-term care (e.g. nursing-home care or home care) and coverage the individual needs (i.e. Medicare, Medicaid or long-term care insurance).

[4] Access these data at: www.cdc.gov/nchs/health_policy/adl_tables.htm

Table 6.1 Projected population of Nantucket, by age

Cumulative Age Group	Census 2010	Projected 2015	Projected 2020	Projected 2025	Projected 2030	Projected 2035	% increase, 2015–2025	% increase, 2015–2030	% increase, 2015–2035
All ages:	10,172	10,667	10,678	10,895	11,371	12,004	2	7	13
Pop.65+	1227	1466	1701	1945	2147	2286	33	46	56
Pop.70+	858	1019	1196	1398	1589	1735	**37**	**56**	70
Pop.75+	535	607	703	831	978	1108	**37**	**61**	**83**
Pop.80+	317	343	372	434	523	617	27	52	**80**
Pop.85+	162	183	185	202	244	291	**10**	**33**	**59**

Source: Massachusetts state data center, the UMass Donahue Institute

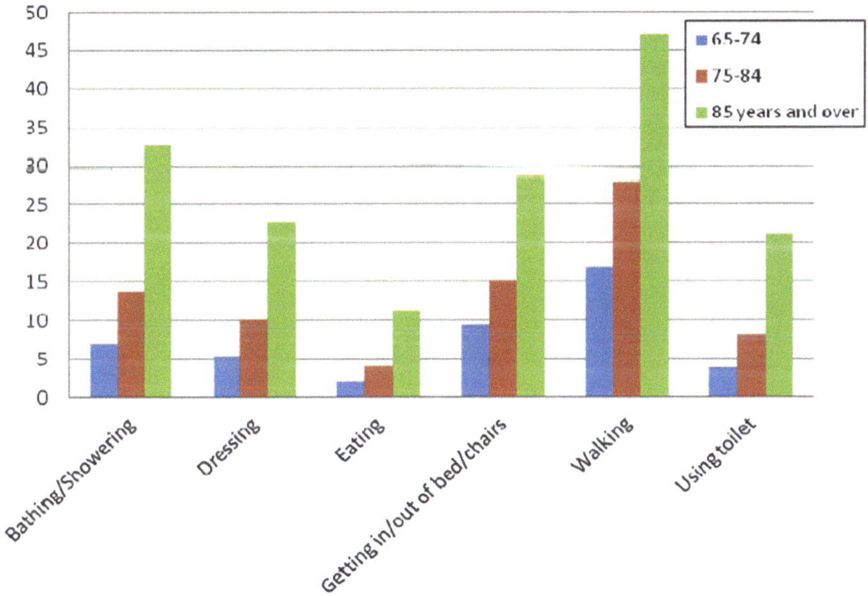

Fig. 6.1 Percent of persons with limitations in activities of daily living by age group: 2009 (Source: DHHS, Administration on Aging, *A profile of older Americans: 2011*, figure 9)

Table 6.2 Period prevalence of limitations in activities of daily living by age: 2003–2007 (civilian noninstitutionalized population)

Limitations in Activities of Daily Living (ADLs)				
	%Distributions of No. of Limitations in ADLs			
Age group	Total	None	1	2+
65–74	100.0 %	97.1 %	0.7 %	2.2 %
75–84	100.0 %	93.9 %	1.4 %	4.7 %
85+	100.0 %	82.2 %	4.7 %	13.2 %

Source: CDC, *Limitations in Activities of Daily Living and Instrumental Activities of Daily Living, 2003–2007*, Table 4

methodology relies upon a key assumption that is open to challenge: that prevalence rates in the future–as of 2035–will be the same as those that prevailed in 2003–2007 (as shown in Table 6.2). With future health improvements, might future elderly cohorts enjoy more "healthy" years of life expectancy, on average, before "nonhealthy" years follow? If so, such improvements would postpone the onset of multiple limitations in ADLs to later years of old age. We can only speculate about this "known unknown," and the conundrum here reminds us that a forecast is no better than the soundness of its underlying assumptions.

The issue here is whether or not such improvements will likely prove consequential relative to the measureable influence of population aging. Here, the applied demographer can only draw upon available evidence to venture an informed guess. We start with two available metrics. The first is life expectancy (LE), or expected years of life at a given age. This is the average remaining years of life a person can

Table 6.3 Projected increase in Nantucket's noninstitutionalized population 65+ with limitations in two or more activities of daily living

Age group	2010 (Census)	Projected 2015	Projected 2020	Projected 2025	Projected 2030	Projected 2035	% Increase		
							2015–2025	*2015–2030*	*2015–2035*
Population									
65 – 74	692	859	998	1114	1169	1178	***30 %***	***36 %***	***37 %***
75 – 84	373	424	518	629	734	817	***48 %***	***73 %***	***93 %***
85+	162	183	185	202	244	291	***10 %***	***33 %***	***59 %***
Total 65+	1227	1466	1701	1945	2147	2286	***33 %***	***46 %***	***56 %***
Estimated Population with Limitations in 2+ ADLs									
65 – 74	15	19	22	25	26	26	***30 %***	***36 %***	***37 %***
75 – 84	18	20	24	30	34	38	***48 %***	***73 %***	***93 %***
85+	21	24	24	27	32	38	***10 %***	***33 %***	***59 %***
Total 65+	54	63	71	81	92	103	***28 %***	***47 %***	***63 %***

Source: Author's calculations applying Table 6.2 prevalence rates to Table 6.3 projected population (derived from Table 6.1 above)

expect to live on the basis of the current mortality rates for the population. The second is "healthy life expectancy" (HLE), the equivalent *healthy* years of LE that a person can expect to live on the basis of the current mortality rates and prevalence distribution of health status in the population.[5] The concepts of "total" and "healthy" life expectancy—and quantitative measures of each—conveniently pinpoint the *average age* at which Nantucket's elderly residents join the target population that OIH serves. In what follows, we utilize the state-specific measures of each for 65-year-old Massachusetts residents: LE is 18.2 years (males) and 20.9 years (females); HLE is 13.8 years (for males) and 15.9 years (for females).

According to these data, a population of 65-year-old Massachusetts males has, on average, 13.8 "healthy" future years of life expectancy. A corollary here is that age 78.8 years (i.e., 65 + 13.8) is the average age at which this 65-year-old male will commence the subsequent "non-healthy" years of life expectancy. For females, the corresponding age is 80.9 years of age (i.e., 65 plus 15.9). In round numbers, then, age 79 (for men) and age 81 (for women) marks the average age at which Nantucket's elderly may begin to need assistance with multiple ADLs. For any particular individual, of course, that onset of need may occur at a younger or older age, and the intensity of need may vary.

The corresponding entire life expectancy (LE) for 65-year-old Massachusetts residents is 18.2 years (males) and 20.9 years (females). These LE's mean that at age 78.8, the average male faces an additional 4.4 non-healthy years of life (i.e., 18.2–13.8); and that at age 80.9, the average female faces an additional 5.0 non-healthy years of life. Again, these durations may be shorter or longer for any particular individual. However, they furnish objective measures of the average anticipated *duration of stay* at OIH by the target population upon admission. In round numbers, then, duration of stay for a would-be OIH resident during the remaining non-healthy years of life expectancy will average about 4.4–5.0 years.

How does this analysis inform our "guess" alluded to above about the prospect that more "healthy" years of life expectancy might delay subsequent "nonhealthy" years? The "guess" here centers on the potential magnitude of gain in "healthy" life expectancy preceding the onset of remaining "nonhealthy" life expectancy. Would the gain be several months? Or could it be several years? To formulate an informed guess here, we can draw upon a body of scientific research to formulate an informed guess.[6] That research suggests that forthcoming gains in "healthy" life expectancy are likely to be modest, adding only a fraction of a "healthy" additional year over the course of an entire future decade. Accordingly, we can treat both HLE and LE as roughly constant in our calculations, subject to updating as more current estimates of HLE and LE become available in the future.

---

[5] CDC, *Morbidity and Mortality Weekly Report*, "State-Specific Healthy Life Expectancy at Age 65 Years—United States, 2007–2009," vol. 62(28), July 19, 2013: pp. 561–566. Access at: www.cdc.gov/mmwr/preview/mmwrhtml/mm6228a1.htm

[6] See: Chernew et al. (2015); Goldman and Olshansky (2013); Jagger (2015), especially Table 4; Olshansky (2015); Easterbrook (2014).

Summing up: We project that Nantucket's noninstitutionalized population 65+ will increase 33 % by 2025, 46 % by 2030, and 56 % by 2035 (Table 6.3, top panel). Assuming constant prevalence rates as shown in Table 6.2, this noninstitutionalized population 65+ will generate 28 % more members with limitations in 2 or more ADLs by 2025, 47 % more by 2030, and 63 % more by 2035 (Table 6.3, bottom panel).

6.4 Interpreting the Results

To appreciate these changes, imagine replacing Nantucket's 2015 resident population of 10,667 with its projected 2035 resident population of 12,004—a 13 % increase. Twenty years hence, many more of these residents will be in their 80s and 90s with, by historic standards, unprecedented longevity ahead. That will mean more Nantucket *families* with members unable to live independently without assistance. More specifically, the 13 % increase in overall population masks a 56 % increase in persons 65 and older (Table 6.1 above, comparing "All ages" with "Pop.65+"). That 56 % increase in elderly residents, in turn, masks a 63 % increase in persons with limitations in 2 or more ADLs (Table 6.3 above, comparing "Total 65+ population" with "Total 65+ Estimated Population with Limitations in 2+ ADLs").

In short, we foresee Nantucket being home to only 13 % more residents by 2035, but among them will be 63 % more persons unable to live independently without assistance. This impending demographic surge foreshadows about 103 prospective candidates for Our Island Home admission by 2035 compared with around 63 candidates now—a current ratio of 1.4 prospects for each of today's 45 beds. That same 1.4 ratio applied to 103 anticipated prospects 20 years from now implies 73 needed beds, not 45.

6.5 Present Capacity vs. Projected Future Need

Against this demographic backdrop, is a 45-bed facility appropriately sized for the future? As of late 2015, 27 of OIH residents were at least 80 years old, roughly congruent with the nonhealthy portion of life expectancy accompanied by the need for assisted living. These 27 residents comprise 8 % of all 343 persons 80 and older enumerated as Nantucket residents on the 2010 federal census and 7 % of all 373 persons 80 and older counted as residents by the Nantucket Town Clerk's 2015 census (see "80+" row in Appendix Table 6.4 ahead). These data imply that OIH figures in the lives of about one of every 14 Nantucket residents 80 and older as of 2015.

Nantucket's population 80 and older is projected to increase 52 % by 2030 and 80 % by 2035 (see Table 6.1 above). A simple back-of-the-envelope calculation

($1.52 \times 27 = 41$; or $1.80 \times 27 = 49$) implies that future needs of the 80+ population alone would absorb virtually the entire capacity of a 45-bed facility.

As an alternative reality check, let us regard the estimated population with multiple ADL limitations (shown in Table 6.3) as a proxy for future need. Here we foresee the population 85 and older increasing 33 % by 2030 and 59 % by 2035. (The population 75–84 evidences much sharper corresponding increases: 73 % by 2030 and 93 % by 2035.) This alternative reality check is broadly consistent with the back-of-the-envelope calculation above. Furthermore, it can enlighten us on an alternative future scenario embodying secular health improvements which might postpone the "unhealthy" span of life expectancy to age 85. Even under this optimistic scenario, we anticipate a 59 % increase by 2035—from 24 persons 85+ with multiple ADL limitations to 38 such persons. Here, too, the data imply that within just 20 years, the needs of the 85+ population alone would fill 38 of the facility's 45 beds. That would leave just 7 beds available to meet the needs of 60 or more persons aged 65–84 with multiple ADL limitations (based on Table 6.3, bottom panel).

To summarize, this analysis offers robust support for an important conclusion: A future facility sized to just 45 beds will eventually find itself turning away many needy elderly Nantucket residents, forcing them and their families to seek some other distant arrangement. "Eventually" could commence within just 15 years; it likely will commence within 20 years. This impending future surely will disrupt local families' capacity to supplement OIH care with frequent personal visits to loved ones, and the universe of families so affected will likely expand with each passing year thereafter.

6.6 Identifying Options for Public Choice

For an isolated island community, local access to assisted living is a manifestly public good that benefits its elderly residents and also their families. OIH affords families the practical possibility of enhancing, through their ongoing companionship, its institutional assistance with activities of daily living. Presently, family members and friends are able to provide most OIH residents such companionship at least weekly and often daily. Institutional residence anywhere off-island would preclude doing so that often.

This analysis underscores the collective need for the community to rethink its priorities for OIH and its future place in the community. Otherwise, there looms the prospect of one needy population segment (e.g., persons 80+) crowding out another, with no explicit public consensus on how to balance the competing needs of families and their elderly members.

At a minimum, a strong case exists for preserving the future option to expand OIH. Looking 20 years ahead, potential demand for OIH beds will intensify sharply, perhaps prompting taxpayers' support for that option. More broadly, the analysis suggests possible ways to preserve OIH's core mission in the future. Our

Island Home's presence affords local family members easy access—a 20-min. drive, at most—to a loved one residing there. That proximity fosters frequent ongoing companionship, enhancing Our Island Home's institutional assistance to a loved one. Family members and friends provide companionship to 2 of every 3 OIH residents at least weekly. Such companionship is a daily event for one of two residents. Off-island institutional residence would preclude this unique enhancement.

A far-sighted vision of OIH would regard local access to assisted living as a *public good*—something that one family can "consume" without reducing its availability to another, and from which no family is excluded. Defining access in this way acknowledges the strength of local family ties in place and values their collective benefits to island families. Fitting OIH to Nantucket's collective future would necessitate new ways of financing it. Just as the existing facility now draws upon those strengths in place, it might capitalize upon the loyalties of close family members, wherever they happen to reside. Such possibilities raise further practical considerations: (1) Taxpayers' appetite for incurring construction and future operating costs of a larger facility; and (2) The feasibility of arrangements whereby those costs might gradually be offset in the distant future through a flow of bequests from Nantucket residents and their family members.

This case study illustrates two distinct ways that applied demographers can inform public choice. First is identifying necessary decisions on the immediate horizon—in this instance, addressing the need to maintain the future option to expand OIH. Second is surfacing unrecognized tradeoffs likely to emerge in the distant future and the hidden costs of overlooking them—here, the prospect of one needy population segment crowding out another and the potential disruption of families' capacity to supplement local institutional care with frequent personal visits to loved ones.

Appendix

The US Census Bureau defines Nantucket's resident population as persons who report Nantucket as their "usual place of residence" as of April 1. That definition excludes many additional Nantucket residents. Nantucket's Town Clerk conducts an annual census by mail from late December-January which identifies a much larger "resident population" based on a more inclusive definition (including eligibility to vote). The substantial differences between these two sources raises the possibility that one or the other definition may prove misleading about the future number of residents—both elderly and nonelderly—who regard themselves as residents of Nantucket. To explore this possibility, I compare age-specific data from each source in Tables 6.4 and 6.5.

Table 6.4 Age distribution of Nantucket residents 55 and older based on census bureau (2010) and town census (2015) enumerations

	2015 Town census		2010 US, Census		
Cumulative age group	No.	% of 55+	No.	% of 55+	Town minus U.S Census
Total 55+	3592	100.0	2512	100.0	1080
60+	2659	74.0	1799	71.6	860
65+	1334	51.1	1227	48.8	607
70+	1155	32.2	858	34.2	297
75+	694	19.3	535	21.3	159
80+	373	10.4	317	12.6	56
85+	179	5.0	162	6.4	17
90	73	2.0	48	1.9	25
95+	12	0.3	11	4.0	1
100+	1	0.0	1	0.0	–

Source: Town Clerk's 2015 census of residents; US Census Bureau, 2010 census, SF1 table QT-P2

Observations:

Town Census (2015) and Census complete count (2010) closely agree in *relative* terms (compare "% of 55+" columns)

Agreement between Town Census (2015) and MA State Data Center's 2015 projected elderly distribution is even closer (data not shown)

Scale differences are apparent (see "Town minus U.S. Census" column) but do not distort relative agreement

Table 6.5 Our Island home residents vs. Town census and U.S. Census counts

				OIH population as % of:	
Cumulative Age Group	2015 Town Census	2010 U.S. Census	2015 Our Island Home Population	2015 Town Census	2010 U.S. Census
55+	3592	2512	41	1.1 %	1.6 %
60+	2659	1799	38	1 4 %	2.1 %
65+	1834	1227	37	2.0 %	3.0 %
70+	1155	858	36	3.1 %	4.2 %
75+	694	535	34	4.9 %	6.4 %
80+	373	317	27	7.2 %	8.5 %
85+	179	162	20	11.2 %	12.3 %
90+	73	48	11	15.1 %	22.9 %
95+	12	11	3	25.0 %	27.3 %
100+	1	1	0	0.0 %	0.0 %

Source: Town Clerk's 2015 census of residents; US Census Bureau, 2010 Census, SF1 Table QT-P2; Our Island Home data *circa* November 2015. Projections by UMass Donahue Institute, "Population Projections for Massachusetts Municipalities" accessed 11/18/2015 at

Observations:

Today, ages 75+ account for 19–21 % of census populations but 83 % of OIH residents

Populations generate OIH residents at noticeable rates starting around ages 75 +

References

Chernew, M., Cutler, D. M., Ghosh, K., & Landrum, M. B. (2015). Understanding the improvement in disability free life expectancy in the U.S. elderly population. In D. A. Wise (Ed.), *Insights in the economics of aging* (University of Chicago Press, forthcoming 2017). Accessed at: http://papers.nber.org/books/wise-21

Esterbrook, G. (2014, October). What happens when we all live to 100?. *The Atlantic*. Accessed at: www.theatlantic.com/magazine/archive/2014/10/what-happens-when-we-all-live-to-100/379338/

Goldman, D. P., & Jay Olshansky, S. (2013). Delayed aging versus delayed disease: A new paradigm for public health. *Public Policy & Aging Report, 23*(4), 16–18.

Jagger, C. (2015). *Trends in life expectancy and healthy life expectancy*. Newcastle University Institute for Ageing and Institute of Health & Society. Accessed at: https://www.gov.uk/government/uploads/system/uploads/attachment_data/file/464275/gs-15-13-future-ageing-trends-life-expectancy-er12.pdf

Olshansky, S. J. (2015). The demographic transformation of America. *Daedalus, 144*(2), 13–19.

Chapter 7
The Growth of Australia's Very Elderly Population: Past Estimates and Probabilistic Forecasts

Tom Wilson and Wilma Terblanche

Abstract Official population estimates in Australia are overstated at the very highest ages, resulting in inaccurate mortality rates, and unreliable forecasts. Official population forecasts are not accompanied by information about their uncertainty, and do not extend into the centenarian ages. The aim of this chapter is to present more accurate estimates of the very elderly population of Australia (those aged 85+) from 1971 to 2014, and probabilistic forecasts out to 2051 by sex and single years of age up to age 110+. Population estimates were calculated from death counts using Extinct Cohort and Survivor Ratio methods, the latter being a newly-refined version. Population forecasts were produced using a probabilistic cohort-component model. The 85+ population of Australia grew from 69,000 in 1971 to 456,000 in 2014, in large part due to mortality reductions. It is forecast to increase to 1.90 million by 2051, with the 95 % prediction interval spanning 1.51 to 2.37 million. The future growth in centenarians is proportionally far greater, but relatively more uncertain. Although the extent of future growth cannot be forecast precisely, huge increases in Australia's very elderly population *will* eventuate.

Keywords Very elderly • Centenarians • Australia • Survivor ratio • Extinct cohort • Probabilistic • Forecasts

Abbreviations

ABS	Australian Bureau of Statistics
ERP	Estimated Resident Population
HMD	Human Mortality Database

T. Wilson (✉)
Northern Institute, Charles Darwin University, Darwin NT 0909, Australia
e-mail: tom.wilson@cdu.edu.au

W. Terblanche
Bond Business School, Bond University, Gold Coast Qld 4229, Australia
e-mail: wterblan@bond.edu.au

D.A. Swanson (ed.), *The Frontiers of Applied Demography*, Applied Demography Series 9, DOI 10.1007/978-3-319-43329-5_7

PROBPOP PROBabilistic POpulation Projections program
RIDR Relative Inter-Decile Range

7.1 Introduction

In common with other industrialised countries, the substantial growth of the elderly population is one of the most prominent features of contemporary Australian demography. Conventionally defined as those aged 65 years and over, the elderly component of the population has increased in number from 1.09 million in 1971 (8.3 % of the total population) to 1.95 million in 1991 (11.3 %) and 3.45 million in 2014 (14.7 %) (ABS 2015). However, for many service providers, planners and policymakers, it is the *very* elderly – often defined as those aged 85 years and above – who are of greatest significance. They are the fastest growing age group in the population and have considerable needs in terms of health, accommodation, income support, and aged-care services (Chomik and MacLennan 2014). They comprise a rapidly growing, and thus increasingly influential, group of consumers and voters.

There is no shortage of research examining population ageing internationally and in Australia, including its characteristics, drivers, and multitude of consequences (e.g. Beard et al. 2012; Cubit and Meyer 2011; Christensen et al. 2009; Productivity Commission 2013; Rowland 2012; Terblanche and Wilson 2014). Neither is there a lack of political awareness of the topic. The Australian Government's Treasury is keenly aware of the implications of population ageing, and every few years produces an 'intergenerational report' focusing primarily on the budgetary consequences of the nation's future population ageing (Treasury 2015). The Productivity Commission, an Australian Government research agency, has also published several major reports on population ageing (e.g. Productivity Commission 2013). But much of this work tends to discuss the 65+ population in fairly homogenous terms. From a policy perspective, those at the very highest ages have greater needs, and there are fewer studies which specifically concentrate on the past and future trends of this important section of the population (but see, for example, McCormack 2010; Terblanche and Wilson 2014).

Many studies which do focus on Australia's very elderly population tend to be limited by at least one of three key issues. First, they often use official population estimates and forecasts. As has been shown for Australia (Terblanche and Wilson 2014) and many other countries (Thatcher et al. 2002), official population estimates at the highest ages are not as accurate as they could be. This is due to mistakes in reporting age on census forms as well as errors introduced in census editing and processing. The published population estimates tend to be overstated from somewhere in the mid-1990s upwards, with the problem increasing in severity with increasing age, and worse for males than females (Terblanche 2015a). Consequently, mortality rates, which use these population estimates as denominators, are understated. Mortality forecasts at the highest ages, and hence population forecasts themselves, are therefore also affected.

Second, existing studies often use a final age group of 85+ or 90+, and sometimes 100+, which is the final age group of official Australian Bureau of Statistics (ABS) Estimated Resident Populations (ERPs). Forecasting a mortality rate for the 90+ population, for example, would not account for the changing age distribution within that age group and its impact on the mortality rate. With a rapidly growing and heterogeneous very elderly population, it is wise to extend the estimates by single years of age up to an age where the final open-ended age group has fewer people in it than the previous single year age group.

Third, few forecasts of the very elderly are accompanied by any indication of their uncertainty, yet many studies have demonstrated that population forecasts always contain some degree of error (e.g. Keilman 2008; Wilson 2012). The traditional way of representing forecast uncertainty is to produce alternative high and low projection variants. But this approach is beset with problems (Lee 1999; Wilson and Bell 2004). There is usually no information supplied on the likelihood of future population lying within the high-low range; the high and low variants are often produced by changing fertility and/or migration assumptions, which will clearly understate uncertainty for the elderly population because of the dominance of mortality at these ages; and the high-low range tends to be inconsistent over time and across variables. Probabilistic population forecasts overcome these limitations, and illustrate the uncertainty of forecasts with internally consistent prediction intervals. Such forecasts have received considerable research attention in demography over the past 20 years (e.g. Alho et al. 2008; Bell et al. 2011; Keilman et al. 2002; Wilson and Rees 2005). But the very top end of the age distribution has not been prominent in this work.

This chapter presents new estimates and forecasts of Australia's very elderly population which attempt to overcome these limitations. New population estimates, extending from 1971 to 2014, were created using a combination of Extinct Cohort and Survivor Ratio methods. These two methods have been shown to yield far more accurate population numbers than census-based estimates (Terblanche and Wilson 2015b; Thatcher et al. 2002; Wilmoth et al. 2007). Our estimates also extend to higher ages than the official ERPs: they are broken down by sex and single years of age up to 109, with a final age group of 110+. The estimates were then used as denominators for mortality rates, which provided the input data to forecast mortality. The future growth and uncertainty of Australia's very elderly population is shown by probabilistic forecasts extending out to 2051. Particular attention is given to future centenarian (100+) and semi-supercentenarian (105+) populations.

Following this introduction, the chapter describes the data and methods used to create the new population estimates dataset and probabilistic forecasts. Section 7.3 describes our new population estimates for the highest ages, whilst Sect. 7.4 presents probabilistic forecasts out to 2051. A summary of key findings and some concluding thoughts form the final section.

7.2 Data and Methods

7.2.1 Population Estimates

Population numbers for 1971 to 2014 by sex and single years of age from 85 to 109 and 110+ were estimated from death counts using the Extinct Cohort (Vincent 1951) and Survivor Ratio methods (Dépoid 1973). Population estimation methods which make use of deaths have been found to produce more accurate very elderly population estimates than those based on census counts because of more accurate information on age in deaths data (Coale and Caselli 1990; Jdanov et al. 2005; Thatcher 1992; Thatcher et al. 2002). These methods are described below.

Extinct Cohort Method The Extinct Cohort method was used to calculate historical population numbers for cohorts for which all members have died. A population estimate for an extinct cohort for any year and age is obtained by summing subsequent cohort deaths (Coale and Caselli 1990; Kannisto et al. 1994; Vincent 1951). The population of the cohort c aged x last birthday on 31st December of year t ($P_{x,t}$) is:

$$P_{x,t} = \sum_{i=1}^{\omega-x} \mathrm{D}_{t+i}^{c}$$

where ω is the age of extinction, and D_{t+i}^{c} is the number of deaths in year $t+i$ from cohort c, which are people born in the year $t-x$. In line with Thatcher et al. (2002) ω is estimated as the highest age at which there was expected to be only one survivor. ω did not exceed 110 in any year. The Extinct Cohort method was used to estimate population numbers from 1971 to 2014 for cohorts born in 1903 and earlier.

Death counts by single year of age for individual calendar years from 1971 to 2014 were obtained from the ABS. Deaths were split between cohorts based on relative birth numbers and estimated survival probabilities (Terblanche and Wilson 2015a). In applying this method an implicit assumption is made that deaths are the only source of population flows, and thus that international migration at these ages is negligible and can be ignored (Thatcher 1992).

Survivor Ratio Method For cohorts that are nearly extinct, population numbers were estimated using the Survivor Ratio method (Dépoid 1973; Thatcher et al. 2002). The Survivor Ratio method was used to estimate population numbers for cohorts aged between 85 and 110+ at 31st December 2014, or cohorts born between 1904 and 1929. Cohort populations in earlier years were then obtained by adding deaths as for the Extinct Cohort method. The survivor ratio is defined as the ratio of a cohort's population at the calculation date to its size k years ago, and can be expressed as:

$$R_x = \frac{P_{x,t}}{P_{x-k,t-k}}.$$

According to the Extinct Cohort method, the number of survivors from a particular cohort k years earlier can be written as:

$$P_{x-k,t-k} = P_{x,t} + \sum_{i=0}^{k-1} \mathrm{D}_{t-i}^{c}$$

so that the survivor ratio for this cohort over k years is:

$$R_x = \frac{P_{x,t}}{P_{x,t} + \sum_{i=0}^{\mathrm{k}-1} \mathrm{D}_{t-i}^{c}}.$$

The estimated population aged x at 31st December of year t is obtained by solving for $P_{x,t}$:

$$P_{x,t} = \frac{R_x}{1 - R_x} \times \sum_{i=0}^{k-1} \mathrm{D}_{t-i}^{c}.$$

In order to avoid variability from year to year, survivor ratios based on the average experience of m older cohorts are used:

$$R_x = \frac{\sum_{j=1}^{m} P_{x,t-j}}{\sum_{j=1}^{m} P_{x-k,t-k-j}}.$$

Based on a retrospective assessment of the accuracy of various nearly extinct cohort estimation methods for Australia, it was found that survivor ratios based on a 5-year age range (k) and 5 older cohorts (m) produced accurate results (Terblanche and Wilson 2015b).

Terblanche and Wilson (2015b) furthermore found that survival improvement across cohorts is best allowed for indirectly by constraining the total nearly-extinct population at the calculation date to the total official population estimates for ages 85+. Official Estimated Resident Populations (ERPs) are provided by the ABS (2015) for single ages 85–99 and in aggregate for ages 100+. ERPs at 30th June were interpolated to 31st December. This method differs slightly from that used by the Human Mortality Database (HMD) in that the HMD applies a constraint of 90+ ERP rather than 85+ ERP. While the Survivor Ratio method with results constrained to 90+ ERP was found to produce very accurate estimates for Australian females, applying an 85+ ERP constraint was found to produce more accurate estimates for Australia on average across age ranges and the sexes (Terblanche and Wilson 2015b).

Period and cohort life tables were constructed using the death data and population estimates at ages 85–110+ for 1971 to 2014 derived using the Extinct Cohort and Survivor Ratio methods, and ERPs for ages below 85 from the ABS (2015). These were supplemented with estimates for earlier periods using Australian death data from the Human Mortality Database (HMD 2015), and, for the younger ages, by historical life tables published by the ABS (2008).

7.2.2 Population Forecasts

Forecasting Model and Program Forecasts of the very elderly population were prepared using a probabilistic cohort-component model. Probabilistic models explicitly incorporate uncertainty about the demographic future and produce forecasts as predictive distributions rather than the single set of numbers which are output by conventional deterministic models. These distributions are created by running a cohort-component model several thousand times with randomly generated sample paths of fertility, mortality and migration, usually from time series models. The thousands of forecast outcomes are then sorted from highest to lowest so that prediction intervals can be marked out. For example, a 95 % prediction interval in a set of 5,000 ranked forecasts extends from the 125th value to the 4,875th.

The forecasting program used in this study was PROBPOP (PROBabilistic POPulation Projections), developed by the first author of the chapter for earlier research (Bell et al. 2011). A minor refinement was made for the present study by adding uncertainty in the initial populations. The model operates in single year intervals with the population disaggregated by sex and single years of age up to 109, ending with age 110+. Time series models are used to generate sample paths for the Total Fertility Rate, life expectancy at birth by sex, total immigration, and total emigration. The varying male and female life expectancy paths, and immigration and emigration trajectories, are correlated in the forecasts to mimic historical trends. The calculations of PROBPOP are performed by a fortran 95 program, though data inputs, assumptions and parameters are read in from an Excel workbook. In this application the jump-off and final dates were set as 30th June 2014 and 30th June 2051 respectively, and 5,000 simulations were chosen. As the simulations are generated, forecasts are progressively written to output files which effectively comprise a large unsorted output database. A separate program then reads and sorts the thousands of forecasts and presents selected outputs in Excel.

Input Data and Assumptions The forecasts begin with the newly-created 30th June 2014 population estimates, calculated using the approach described above in Sect. 7.2.1. These initial jump-off populations were subject to random variation in each of the 5,000 forecast simulations because in Australia these numbers are population *estimates*, not precise counts.

The long-run Total Fertility Rate (TFR) assumption used was 1.90, approximately the level of fertility recorded in Australia for the last few years (ABS 2015). Sample paths for the TFR were created from a random walk with drift model subject to ceiling and floor limits.

Immigration and emigration sample paths were generated with ARIMA(1,0,0) time series models, with correlated random numbers for errors obtained by Cholesky decomposition of the variance-covariance matrix. For the first few years of the forecasts short-term predictions from the Department of Immigration and Border Protection (2015) were used. From 2019 to 2020 onwards slight increases in both immigration and emigration totals were assumed, resulting in Net Overseas Migration rising from 247,000 in 2019–2020 to 267,000 by 2050–2051.

Past mortality rates by age and sex were calculated using the newly created population estimates as denominators. Mortality and life table forecasts were created by geometric extrapolation of age-specific death rates from 1971 to 2014. Rates were smoothed across most ages using cubic splines and in the centenarian ages with a fitted logistic curve. A detailed evaluation by Terblanche (2015a) concluded that the geometric method yields forecasts which are just as accurate as far more complex approaches. Life expectancy at birth was forecast to increase at a gradually decelerating pace, rising from 84.7 years in 2013–2014 to 90.0 in 2050–2051 for females, and from 80.6 years in 2013–2014 to 88.0 in 2050–2051 for males. Sample paths for life expectancy at birth were generated using random walk models. Correlated random numbers for male and female life expectancy at birth errors were produced by Cholesky decomposition of the variance-covariance matrix. For each life expectancy at birth value generated by the random walk model the forecasting program interpolated the forecast life tables to obtain a corresponding set of age-specific death rates. By way of illustration, Fig. 7.1 shows the median female life expectancy at birth forecast, 95 % prediction interval bounds, and a small selection of the 5,000 life expectancy sample paths that were created.

7.3 Estimates of Australia's Very Elderly Population

7.3.1 85+ Population

Australia's very elderly population has grown almost seven-fold over the 1971–2014 period, from 69,000 to 456,000. This exponential growth is clear from Fig. 7.2, which shows male and female very elderly population estimates over the estimation period. The very elderly population has grown at an average annual rate of 4.5 % compared to 1.4 % per annum for the total population. As a result, the very elderly increased their share of the total population from 0.5 % in 1971 to 1.9 % in 2014.

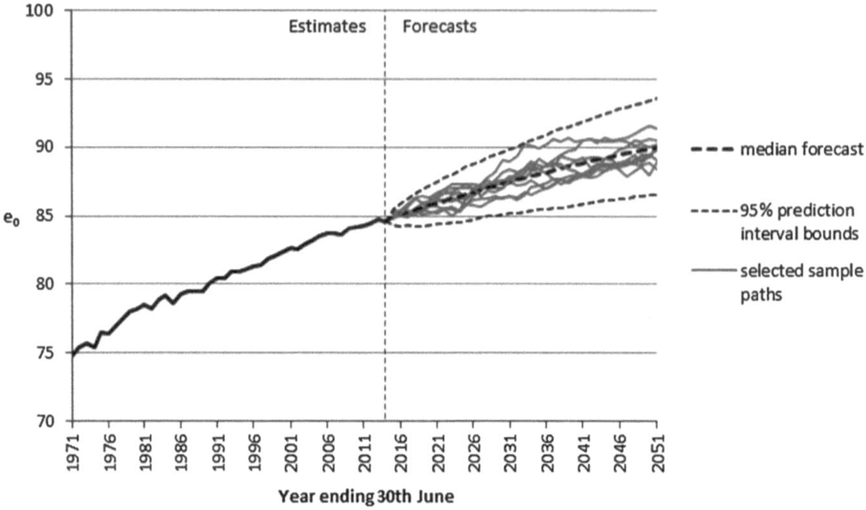

Fig. 7.1 Female life expectancy at birth estimates and forecasts, 1970–71 to 2050–51 (Source: authors' estimates and forecasts)

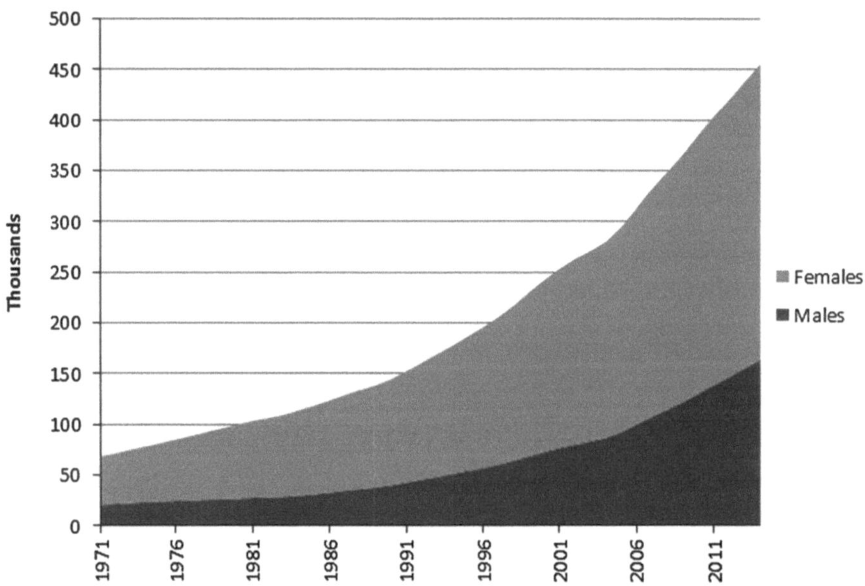

Fig. 7.2 Very elderly population estimates by sex, 1971 to 2014 (Source: authors' calculations)

7.3.2 Age-Sex Composition

During the past four decades, the very elderly population itself has aged, as can be seen from Table 7.1. This table shows the proportions aged 85–89, 90–99 and 100+ relative to the 85+ population in selected years between 1971 and 2014. Over the period from 1971 to 2014, nonagenarians (ages 90–99) and centenarians (ages 100+) respectively increased from 24.5 % and 0.3 % of the total 85+ population to 34.2 % and 0.8 %. The proportion of those aged 85–89 decreased correspondingly, from 75.2 % in 1971 to 65.0 % in 2014.

While the very elderly population was becoming more heavily weighted to older persons, its sex composition also changed. For example, in 1971, 31 % of the very elderly population consisted of males. By 2014, this figure had increased to 36 %. Figure 7.3 shows sex ratios by cohort for ages 85–89, 90–99 and 100+. Changing sex ratios showed a strong cohort effect, with a decreasing trend for birth cohorts from 1871 to 1895, and an increasing trend for cohorts born from 1896 to 1929. For the 1896 cohort, there were around 37 males per 100 females aged 85–89, 27 males per 100 females aged 90–99, and 14 males per 100 females aged 100+. For the 1914 cohort, the number of males per 100 females increased to 49 (ages 85–89), 37 (ages 90–99) and 24 (ages 100+). The increasing sex ratio was largely due to a much greater improvement in male survival relative to females at ages 65–85 for cohorts born after 1895. Male survival beyond age 85 for these cohorts also improved more than that of females. An average of 14.3 % of males and 28.7 % of females born in 1924 survived to at least age 90, compared to 2.5 % of males and 5.9 % of females born in 1874. Males born in 1924 were therefore almost six times more likely to survive to age 90 compared to those born 50 years ago.

The changing relative prevalence of smoking among males and females has been suggested as one of the major drivers of different observed mortality improvements (Thatcher 1999; Trovato 2005). In low mortality countries the male death rate attributable to smoking declined from the mid-1980s after increasing substantially since the 1950s (Bongaarts 2014). In contrast, the smoking-related death rate for females continued to increase from the 1950s to 2010, although it was still only around half the level for males at 2010 (Bongaarts 2014). This pattern was also observed in Australia. According to Peto et al. (2006), female deaths attributed to smoking, in particular lung cancer deaths, increased since the 1970s despite an

Table 7.1 Age distribution (%) within the very elderly age group, 1971–2014

Year	Age group 85–89	90–99	100+	85+
1971	75.2	24.5	0.3	100.0
1981	70.8	28.7	0.5	100.0
1991	70.1	29.2	0.7	100.0
2001	69.5	29.8	0.6	100.0
2011	67.6	31.7	0.7	100.0
2014	65.0	34.2	0.8	100.0

Source: authors' calculations

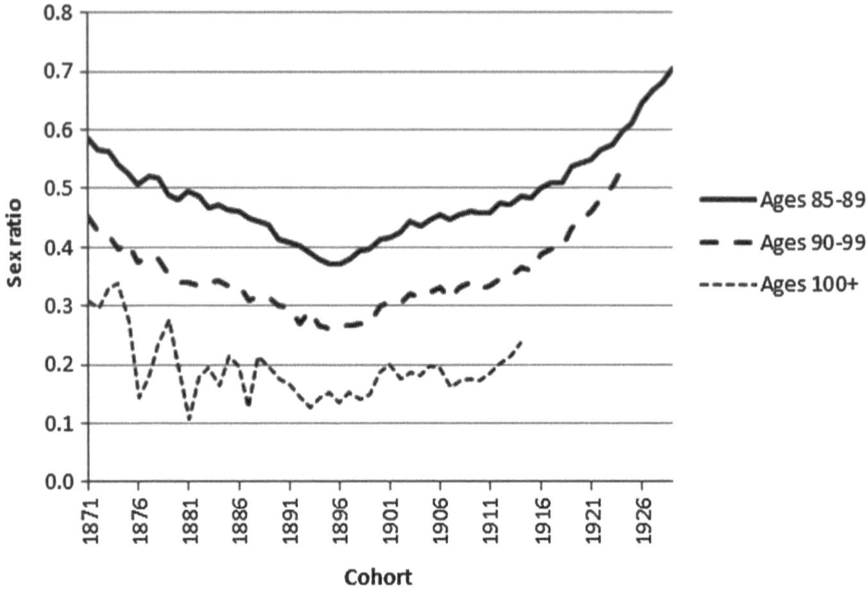

Fig. 7.3 Very elderly population sex ratios, cohorts born 1871–1929 (Source: authors' calculations)

overall decline in death rates, while male smoking-attributed deaths declined since the 1980s.

Although the growth in male centenarians started to exceed that of females at the start of the twenty-first century, sex ratios at ages 100+ did not increase to the same extent as at younger ages. In 1971, 78 % of centenarians were female. This proportion increased to around 84–85 % in the next 10 years, and remained around that level for the next three decades.

Estimated numbers of males and females at single ages 85–104 and in aggregate for 105+ in 1971, 1992 and 2014 are shown in Fig. 7.4. Figure 7.5 shows the total increase at each age from 1971 to 1992 and from 1992 to 2014. It is clear from these figures that male and female numbers at all ages increased substantially. Between 1971 and 1992, growth in female numbers increased with age and exceeded that of males at all ages. Over this period, growth in male numbers at ages up to 100 did not show a similar pattern of increase with age. Male numbers at most ages increased much more over 1992 to 2014 compared to the period 1971 to 1992, and their growth increased with age up to 93.

Centenarian numbers increased more than 17-fold, from 210 in 1971 to around 3,650 in 2014. In 1971 there were 16 centenarians per million of the total population. By 2014, this has increased to 156. The number of semi-supercentenarians (ages 105+) increased from only 3 in 1971 (all female) to an estimated 155 in 2014 (137 female and 18 male).

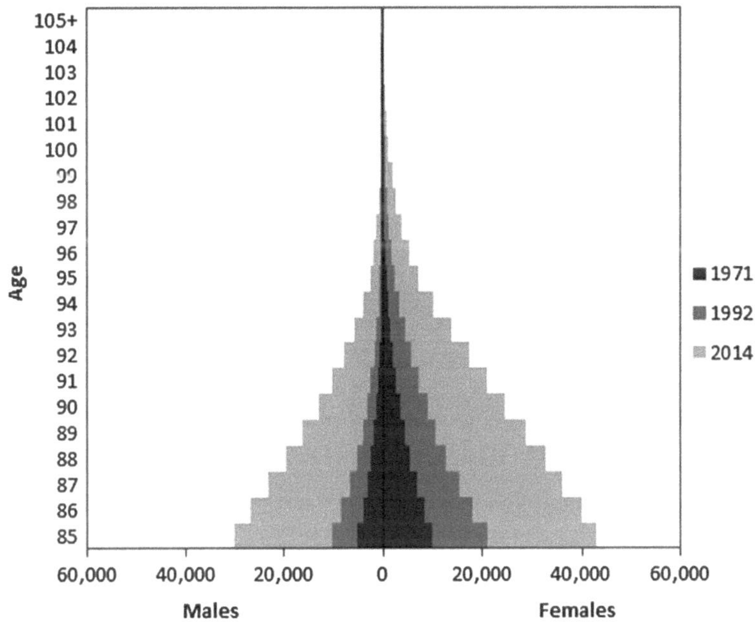

Fig. 7.4 Very elderly population age-sex structure, 1971, 1992 and 2014 (Source: authors' calculations)

7.3.3 Drivers of Growth

The relative contributions of changes in births, survival and net migration to the growth in nonagenarian numbers from 1971 to 2014 are shown in Fig. 7.6. Figure 7.7 shows the contribution of these factors to the growth in centenarian numbers from 1981 to 2014. An initial year of 1981 had to be chosen because of data constraints relating to the earliest cohorts for which cohort life tables could be constructed. The numbers of centenarians in 1981 and 2014 were expressed as the product of births, cohort survival from birth to age 65, age 65 to 85, age 85 to 100 and beyond age 100, and a factor representing the impact of net migration (Vaupel and Jeune 1995; Thatcher 1999; Terblanche 2015b). The factor increases shown in Figs. 7.6 and 7.7 represent the ratios of these factors at the two dates. For example, the factor increase in male births of 1.7 shown in Fig. 7.7 represents the ratio of birth counts in 1905–1914 compared to 1872–1881. Survivors from these cohorts were aged 100+ in 2014 and 1981 respectively. For nonagenarians, improvements in cohort survival between ages 65–85 contributed most to the increase in their numbers from 1971 to 2014, followed by an increase in birth numbers. By far the largest driver of centenarian growth between 1981 and 2014 was an improvement in cohort survival from age 85 to 100.

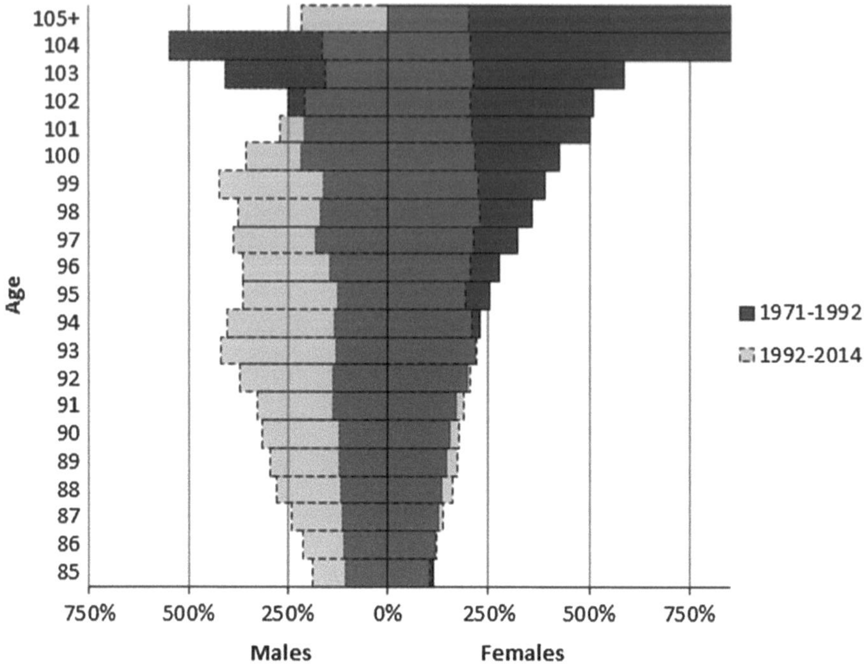

Fig. 7.5 Very elderly population increases by age and sex, 1971–1992 and 1992–2014 (Source: authors' calculations)

7.3.4 Comparison of Centenarian Numbers with Those of ABS

The differences between our estimates and those of the ABS are most marked in the centenarian ages. A comparison of the two sets of figures from 1971 to 2014 is shown in Fig. 7.8. Recall that ABS ERPs are derived from census counts, while the authors' estimates are derived from death counts using the Extinct Cohort and Survivor Ratio methods. ERPs show a more volatile and less plausible pattern over time. Up to around 1985, 100+ ERPs for females were lower than our estimates but since then, with the exception of 1992–1995, they have exceeded our estimates by around 20 %. Since around 2006, the gap between female ERPs and our estimates has gradually reduced and in 2014 the two sets of numbers were very close. In the case of males, ERPs were generally higher than our estimates, especially during 1997–2003. The largest difference was observed in 2001, with ERP of 610 compared to our estimate of 223. At 30th June 2014 the ERP for males aged 100+ was 842, compared to our figure of 600.

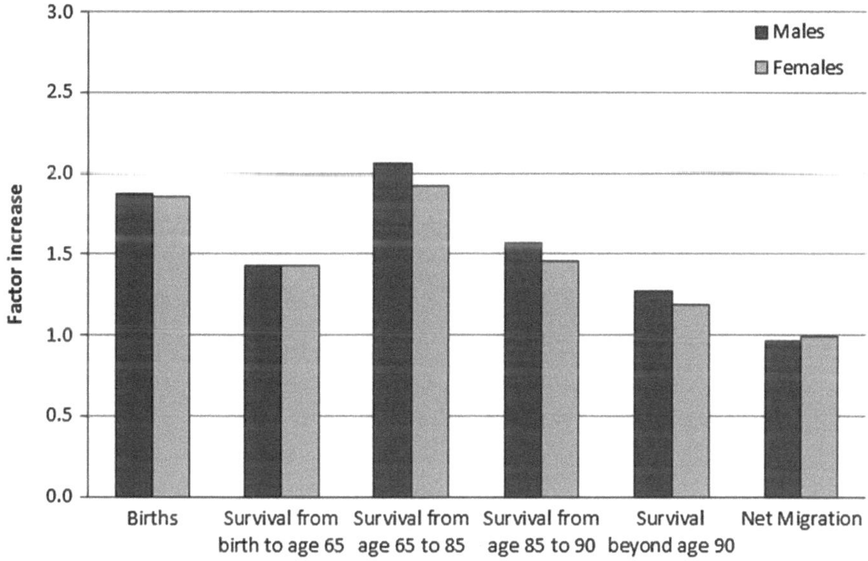

Fig. 7.6 Factor increases in births, survival and net migration, explaining growth in population aged 90–99 from 1971 to 2014, by sex (Source: authors' calculations based on own estimates, ABS (2008, 2015) and HMD (2015))

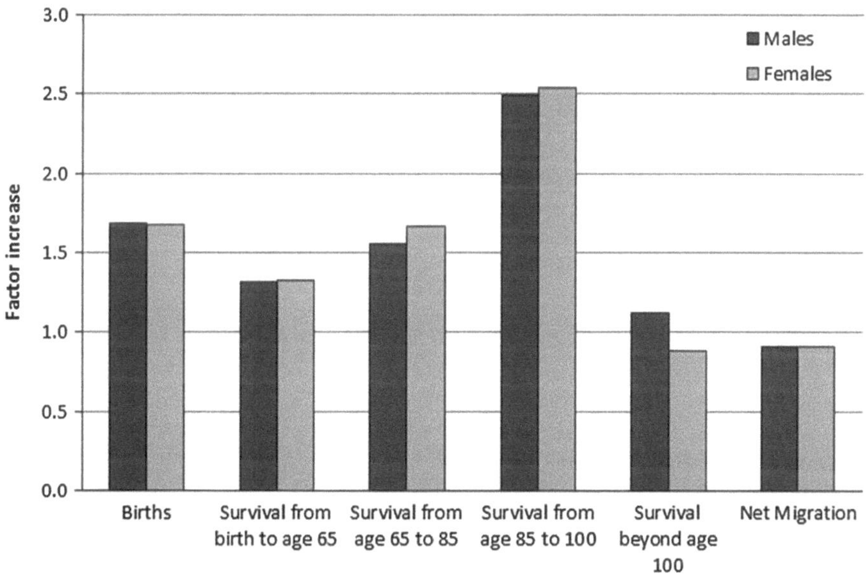

Fig. 7.7 Factor increases in births, survival and net migration, explaining growth in population aged 100+ from 1981 to 2014, by sex (Source: authors' calculations based on own estimates, ABS (2008, 2015) and HMD (2015))

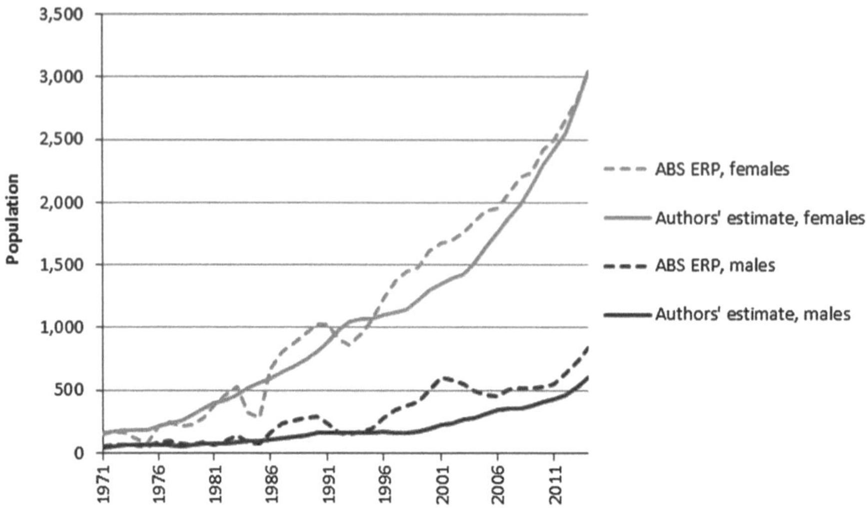

Fig. 7.8 Comparison of ABS and authors' estimates of the centenarian population, 1971 to 2014 (Source: authors' calculations and ABS (2015))

7.4 Forecasts of Australia's Very Elderly Population

7.4.1 85+ Population

Australia's very elderly population will increase substantially in the coming decades, as Fig. 7.9 shows. From an estimated 456,000 in 2014, the 85+ population is forecast to increase substantially over the next 16 years, with 95 % of simulations lying between 778,000 and 975,000 by 2031, and with the median of the distribution at 871,000. Over the following 20 years growth will accelerate as the 1946–65 baby boom generation enters the 85+ age group. By 2051 the median of the predictive distribution is set to reach 1.90 million with the 95 % prediction interval spanning 1.51–2.37 million.

The strength of these probabilistic forecasts is that they show how the demographic future of the 85+ age group is quite certain for the next decade or so but becomes increasingly uncertain further into the future. But they also demonstrate that huge growth is assured: by 2051 even the lower bound of the 95 % prediction interval is more than triple the population in 2014. Prediction intervals in probabilistic forecasts are generally calculated for population ranges, but they may also be used to describe time periods for the passing of particular population milestones. For the 85+ population, the 95 % interval for achieving one million extends from 2032 to 2037.

How do these forecasts compare to the latest 2012-based population projections for Australia prepared by the ABS (2014)? Although the ABS produce a large number of projection variants their publications focus mostly on just three, which they label Series A, B and C. Series A contains high fertility, high life expectancy

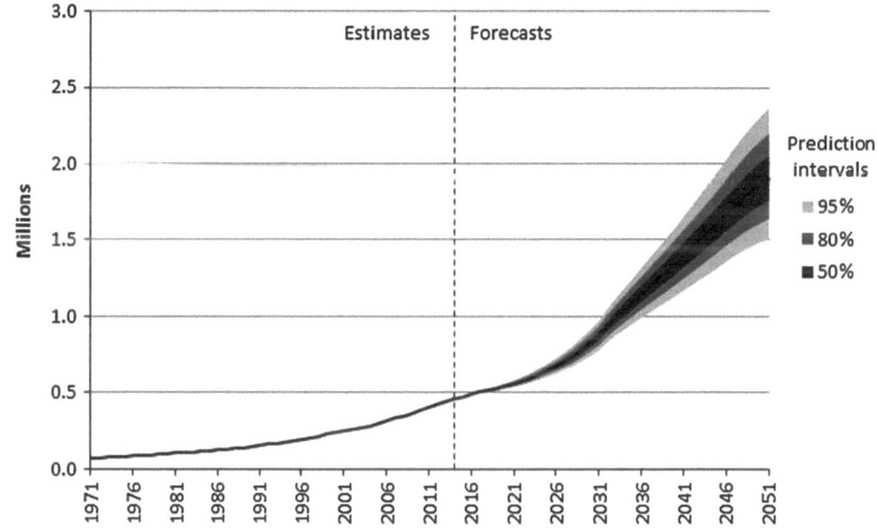

Fig. 7.9 The growth of Australia's very elderly (85+) population, 1971–2051 (Source: authors' calculations)

and high net overseas migration assumptions, Series B has medium assumptions for all components and is widely interpreted as the principal series, and Series C contains low fertility, medium life expectancy and low net overseas migration assumptions. For the 85+ population Series B and C are almost identical and project a very elderly population of 805,000 by 2031 and 1.58 million by 2051. Series A projects 843,000 by 2031 and 2.15 million by 2051. The range between the highest and lowest of these series covers about a quarter of our predictive distribution by 2031, but about 80 % of it by 2051 with Series A close to the upper bound of the 80 % prediction interval and Series B and C close to its lower bound.

7.4.2 *Centenarians*

For Australia's centenarian population, forecast growth is far more dramatic than for the 85+ population as a whole, as Fig. 7.10 shows. Centenarian numbers are forecast to grow from about 3,600 in 2014 to between 9,000 and 11,700 by 2031 (95 % interval) and to between 19,500 and 39,500 by 2051 (95 % interval). The medians of the forecast distribution for these two dates are, respectively, 10,200 and 27,200. The milestone of 10,000 centenarians is expected to be passed sometime between 2028 and 2037 (95 % interval).

Compared to the 85+ population as a whole, there is more uncertainty about the future of the centenarian group. Relative uncertainty can be measured with the Relative Inter-Decile Range (RIDR) which is defined as the 80 % prediction

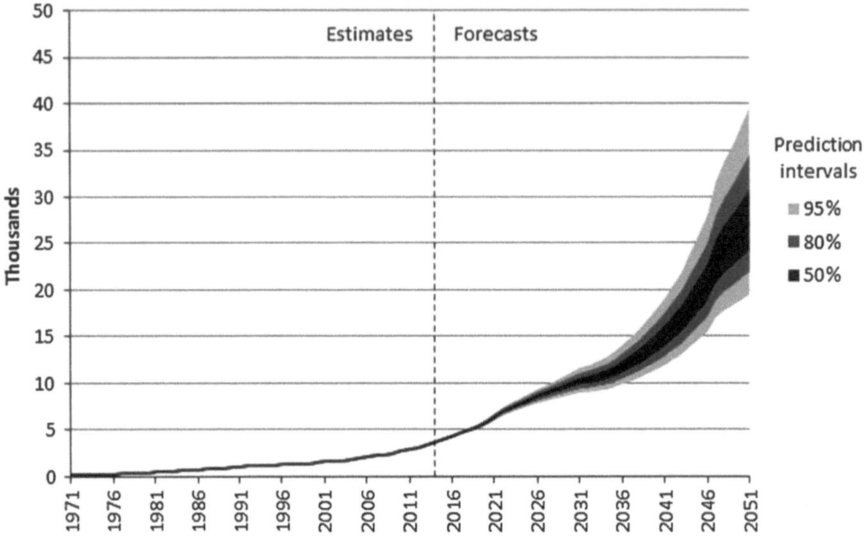

Fig. 7.10 The growth of Australia's centenarian (100+) population, 1971–2051 (Source: authors' calculations)

interval divided by the median (Lutz et al. 2004). By 2051 the RIDR for the 85+ population forecast is 0.29 whilst for the centenarian population it is 0.46. The greater uncertainty about the centenarian population is the result of higher death rates than the 85+ population as a whole.

How do our forecasts compare to those of the ABS (2014)? The latest ABS projections of the centenarian population begin higher than ours and remain above the upper 95 % prediction interval bound of our forecasts through the forecast horizon. Series B and C track a few thousand above the upper 95 % bound but move in line with it. Until the late 2030s, the high life expectancy Series A projection does the same, but after this time it increases exponentially, soaring to 94,000 by 2051. The ABS projections are much higher than our forecasts because they are based on forecast mortality rates which are too low, which in turn is the result of historical mortality rates being calculated using population estimates which are too high, as mentioned earlier.

One significant feature of the centenarian population which will change over coming decades is its overwhelming female dominance. In 2014, 84 % of centenarians were female. Figure 7.11 shows how the female percentage is forecast to decline out to mid-century. By 2051 the 95 % prediction interval for percentage female ranges from 59 to 72 %, so although females are likely to remain more numerous than males in the centenarian ages there will be more of a balance between the sexes. The cause of this shift in sex distribution is the reduction in differences between male and female death rates at the highest ages due to the greater decline of male death rates.

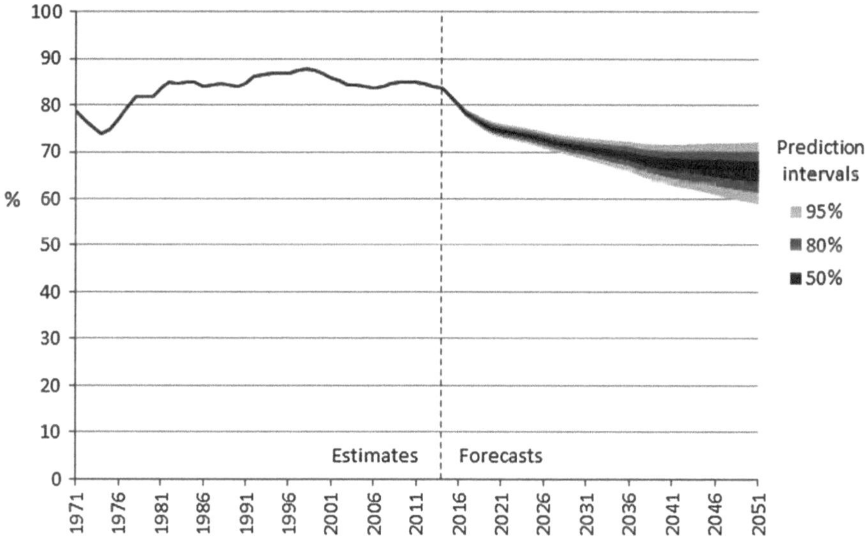

Fig. 7.11 The changing female percentage of Australia's centenarian population, 1971 to 2051 (Source: authors' calculations)

7.4.3 Semi-supercentenarians

People aged 110 years and above are described as supercentenarians. Supercentenarians in Australia are very few in number even at the end of the forecast horizon, so instead we focus here on semi-supercentenarians, those aged 105 and above. Figure 7.12 shows their past and forecast growth out to 2051. From a population of about 150 in 2014, semi-supercentenarian numbers are forecast to increase to between 480 and 590 by 2031 (95 % prediction interval) and then on to between 950 and 1,800 by 2051 (95 % prediction interval). The medians of the forecast distribution for these two dates are, respectively, 530 and 1,290.

7.4.4 Population Pyramids

Figure 7.13 displays the forecasts in the form of a population pyramid for ages 85 and above for 2014, 2031 and 2051. They clearly demonstrate the huge growth and shift in sex distribution which will eventuate. Unlike population pyramids covering the whole age range, they reveal a rapid decline in population with age without any peaks and troughs due to cohort waves. Although cohort sizes in the younger adult ages have varied over time due to births and overseas migration, the mortality rates at these advanced ages remain sufficiently high to deplete a large proportion of each cohort every year and thus create the tapered shape shown in the graphs.

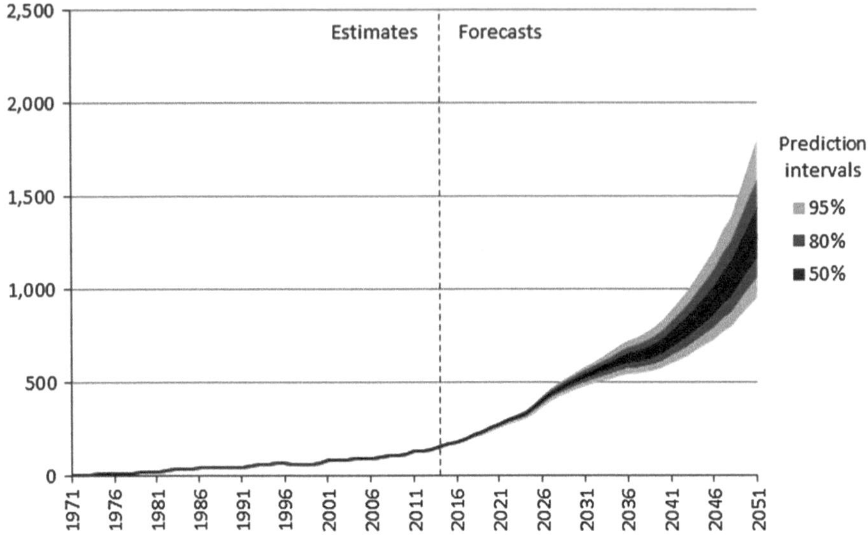

Fig. 7.12 The growth of Australia's semi-supercentenarian (105+) population, 1971 to 2051 (Source: authors' calculations)

7.5 Summary and Conclusions

This chapter has presented updated estimates of Australia's very elderly population and the first set of probabilistic forecasts focusing on the very highest ages. Population estimates, created using the Extinct Cohort and Survivor Ratio methods, are far more accurate than those derived from census counts, especially at the centenarian ages. These more accurate estimates were used to calculate more reliable death rates, which in turn permitted the preparation of more accurate mortality forecasts. Our probabilistic forecasts illustrate the extent to which growth in Australia's very elderly population is uncertain, and quantify it through prediction intervals. These intervals reveal there to be very little uncertainty in the coming decade, but increasing amounts the further into the future one goes, and relatively more for some variables (e.g. centenarians) than others (e.g. the 85+ population as a whole). What is certain, however, is that huge growth in population numbers at the highest ages *will* eventuate. Australia's demography is now entering an era in which the very elderly will form a sizeable section of the population for the first time.

Our estimates and forecasts also indicate that the official ABS estimates of the very elderly are not as accurate as they could be, and that their projections are also problematic. As Fig. 7.8 shows, ABS centenarian ERPs exhibit volatile and implausible trends. Their centenarian projections in Series A, B and C exceed the upper bound of our 95 % prediction interval because of forecast death rates which are too low, themselves the result of inflated ERPs at the highest ages. We encourage the ABS to adopt the Extinct Cohort and Survivor Ratio methods for estimating the very elderly population as described in this chapter. The UK's Office

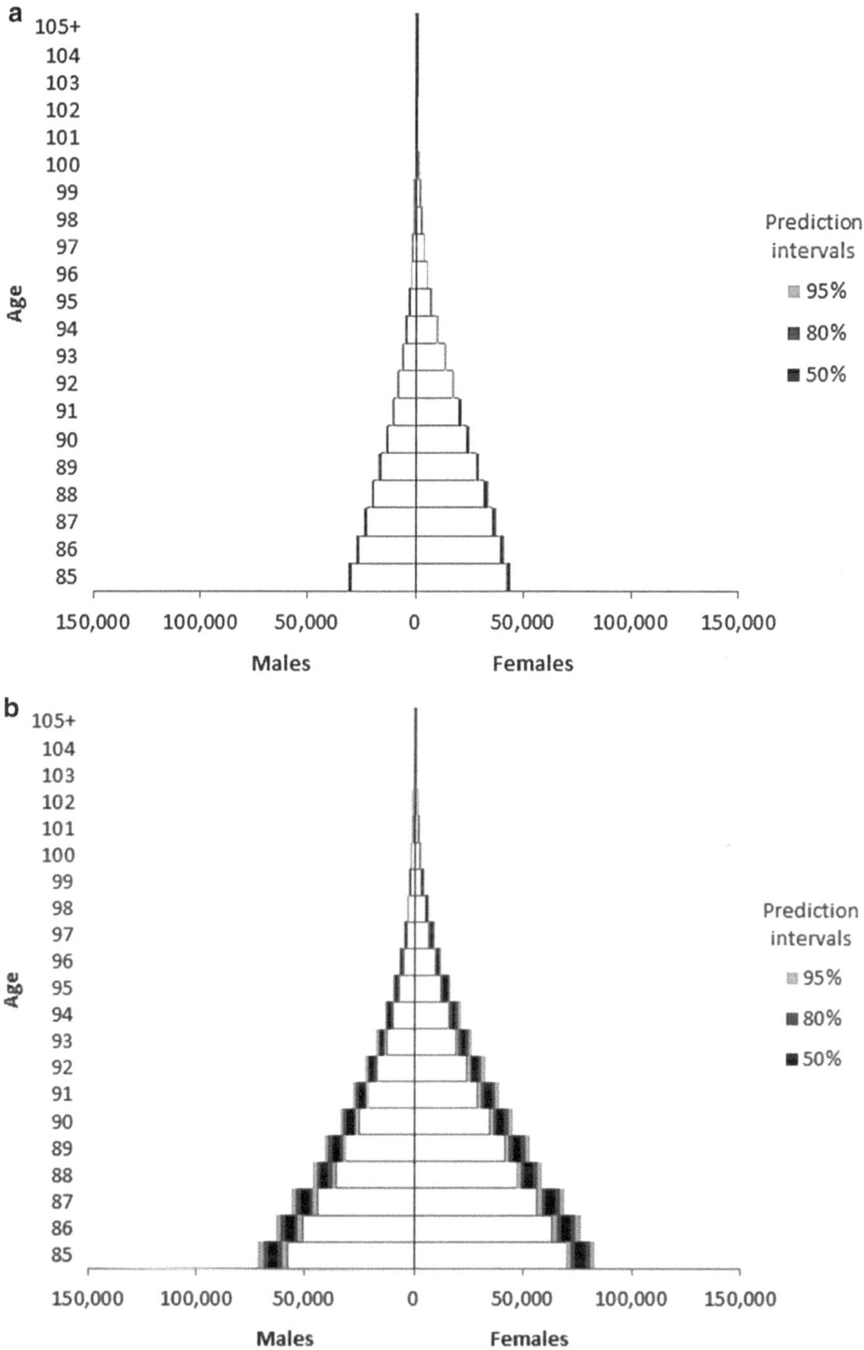

Fig. 7.13 The age-sex structure of Australia's very elderly population, 2014, 2031 and 2051. (**a**) 2014. (**b**) 2031. (**c**) 2051 (Source: authors' calculations)

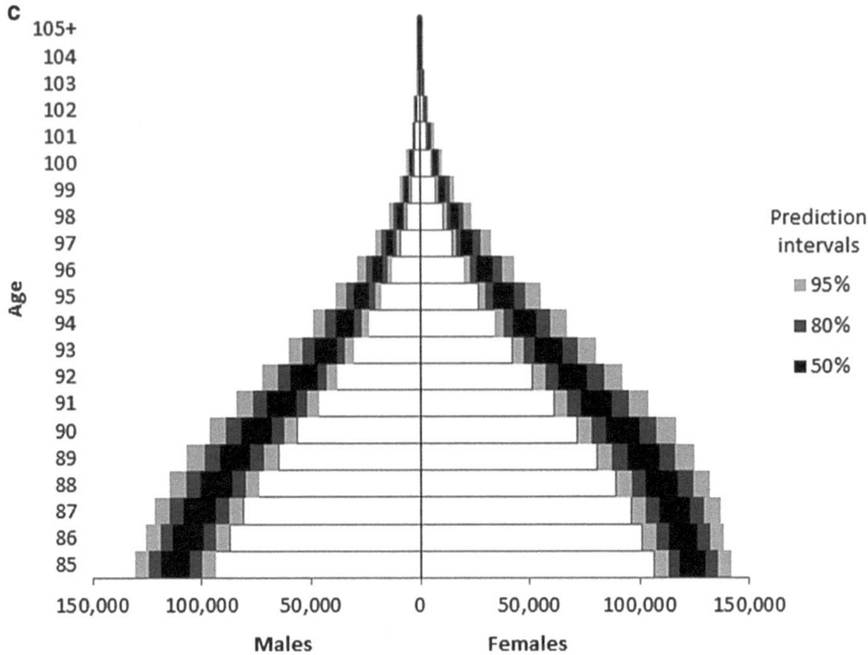

Fig. 7.13 (continued)

for National Statistics has applied these methods for estimating very elderly populations for several years now (ONS 2015). Furthermore, uncertainty about the future of the very elderly population is not reliably indicated by high-low ranges from the ABS Series A, B and C projection variants. It would be far better if the ABS produced just one forecast and then created prediction intervals around it from a probabilistic model.

References

ABS. (2008). *Australian historical population statistics, 2008*. Canberra: ABS.

ABS. (2014). *Population projections, Australia, 2012 (base) to 2101*. Canberra: ABS.

ABS. (2015). *Australian demographic statistics* (March quarter 2015). Canberra: ABS.

Alho, J. M., Hougaard Jensen, S. E., & Lassila, J. (Eds.). (2008). *Uncertain demographics and fiscal sustainability*. Cambridge: Cambridge University Press.

Beard, J., Biggs, S., Bloom, D., Fried, L., Hogan, P., Kalache, A., & Olshansky, J. (2012). *Global population ageing: Peril or promise?* (Working Paper No. 89). Program on the Global Demography of Aging, Harvard University.

Bell, M., Wilson, T., & Charles-Edwards, E. (2011). Australia's population future: probabilistic forecasts incorporating expert judgement. *Geographical Research, 49*(3), 261–275.

Bongaarts, J. (2014). Trends in causes of death in low-mortality countries: Implications for mortality projections. *Population and Development Review, 40*(2), 189–212.

Coale, A. J., & Caselli, G. (1990). Estimation of the number of persons at advanced ages from the number of deaths at each age in the given year and adjacent years. *Genus, 46*(1/2), 1–23.
Chomik, R., & MacLennan, M. (2014). *CEPAR research brief 2014/01 and 2014/02*. Sydney: The University of New South Wales. Aged care in Australia.
Christensen, K., Doblhammer, G., Rau, R., & Vaupel, J. W. (2009). Ageing populations: The challenges ahead. *The Lancet, 374*, 1196–1208.
Cubit, K. A., & Meyer, C. (2011) Aging in Australia. *The Gerontologist, 51*, 583 589.
Department of Immigration and Border Protection. (2015). *The outlook for net overseas migration as at June 2015*. Canberra: DIBP.
Dépoid, F. (1973). La mortalité des grands vieillards (French Edition). *Population, 28*(4/5), 755–792.
Human Mortality Database. (2015). University of California, Berkeley (USA), and Max Planck Institute for Demographic Research (Germany). Available at www.mortality.org
Jdanov, D. A., Scholz, R. D., & Shkolnikov, V. M. (2005). Official population statistics and the Human Mortality Database estimates of populations aged 80+ in Germany and nine other European countries. *Demographic Research, 13*, 14–14.
Kannisto, V., Lauritsen, J., Thatcher, A. R., & Vaupel, J. W. (1994). Reductions in mortality at advanced ages: Several decades of evidence from 27 countries. *Population and Development Review, 20*(4), 793–810.
Keilman, N. (2008). European demographic forecasts have not become more accurate over the past 25 years. *Population and Development Review, 34*, 137–153.
Lee, R. D. (1999). Probabilistic approaches to population forecasting. In W. Lutz, J. Vaupel, & D. A. Ahlburg (Eds.), *Frontiers of population forecasting* (pp. 156–190). New York: Population Council.
Lutz, W., Sanderson, W. C., & Scherbov, S. (2004). The end of world population growth. In W. Lutz, W. C. Sanderson, & S. Scherbov (Eds.), *The end of world population growth in the 21st century* (pp. 17–83). London: Earthscan.
McCormack, J. (2010). Being very old in a young country: Centenarians and supercentenarians in Australia. In H. Maier, J. Gampe, B. Jeune, J. W. Vaupel, & J.-M. Robine (Eds.), *Supercentenarians* (pp. 101–117). Dordrecht: Springer.
ONS. (2015). *Estimates of the very old (including Centenarians), England and Wales, and United Kingdom, 2002 to 2014*. Statistical Bulletin. London: ONS.
Peto, R., Lopez, A. D., Boreham, J., & Thun, M. (2006). *Mortality from smoking in developed countries 1950–2000*. Australia, www.deathsfromsmoking.net.
Productivity Commission. (2013). *An ageing Australia: Preparing for the future*. Canberra: Productivity Commission.
Rowland, D. T. (2012). *Population aging*. Dordrecht: Springer.
Terblanche, W. (2015a). Population estimates and projections for Australia's very elderly population at state and national level. PhD thesis, University of Queensland, Brisbane, Australia.
Terblanche, W. (2015a). Explaining the substantial growth of centenarian populations in Australia's states, 1981 to 2012'. *Journal of Population Research, 32*(2), 81–93.
Terblanche, W., & Wilson, T. (2014). Understanding the growth of Australia's very elderly population, 1976 to 2012. *Journal of Population Ageing, 7*(4), 301–322.
Terblanche, W., & Wilson, T. (2015a). Accuracy of nearly-extinct-cohort methods for estimating very elderly sub-national populations. *International Journal of Population Research.*
Terblanche, W., & Wilson, T. (2015b). An evaluation of nearly-extinct cohort methods for estimating the very elderly populations of Australia and New Zealand. *PLOS ONE, 10*(4).
Thatcher, A. R. (1992). Trends in numbers and mortality at high ages in England and Wales. *Population Studies, 46*(3), 411–426.
Thatcher, A. R. (1999). The demography of centenarians in England and Wales. *Population Trends, 96*(Summer 1999), 5–12.
Thatcher, A. R., Kannisto, V., & Andreev, K. F. (2002). The survivor ratio method for estimating numbers at high ages. *Demographic Research, 6*, 1–18.

Trovato, F. (2005). Narrowing sex differential in life expectancy in Canada and Austria: Comparative analysis. *Vienna Yearbook of Population Research, 2005*, 17–52.

Treasury. (2015). *2015 Intergenerational report: Australia in 2055*. Canberra: Commonwealth of Australia.

Vaupel, J. W., & Jeune, B. (1995). The emergence and proliferation of centenarians. In B. Jeune & J. W. Vaupel (Eds.), *Odense monographs on population aging vol 2, Exceptional longevity: From prehistory to the present* (Vol. 2). Odense: Odense University Press.

Vincent, P. (1951). La mortalité des vieillards. *Population* (French Edn), 181–204.

Wilmoth, J. R., Andreev, K. F., Jdanov, D. A., Glei, D. A., Boe, C., Bubenheim, M., Philipov, D., Shkolnikov, V., & Vachon, P. (2007). Methods protocol for the Human Mortality Database. University of California, Berkeley, and Max Planck Institute for Demographic Research, Rostock. http://mortality.org [version 31/05/2007].

Wilson, T., & Bell, M. (2004). Australia's uncertain demographic future. *Demographic Research, 11*(8), 195–234.

Wilson, T., & Rees, P. (2005). Recent developments in population projection methodology: A review. *Population, Space and Place, 11*(5), 337–360.

Wilson, T. (2012). Forecast accuracy and uncertainty of Australian Bureau of Statistics state and territory population projections. *International Journal of Population Research, 2012*.

Part II
Data: Issues and Analyses

Chapter 8
An Assessment of Historical Demographic Analysis Estimates for the Black Male Birth Cohorts of 1935–39

Kirsten West, Jason Devine, and J. Gregory Robinson

Abstract A birth cohort consists of persons born during the same year or years. Historical Demographic Analysis (DA) estimates reveal that in census after census, Black males born in the decades after the vital registration system was established are undercounted at higher rates than Black females and non-Black males and females. A cohort pattern of coverage error by race and sex does not appear logical. One might expect certain ages to be more difficult to enumerate than other ages. One might also expect period effects to show up in data collections that occur decades apart. Age and period effects interact to create their own effects. However, it is difficult to understand how cohorts of a given race and sex are missed persistently every 10 years. The anomaly observed in the coverage estimates may stem not solely from problems of census enumeration of this subpopulation, but from aspects inherent to the DA method. This analysis includes an examination of the factors used to correct for incomplete and missing data for Black male births in the early years of the vital registration system and the sensitivity of the estimates to assumptions about completeness. Race classification errors are also discussed.

Keywords Black males • Census coverage error • Cohort analysis

8.1 Introduction

Every decade, the U.S. Census Bureau prepares Demographic Analysis (DA) estimates of the size, age, sex, and race composition of the national population. Unlike annual population estimates, the DA estimates do not use the decennial census as its base. They are developed from historical data on births, deaths, and

---

Kirsten West and Gregg Robinson are now retired.
This paper is released to inform interested parties of ongoing research and to encourage discussion of work in progress. Any views expressed on statistical, methodological, technical, or operational issues are those of the authors and not necessarily those of the U.S. Census Bureau.

K. West (✉) • J. Devine • J.G. Robinson
Population Division, U.S. Census Bureau, Washington, DC, USA
e-mail: kwestiaos@gmail.com

D.A. Swanson (ed.), *The Frontiers of Applied Demography*, Applied Demography Series 9, DOI 10.1007/978-3-319-43329-5_8

international migration. Thus, they are independent of the most recent decennial census and are ideal benchmarks for assessing coverage errors for the total population, and the population by age, sex and limited race categories (Black and all others).

A number of assessments show that the counts for Black males are persistently lower than the benchmarks. In the 1940s, Price (1947) used data from the First Selective Service Registration for June 30, 1941 to assess the 1940 Census count for males. He found that the census under-enumerated all males by about 3 % and Black males in the age group 21–35 by about 13 %. Price speculated that there is something unique about these ages. He suggested that the census misses young Black males because they are a highly mobile population with weak ties to a household. Passel and others identified a suspicious cohort effect for the Black 1935–1950 birth cohorts in the 1940–1990 censuses (Passel 1991, 1992). Robinson and his co-authors also identified a peculiar undercount pattern for Black adult males aged 35–54 in 1980 (Robinson et al. 1990). In more recent assessments, without attributing cause, Robinson used DA estimates as the benchmark to point to a persistent high undercount for adult Black males—5.0–12.0 % nationally in every census since 1940 (Robinson 1997, 2012). Historical survey-based estimates of the coverage of Black males support the finding that the Black males are difficult to enumerate, but do not suggest the levels of error implied by DA.

This study investigates if the DA estimates of the Black males are too high. To simplify the discussion, the focus is on the 1935–1939 birth cohort. However, the phenomenon is not unique to this cohort. The differential coverage pattern for Black males is evident in adjacent cohorts.

Table 8.1 presents the differences between the census and either the survey-based estimates or the cohort-component-based DA estimates for the Black population aged 50 and older by sex. The results are for the 1990 Census, Census 2000, and the 2010 Census.

In 1990, the DA method estimated a 5.3 % net undercount for Black males aged 50 and older compared to the Post Enumeration Survey (PES) estimates of a 0.4 % net overcount. The DA showed a net overcount of 0.5 % for Black females in this age group. In comparison, the PES showed a net overcount of 1.2 %.

In Census 2000, the DA and the initial Accuracy and Coverage Evaluation (A.C.E.) estimates were again different, with the DA measuring a 3.9 % net undercount for Black males compared to the A.C.E. estimate of a 2.5 % net overcount. The final revised A.C.E. Revision II estimates for Black males were adjusted for correlation bias using DA sex ratios.[1] This resulted in estimates of a 2.4 % net undercount—the same direction as the DA estimates. Both methods estimated a net overcount for the Black females aged 50 and older amounting to 1.0 % in the DA and 2.5 % in the A.C.E.

In 2010, the post-enumeration survey results were computed for two Black groups: Black Alone (BA) and Black Alone or in Combination (BAIC). Census

[1] The accuracy of estimates of coverage from post-enumeration surveys are affected by correlation bias. Correlation bias occurs if the census and the survey miss the same type of people. Sex ratios calculated from the DA correct for this bias.

Table 8.1 Percent difference for the black population aged 50 and older by sex: results from post enumeration surveys and demographic analysis: 1990, 2000 and 2010

	Black population aged 50 and older	
Category	Male (1)	Female (2)
1990		
Post Enumeration Survey (PES)	0.4	1.2
Demographic analysis	−5.3	0.5
2000		
Accuracy and Coverage Evaluation (A.C.E.)		
No adjustment for correlation bias	2.5	2.5
With adjustment for correlation bias	−2.4	2.5
Demographic analysis	−3.9	1.0
2010		
Census Coverage Measurement (CCM)		
Black Alone	−2.3	3.1
Black Alone or in Combination	−2.3	3.1
Demographic Analysis		
Black Alone	−4.6	0.8
Black Alone or in Combination[a]	−2.3	3.3
Average[b]	−3.4	2.0

Note: Percent difference = ((census count – DA estimate)/census count). A negative sign indicates an undercount

[a]The Census counts are classified as Black Alone or Black Alone or in Combination, whereas the Demographic Analysis estimates can only calculate Black Alone or in Combination estimates for the population aged 0–30. Thus, the difference is not a pure estimate of Black Alone or in Combination

[b]The average is based on the Black Alone or Black Alone or in Combination percent differences

Source: 1990: U.S. Census Bureau, 2001a. "ESCAP II Demographic Analysis Results "by J. Gregory Robinson. Executive Steering Committee on Accuracy and Coverage Evaluation Policy II, Report number 1, Oct. 13, 2001

2000: U.S. Census Bureau, 2001a. "ESCAP II Demographic Analysis Results "by J. Gregory Robinson. Executive Steering Committee on Accuracy and Coverage Evaluation Policy II, Report number 1, Oct. 13, 2001 and U. S. Census Bureau. 2001b. "Accuracy and Coverage Evaluation: Demographic Analysis Results" by J. Gregory Robinson, DSSD Census 2000 Procedures and Operations Memorandum Series B-4, March 2. 2002

2010: http://www.census.gov/coverage_measurement

Coverage Measurement (CCM) estimated a 2.3 % net undercount for both race categories for the male population.

In Table 8.1, three categories are shown for the DA estimates: BA, BAIC and an average of the two. Whereas the census tabulates two Black race categories for all ages, the DA can only produce BAIC estimates for the population aged 0–30. The CCM and the DA results for the BAIC male population is the same (a 2.3 % net undercount). For the BA population DA measures a net undercount of 4.6 %

compared to the CCM estimate of 2.3 % CCM. The net undercount based on the average of BA and BAIC is 3.4 %.

For the BA females aged 50 and older, the DA and the CCM results are different, but in the same direction in 2010. Both show a net overcount (3.1 % according to CCM and 0.8 % according to DA). However, for the BAIC population, both show a net overcount around 3.0 %. The averaged DA difference is a 2.0 % net overcount (Table 8.1).

Studies of differential coverage by race have focused either on documenting the differences for specific age groups (age effects) or on tracking the improvements in coverage from one census to the next (period effects). Consequently, the focus has been on specific population subgroups such as young children, the college age population, young adult males, or the population aged 65 and over (U.S. Census Bureau 1973; Passel et al. 1977; Passel and Robinson 1988; West and Robinson 1998). Since the DA methodology is a cohort-component approach with births, deaths, and net international migration used as the building blocks, cohort effects can also be gleaned from the DA estimates though only a few studies have done so (Passel 1991; Robinson et al. 1990). In these studies, the emphasis has been primarily on creating the best possible DA estimates for the total population rather than on the differential coverage of a specific subpopulation such as Black males.

This study examines cohorts from the early years of the formation of the vital registration system, i.e., the cohorts born between 1935 and 1940. In 2010, these cohorts are carried forward in the estimates for 70–74 years providing the opportunity for a comparison of the DA estimates to counts from seven different censuses. If one of the components is biased in the cohort-component approach, the method results in a cohort that is either too small or too large causing bias in the measurement of coverage.

Since the DA estimates are based on components of change (births, deaths, and estimates of net international migration), the method starts with the population born since 1935. This encompasses the population under age 75 in 2010, under age 65 in 2000, under age 55 in 1990, under age 45 in 1980, ..., under age five in 1940. For the ages not covered in the administrative records, other techniques and data sources come into play. Since the 1970 Census, Medicare has been available to estimate the population aged 65 and over. In 1990, the cohort component method was used for the population aged 0–55 and the Medicare data for the population aged 65 and over. The estimates for ages 55–65 are obtained by reverse surviving the Medicare population with the components of population change. In 2000, the population aged 0–65 are from the cohort–component approach and the population aged 65 and older from Medicare. In 2010, it was possible to estimate the population aged 0–75 using the cohort component approach and Medicare (adjusted for underenrollment) for the population aged 75 and over.

The sections below describe each component of the DA estimate in 2010 and the underlying assumptions of the estimation. The corrections for birth registration completeness are discussed along with corrections to the mortality data for Black infants to understand potential impacts on the magnitude of census coverage errors. The possible impact of international migration of Black males is also presented.

Finally, to produce estimates by race characteristics it is necessary to assign race to the vital registration data. The potential sensitivity in the estimates to different race classification schemes is considered.

8.2 Background

The DA estimates of the population born between 1935 and 1940 are based on registered births corrected for under-registration, registered deaths corrected for under-registration of infant deaths, and estimates of net international migration. The DA method starts with the births. It then follows each birth cohort through its life course, subtracts deaths and those who leave the country, and adds new immigrants. Since births are much more numerous than deaths or migrants for a given cohort, errors in the number of births have the largest potential impact on the estimates.

The quality of the birth registration data have been researched a number of times. Tests of birth registration completeness were conducted for 1940, 1950, and 1964–1968. Over the years, the methods were modified, based on emerging research that identified inconsistencies in the time series of DA coverage estimates that could be related to deficiencies in the test results or the method of interpolation/ extrapolation to years between the tests. For the 1970, 1980, and 1990 DA estimates, the time series of births since 1968 were based on the extrapolation of the 1964–1968 Birth Registration Test (BRT) results. The 2000 DA operation also incorporated methodological corrections for under-registration in the vital statistics data. Subsequent evaluations of the DA assumptions about birth and death registration completeness concluded that the births were correctly adjusted (U.S. Census Bureau 2001). The births for 1980–2000 were modified slightly based on the individual record file. Otherwise, the 2000 DA output data were used as input for the 2010 DA without any further adjustments (Devine et al. 2011). However, post-2010 assessments of the cohort-component-based estimates of the Black males suggest that the earlier adjustments may not have addressed the registration issue sufficiently (West 2012, 2014).

Table 8.2 shows the differential coverage error pattern by age, race and sex in 1990, 2000 and 2010. The differences are based on averaged BA and BAIC for the census counts and averaged BA and BAIC DA estimates for ages under 30 and 'race of father' for ages 30–74. With exception of the age group 10–19 in 2000 and 2010 and 15–19 in 1990, the DA estimates for Black males are always higher than the census counts. The pattern for the non-Black males look more similar to the pattern for Black males, but the difference from the census is much smaller. A pattern of census overcount starts to emerge for Black females when they reach age 50 and for non-Black females starting at age 35.

Figures 8.1, 8.2, 8.3 and 8.4 show the results for selected age groups in 1990, 2000 and 2010 by race and sex. Figure 8.1 is for Black males. There is a persistent pattern of undercount regardless of age and year of the census. The 1935–1939 birth

Table 8.2 Percent difference between 1990, 2000 and 2010 census counts and cohort-component-based demographic analysis estimates by race, sex, and 5-year age groups: Ages 0–74

		% Diff. Black				% Diff. Non-Black			
		Male		Female		Male		Female	
Age at time of census	Census	BA	Averaged	BA	Averaged	BA	Averaged	BA	Averaged
	2010	−5.9	−4.9	1.5	2.6	0.3	0.2	2.2	2.1
70–74	2000	n.a.	n.a.	n.a.	n.a.	n.a.	n.a.	n.a.	n.a.
	1990	n.a.	n.a.	n.a.	n.a.	n.a.	n.a.	n.a.	n.a.
	2010	−3.9	−2.9	1.5	2.7	1.1	1.0	2.9	2.8
65–69	2000	n.a	n.a.	n.a.	n.a.	n.a.	n.a.	n.a.	n.a.
	1990	n.a	n.a.	n.a.	n.a.	n.a.	n.a.	n.a.	n.a.
	2010	−1.4	−0.3	2.6	3.8	1.9	1.8	3.5	3.4
60–64	2000	−6.0	n.a.	1.5	n.a.	−1.1	n.a.	0.8	n.a.
	1990	-n.a	n.a.	n.a	n.a.	n.a	n.a.	n.a	n.a.
	2010	−5.2	−4.0	0.5	1.8	0.9	0.7	2.9	2.8
55–59	2000	−8.1	n.a.	−0.9	n.a.	−1.4	n.a.	0.5	n.a.
	1990	-n.a	n.a.	n.a	n.a.	n.a	n.a.	n.a	n.a.
	2010	−7.2	−6.0	0.0	1.4	0.5	0.4	2.6	2.4
50–54	2000	−5.5	n.a.	0.7	n.a.	−0.2	n.a.	1.4	n.a.
	1990	−12.7	n.a.	−2.2	n.a.	−3.6	n.a.	−1.5	n.a.
	2010	−9.1	−7.8	−1.3	0.3	−1.0	−1.2	2.0	2.1
45–49	2000	−9.1	n.a.	−1.6	n.a.	−0.8	n.a.	0.8	n.a.
	1990	−12.8	n.a.	−2.9	n.a.	−3.3	n.a.	−1.6	n.a.
	2010	−9.0	−7.4	−0.5	1.4	−0.7	−1.0	2.1	1.8
40–44	2000	−10.8	n.a.	−2.0	n.a.	−0.6	n.a.	0.9	n.a.
	1990	−11.2	n.a.	−1.9	n.a.	−1.8	n.a.	0.1	n.a.
	2010	−10.7	−8.8	−0.6	1.7	−1.0	−1.3	2.4	2.0
35–39	2000	−10.3	n.a.	−1.2	n.a.	−0.8	n.a.	1.2	n.a.
	1990	−12.5	n.a.	−2.4	n.a.	−1.9	n.a.	−0.2	n.a.

	2010	−11.3	−8.9	−2.2	0.6	−1.5	−1.9	2.2	1.7
30–34	2000	−9.2	n.a.	0.4	n.a.	−0.2	n.a.	1.3	n.a.
	1990	−14.8	n.a.	−4.1	n.a.	−2.0	n.a.	−0.1	n.a.
	2010	−7.4	−8.2	0.3	−0.2	−0.5	−0.3	1.5	1.6
25–29	2000	−8.8	n.a.	0.0	n.a.	−0.2	n.a.	1.0	n.a.
	1990	−13.4	n.a.	−5.7	n.a.	−3.6	n.a.	−1.8	n.a.
	2010	−1.9	−2.9	1.2	0.6	1.9	2.2	1.2	1.3
20–24	2000	−5.3	n.a.	0.4	n.a.	0.8	n.a.	2.0	n.a.
	1990	−6.2	n.a.	−2.9	n.a.	−0.2	n.a.	0.4	n.a.
	2010	4.6	3.9	3.6	3.2	1.8	1.9	1.4	1.5
15–19	2000	1.7	n.a.	1.8	n.a.	2.6	n.a.	2.9	n.a.
	1990	0.2	n.a.	−0.3	n.a.	1.8	n.a.	2.0	n.a.
	2010	1.2	1.2	0.7	0.9	−0.4	−0.4	0.7	−0.7
10–14	2000	1.5	n.a.	1.2	n.a.	1.6	n.a.	1.2	n.a.
	1990	−3.4	n.a.	−3.4	n.a.	0.7	n.a.	0.7	n.a.
	2010	−1.2	−1.6	−1.2	1.3	−2.4	−2.1	−2.3	−2.3
5–9	2000	−1.4	n.a.	1.9	n.a.	−1.1	n.a.	−1.5	n.a.
	1990	−6.9	n.a.	−6.7	n.a.	−2.4	n.a.	−2.2	n.a.
	2010	−4.5	−5.6	−4.3	5.3	−4.8	−4.5	−4.6	−4.5
0–4	2000	−5.3	n.a.	−5.5	n.a.	−3.3	n.a.	−3.8	n.a.
	1990	−7.8	n.a.	−7.4	n.a.	−2.9	n.a.	−3.0	n.a.

Note: Percent difference = ((census count − DA estimate)/census count). A *negative sign* indicates an undercount

The Averaged 2010 comparisons are 'race of father' births-based DA estimates compared to an average of Black Alone (BA) and Black Alone or In Combination (BAIC) census counts. The 2010 DA operation produced separate BA and BAIC estimates for the population under age 30, but not for older ages

The results for the Non-Black population are residuals, calculated by subtraction (total population minus Black population)

[a]In 2000, it was not possible to produce DA estimates from the cohort-component method for the population aged 65–69 and 70–74

[b]In 1990, it was not possible to produce DA estimates from the cohort-component method for the population aged 55–59, 60–64, 65–69 and 70–74

Sources: 2010 DA estimates are consistent with May 2012 release of the Middle Series (www.census.gov/popest/research/demo-analysis.html); the 2000 and 1990 DA estimates are consistent with ESCAP II releases. See Appendix Table B3 in http://www.census.gov/dmd/www/pdf/Report1.PDF. The estimates shown here are consistent with DA estimates published in Robinson et al. (1993)

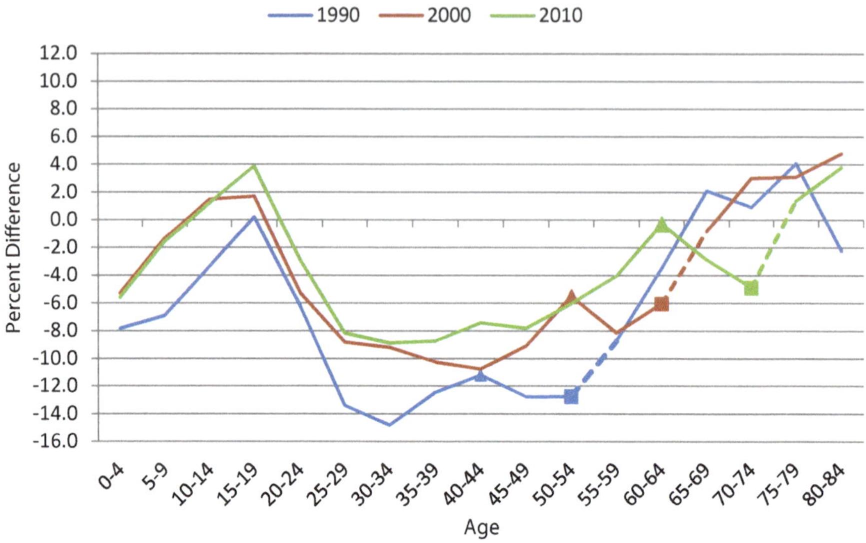

Fig. 8.1 Percent differences between cohort-component-based demographic analysis estimates and 1990, 2000, and 2010 censuses: Black male (Source: Census Bureau Staff calculations)

cohort is marked with squares. The age group 50–54 is particularly low in 1990. Age groups 60–64 in 2000 and 70–74 in 2010 are more similar.

Figure 8.2 shows the differences for Black females. Black females have a smaller percent undercount compared to the Black males (except for children) and make a transition to census overcount in the older age groups. (In 2000 and 2010, the squares are above the 0.0 % line.)

Figures 8.3 and 8.4 show the results for the non-Black population. In this sub-population, no estimate exceeds 5 % and an overcount occurs for females in 2000 and both males and females in 2010 for the birth cohort 1935–1939. In the figures, the 1935–1939 cohort is indicated by squares.

The focus of this paper is the 1935–1939 birth cohorts (shown as squares in Figs. 8.1, 8.2, 8.3, and 8.4). Table 8.3 follows these cohorts showing the differences between the census count and the estimates by race and sex. Black males persistently have differences that are much higher than those for Black females or the non-Black population (males and females).

In general, the census counts are lower than the DA estimates by race and sex until age 60. At that age, the counts for the females (Black and non-Black) are higher than the estimates. The counts for the males (Black and non-Black) remain lower for the 60–64 age group (in 2000), but the differences for non-Black males are 1.1 % compared to 6.0 % for Black males. By age 70 (in 2010), the counts for the non-Black males are no longer lower. However, the Black males aged 70 and older continue to experience a count that is 5.9 % lower than the estimate.

As previously stated, for simplification of the issue, this paper focuses on the 1935–1939 cohorts. The percent differences for the cohort of 1935–1939 are in

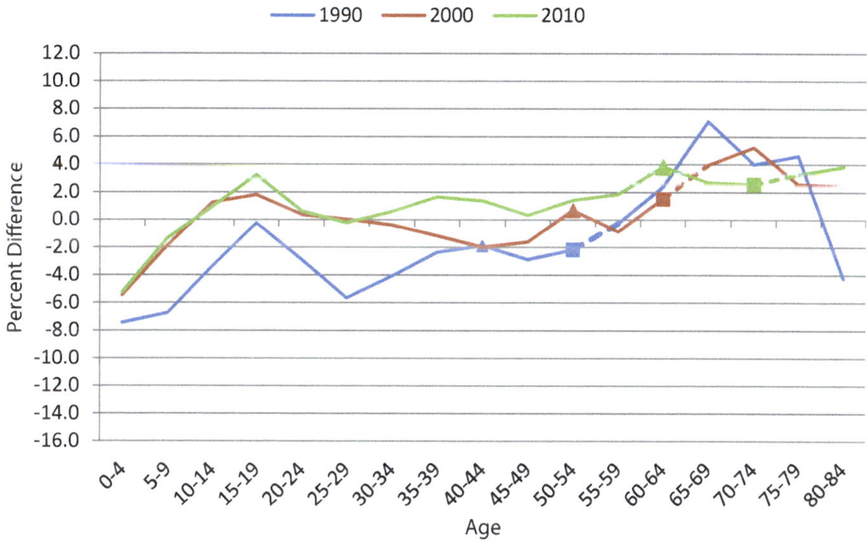

Fig. 8.2 Percent differences between cohort-component-based demographic analysis estimates and 1990, 2000, and 2010 censuses: Black female (Source: Census Bureau Staff calculations)

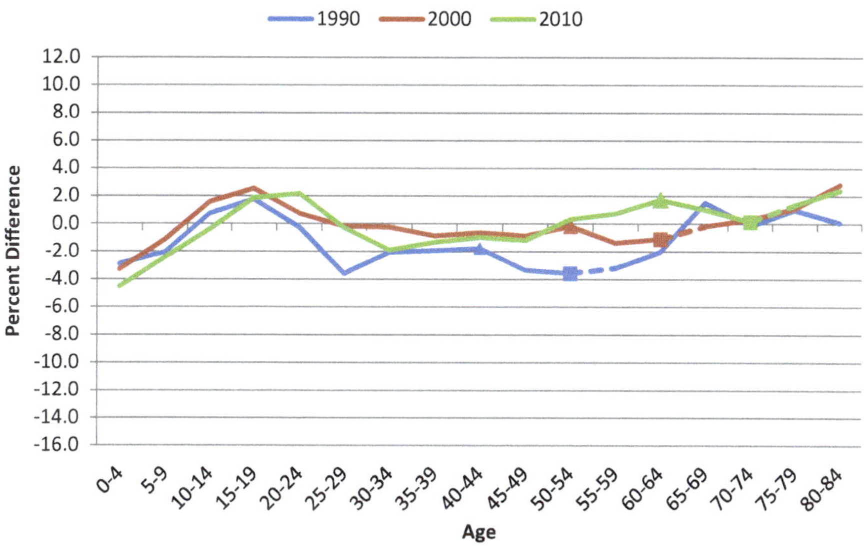

Fig. 8.3 Percent differences between cohort-component-based demographic analysis estimates and 1990, 2000, and 2010 censuses: Non-Black male (Source: Census Bureau Staff calculations)

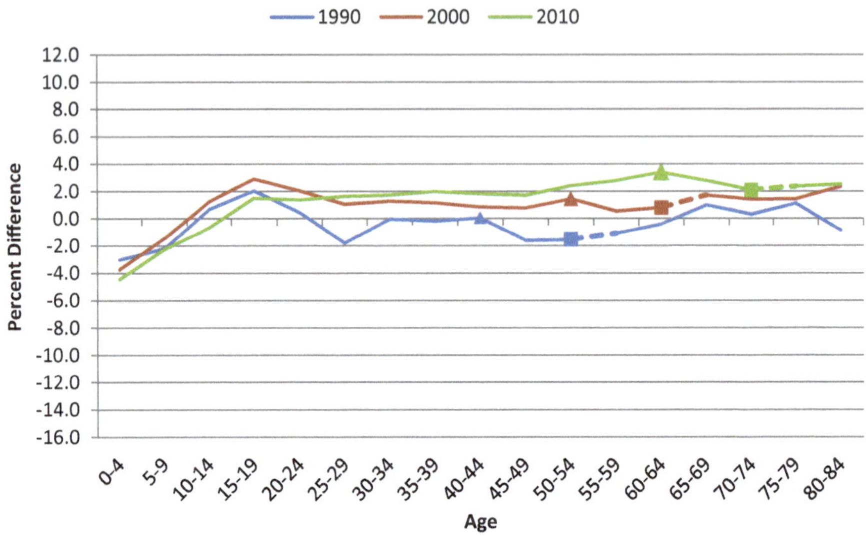

Fig. 8.4 Percent differences between cohort-component-based demographic analysis estimates and 1990, 2000, and 2010 censuses: Non-black female (Source: Census Bureau Staff calculations)

Table 8.3 Percent differences between demographic analysis estimates and census counts by race, sex and census year for the population born between April 1, 1935 and April 1, 1940

		Black		Non-Black	
Census year	Age at census	Male (1)	Female (2)	Male (3)	Female (4)
1940	0–4	−13.4	−11.8	−7.8	−7.3
1950	10–14	−3.6	−3.3	−2.3	−2.3
1960	20–24	−16.1	−6.6	−5.2	−2.8
1970	30–34	−15.9	−1.8	−3.7	−1.1
1980	40–44	−13.2	−0.0	−2.8	0.0
1990	50–54	−12.7	−2.2	−3.6	−1.5
2000	60–64	−6.0	1.5	−1.1	0.8
2010	70–74	−5.9	2.6	0.2	2.1

Note: Percent difference = ((census count − DA estimate)/census count). A *negative sign* indicates an undercount

Sources: 2010 DA estimates are consistent with May 2012 release of the Middle Series (www.census.gov/popest/research/demo-analysis.html); the 2000 and 1990 DA estimates are consistent with ESCAP II releases. See Appendix Table B3 in http://www.census.gov/dmd/www/pdf/Report1.PDF. The estimates shown here are consistent with DA estimates published in Robinson et al. (1993). The 1940–1980 DA are unpublished and consistent with revised 1990 and 2000 DA estimates (from ESCAP II)

Fig. 8.5. Figs. 8.6, 8.7, and 8.8 cover the birth cohorts of 1945–1949, 1955–1959 and 1965–1969.

As first seen in Fig. 8.5, the coverage of Black males born between 1935 and 1940 look close to Black females during the first 5 years (appearing first in the 1940 Census) and to the other race/sex groups only for age group 10–14 (1950 Census). The counts for young Black males are particularly low in the 1960 and 1970 Censuses at ages 20–24 and 30–34. As the population ages, the percent differences diminish, but unlike the Black females, and the non-Black males and females, Black males are never overcounted.

For the cohorts born between 1945 and 1950, the same pattern emerges (Fig. 8.6). The coverage of Black males comes close to females in the early ages and in age group 10–14, but an overcount is never achieved, and the differential undercount of Black men between ages 20–24 and 50–54 remains wide (though not as wide as for the 1935–1939 cohort).

Figure 8.7 starts in 1960 with the age group 0–4 estimates for the 1955–1959 birth cohorts. Again, the lines in the graphs are similar to the lines in Figs. 8.5 and 8.6. The census counts for the Black males are always lower than the estimates.

Finally, Fig. 8.8 shows the percent difference of DA estimates and census counts for the 1965–1969 birth cohort. The Black males and females are similar up to age 10. At that point, the Black females start looking more like their non-Black counterparts whereas the differences for young Black males grow larger. The differences for the Black males in their 20s has a percentage similar to that observed for the 0–4 age group.

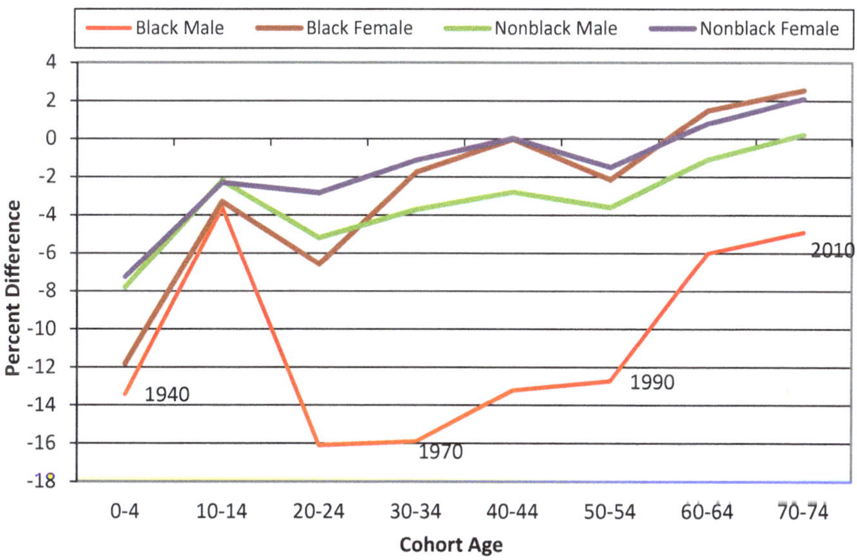

Fig. 8.5 Percent differences between cohort-component-based demographic analysis estimates and the censuses for birth cohorts of 1935–1939 (Source: Census Bureau Staff calculations)

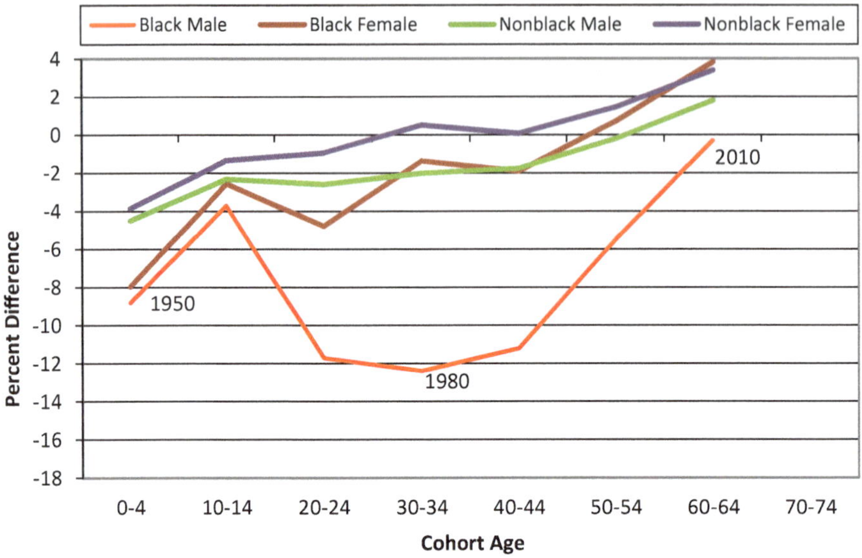

Fig. 8.6 Percent differences between cohort-component-based demographic analysis estimates and the censuses for birth cohorts of 1945–1949 (Source: Census Bureau Staff calculations)

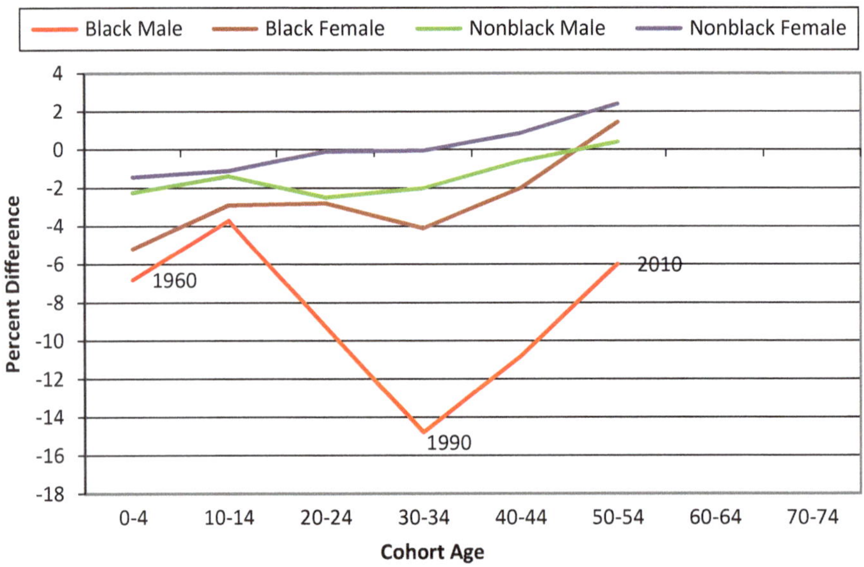

Fig. 8.7 Percent differences between cohort-component-based demographic analysis estimates and the censuses for birth cohorts of 1955–1959 (Source: Census Bureau Staff calculations)

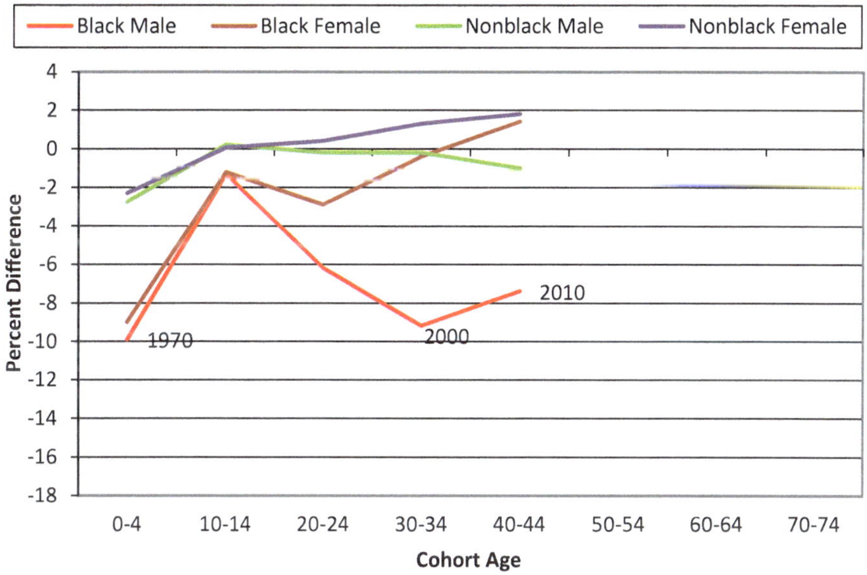

Fig. 8.8 Percent differences between cohort-component-based demographic analysis estimates and the censuses for birth cohorts of 1965–1959 (Source: Census Bureau Staff calculations)

A cohort pattern of coverage error by race and sex does not appear logical, but the pattern is seen for every Black male cohort starting at age 20–24. One might expect certain ages to be more difficult to enumerate. One might also expect period effects, i.e., when strategies for collecting and processing the data change or the respondents are influenced by a current event, differential outcomes may follow. It is also perceivable that age and period factors interact to affect the successful implementation of the census process for all populations, but it is difficult to understand how a cohort of a given race and sex is missed persistently every 10 years. So are the observed differences too big? The problem may lie in the DA methods and the sensitivity of the assumptions used to estimate the components.

8.3 Demographic Analysis Component Method

In the following sections, we search for possible explanations of the persistent differences between the count and the estimate by reviewing each of the components, starting with births. The assessment includes an examination of the factors used to correct for incomplete and missing vital registration data. The possible impact of race classification errors is also discussed. As stated above, the focus is on the 1935–1939 Black male birth cohorts—the first cohorts for which records were used from the vital registration system.

8.3.1 Births—Incomplete Birth Registration

Since the 1970 DA operation, the 1935–1939 cohorts have been corrected for under-registration based on estimates of birth registration completeness. The correction factors for these cohorts are derived from birth registration completeness tests conducted for the year 1940. Later tests, in 1950 and for the period 1964–1968, provide correction factors for the cohorts in those years. For years not in the tests, the factors are obtained through interpolation and extrapolation.

The 1940 Birth Registration Completeness Test (BRT) concluded that the registration of Black births in 1940 was 82 % complete. In comparison, the completeness for White births was 94 %. However, Robinson (1991) speculated that the 1940 test overstated the under-registration for the Black population. The birth registration test involved matching all infants under 4 months of age who were counted in the 1940 Census to their birth certificates. Robinson and his co-authors hypothesized that there were a number of false non-matches especially for Black infants, because field follow-up procedures were not implemented uniformly in all geographical areas when the data for the test were collected. Furthermore, the test implied that Black births occurring out of hospitals were registered at the same rate as White births occurring out of hospitals. The subsequent birth-registration tests did not support this finding.

A recalculation of the completeness rates for Black births using the factors for White births not occurring in hospitals (88.2 %) brought the rate up to 89.9 %—7.9 percentage points higher than the test result.

Research supports a higher estimate of completeness. Whelpton estimated the completeness of the birth registration for Blacks births in 1940 to be 89.5 %—7.5 percentage points higher than the estimate derived from the test (Whelpton 1950). However, Robinson and others (1990) conducted further analyses using different rates of completion and concluded that an 86 % completion was probably the most demographically plausible level.

Passel (1992) agreed with the assessment that the 1940 Test indicated biased estimates of the Black 1935–1939 birth cohort. He too concluded that the registration was more complete than the test showed and concurred that Black births are more affected than non-Black births "because of the geographic concentration of Blacks in states with less sophisticated statistical agencies and higher rates of under-registration" (Passel 1992:3).

Subsequently, he applied a regression approach to develop new estimates of corrected Black births for the period 1935–1980. He used odds ratios of registration to adjust the original corrected births. The adjustment raised the number of births. The overall completeness of registration using the modified regression equation is 86.6 % or 4.6 percentage points higher than the percent completeness derived from the 1990 Test. Once the 1940 values for the regression constants were set, Passel determined the values for 1935–1940 (Passel 1992).

Table 8.4 summarizes the range of percent completion for Black births based on the 1940 Test and the assessments of the National Center for Health Statistics

Table 8.4 Percent completeness for registration of black births in 1940 by source

Source	Percent completeness
National Center for Statistics (NCHS), original	82.0
Robinson et al. (1990), revised	86.0
Passel (1992)	86.6
Whelpton (1950)	89.5
Robinson et al. (1990), original	89.9

Source: See text

Table 8.5 Assumed birth registration completeness for black birth cohort 1935–1939

	Assumptions about birth registration completeness in percent				
		DA 2010			
	Original	Low	Middle	High	Alt. 1
Birth year	(1)	(2)	(3)	(4)	(5)
Average	79.9	80.4	84.2	88.4	89.6
1935	78.9	79.4	83.3	87.7	89.0
1936	79.3	79.8	83.7	88.0	89.2
1937	79.8	80.3	84.1	88.3	89.5
1938	80.4	80.8	84.5	88.6	89.9
1939	81.3	81.7	85.3	89.2	90.3

Source: Column 1: Devine et al. (2011)
Columns 2–4: (www.census.cov/popest/research/demo-analysis.html)
Columns 5: Census Bureau Staff calculations

(NCHS 1942–1943, 1946–1947, 1950–1952, 1957–1962), and three demographers: Passel, Robinson and Whelpton. The range is from a low of 82 % to a high of 89.9 %.

The managers of the 2010 DA operation recognized the sensitivity of the components to varying assumptions and chose to produce a range of estimates. For the birth registration completeness, the 2010 DA range was set to a low of 80.4 % and a high of 88.4 % with a middle series representing an assumed 84.2 % completeness.[2] Table 8.5 shows the percent range used for the 2010 DA for the Black 1935–1939 birth cohorts.

The last column of Table 8.5 offers an additional assumption. Here, it is the hypothesis that for these cohorts, the completeness of the registration of Black births is about the same as the completeness for White births and the completeness is set to 90 % on average. This alternative is included mainly to demonstrate the impact of the number of additional births needed to account for under-registration.

[2] The low and high estimates represent the outcome obtained by multiplying the original estimated number of unregistered births by age, sex, and race by 0.30 and then increasing or decreasing the original number of births added due to completeness. This range looked plausible when compared to the Census 2000 count of the native born population (Devine et al. 2012).

Table 8.6 Number of black male births using different assumptions about birth registration completeness for birth cohorts 1935–1939

		Assumptions about birth registration completeness			
	Registered births	80 %	84 %	88 %	90 %
Birth year	(1)	(2)	(3)	(4)	(5)
1935–1939	662,943	829,143	787,344	749,936	740,048
1935	128,725	163,150	154,476	146,779	144,624
1936	127,101	160,279	151,871	144,433	142,454
1937	133,677	167,515	158,950	151,390	149,371
1938	135,439	168,456	160,264	152,866	150,784
1939	138,001	169,743	161,782	154,710	152,816

Source: Column 1: Table 4 in Robinson (2011)
Columns 2–5: Table 5 in Robinson (2011)

Table 8.7 Numeric difference in the number of black male births using different assumptions about birth registration completeness for cohort 1935–1939

	Assumptions about birth registration completeness			
	80 %	84 %	88 %	90 %
Birth year	(1)	(2)	(3)	(4)
1935–1939	166,200	124,401	86,993	77,105
1935	34,425	25,751	18,054	15,899
1936	33,178	24,770	17,332	15,353
1937	33,838	25,273	17,713	15,694
1938	33,017	24,825	17,427	15,345
1939	31,742	23,782	16,709	14,815

Source: Table 6 in Robinson (2011)

Table 8.6 presents hypothetical numbers of Black births. An assumption that 20 % of the Black male births are not registered would result in 829,143 Black male births in the period 1935–1939. If 16 % were missed, the total for the period would be 787,344; if 12 % were missed the total number is close to 750,000, and if only 10 % were assumed missed, the total number would be closer to 740,000.

The sensitivity of the numbers to varying assumptions is captured again in the next table (Table 8.7). It shows the numeric difference in the number of Black male births using different assumptions. All differences are based on the original number of registered births. Over the 5-year time span, the number of births to be added range from about 77,000–166,000. Annually, the range is from a low of around 14,800 to a high of around 34,400 births.

8.3.2 *Deaths—Incomplete Infant Death Registration*

Mortality is another factor to consider in the cohort-component approach. In the end, if a birth cohort has been inflated to a level that is too high, the cohort will remain too large as it ages. There can be no decrements from those members of the cohort who exist only on paper. However, focusing on the birth event and differential outcomes by race, discrepancies in adjustments for underreporting of infant deaths are of interest. If deaths to Black births were understated, the births will remain in the cohort as it ages.

There have been no tests of the completeness of the national death registers. The National Center for Health Statistics (NCHS) has never adjusted registered deaths, including infant deaths, for an assumed percentage of under-registration (McDevitt 2001a, b). In the DA estimation method, the deaths statistics for adults are considered complete as of 1935. There is no adjustment for under-registration of deaths in the preparation of the demographic estimates for the population aged 1 and older. Adjustments are made to infant deaths occurring prior to 1960. These deaths are adjusted upward by one-half the percentage of the correction for the under-registration of births. For example, in 1940, if the birth under-registration were estimated at 18 %, registered infant deaths would be adjusted upward by 9 %. Within the Census Bureau, there is no institutional memory about the origin of the infant deaths under-registration assumption of half the race-specific level of incompleteness of births (McDevitt 2001a). The assumption could be too low rendering these early birth cohorts too high. In their research, Preston and his co-authors assume that Black infant deaths have been subject to the same level of under-registration as births from 1935 to 1990 (Preston et al. 1998).

In the context of the evaluations following the 1990 Census, Robinson et al. (1990) set the correction factor for infant mortality occurring between 1935 and 1960 equal to the factor used to correct for birth under-registration for those years to gauge the possible impact on the estimated census undercount for those cohorts. They found that the undercount would have been smaller, but that overall the magnitude of the impact from this factor was far smaller than the potential impact from the correction of births. If instead of assuming a 1/2 to 1 ratio of deaths to birth, a ratio of 1 to 1 is applied, the number of infant deaths would increase by about 2000–3000 per year for all births.

Evaluations of the 2000 DA estimates found that assumptions regarding infant death registration were reasonable though it left open the door for further research on this topic (McDevitt 2001b).

The 2010 DA operation used the 2000 DA results as input. After the 2010 DA production, Condon (2010) analyzed the assumptions regarding the completeness of the death registration for infants in the birth cohorts 1936–1946. Her analysis concluded that differential under-registration of infant deaths would impact Blacks more than Whites.

The analysis expanded on the research done by McDevitt (2001b) by focusing on the differential impact on the Black population compared to the White population. Furthermore, the notion that deaths for young children (the 1–4 and 5–9 year-olds) might be similarly misreported was introduced (the infant deaths are adjusted by half the race-specific level of incompleteness of births for these birth cohorts). Sensitivity analyses were performed to show the impact of setting infant death under-registration equal to the race-specific level of incompleteness (100 %), as

Table 8.8 Assumptions about death registration completeness and numeric difference in infant deaths for black male births 1935–1939

Percent of birth registration completeness correction	Number of deaths	Difference from reported infant deaths	
		Amount	Percent
	(1)	(2)	(3)
If 100 % of correction factor	73,270	14,716	25.1
If 50 % of correction factor (DA 2010)	65,089	6535	11.2
Reported infant deaths 1935–1939	58,554		

Source: Table 4 in Robinson (2011)

well as 25 and 75 %. Race-specific adjustments for the deaths of children aged 0–9 were also tested.

Appendix Table 8.11 shows results from previous work by Condon that applies to the 1935–1940 cohorts for Whites and Blacks. The extract indicates that the sensitivity is substantial for the Black population. The current assumption uses 50 % of the birth correction factor. If that were to be changed to 100 %, the Whites would see an increase in deaths that ranges from 0.2 to 0.3 %. For Blacks, the increase would range from 1.3 to 2.1 % (as a percent of the DA population aged 70–74 in 2010).

Table 8.8 summarizes the results for the Black births 1935–1940 (male and female). Additional deaths would be either subtracted or added to these birth cohorts depending on the assumption about the completeness of the registration of infant deaths. The 50 % assumption increases the number of deaths by 11.2 %. An assumption of 100 % of the birth correction factor would increase the number of deaths by 25.1 %.

8.3.3 Net International Migration—Incomplete Immigration Data

A number of research projects have focused on net international migration and ways to measure immigrants in the DA estimates as well as the census (see e.g. Robinson 2011). Most research has examined migration of the Hispanic origin population. There is a void in the research regarding the impact of this component on the estimates for the Black population, Black males, and specifically older Black males—the focus of this paper. In the estimation process, the net international immigration component generally is assumed small for these cohorts (1935–1939) and it is believed that immigration could not disproportionately change the size of this population. If this assumption is false, and/or if Black immigrants in the age groups are classified erroneously as non-Black immigrants, the size of the Black cohorts would increase.

8.3.4 Net International Migration—Incomplete Emigration Data

Emigration lowers the cohort size. It is assumed that there is no significant attrition in the Black population in these cohorts due to emigration. If this assumption is false, the cohorts would become smaller and subsequently the census undercount would be smaller. However, the number of Black emigrants would have to be substantial before an impact would be noticeable in the size of the cohort.

8.3.5 Race Classification in DA Processing

When constructing the demographic estimates of the population, births, deaths, and immigration events must be assigned to a specific race group. Different numbers are produced depending on the algorithm used to assign race to a birth, death or an immigrant. The potential impact of race classification errors on the DA estimates is limited because the DA processing allows for only two race categories.

In the birth component, race can be ascribed to each birth based on the recorded race of the mother or the father, or a combination of both. In the DA processing for the birth cohorts 1935–1939, it is assumed that the child is the same race as the father. If the parents belonged to different race groups, it would not be apparent from the birth certificate. For the birth cohorts in this analysis, the birth certificate did not provide the option to record more than one race.

The death certificate provides the race of a deceased person. This information, in turn, comes from a proxy such as a family member, or a funeral director. There is no reason to believe the process produces a substantial amount of classification error for the Black/Non-Black distinction.

Most of the data on legal immigration for the early birth cohorts come from the Immigration and Naturalization Service (INS). These data are classified by country of birth, but not by race. The Census Bureau assigns the legal immigrants to race groups based on responses from the previous census or more recent data, if available. The race classification is not based on responses from the immigrants themselves. It is possible that the assigned race used in the DA processing is different from the race category chosen by this group in their responses to the census.

8.3.6 Classification of the Census Count by Race in DA Processing

In order to assess the completeness of the census counts by race, the DA estimates and the counts must be presented in identical race categories. On the census side, race is either self reported or reported by a proxy—a person completing the census questionnaire on behalf of the members of the household. It is possible that a person is reported with a particular race on the vital registration record and then self-identifies differently in the census. However, it is unlikely that the impact is significant for the cohorts included in this analysis (Robinson 2011). The 2000 and 2010 censuses included an option to report multiple races and both censuses provided an option to report one's race as "Some Other Race." It is possible for a person to be classified in a census race category that is inconsistent with the DA race classification. However, it is unlikely that the impact is significant for the cohorts in question.

8.4 Discussion of the DA Methodology

The previous section discussed the cohort-component method used to produce the DA estimates. The major observations regarding the sensitivity of the cohort component methodology to assumptions are summarized in this section. The focus is on factors that might ultimately explain why a subgroup of the population is persistently being missed in census enumerations. Since the DA estimates serve as the benchmark for measuring net undercount, the emphasis is on the DA method, rather than the census operation.

8.4.1 Sensitivity of Cohort Components to Assumptions

The sensitivity of the birth components to altering assumptions about the completeness should be considered. The correction for incomplete registration when the national vital registration system was newly established is important for understanding the persistent difference between the DA estimate and the census counts. Depending on assumptions used, the 5-year cohort of Black males (1935–1939) could have started out as much as much as 829,143 (assumed under-registration of 20 %) or as small as 740,048 (assumed under-registration of 10 %). These numbers were presented in Table 8.6.

The sensitivity of deaths to altering assumptions about completeness of the registration might also be a factor. If birth registration is more complete than originally thought it follows that infant death registration would also be more complete. Were this to be the case, the size of the estimated cohort would be too small because the correction created too many deaths to be subtracted. If the completeness was assumed to be 100 % of the percent used to correct for under-registration of births rather than for example 50 %, there could be as many as 8181 more Black male deaths. This would lower the DA population for Black men aged 70–74 in 2010 by 2.1 %.

Finally, with regard to immigration, failure to add the proper number of immigrants that belongs to the 1935–1939 birth cohorts would render the estimated size of the cohort too small. In all likelihood, emigration would be too small a factor for the 1935–1939 Black male birth cohorts to have an impact.

The estimated cohort size resulting from the manipulations of the components could be further complicated by the method used to classify a birth, a death or an immigration event as Black. But the impact would be small.

When comparing the DA estimates to the census by race, the estimates and the count must be classified in the same conceptual race categories. The possible impact of race classification errors was considered. Since the larger than expected undercount is observed in every census for Black men, in order to create this outcome, the respondents must self-report or be recorded persistently as a race other than Black on the census form though they were classified as Black on the birth certificate. Given the unique patterns observed for Black men, Black males must do so more frequently than Black females.

The literature provides some insight about the impact of this particular issue. Robinson et al. (1990) concluded that classification errors attributable to a shift in the reporting of Black as one's race over time could not have a large enough impact to account for the observed differential by race in the estimate of undercount, especially in the older birth cohorts such as 1935–1939.

In summary, the scrutiny of the assumptions associated with each of the components used to create the DA estimate suggests that the birth registration factors applied to the Black births and the related corrections for infant mortality have a potential impact on the size of the Black male births cohorts 1935–1939. Though they would pull in opposite directions, the likely corrections to both would lower the estimated DA population. Immigration and race classification errors probably do not have a significant impact.

8.4.2 Gender Differences

Throughout this paper comparisons have indicated that Black females do not follow the same pattern of census undercount as Black men. It is difficult to provide logical reasons why the DA estimates should be constructed differently for females. In the absence of empirical guidance on this issue, the males and females are treated the same in the DA processing. The corrections for births are not gender specific under the assumption that in the United States in the 1930s, a birth of a Black female would be just as likely to be registered, or not, as the birth of a Black male. The same assumption applies to the completeness of infant death registration. However, historically, infant mortality is higher for males than females (Singh and Van Dyck 2010). Thus, it is could be argued that Black male and female infant deaths should receive different corrections.

Similarly, the literature does not address assignment of race by sex to the vital records nor the immigration components. Logically, there is no reason to suspect gender differences in ascribed race in any of the DA components.

On the census side, the literature is also silent on gender differences in identification with a specific race. It is unknown if males identify differently than females or if this applies to the older Black population. Race classification errors from such sources have not been researched, nor their potential impact on the measurement of coverage. However, in all likelihood such errors would be small.

8.4.3 *Implications for the Measurement of Census Coverage Error*

The size of a cohort matters because ultimately the DA estimates are benchmarks to gauge the quality of the census. Annual population estimates in turn take the most recent census as its starting point. It is important to know the size of the effect carried forward.

8.4.4 *Creating Optional Series of Demographic Analysis Estimates*

This paper presents the percent differences between the count and the DA estimates for the three most recent censuses. The calculations are based on the middle series of DA estimates released in May 2012. Post-2010 DA assessments deemed that for the total population, the estimates in this series were more consistent with the Census 2000 native born counts than the high and low series estimates of births (Devine et al. 2011). However, the assessments did not address what would be the most appropriate estimate within the range of estimates for a single subpopulation such as Black males, nor for a specific age group such as those aged 70–74 in 2010.

Table 8.9 shows the outcome for the assessment of census coverage errors if a different set of estimates were produced and chosen as the benchmarks for this subgroup. If the birth registration is assumed to be 88 % complete rather than 84 %, and all other components stayed constant, the 13.4 % undercount of Black male children in age group 0–4 in the 1940 Census drops to 8.2 % The 1950 Census

Table 8.9 Percent differences between demographic analysis estimates and census counts by census year for the black male population born between April 1, 1935 and April 1, 1940 by assumptions about birth registration

		Assumptions about birth registration completeness		
	Age at census	84 %	88 %	90 %
Census year	(1)	(2)	(3)	(4)
1940	0–4	−13.4	−8.2	−6.8
1950	10–14	−3.6	1.7	3.1
1960	20–24	−16.1	−10.6	−9.2
1970	30–34	−15.9	−10.3	−8.9
1980	40–44	−13.2	−7.5	−6.0
1990	50–54	−12.7	−6.6	−5.0
2000	60–64	−6.0	1.3	3.2
2010	70–74	−4.9	4.0	6.5

Source: U.S. Census Bureau staff calculations

overcounts the Black males in age group 10–14. The next four censuses again undercount this cohort. When the cohort reaches age group 60–64 in 2000 and 70–74 in 2010, the census counts are again higher than the estimates.

If instead the assumptions about completeness are 90 %, with all other components remaining constant, the census and the DA estimates are even closer: the undercount is reduced and the overcount is increased. However, the overcount for the Black men aged 70–74 in 2010 reaches 6.5 %, which suggests that this level of correction might not be plausible. The exercise is hypothetical, but points to the sensitivity of the DA estimates to the assumptions about the completeness of the birth registration.

Figures 8.9, 8.10 and 8.11 depict the outcome for the 1990 Census, Census 2000, and the 2010 Census of various assumptions about the birth registration completeness. The 84.2 % inherent in the 2010 Middle series estimates is shown in Fig. 8.9. The 1935–1939 cohort of Black males (shown by squares) are undercounted in all three censuses as discussed.

If a more complete registration were assumed such as the one used in the 2010 DA Low series (30 % fewer births than in the Middle series) the outcome would look like that shown in Fig. 8.10. The 1935–1939 cohort of Black males (shown by squares) are undercounted in the 1990 Census when they are in the age group 50–54. In both 2000 and 2010, the cohort would show an overcount.

Finally, Fig. 8.11 shows the hypothetical case of assuming 40 % fewer births than the DA 2010 Middle series. This series is not demographically plausible. The implied overcount of the cohort in 2010 is not reasonable.

8.4.5 Use of Sex Ratio to Select Most Plausible Series

Demographers often use the sex ratio or the proportion of males to females to decide the plausibility of an estimate. The ratios shown in Table 8.10 are for the

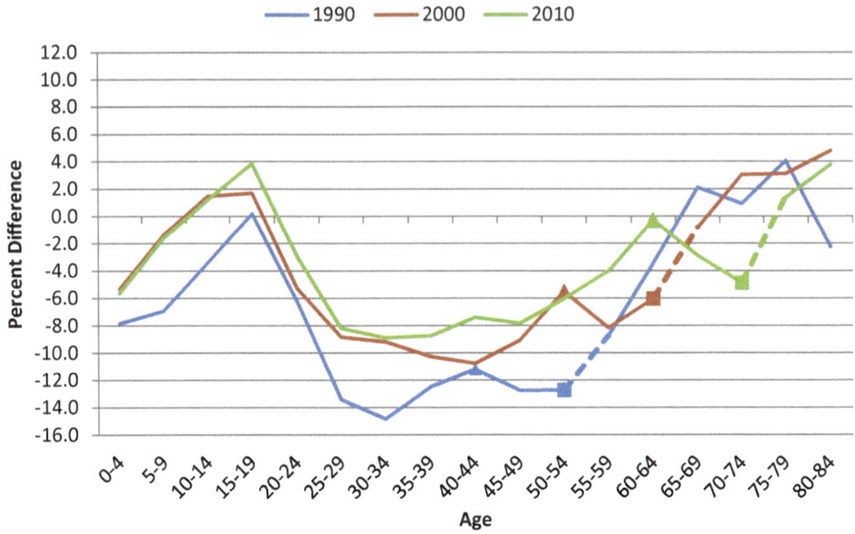

Fig. 8.9 Percent differences between cohort-component-based demographic analysis estimates and 1990, 2000, and 2010 censuses, based on 84 % BRC assumption: Black male (Source: Census Bureau Staff calculations)

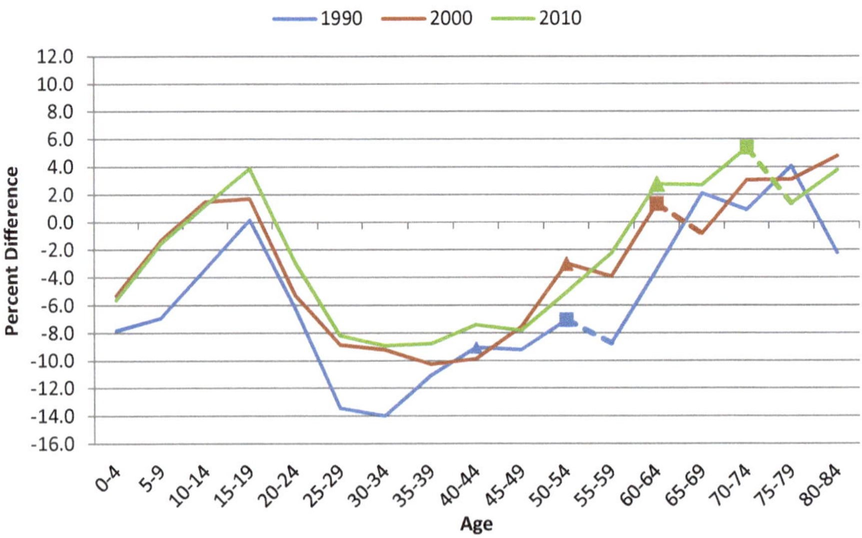

Fig. 8.10 Percent differences between cohort-component-based demographic analysis estimates and 1990, 2000, and 2010 censuses, based on 88 % BRC assumption: Black male (Source: Census Bureau Staff calculations)

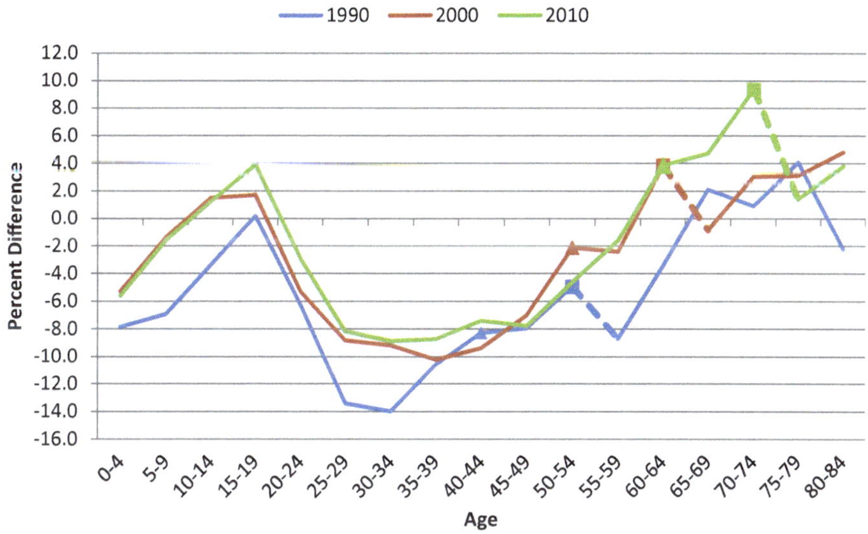

Fig. 8.11 Percent differences between cohort-component-based demographic analysis estimates and 1990, 2000, and 2010 censuses, based on 90 % BRC assumption: Black male (Source: Census Bureau Staff calculations)

Table 8.10 Sex ratios for the black population by census year, age at census, and assumptions about birth registration completeness for males and females

		Assumptions about birth registration completeness		
	Age at census	84 %	88 %	90 %
Census year	(1)	(2)	(3)	(4)
1940	0–4	100.9	101.0	101.0
1950	10–14	100.5	100.5	100.5
1960	20–24	98.0	97.9	97.8
1970	30–34	96.8	96.7	96.6
1980	40–44	95.3	95.0	95.0
1990	50–54	91.9	91.4	91.2
2000	60–64	85.2	84.3	84.0
2010	70–74	76.7	74.8	74.3

Source: U.S. Census Bureau staff calculations
Note: Sex ratio = (male/female)*100

range of completeness (84–90 %) presented in Table 8.9 and the additional assumption that the number of Black females added to compensate for incomplete birth registration would be reduced by the same amount as the Black males under the varying completeness assumptions. The variability is too small to provide any guidance as to the most plausible series, but can assess plausibility of age groups of cohorts.

8.5 Conclusion

The DA estimates are benchmarks for measuring coverage errors in the census. The census in turn serves as the base for producing intercensal population estimates. With multiple observations of coverage across several censuses, it is now possible to judge the quality of the demographic estimates and to identify and correct for anomalous patterns for early cohorts. In this paper, we focused on the Black male cohorts of 1935–1939, but subsequent cohorts appear to exhibit the same patterns. The census count is lower than the estimates for the Black male population.

The data quality of each component used to construct the estimates and the assumptions of the method were analyzed. The analysis shows that the birth registration factors applied to the Black births and the related corrections for infant mortality impact the size of the Black male births. Manipulations of these factors could lower the estimate of the DA population. Immigration and race classification errors probably do not have a significant impact.

If the registration of the Black male birth cohorts of 1935–1939 were more complete than anticipated, the DA estimates of Black men are too high. In all likelihood, the net undercount of adult Black men would still be disproportionally high relative to other race-sex groups, but the net undercount might have been overstated in the past seven censuses. Research efforts leading up to the 2020 DA should revisit the range of uncertainty to be expected in the components. It is the recommendation that the analyses focus on cohorts as well as age or period effects.

Appendix

Table 8.11 Effects of alternative adjustments for under-registration of infants deaths on demographic analysis estimates by race, 1935–1940[a]

Year of birth	DA 2010 estimate of population (1)	Reported deaths			Adjustment for under-registration[b]				Implied difference from 50 %				Difference as % of		
		Male (2)	Female (3)	Total (4)	Equal to (5)	75 % of (6)	50 % of (7)	25 % of (8)	With 100 % (9 = 5–7)	With 75 % (10 = 6–7)	With 25 % (11 = 8–7)	With reported (12 = 4–7)	Current deaths (13 = 9/7)	2010 DA (14 = 9/1)	Reported deaths (15 = 9/4)
White															
1935	1,500,137	56,517	41,614	98,131	106,803	104,495	102,284	100,165	4519	2211	−2119	−4153	4.4	0.3	4.6
1936	1,546,272	57,054	42,674	99,728	108,177	105,933	103,781	101,714	4396	2152	−2067	−4053	4.2	0.3	4.4
1937	1,666,679	55,623	41,647	97,270	105,157	103,068	101,060	99,129	4097	2008	−1931	−3790	4.1	0.3	4.2
1938	1,755,208	54,202	40,468	94,670	101,861	99,963	98,134	96,371	3727	1829	−1763	−3464	3.8	0.2	3.9
1939	1,808,771	50,306	37,758	88,064	94,085	92,504	90,975	89,496	3110	1529	−1479	−2911	3.4	0.2	3.5
Black															
1935	151,617	11,700	9263	20,963	26,569	24,904	23,435	22,130	3314	1469	−1305	−2472	13.4	2.1	15.0
1936	154,887	12,066	9605	21,671	27,321	25,649	24,170	22,852	3151	1479	−1318	−2499	13.0	2.0	14.5
1937	169,682	11,951	9613	21,564	27,016	25,410	23,984	22,710	3032	1426	−1274	−2420	12.6	1.8	14.1
1938	178,560	11,636	9269	20,905	26,008	24,512	23,179	21,983	2829	1333	−1196	−2274	12.2	1.6	13.5
1939	192,520	11,201	8598	19,799	24,359	23,033	21,844	20,772	2515	1189	−1072	−2045	11.5	1.3	12.7

[a]Modified from McDevitt (2001a) Table 1a and Condon (2010) Table 1

[b]Under-registration of deaths using under-registration of births as guideline. Current adjustment uses 50 %.

References

Condon, K. M. (2010). *2010 Demographic analysis technical documentation series: Sensitivity analysis of DA population estimates in 2010 to alternative assumptions about completeness of infant and young children (1–4 years, and 5–9 years) death registration*. U.S. Census Bureau. 2010 Demographic Analysis Technical Documentation Memorandum (A-8). Issued September 22, 2010.

Devine, J., Sink, L., DeSalvo, B., & Cortes, R. (2011). *The use of vital statistics in the 2010 demographic analysis estimates* (Population Division working paper #88). Issued October 2011.

Devine, J., Bhaskar, R., DeSalvo, B., Robinson, J. G., Scopilliti, M., & West, K. K. (2012). *The development and sensitivity analysis of the 2010 demographic analysis estimates* (Population Division working paper #93). Issued February 2012.

McDevitt, T. (2001a). *Sensitivity of estimated Cohort size to alternative assumptions about completeness of infant death registration, and proposed acceptance of current DA assumptions*. Memorandum #7 to DAPE Core Team. Issued June 25, 2001.

McDevitt, T. (2001b). *Sensitivity of births to alternative assumptions about completeness of registration, implied impact on net undercount for census 2000 and proposed change to demographic analysis assumptions on completeness of birth registration*. Memorandum # 5 to DAPE Core Team. Issued June 8, 2001.

National Center for Health Statistics. (1942–1943). *Volume I-Natality*. Department of Health and Human Services, Public Health Service, Center for Disease Control and Prevention.

National Center for Health Statistics. (1946–1947). *Volume I-Natality*. Department of Health and Human Services, Public Health Service, Center for Disease Control and Prevention.

National Center for Health Statistics. (1950–1952). *Volume I-Natality*. Department of Health and Human Services, Public Health Service, Center for Disease Control and Prevention.

National Center for Health Statistics. (1957–1962). *Volume I-Natality*. Department of Health and Human Services, Public Health Service, Center for Disease Control and Prevention.

Passel, J. S. (1991). *Age-period-cohort analysis of census undercount rates for race-sex group, 1940:1980: Implications for the method of demographic analysis*. In Proceedings of the American Statistical Association Social Statistics Sections, 326–331.

Passel, J. S. (1992). *Alternative estimates of black births corrected for under-registration: 1935–1980*. Washington, DC: The Urban Institute.

Passel, J. S., & Robinson, J. G. (1988). Methodology for developing estimates of coverage in the 1980 census based on demographic analysis: Birth and death statistics, 1935–1980. Preliminary Evaluation Results Memorandum No. 112, February 24, 1988 (Revised September 1988). Washington, DC: U.S. Bureau of the Census.

Passel, J. S., Rives, N. W. Jr., & Robinson, J. G. (1977). Estimating the completeness of registration of white births for states: A reassessment of the 1940, 1050, and 1964–68 birth registration tests. Paper presented at the Annual Meeting of the Population Association of America, St. Louis, Missouri, April.

Preston, S. H., Elo, I. T., Foster, A., & Haishan, F. (1998). Reconstructing the size of the African American population by age and sex, 1930–1990. *Demography, 35*(1), 1–21.

Price, D. O. (1947). A check on under-enumeration in the 1940 census. *American Sociological Review, 12*(1), 44–49.

Robinson, J. G. (1991). *Demographic analysis evaluation project D1: Error in the birth registration completeness estimates*. 1990 Decennial Census Preliminary Research and Evaluation Memorandum No. 74. Washington, DC: U.S. Bureau of the Census.

Robinson, J. G. (1997). *The differential undercount of adult black men: Is it a myth?* Internal Census Bureau Memorandum. November 3, 1997.

Robinson, J. G. (2011). *Coverage of population in census 2000 based on demographic analysis: The history behind the numbers* (Population Division working paper #91). Issued October 2011.

Robinson, J. G., Gupta, P. D., & Ahmed, B. (1990). *A case study in the investigation of errors in estimates of coverage based on demographic analysis: black adults aged 34 to 54 in 1980*. Paper presented at the American Statistical Association Annual Meeting, Anaheim, California, August 1990.

Robinson, J. G., Ahmed, B., Gupta, P. D., & Woodrow, K. A. (1993). Estimating coverage in the 1990 United States census based on demographic analysis. *Journal of the American Statistical Association, 88*(423), 1061–1071.

Singh, G. K, & van Dyck, P. C. (2010). *Infant mortality in the United States, 1935–2007: Over seven decades of progress and disparities*. Rockville: U.S. Department of Health and Human Services, Health Resources and Services Administration, Maternal and Child Health Bureau. 2010.

U.S. Bureau of the Census. (1973). *Census of Population and Housing: 1970 Evaluation and Research Program PHC(E)-4. Estimates of coverage of population by sex, race, and age: Demographic analysis*. Washington, DC: U.S. Government Printing Office. Table A12.

U.S. Bureau of the Census. (2001). *Births and deaths evaluation. Census 2000 Demographic Analysis-Population Estimates research project (DAPE)*. Final Executive Summary. Issued November 29, 2001.

West, K. (2012). *Is the elderly a hard to count population? Evidence from the 2010 demographic analysis estimates*. Paper presented at the International Conference on Methods for Surveying and Enumerating Hard-To-Reach Populations, New Orleans, Louisiana, October 3 – November 2, 2012.

West, K. (2014). *An assessment of the cohort-component-based demographic analysis estimates of the population aged 55 to 64 in 2010*. Paper prepared for the Applied Demography Conference in San Antonio, Texas, January 8–10, 2013.

West, K., & Robinson, J. G. (1998). *What do we know about the undercount of children?*. Population Division working paper #39. Issued August 1999.

Whelpton, P. (1950). *Birth and birth rates in the entire United States, 1909 to 1949*. National Center for Health Statistics. Vital statistics special reports, *33*(8).

Chapter 9
Reporting of Race Among Hispanics: Analysis of ACS Data

Howard Hogan

This report is released to inform interested parties of ongoing research and to encourage discussion of work in progress. Any views expressed on the statistical and methodological issues in this report are those of the author and not necessarily those of the US Census Bureau.

Abstract The Census Bureau collects information concerning race and Hispanic origin using two separate questions. Hispanics may report any race or combination of races. The majority select a specific race, however a substantial minority report a nationality, a group or a general category as their "race." Many researchers choose to ignore the reported race among Hispanics, treating Hispanic origin as a category equivalent to race, the justification often being that "race is not a meaningful concepts among Hispanics." This paper examines the reporting of race among Hispanics. It looks at how reported race relates to specific Hispanic origin, nativity and other predictors. It analyzes whether including race adds additional power to explain social outcomes among Hispanics. It also looks at two small but significant groups. One groups is Hispanics who report both a specific race and an additional response. The other group is Black Hispanics. The analysis is based on the 2008–2012 "Five Year" American Community Survey files. By examining race within national group, it allows a better understanding of the separate contribution of race to social analysis. It shows that race is a meaningful variable, but interpretation may be complex.

Keywords Race • Hispnic origin • American community survey

H. Hogan (✉)
US Census Bureau, Washington, DC, USA
e-mail: Howard.R.Hogan@census.gov

D.A. Swanson (ed.), *The Frontiers of Applied Demography*, Applied Demography Series 9, DOI 10.1007/978-3-319-43329-5_9

9.1 Introduction

The U.S. Census Bureau, including the American Community Survey collects data on race and Hispanic origin in two separate questions. "People who identify their origin as Hispanic, Latino, or Spanish may be of any race." (US Census Bureau 2013)

> U.S. federal government agencies must adhere to standards issued by the Office of Management and Budget (OMB) in October 1997, which specify that race and Hispanic origin (also known as ethnicity) are two separate and distinct concepts. These standards generally reflect a social definition of race and ethnicity recognized in this country, and they do not conform to any biological, anthropological, or genetic criteria. The standards include five minimum categories for data on race: "American Indian or Alaska Native," "Asian," "Black or African American," "Native Hawaiian or Other Pacific Islander," and "White." There are two minimum categories for data on ethnicity: "Hispanic or Latino" and "Not Hispanic or Latino." The concept of race reflects self-identification by people according to the race or races with which they most closely identify. Persons who report themselves as Hispanic can be of any race and are identified as such in our data tables. (http://www.census.gov/population/hispanic/about/faq.html)

In spite of this, race data are often ignored with analyzing the Hispanic population. Clearly, when the sample size is small, for example in the Current Population Survey, there is often no choice. However, ignoring race among Hispanics is often done based on a belief that race is not a meaningful concept for Hispanics. Outside the Census Bureau, "Hispanic" is often treated as if it were a distinct race, with the term "White" defined to mean only non-Hispanic White.

Data from the 2008–2012 "5-Year" American Community Survey are sufficiently rich as to allow us to ask whether the reporting of race among Hispanics is more or less random or whether one can gain insights by including this information. While the meaning of reported race among Hispanics presents complex issues, important insights can be gained by analyzing these data.

Some background is important to understanding the data. Most readers are acquainted with the OMB standards on race (OMB 1997), which define five race groups, but allow respondents to choose one or more among these. Each of these is defined as, essentially self-identification, except for American Indians and Alaska Natives. Here, the standards define:

> American Indian or Alaska Native
> A person having origins in any of the original peoples of North and South America (including Central America), and who maintains tribal affiliation or community attachment.

Because of this, there is no category for those who might see their origins, in whole or part, in the original peoples of South and Central America but who do not maintain tribal affiliation or attachment an indigenous community. The Census Bureau does not emphasis this rule in it materials. However, the terms used on the questionnaire may not seem appropriate to many people with South or Central American racial origin:

> American Indian or Alaska Native: Print name of enrolled or principal tribe

Because of this, it is not possible from the data to identify those who might see their origins in the original peoples of South and Central America but who do not have an enrolled or principal tribe or who do not identify with the term "American Indian." Many Hispanics with indigenous or mixed racial heritage, for example those who might identify with the term "Mestizo" are difficult to identify from these official categories.

In addition to the five specific race categories, the Census and ACS allow respondents to write in other responses under "Some Other Race." It is important to recognize that this is a catch-all, residual, "not-elsewhere-classified" category. It should not be interpreted as representing one particular and distinct "other race." Different respondents wrote in different words representing different understandings of race and their individual background. Most responses listed under "Some Other Race" that did not use a South American, Central American or other "Hispanic" term, were reassigned by the Census Bureau to a specific "OMB" race, so that nearly all responses left in this rubric are Hispanic terms and groups.

The 2010 Census and the ACS collect these data in essentially the same manner. For most of the country, a paper questionnaire is first mailed to the housing unit asking for self-response and mail back. Housing units which do not respond are followed up, largely by personal visit in both systems. The difference is that the 2010 Census used a paper questionnaire and a temporary field staff. The ACS uses a computer assisted personal interview (CAPI) and a permanent field staff.

It is important to note for the purposes of this analysis, that the ACS results are statistically controlled to the annual population estimates, which are based on the 2010 Census. The data are controlled, at the local level to agree on "Hispanic Origin," but not on race within Hispanic Origin. Thus, when analyzing race reporting among Hispanics, the reporting is essentially not controlled to these 2010 Census based estimates. Finally, the data reported here reflect standard coding, editing, weighting and imputation of the ACS data. "Reported" then is "reported" by the Census Bureau and not necessarily by the respondent.

Table 9.1 presents the distribution by race among Hispanics in the 2010 Census and the ACS. Three facets are evident. "White Alone" is by far the biggest race group in both the Census and the ACS. However, in both collections, a substantial proportion provided responses to the race question other than one (or more) of the five specific categories. Interestingly, more Hispanics are reported as "White Alone" in the ACS than in the 2010 Census. The reasons are unclear, but the better training of the ACS interviewers is a possible explanation. In the remainder of this paper, we look below these broad categories to see what can be learned, looking in detail at the ACS.

Table 9.1 Hispanics by standard race groups: 2010 census vs ACS

	2010 Census		2008–2012 ACS			
	Number	Percent	Number	Percent	MOE	Difference
Hispanic or Latino	50,477,594		50,545,275			
White alone	26,735,713	53.0	32,394,938	64.1	0.1	−11.1
Black or African American alone	1,243,471	2.5	1,039,257	2.1	0.0	0.4
American Indian and Alaska Native alone	685,150	1.4	478,334	0.9	0.0	0.4
Asian alone	209,128	0.4	167,001	0.3	0.0	0.1
Native Hawaiian & Other Pacific Islander alone	58,437	0.1	34,339	0.1	0.0	0.0
Some Other Race alone	18,503,103	36.7	14,198,178	28.1	0.1	8.6
Two or More Races	3,042,592	6.0	2,233,228	4.4	0.0	1.6

MOE: indicates a measure of sampling error using a 90 % confidence interval

9.2 Race by Detailed Hispanic Categories

Hispanics in the United States constitute a diverse group. However, the ACS collects and allows analysis of detailed Hispanic groups. Of most interest to this analysis are the different responses within the Hispanic category. Given the very different patterns of immigration, colonization, intermarriage and culture represented among Hispanics, it should not be surprising that different groups view race differently. The ACS reports the detailed Hispanic response. Many of these are clear and, additionally, large enough to analyze separately. A few categories needed to be combined. The collapsing standard in many Census Bureau publications will not serve this analysis well. For this analysis, I have created 17 groups, based on a general understanding of the culture and history. Other groupings are certainly possible, and may be better. Further, no special merit should be attributed to the short-hand variable name given to each group. The important point for this analysis is that the groups were constructed independently of the ACS data reported in the following sections.

Total	
MEXICAN	Mexican, Mexicano, La Raza, Mexican American Indian, etc.
PUERTO RICAN	PUERTO RICAN: write-in or check box
CUBAN	CUBAN: write-in or check box
COLOMBIAN	COLOMBIAN
DOMINICAN	DOMINICAN
ECUADORIAN	ECUADORIAN
GUATEMALAN	GUATEMALAN
HONDURAN	HONDURAN
PERUVIAN	PERUVIAN
SALVADORAN	SALVADORAN

(continued)

SO SOUTH AMER	Argentine, Chilean, Uruguayan
OTHER SO AMER	Bolivian, Paraguayan, Venezuelan, etc.
OTHER CENT AMER	Costa Rican, Nicaraguan, Panamanian, Canal Zone
SPANIARD	Spaniard, Castilian, Spanish Basque, etc.
GENERIC Hispanic	"Hispanic"
GENERIC Spanish	"Spanish"
ALL OTHER	Latin, Latino, Californio, Spanish American, Mestizo, Multiple origins, Other Spanish/Hispanic n.e.c.

Appendix A gives the detailed codes associated with each of these groups.

Table 9.2 gives the estimated totals for each of these groups.

Additionally, it is necessary to collapse race groups to maintain sample size. Four groups were chosen:

1. White Alone
2. Black Alone
3. "Some Other Race" Alone
4. "Some Other Race" in combination
5. All other.

Again, other groupings are possible. However, these allow us to focus on some interesting issues in race reporting. Table 9.3 gives the totals for these five groups.

Table 9.4 presents the percent by race group within each of the Hispanic Origin groupings. This table is sorted by the estimated percentage White Alone. It is clear that some groups (Cuban, South-South-American, Spaniard) have a high percent reported as White alone, where Dominicans have a low percentage reported as White Alone. Dominicans and "Other Central Americans" also report a high percent Black Alone. Similar findings have been reported from other data sets (See Humes and Hogan 2013). This table, together with a general understanding of the histories of the donor countries, should convince us that reporting of race among Hispanics is neither random nor arbitrary but reflects complex historical, national and cultural understandings. In the next section, we seek to better understand these findings by relating them to age and place of birth.

Table 9.2 Detailed Hispanic groups: counts and percents

	Number	MOE	Percent	MOE
Total	50,545,275		100	
Mexican	32,695,701	188,399	64.7	0.58
Puerto Rican	4,739,207	2418	9.4	0.05
Cuban	1,821,626	1561	3.6	0.09
Colombian	973,448	992	1.9	0.10
Dominican	1,492,849	2564	3.0	0.17
Ecuadorian	636,933	582	1.3	0.09
Guatemalan	1,143,197	2132	2.3	0.19
Honduran	696,668	760	1.4	0.11
Peruvian	578,583	738	1.1	0.13
Salvadoran	1,827,447	4489	3.6	0.25
So south Amer	433,741	296	0.9	0.07
Other So Amer	363,386	467	0.7	0.13
Other Cent Amer	694,756	1343	1.4	0.19
Spaniard	689,392	113	1.4	0.02
Generic Hispanic	512,422	616	1.0	0.12
Generic Spanish	498,282	688	1.0	0.14
All Other	747,637	1280	1.5	0.17

Table 9.3 Hispanics by race groups

		Number	MOE	Percent	MOE
Total		50,545,275			
White Alone	White-A	32,394,938	21,882	64.1	0.07
"Some other Race" Alone	SOR-A	14,198,178	8985	28.1	0.06
Black Alone	Black-A	1,039,257	208	2.1	0.02
"Some other Race" in Combination	SOR-C	1,272,046	281	2.5	0.02
All other	ALL OTHER	1,640,856	409	3.2	0.02

9.3 Race by US Born and Age

Many immigrants to the United States, non-Hispanic as well as Hispanic, are confused by the race concepts of American (U.S.) society in general and Federal surveys in particular. Faced with a questionnaire that lists several nationalities (e.g. Chinese, Asian Indian) many people assume that the government wishes to know their place of origin.[1] However, one might hypothesis that those who have lived in the US all their life may be better acquainted with the racial concepts used in Federal surveys.

[1] The Census Bureau's editing procedures reassign most national origin responses to a specific race. For example, people who write in Portuguese or Iranian are re-coded as White; people who write in Haitian or Jamaican are recoded as black. Hispanic national write-in responses are not re-coded to a specific race.

Table 9.4 Hispanic group by race group

Group	White-A	SOR-A	Black-A	SOR-C	ALL OTHER
CUBAN	87.6	5.6	3.8	1.3	1.7
SO SOUTH AMER	84.4	11.2	0.5	2.4	1.6
SPANIARD	75.5	8.3	1.2	3.6	11.3
COLOMBIAN	75.2	18.4	1.5	2.8	2.1
OTHER SO AMER	73.4	17.9	2.0	3.5	3.3
MEXICAN	66.1	28.3	0.8	2.1	2.8
GENERIC spanish	65.9	16.7	1.1	5.0	11.2
Total	**64.1**	**28.1**	**2.1**	**2.5**	**3.2**
PERUVIAN	62.5	28.7	0.9	3.8	4.2
OTHER CENT AMER	60.3	21.8	10.3	4.1	3.5
ECUADORIAN	58.6	34.8	1.2	3.0	2.5
All Other	57.5	27.2	4.1	4.5	6.7
PUERTO RICAN	57.3	27.0	7.0	3.8	4.9
HONDURAN	55.5	35.0	4.4	2.4	2.7
SALVADORAN	53.3	41.2	0.7	2.5	2.3
GUATEMALAN	50.7	41.3	1.2	2.7	4.2
GENERIC Hispanic	46.8	43.0	1.4	6.3	2.5
Dominican	34.3	46.3	11.6	3.4	4.3

Sorted by Percent White alone

Table 9.5 presents the percentage who provided only a non-standard race response in the "Some Other Race" space. It classifies the people by whether they were born in the United States versus all others, including all immigrants and those born in Puerto Rico or abroad of US citizens. Since many US born children may have had their race reported by a foreign born adult, the tabulation is repeated for only those 18 and over. The last column give the difference in those not choosing a standard race between US born and all other.

The general trend is clear. Adult US born Hispanics are less likely to write in only a non-standard response. The exceptions, Cuban and Spaniard, are groups with a low percentage of both groups reporting a non-standard race.

It is interesting to note that in the mainland, 27 % of Puerto Ricans report some other race and 57 report white alone. In Puerto Rico itself, only 10 % of the population reports as "Some Other Race." 71 % report as "White Alone" and another 11 % report "White in Combination." (PRCS: DP-05 2008–2012).

9.4 Social and Economic Conditions

This section examines several social and economic variables related to Hispanic origin and race. In each case the analysis will cross classify by origin group. Additionally, we will largely focus on the adult US Born population. This will

Table 9.5 Percent not reporting a standard racy by US born

	Total	US BORN Total	All other Total	US BORN Adult	All other Adult	Adult Difference	MOE difference
Total	28.1	25.1	32.5	24.0	32.5	−8.6	0.2
MEXICAN	28.3	25.1	34.0	23.9	34.1	−10.1	1.0
PUERTO RICAN	27.0	25.6	30.1	26.9	30.1	−3.2	0.2
Cuban	5.6	5.8	5.4	6.2	5.4	0.8	0.6
Colombian	18.4	15.9	19.7	17.5	19.7	−2.1	0.5
DOMINICAN	46.3	43.2	48.4	42.2	48.2	−6.0	1.2
ECUADORIAN	34.8	30.8	36.9	29.4	37.1	−7.7	1.3
GUATEMALAN	41.3	36.8	43.4	36.0	43.6	−7.6	1.9
HONDURAN	35.0	30.6	37.0	27.1	37.0	−9.9	1.8
PERUVIAN	28.7	21.8	31.8	22.8	31.6	−8.8	2.2
SALVADORAN	41.2	38.6	42.8	40.7	42.9	−2.2	2.0
SO SOUTH AMER	11.2	9.6	12.0	10.6	11.9	−1.3	1.3
OTHER SO AMER	17.9	15.0	19.0	16.0	19.0	−3.0	1.4
OTHER CENT AMER	21.8	18.3	24.1	18.2	24.2	−6.0	2.2
SPANIARD	8.3	8.6	6.8	9.2	6.9	2.3	1.4
GENERIC Hispanic	43.0	41.0	57.3	42.0	57.4	−15.4	2.5
GENERIC SPANISH	16.7	15.6	25.4	15.6	25.1	−9.5	2.1
All other	27.2	24.9	34.0	22.5	34.0	−11.5	1.5

MOE: indicates a measure of sampling error using a 90 % confidence interval

allow us to consider the information value of the race variable for Hispanics. In analyzing these data, one must keep in mind that the direction of the causality is not only one way. People who are perceived to be of a certain race in American society may gain or lose economic and social benefits. However, those who are relatively well off or relatively disadvantaged may, as a consequence, see themselves and be perceived as members of a particular race.

Table 9.6 presents the percentage of US Born adults who speak only English for those reporting White Alone and those who provide a response in "some other race." Those reporting White Alone are more likely to speak only English. The pattern is consistent with the exception of "South South Americans."

Table 9.7 presents the percentage of US Born adults who have education less than High School Diploma., again cross tabulated by group. It also presents the difference between the percentage among "White Alone" and those responding only in "Some Other Race." Overall, there is a 4 %age point difference. The difference is positive for all groups, but not statistically significant for several of

Table 9.6 Speaks English only

	White alone	Only some other race	Difference	MOE
Total	43.4	31.2	−12.2	0.7
MEXICAN	74.4	32.5	−41.9	5.5
PUERTO RICAN	41.7	27.9	−13.8	1.2
CUBAN	46.4	34.8	11.6	1.2
COLOMBIAN	36.6	21.0	−15.6	2.6
DOMINICAN	30.9	9.7	−21.2	8.2
ECUADORIAN	20.4	14.5	−5.9	2.4
GUATEMALAN	30.9	13.6	−17.4	11.2
HONDURAN	26.1	15.8	−10.3	9.7
PERUVIAN	30.9	25.6	−5.3	5.5
SALVADORAN	43.2	11.8	−31.4	13.1
SO SOUTH AMER	22.4	33.0	10.7	8.3
OTHER SO AMER	46.8	29.6	−17.2	19.3
OTHER CENT AMER	42.4	32.5	−9.9	9.6
SPANIARD	42.6	61.2	18.6	3.1
GENERIC Hispanic	47.2	40.6	−6.5	7.7
GENERIC Spanish	70.6	49.9	−20.7	7.1
ALL OTHER	51.9	38.5	−13.5	3.7

MOE: indicates a measure of sampling error using a 90 % confidence interval

the groups. For example, the difference among those reporting "Mexican," the largest group, is 2.9 percentage points.

Table 9.8 presents the data for the percent in poverty. The universe here is US Born Adults in households. Overall, 3 percentage points more Hispanics who report only within the "Some other race" category are in poverty than those reporting White Alone. For some groups, the difference is quite large. However, again, the difference for other groups is small and not statistically significant. In two groups, the difference is negative, but well within sampling error. However, remember that the point of this analysis is not to show that those of a particular race are consistently advantaged or disadvantaged. Rather it is simply to show that race is a variable that can be usefully considered.

Finally, Table 9.9 presents the percentage of women who had a birth in the past year. Here the universe is US born women 18–50. The differences for most groups are small and not statistically significant. Race has only modest explanatory power for this variable. This may be attributed to that fact that the event (having a child in the previous year) is relatively rare, the group "at risk" (here restricted to US Born Women) is relatively small, and the differences, if they exist, are probably moderate.

Of course, many of the US born adults are children of immigrants. Since the ACS lacks the ability to separate second generation adults from other native born population, this group cannot be separated. However, as a possibly related

Table 9.7 Percent less than high school diploma among US born adults

						SOR-A - White-A		
	White-A	SOR-A	BLACK-A	SOR-C	ALL OTHER	DIFF	MOE	DIFF not Significant
Total	18.9	22.9	17.7	16.7	17.0	4.0	0.6	
MEXICAN	21.1	24.0	20.7	17.4	19.2	2.9	1.3	
PUERTO RICAN	15.6	22.8	19.2	18.4	15.1	7.2	1.1	
CUBAN	8.8	12.8	12.6	9.6	10.8	4.0	2.8	
COLOMBIAN	7.4	11.0	15.7	9.8	5.3	3.6	6.8	X
DOMINICAN	13.4	17.3	13.1	14.5	11.4	3.9	2.0	
ECUADORIAN	8.0	12.4	12.2	8.0	5.3	4.4	8.5	X
GUATEMALAN	17.8	21.4	7.1	17.5	14.6	3.7	5.8	
HONDURAN	15.1	24.4	17.1	12.6	24.3	9.3	5.3	
PERUVIAN	7.1	11.0	14.6	7.5	6.9	3.9	10.6	X
SALVADORAN	18.2	18.7	15.4	12.6	14.1	0.6	7.2	X
SO SOUTH AMER	5.8	7.8	21.0	7.0	5.6	2.0	17.7	X
OTHER SO AMER	7.3	8.9	4.4	6.3	5.4	1.6	6.1	X
OTHER CENT AMER	9.2	10.7	9.2	9.3	11.5	1.4	2.6	X
SPANIARD	9.7	14.2	13.2	13.1	11.4	4.5	5.3	X
GENERIC Hispanic	23.8	27.0	24.0	22.5	25.3	3.3	7.3	X
GENERIC Spanish	13.8	22.1	15.3	15.1	13.7	8.2	6.6	
ALL OTHER	17.9	21.0	22.5	19.8	19.1	3.2	3.8	X

MOE: indicates a measure of sampling error using a 90 % confidence interval

Table 9.8 Percent in poverty among US born adults by race and Hispanic origin groups

						SOR-A - White-A		
	White-A	SOR-A	BLACK-A	SOR-C	ALL OTHER	DIFF	MOE	DIFF not Significant
Total	15.8	18.8	24.9	16.6	19.0	3.1	0.9	
MEXICAN	16.2	17.6	24.7	15.4	18.8	1.4	1.7	X
PUERTO RICAN	17.5	25.4	27.9	21.7	21.7	7.9	1.5	
CUBAN	10.8	21.0	24.3	24.2	16.9	10.1	4.3	
COLOMBIAN	10.3	14.7	21.8	11.6	15.1	4.4	9.6	X
DOMINICAN	17.6	20.9	23.4	24.6	20.8	3.3	3.0	
ECUADORIAN	10.8	14.4	13.3	11.1	14.4	3.6	10.9	X
GUATEMALAN	16.1	17.4	22.3	13.2	26.8	1.4	11.1	X
HONDURAN	19.3	23.4	19.8	8.9	16.7	4.1	6.7	X
PERUVIAN	10.5	12.7	9.7	11.7	15.2	2.3	11.2	X
SALVADORAN	14.2	15.0	26.5	13.1	11.7	0.8	10.8	X
SO SOUTH AMER	9.7	8.8	33.9	8.4	18.3	–0.8	25.4	X
OTHER SO AMER	12.6	12.2	16.8	12.8	12.7	–0.5	12.3	X
OTHER CENT AMER	13.3	16.6	13.8	11.3	19.4	3.4	3.9	X
SPANIARD	11.3	13.7	16.6	14.4	16.1	2.5	7.1	X
GENERIC Hispanic	19.6	24.0	32.5	18.6	22.3	4.4	9.9	X
GENERIC Spanish	13.3	16.7	22.1	14.7	17.8	3.4	9.3	X
ALL OTHER	15.4	19.9	24.6	19.0	19.7	4.6	4.9	X

Table 9.9 Percent in birth last year among US born women by race and Hispanic origin groups

						SOR-A - White-A		
	White-A	SOR-A	BLACK-A	SOR-C	ALL OTHER	DIFF	MOE	DIFF not Significant
Total	7.4	7.8	7.8	7.3	6.9	0.4	0.6	X
MEXICAN	7.9	8.3	9.7	7.9	7.1	0.5	1.2	X
PUERTO RICAN	6.9	7.1	8.0	7.4	7.7	0.2	0.9	X
CUBAN	6.0	5.7	5.7	3.7	6.0	−0.2	2.5	X
COLOMBIAN	6.1	5.9	4.9	5.4	7.2	−0.2	5.1	X
DOMINICAN	7.4	6.4	7.7	7.2	6.0	−1.0	1.9	X
ECUADORIAN	6.5	6.8	0.0	8.5	5.3	0.3	1.9	X
GUATEMALAN	6.2	8.3	5.3	7.1	4.8	2.1	5.4	X
HONDURAN	6.3	8.8	7.3	8.9	8.1	2.5	4.3	X
PERUVIAN	5.2	6.4	11.6	7.6	3.9	1.2	11.3	X
SALVADORAN	8.0	7.6	10.1	5.3	3.8	−0.4	7.3	X
SO SOUTH AMER	6.4	3.8	26.7	2.2	8.5	−2.5	22.9	X
OTHER SO AMER	5.6	5.9	5.5	5.7	5.5	0.2	8.0	X
OTHER CENT AMER	4.9	6.6	5.7	2.8	6.8	1.6	2.6	X
SPANIARD	5.1	5.1	2.2	4.4	6.8	0.0	3.1	X
GENERIC Hispanic	7.6	6.3	0.7	9.5	3.4	−1.2	2.1	X
GENERIC Spanish	5.8	4.7	14.1	4.5	5.7	−1.1	9.4	X
ALL OTHER	5.2	6.6	3.6	5.3	6.1	1.4	2.5	X

Table 9.10 Percent less than high school diploma among US born adults who speak english only by race and Hispanic origin groups

						SOR-A - White-A		
	White-A	SOR-A	BLACK-A	SOR-C	ALL OTHER	DIFF	MOE	DIFF not Significant
Total	13.9	18.3	16.8	14.6	14.6	4.4	0.8	
MEXICAN	15.2	18.8	19.5	14.3	16.0	3.6	1.5	
PUERTO RICAN	13.1	18.8	17.8	17.6	13.2	5.7	1.6	
CUBAN	8.3	11.3	12.7	10.2	12.2	3.0	3.9	X
COLOMBIAN	6.7	9.0	11.8	10.4	4.7	2.2	8.8	X
DOMINICAN	10.0	12.4	10.6	12.6	10.0	2.4	4.2	X
ECUADORIAN	8.9	13.4	18.5	9.0	5.5	4.5	16.7	X
GUATEMALAN	13.9	19.0	5.2	22.2	10.7	5.0	8.3	X
HONDURAN	10.9	11.9	10.9	15.0	17.0	1.1	7.7	X
PERUVIAN	5.4	12.8	14.3	9.9	8.4	7.4	17.2	X
SALVADORAN	15.3	20.4	20.4	16.9	18.2	5.2	14.4	X
SO SOUTH AMER	5.8	4.6	10.9	7.6	8.9	−1.2	18.2	X
OTHER SO AMER	6.7	5.9	4.9	9.6	3.2	−0.8	7.8	X
OTHER CENT AMER	5.7	8.5	8.5	11.0	13.3	2.8	3.6	X
SPANIARD	8.2	10.9	12.3	11.7	10.5	2.7	5.8	X
GENERIC Hispanic	18.3	21.5	22.0	17.8	22.8	3.2	9.4	X
GENERIC Spanish	10.8	15.2	14.8	12.9	11.5	4.4	7.6	X
ALL OTHER	14.9	17.3	22.1	16.1	16.3	2.3	4.7	X

Table 9.11 Percent in poverty among US born adults who speak English only by race and Hispanic origin groups

						SOR-A - White-A		
	White-A	SOR-A	BLACK-A	SOR-C	ALL OTHER	DIFF	MOE	DIFF not Significant
Total	14.5	17.5	24.6	15.8	18.9	3.1	0.5	
MEXICAN	14.7	16.4	24.1	14.4	18.6	1.7	0.5	
PUERTO RICAN	15.8	22.6	27.5	21.3	21.7	6.7	1.4	
CUBAN	13.0	25.6	23.6	23.3	17.2	12.7	5.5	
COLOMBIAN	12.1	16.1	31.1	14.2	15.5	4.0	6.4	X
DOMINICAN	15.2	21.2	25.7	32.8	19.2	6.0	5.8	
ECUADORIAN	13.7	16.6	17.2	16.1	22.6	2.9	8.0	X
GUATEMALAN	14.5	16.9	20.3	13.1	33.5	2.4	7.2	X
HONDURAN	17.1	20.1	16.7	10.4	18.9	3.0	10.3	X
PERUVIAN	9.4	18.4	1.4	16.1	15.4	9.0	7.7	
SALVADORAN	13.9	15.6	18.8	14.3	17.1	1.7	5.0	X
SO SOUTH AMER	11.4	8.7	44.4	14.8	25.6	**−2.7**	6.9	X
OTHER SO AMER	15.0	9.9	21.7	12.2	8.0	**−5.1**	8.9	X
OTHER CENT AMER	12.7	17.2	13.7	11.3	20.8	4.5	5.5	X
SPANIARD	10.9	13.1	16.7	13.4	15.5	2.2	2.6	X
GENERIC Hispanic	17.9	24.0	34.1	15.8	20.9	6.1	2.9	
GENERIC Spanish	12.4	15.6	18.8	14.9	18.0	3.2	2.9	
ALL OTHER	15.3	18.6	26.2	14.6	19.9	3.3	3.1	

Universe: household population only
MOE: indicates a measure of sampling error using a 90 % confidence interval

surrogate, we can analyze the data among US born adults who speak English only. This group would seem to the most culturally aware of U.S. racial concepts.

Table 9.10 gives the percent with less than High School diploma among US born adults who speak English only. Overall, there is a 4.4 percentage point difference between those reporting only in the "some other race" box and those reporting white alone. Most of the differences are positive, although not statistically significant. A few are negative but also not statistically significant. Again, remember that the point of this analysis is not to show that those of a particular race are consistently advantaged or disadvantaged. Rather it is simply to show that race is a variable that can be usefully considered.

Table 9.11 presents the data on poverty for the same group. Those reporting only in the "some other race" box are more likely to be in poverty. Differences that are clearly statistically significant also measured for the Mexican, Puerto Rican and Cuban groups, as well as for the group responding with generic Hispanic terms. The differences for the other groups are generally "positive" (lower poverty for the white alone), but not statistically significant. A few are "negative" but also not significant.

In summary, racial differences in social and economic outcomes can be detected for Hispanics. These differences cannot be explained by Hispanic national or other response categories.

9.5 "Some Other Race" in Combination

A small group of particular interest is those Hispanics who reported a specific race as well as writing in a response in the "Some Other Race" box. Some of these people may be trying to report that, say, their mother was Non-Hispanic white, while their father was a Mexican of Central American indigenous origin. In other words, they were seeking to report two distinct races. On the other hand, they may simply be reporting a race and a national origin, say white Cuban. The total for "Some other Race in Combination" in the ACS is relatively small, around 1.3 million. When cross tabulated by the Hispanic groups used above, the numbers become quite small indeed. The only groups large enough for separate analysis are those reporting either Mexican or Puerto Rican origins.

Figure 9.1 presents box plots of the percent in that speak only English by race groups, with the units of analysis being the individual Hispanic sub-groups. Although one might expect the distribution of the "in combination" groups to lie between the White alone and the "some other race" alone groups, this is not the case. (Of course, not all of the "in combination" are with white.) There is a general tendency for those who report "in combination" to be more likely to speak only English. Table 9.12 breaks out the data for the total and the three larger groups (Mexican, Puerto Rican and All Other). It also computes the differences between "some other race" alone and white alone and the combination group. It is clear from this table that the "in combination" group is distinct and not particularly similar to

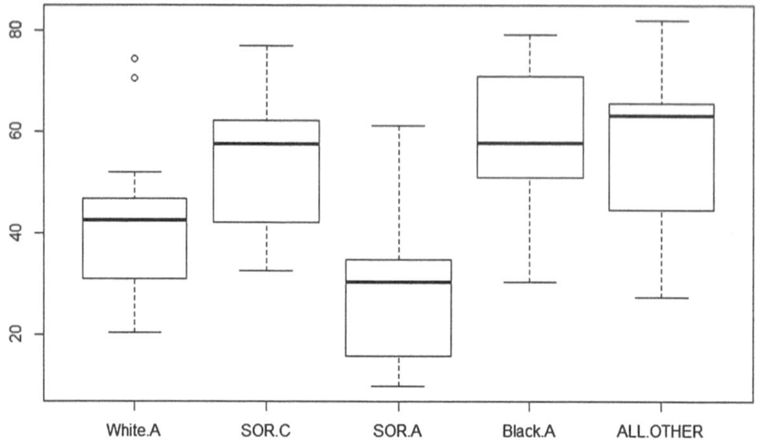

Fig. 9.1 Boxplot of the percent in that speak only English by race groups

either the White alone or the "some other race" alone groups, at least with respect to this variable.

Figure 9.2 repeats this analysis for those reporting less than a high school diploma. Here, the differences between the three groups is smaller. Table 9.13 presents the figures. The results are mixed. For the total and Mexican, the percentage is not "in between" white and "some other race "alone. For the other groups, it does lie between.

Figure 9.3 presents the data for poverty. It suggests that the "in combination" group is distinct and perhaps somewhat closer to the White alone category than the "some other race" alone category. Table 9.14 presents the numeric analysis

Altogether, this analysis suggests than when sample size is not sufficient to form a separate group, which will often be the case, those reporting "some other race" in combination be included with the other specific race mentioned, rather than with those reporting "some other race" alone.

9.6 Black Hispanics

Another group of interest is Black Hispanics. When sample size is limited, many researchers face the question of whether to combine Black Hispanics with other Hispanics or with other Blacks. Blacks constitute 3.4 percent of Hispanics and Hispanics constitute 4.0 % of Blacks, including both those who report only Black and those who report Black in combination. Dominicans and Panaians include a significant percent reporting Black. However the percent for other groups can be quite small. Here we will look at only the totals, regardless of Hispanic sub-group. As will be clear, the answer depends on the research question being asked.

Table 9.12 Speaks English only US born adults by race and Hispanic origin groups

						White-A - SOR-A		SOR-C - SOR-A	
	White-A	SOR-A	BLACK-A	SOR-C	ALL OTHER	Diff	MOE	Diff	MOE
Total	43.4	31.2	61.7	60.9	65.6	12.2	0.7	29.7	0.7
MEXICAN	41.7	32.5	74.9	62.3	65.3	9.2	1.2	29.7	0.9
PUERTO RICAN	46.4	27.9	58.6	60.8	64.0	18.4	1.2	32.8	1.7
CUBAN	47.8	28.6	55.3	58.0	67.3	19.2	1.3	29.4	1.4

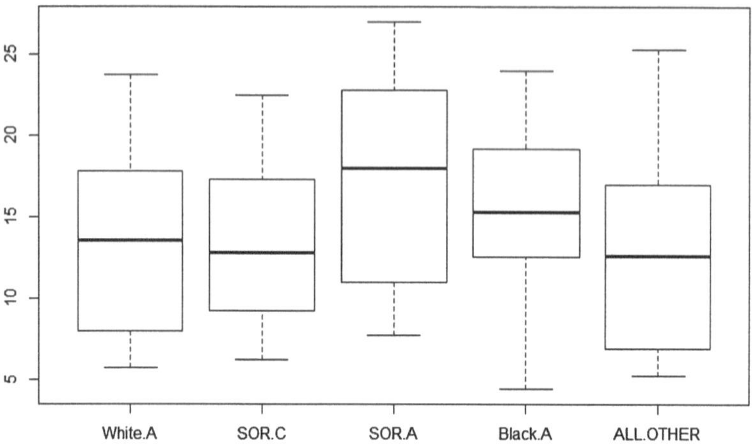

Fig. 9.2 Boxplots of the percent with less than high school by race groups

Table 9.13 High School Diploma or more: US born adults by race and Hispanic origin groups

	Comparisons with Some Other Race in Combination								
						White-A - SOR-A		SOR-C – SOR-A	
	White-A	SOR-A	BLACK-A	SOR-C	ALL OTHER	Diff	MOE	Diff	MOE
Total	81.1	77.1	82.3	83.3	83.0	4.0	0.6	6.2	0.6
Mexican	78.9	76.0	79.3	82.6	80.8	2.9	1.3	6.6	0.8
PUERTO RICAN	84.4	77.2	80.8	81.6	84.9	7.2	1.1	4.4	1.5
All Others	87.5	80.9	86.2	85.7	86.6	6.6	1.0	4.7	1.1

Table 9.15 presents the data on those who speak English only. This table is for the all those in the ACS for whom the question is asked. As would be expected, an overwhelming percentage of non-Hispanic Blacks speak only English, but only about a quarter of Hispanics speak only English. Black Hispanics are far less likely to speak only English than are other Blacks, but the percentage is higher than for non-Black Hispanics.

Table 9.16 presents data on the percent with less than a high school diploma. On this measure Black Hispanics resemble more closely non-Black Hispanics than they do non-Hispanic Blacks.

Table 9.17 is perhaps the most interesting. It presents the data on poverty. Poverty among Black Hispanics is higher than among either non-Hispanic Blacks or non-Black Hispanics. This may be attributable to their distinct origin (Panama, Dominican Republic, etc.) or perhaps more recent immigration status, or perhaps

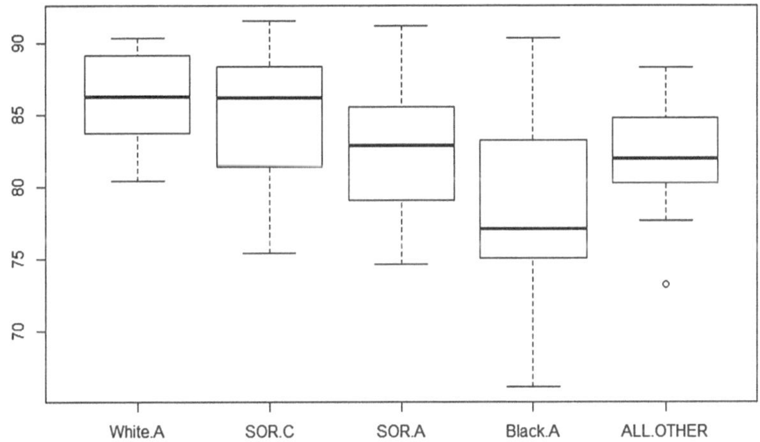

Fig. 9.3 Boxplots of the in poverty by race groups

Table 9.14 Poverty US born adults by race and Hispanic origin groups

	Comparisons with Some Other Race in Combination								
						SOR-A - White-A		SOR-A - SOR-C	
	White-A	SOR-A	BLACK-A	SOR-C	ALL OTHER	Diff	MOE	Diff	MOE
Total	15.8	18.8	24.9	16.6	19.0	3.1	0.9	2.3	0.7
Mexican	16.2	17.6	24.7	15.4	18.8	1.4	1.7	2.2	0.9
PUERTO RICAN	17.5	25.4	27.9	21.7	21.7	7.9	1.6	3.7	2.1
All Others	13.2	18.6	21.5	16.2	17.8	5.4	1.5	2.4	1.5

Table 9.15 Speaks English only

	TOTAL	YES	NO	% YES	MOE
Not Hispanic Black	37,015,414	34,422,583	2,592,831	93.0	0.035
Black Alone	35,054,970	32,607,790	2,447,180	93.0	0.036
Black in Combo	1,960,444	1,814,793	145,651	92.6	0.155
Hispanic not Black	43,994,926	10,604,302	33,390,624	24.1	0.053
Total Black Hispanics	1,443,450	674,428	769,022	46.7	0.343
Hispanics Black Alone	922,327	407,103	515,224	44.1	0.428
Hispanics Black in Combo	521,123	267,325	253,798	51.3	0.573

due to the combined social and economic disadvantages of race and immigrant status.

Race and ethnicity are both relevant in many legal issues, such as voting rights. It is therefore useful to examine the characteristics of Black Hispanics among the US Citizen voting age population. Unlike earlier tables in this paper, citizen includes

Table 9.16 High school diploma or more

	TOTAL	YES	NO	% YES	MOE
Not Hispanic Black	38,341,801	14,936,867	23,404,934	39.0	0.1
Black Alone	36,182,607	13,664,699	22,517,908	37.8	0.1
Black in Combo	2,159,194	1,272,168	887,026	58.9	0.3
Hispanic not Black	45,979,506	25,033,314	20,946,192	54.4	0.1
Total Black Hispanics	1,542,139	837,437	704,702	54.3	0.4
Hispanics Black Alone	969,401	482,501	486,900	49.8	0.5
Hispanics Black in Combo	572,738	354,936	217,802	62.0	0.6

Table 9.17 Percent in poverty: blacks and Hispanics

	TOTAL	YES	NO	% NO	MOE
Not Hispanic Black	38,504,726	28,352,597	10,152,129	26.4	0.1
Black Alone	36,136,442	26,586,463	9,549,979	26.4	0.1
Black in Combo	2,368,284	1,766,134	602,150	25.4	0.3
Hispanic not Black	47,873,498	36,409,806	11,463,692	23.9	0.1
Total Black Hispanics	1,633,071	1,176,178	456,893	28.0	0.4
Hispanics Black Alone	997,641	711,620	286,021	28.7	0.5
Hispanics Black in Combo	635,430	464,558	170,872	26.9	0.6

Table 9.18 Speaks English only or very well among adult US citizens

	TOTAL	YES	NO	% NO	MOE
Not Hispanic Black	27,145,257	26,802,615	342,642	1.26	0.02
Black Alone	26,175,244	25,847,730	327,514	1.25	0.02
Black in Combo	970,013	954,885	15,128	1.56	0.14
Hispanic not Black	20,938,674	16,281,291	4,657,383	22.24	0.10
Total Black Hispanics	747,574	617,216	130,358	17.44	0.48
Hispanics Black Alone	522,063	427,798	94,265	18.06	0.58
Hispanics Black in Combo	225,511	189,418	36,093	16.00	0.84

those born in the U.S., naturalized citizens and those born in Puerto Rico and other territories.

Speaking English less than very well is a characteristic important in voting rights, notably the provision of non-English ballots. Table 9.18 presents these data. Among citizens, the proportion of Black-Hispanics who speak English less than very well is lower than that for non-Black Hispanics, but still closer to that group than to non-Hispanic Blacks.

Table 9.19 presents the proportion of adult US Citizen Black Hispanics with less than a High School diploma. It lies between that for the two other groups.

Finally, Table 9.20 presents the data for the percent in poverty. Here, Black-Hispanics are more similar to non- Hispanic Blacks than to non-Black Hispanics.

Table 9.19 Percent with high school diploma among adult US citizens

	TOTAL	YES	NO	% YES	MOE
Not Hispanic Black	27,145,257	4,875,132	22,270,125	18.0	0.1
Black Alone	26,175,244	4,753,155	21,422,089	18.2	0.1
Black in Combo	970,013	121,977	848,036	12.6	0.4
Hispanic not Black	20,938,674	5,144,545	15,794,129	24.6	0.1
Total Black Hispanics	747,574	153,572	594,002	20.5	0.5
Hispanics Black Alone	522,063	111,718	410,345	21.4	0.6
Hispanics Black in Combo	225,511	41,854	183,657	18.6	0.9

Table 9.20 Poverty among US citizens 18 plus

	TOTAL	YES	NO	% NO	MOE
Not Hispanic Black	25,667,062	19,965,706	5,701,356	22.2	0.1
Black Alone	24,756,713	19,245,079	5,511,634	22.3	0.1
Black in Combo	910,349	720,627	189,722	20.8	0.5
Hispanic not Black	20,385,243	17,088,239	3,297,004	16.2	0.1
Total Black Hispanics	707,331	546,663	160,668	22.7	0.6
Hispanics Black Alone	493,421	377,202	116,219	23.6	0.7
Hispanics Black in Combo	213,910	169,461	44,449	20.8	1.0

So, there is no uniform answer to the question of which way to combine Black-Hispanics when sample size is too small to consider this group separately. The user is advised to examine variables of interest for this group at a more aggregate level where sample size would be sufficient to make informed decision. These results should then guide the collapsing for the analysis at the smaller geographic or other domains.

9.7 Conclusion

Using the richness of the ACS, this paper examined the meaning of reported race among Hispanic groups. It found, not surprisingly, that different Hispanic groups reported quite different racial composition, even when all the responses in the residual "Some Other Race" category are lumped together. More interestingly, it also showed that, even controlling for not just Hispanic, but also Hispanic sub-groups, and even when the analysis was limited to adults born in the United States. meaningful differences in social and economic conditions were measured for different races.

One of the major social questions for America's future is whether the grand children and great grandchildren of current Hispanic immigrants will assimilate into current racial categories or form a separate and distinct racial or social group.

Will the great grand-children of White immigrants from Cuba or Chile be seen and treated the same as non-Hispanic whites or will they be treated as Hispanics? Will the great grand-children of Black immigrants from Panama or the Dominican Republic be seen and treated the same as non-Hispanic African Americans or the descendants of immigrants from Africa or Haiti, or will they become part of a Hispanic social or racial group? Will "Mestizo" form a separate racial or ethnic group? Of course, other possibilities can be imagined. This paper cannot answer these large questions. However, it points to a rich data set from which these questions can begin to be answered.

Appendix

Group name	Census code	GP numb	Detailed group
Spain	200	1	SPANIARD
Spain	201	1	ANDALUSIAN
Spain	202	1	ASTURIAN
Spain	203	1	CASTILLIAN
Spain	204	1	CATALONIAN
Spain	205	1	BALEARIC ISLANDER
Spain	206	1	GALLEGO
Spain	207	1	VALENCIAN
Spain	208	1	Canarian
Spain	209	1	Spanish Basque
Mexico	210	2	Mexican (check box)
Mexico	211	2	Mexican
Mexico	212	2	Mexican American
Mexico	213	2	Mexicano
Mexico	214	2	Chicano
Mexico	215	2	La Raza
Mexico	216	2	Mexican American Indian
Mexico	218	2	Mexico
Mexico	219	2	MEXICAN INDIAN
PUERTO RICAN	260	3	Puerto Rican (check box)
PUERTO RICAN	261	3	PUERTO RICAN
CUBAN	270	4	Cuban (check box)
CUBAN	271	4	CUBAN
COLOMBIAN	234	5	COLOMBIAN
DOMINICAN	275	6	DOMINICAN
ECUADORIAN	235	7	ECUADORIAN
GUATEMALAN	222	8	GUATEMALAN
HONDURAN	223	9	HONDURAN
PERUVIAN	237	10	PERUVIAN

(continued)

Group name	Census code	GP numb	Detailed group
SALVADORAN	226	11	SALVADORAN
AR-CH-UR	231	13	ARGENTINEAN
AR-CH-UR	233	13	CHILEAN
AR-CH-UR	238	13	URUGUAYAN
Other S.America	232	14	BOLIVIAN
Other S.America	236	14	PARAGUAYAN
Other S.America	239	14	VENEZUELAN
Other C America	221	15	COSTA RICAN
Other C America	224	15	NICARAGUAN
Other C America	225	15	PANAMANIAN
Other C America	229	15	CANAL ZONE
Generic Spanish	282	16	Spanish
Generic Terms	281	18	Hispanic
Generic Terms	227	19	CENTRAL AMERICAN
Generic Terms	242	19	South American
Generic Terms	250	19	LATIN AMERICAN
Generic Terms	251	19	LATIN
Generic Terms	252	19	LATINO
Other Groups	228	19	Central American Indian
Other Groups	240	19	South American Indian
Other Groups	241	19	Crollo
Other Groups	283	19	Californio
Other Groups	284	19	Tejano
Other Groups	285	19	Nuevo Mexicano
Other Groups	286	19	Spanish American
Other Groups	287	19	Spanish American Indian
Other Groups	288	19	Meso American Indian
Other Groups	289	19	Mestizo
Other Groups	290	19	Caribbean
Other Groups	291	19	Multiple Hispanic origin
Other Groups	299	19	Other Spanish/Hispanic, n.e.c.
Other Spanish	280	19	Other Spanish/Hispanic (check box)

References

Humes, K., & Hogan, H. R. (2013). Measurement of race and ethnicity in a changing, multicultural America. *Race and Social Problems*. September 2009, *1*(3), 111–131.http://link.springer.com/article/10.1007%2Fs12552-009-9011-5#page-2

Office of Management and Budget. (1997). Standards for the classification of federal data on race and ethnicity, Federal Register. http://www.whitehouse.gov/omb/fedreg_1997standards

US Census Bureau. (2013). *Hispanic origin main*. http://www.census.gov/population/hispanic/

Chapter 10
Promises and Pitfalls of the Puerto Rico Community Survey: Lessons from Persons-Per Household and Household Distributions

Matt Kaneshiro and Christine Pierce

Abstract Previous research has highlighted the difficulties of using the American Community Survey (ACS) to produce stable estimates of persons-per-household (PPH) at the county-level (Swanson and Hough 2012). Estimates of PPH and other demographic characteristics are made even more difficult using the Puerto Rico Community Survey (PRCS) due to the questionable quality of the base population and household estimate controls for Puerto Rico. This research demonstrates that treating the PRCS as a probability survey of households can produce relatively stable PPH. Specifically, PPH estimates can be determined by computing a weighted average of household size (and ignoring persons' weights and household sums-of-weights altogether). Taking this finding one step further, this research evaluates the stability of other household distributions from the PRCS and notes that the slight differences in the universes captured between the PRCS and Decennial Census can produce biased estimates when using the PRCS as a direct replacement for the long form of the Decennial Census. This chapter concludes with comments on the "promises and pitfalls" of the PRCS, highlighting the roles that the PRCS can play in future estimates while also being cautious of the biases that are inherent in the sampling structure and errors in previous estimates.

Keywords Puerto Rico community survey • American community survey • Puerto Rico • Persons-per-household • Census Bureau

M. Kaneshiro (✉)
Customer Analytics, LLC, Verona, WI, USA
e-mail: matt.kaneshiro@customeranalytics.com

C. Pierce
Nielsen, New York, NY, USA
e-mail: Christine.Pierce@nielsen.com

D.A. Swanson (ed.), *The Frontiers of Applied Demography*, Applied Demography Series 9, DOI 10.1007/978-3-319-43329-5_10

10.1 Introduction

As the dust settles from the 2010 Decennial Census, the U.S. Census Bureau Population Division is – as is not unusual – faced with the threat of budget cuts. Such budgetary limitations are troublesome for those who depend on Census releases since the quality of the data that are produced by the Population Estimates Division are put at risk. The quality of the Population Estimates in turn affects the quality of the American Community Survey (ACS) and Puerto Rico Community Survey (PRCS) due to these surveys' reliance on the Population Estimates for population and household controls. The quality of the ACS and PRCS estimates in turn affect the decisions of policy-makers and other users of ACS/PRCS data as the ACS and PRCS serve as the replacement for the Decennial Census' long-form which many institutions had relied upon for policy decisions. In short, the releases of the Population Division have far-reaching impacts.

The Puerto Rico Community Survey serves as a useful case study for determining the possible negative impacts of data quality that can result from budget cuts. Perhaps the most notable limitations of the PRCS include the lack of Master-Address-File updates and the high errors of the Population and Household Estimate updates of the 2000s. Under such circumstances, however, this research demonstrates that one could still leverage the strengths of the PRCS to produce useable household distributions. Specifically, trends in persons-per-household as well as in household distributions could be detected when treating the PRCS simply as a weighted household survey. However, one must be cognizant of the biases that are inherent in the PRCS universe as the PRCS is highly sensitive to mobile populations. Larger implications of this research highlight the lessons that the PRCS can teach us – namely in the "promises and pitfalls" of the ACS in a hypothetical world of declining budgetary resources and estimate quality.

10.2 The Sample

Sampling for the ACS begins with the Master Address File (MAF) which is the Census Bureau's most trusted resource for an exhaustive list of occupied households. The MAF can simply be thought of as the master frame from which household samples can reliably be drawn. Based on the MAF, housing units are assigned to one of five independent subsamples per county that are intended to be representative of the county. Each year, sampling for the ACS only occurs within one of the subsamples. The end product is a household survey of approximately 250,000 households per month that becomes aggregated to become part of a 1, 3, or 5-year sample of all households in the U.S. (U.S. Census Bureau 2009a). Serving as the foundation of the ACS samples, a frequently-updated MAF is a quintessential component for the drawing of representative samples.

One challenge faced by the PRCS is the limited updates of the MAF since the Puerto Rican component of the MAF is not updated by the U.S. Postal Service's Delivery Sequence Files as is done for the continental U.S. The MAF for Puerto Rico is, however, modified through "Census Bureau field operations" (U.S. Census Bureau 2011b) which would theoretically offset some of the biases that result from the limited MAF updates. In general, however, the sampling frame for the PRCS is dependent on address files that become increasingly outdated throughout the decade, effectively biasing the sample toward older occupied housing units.

10.3 Weights, Controls, and Demographic Estimates

Table 10.1 displays estimates of 1-Year PRCS estimates of households, the 2009 Vintage Population Estimates (which serve as the PRCS person weight controls), and 1-Year PRCS estimates of persons per household. Each of these figures is compared to a respective figure which is linearly interpolated from the 2000 to 2010 Census. The differences between the PRCS/Population Estimate and the interpolated Census counts are displayed in the "Diff, %" columns. Each of these three sets of estimates will be expanded upon in the sections below. As displayed in the table (and in the coming sections), the differences for each of the measures from the interpolated Census estimates became progressively larger throughout the decade. In brief, the problematic household estimates is likely the result of the limited updates of the Master Address File and attendant PRCS household weighting procedures while the problematic Population Estimates is likely the result of inaccurate migration estimation. The problematic PPH estimates arose due to the numerators and denominators of PPH going in the opposite direction of what was interpolated from the Decennial Censuses.

To briefly summarize the weighting procedures implemented for the ACS, household weights are initially weighted by selection probabilities. These weights are then adjusted by non-responding housing units, and the weights are lastly controlled to Household Estimates by county or groups of less-populous counties. Population weights are initially assigned the respective household weight. These weights are then adjusted by the sub-county control areas, household spousal relationships, householder status, race, Hispanic origin, sex, and age (U.S. Census Bureau 2009a). In short, weights are determined by selection probabilities, Housing Unit Estimates, and Population Estimates.

There are a few differences that distinguish the weighting procedure for the PRCS from that of the ACS. Perhaps most notably, the PRCS does not include a process that weights on housing unit controls. Instead, housing unit weights are determined by sampling rates and are adjusted based on a number of data-collection mode-of-administration biases (U.S. Census Bureau 2011a, b, 2009b). As noted in the previous section, the limited updates of the MAF add to the problem of producing Housing Unit weights since sampling for newly constructed units

Table 10.1 Estimates of households, population, and persons per household by source

	Households			Total population			Persons per household		
Year	Census (Interpolated)	PRCS (SOW)	Diff, %	Census (Interpolated)	2009 Pop Estimate	Diff, %	Census (Interpolated)	PRCS (SOW)	Diff, %
2005	1319	1254	−4.9	3768	3911	3.8	2.83	3.08	8.9
2006	1331	1240	−6.8	3759	3927	4.5	2.80	3.13	11.7
2007	1342	1204	−10.3	3751	3941	5.1	2.77	3.24	16.9
2008	1354	1186	−12.4	3743	3955	5.7	2.74	3.29	20.2
2009	1365	1181	−13.5	3734	3967	6.2	2.71	3.32	22.6

Note: "Census (Interpolated)" represents linearly-interpolated estimates from the 2000 and 2010 Decennial Censuses. "PRCS (SOW)" represents estimates computed using Puerto-Rico wide sums of weights. "2009 Pop Estimate" is gathered from the official Census Bureau 2009 vintage Population Estimates

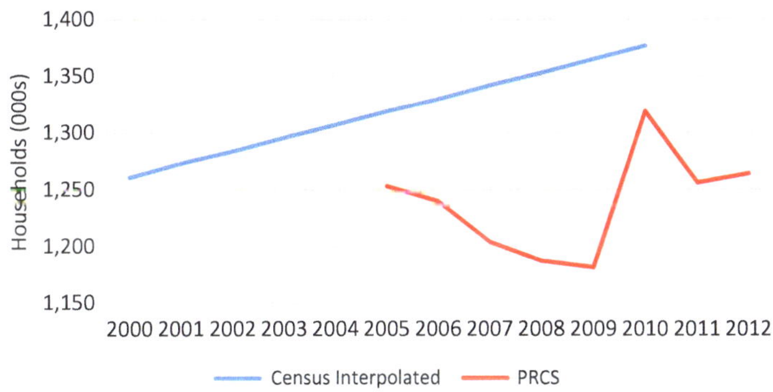

Fig. 10.1 Household estimates for Pureto Rico: census interpolated vs. PRCS

becomes understated. The resulting household estimates derived through the PRCS are thus volatile as they are not benchmarked to existing household controls.

Figure 10.1 illustrates interpolated Decennial Census (2000 and 2010) and annual PRCS estimates of households. As depicted, the PRCS estimated a decline in households through the 2000s when the opposite was more likely to be the case. As displayed on Table 9.1, the difference between the two estimates reaches −13.5 % in 2009. The large spike upward in 2010 reflects the PRCS' alignment with the 2010 Census, although even this alignment falls short of the Census enumeration. Following 2010, however, the PRCS estimate demonstrates another downward estimate, which suggests that the PRCS may continue to have problems in producing accurate housing unit weights.

Population estimates for Puerto Rico were estimated using a component method that relied on estimates of births, deaths, and migration (U.S. Census Bureau 2011a, b). The most questionable input for the population estimates was likely the net migration input which was determined using a residual technique for the previous decade. Specifically, the difference between the 2000 enumerated population and the survived 1990 enumerated population was used to determine migration rates. This method works under the assumptions that the net coverage rates for both censuses were equal and that the migration rates for the 1990s were the same as those of the 2000s.

Figure 10.2 illustrates the 2009 Vintage Population Estimates for Puerto Rico and the interpolated Decennial Census populations, displaying that the Population Estimate growth went in the opposite direction from the enumerated Decennial Censuses. By 2009, the differences between the Population Estimates and the interpolated Census Estimate reached 6.2 %. As the large discrepancies were determined to be a result of the migration component, the method used to determine net migration was changed to utilize the "residence one year ago" item on the ACS and PRCS for the 2010 decade (U.S. Census Bureau 2011a, b, 2012) (Fig. 10.3).

With inflated Population Estimates and deflated Household Estimates, the resulting persons-per-household estimates were greatly inflated with errors exceeding 20 % for 2009. While this problem was particularly acute for the PRCS, the

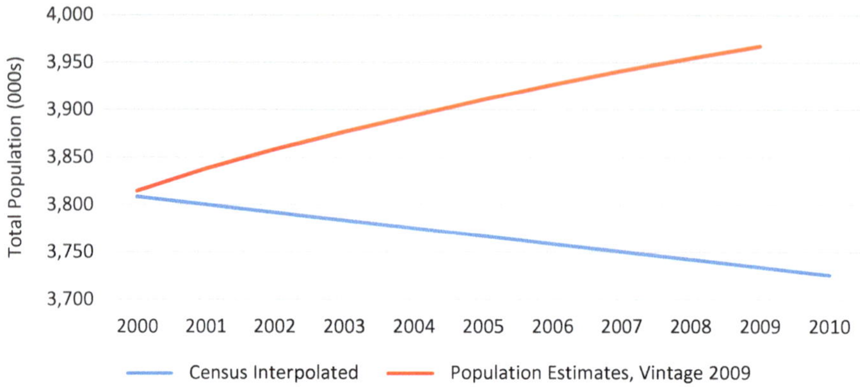

Fig. 10.2 Total population estimates for Pueto Rico: census interpolated vs. Population estimates

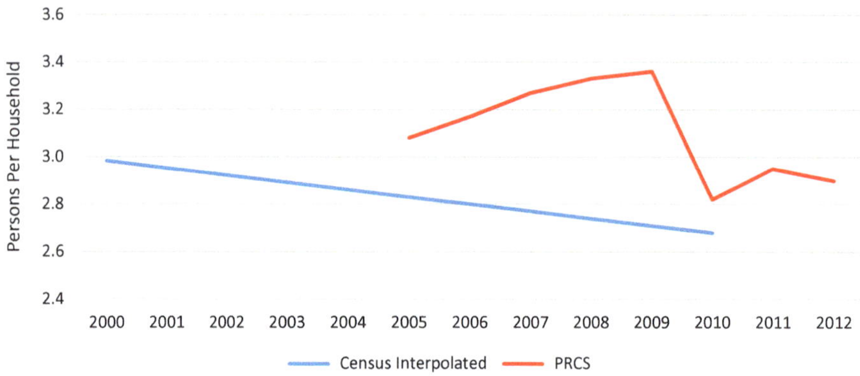

Fig. 10.3 Persons per household estimates for Puerto Rico: census interpolated vs. PRCS

ACS was also notorious for producing volatile PPH at the county-level. Swanson and Hough (2012) evaluated the stability of the 1-year ACS persons per household (PPH) estimates and noted that there was tremendous annual fluctuation that did not align with demographic theory in that the numerous dynamics that drive PPH changes would not be likely to produce annual "bounces" in PPH. These authors highlight that the ACS' weighting design is inherently representative of a "statistical perspective" of population dynamics where changes are represented through sample variation. This contrasts with the vision of data through a "demographic perspective" which views slow-moving population dynamics as changing the population structure over time.

10.4 Current Study: Method and Data

This study explores techniques used to produce annual persons-per-household estimates using PRCS microdata. This approach allows the user to bypass the errors that may be present for the Population and Household control estimates to potentially produce household estimates (if one were to have a reliable estimate of the total population). The quality of this estimate instead depends on the relationship between the representativeness of the household sample and the quality of the household weights. The method simply computes persons-per-household as a weighted average of household size. In other words, PPH is determined as:

$$PPH = \frac{\sum (HHSize_i \times HHWeight_i)}{\sum HHWeight_i}$$

...where i is a PRCS-sampled household. The estimates created using this method will be termed the "Weighted Distribution Estimates."

The counterpart of the Weighted Distribution PPH estimates will be termed the "PRCS (SOW – sums of weights)" estimates, which is simply the result of determining PPH based on sums of household and persons weights for the PRCS microdata. The proxy for the "accurate" PPH estimate is a linearly-interpolated PPH estimate between the 2000 and 2010 Decennial Census which will be called the "Census Interpolated" estimate.

This study draws on 1-year PRCS Public-Use Microdata Samples for 2005–2012 for Total Puerto Rico Estimates. Public Use Microdata Areas (PUMAs) were allocated into municipios using the Census Bureau's "Equivalency Files" for 2000. The only municipios that were included in this analysis were those for which each PUMA could be mapped to one-and-only-one municipio, thus ensuring the accurate representation of municipios from the PUMS data. These select municipio-level estimates draw from the 1-year PRCS PUMS data from 2005 to 2012 and were limited by the lack of the Equivalency Files for the 2010 PUMA Definitions (which were used in the 2012 PRCS data).

10.5 Results

Table 10.2 displays the interpolated Census PPH estimates in the column titled "Census Interpolated" and compares these figures with the PPH estimates using raw sums of persons and household weights (PRCS SOW) and the household-distribution-based method described above (Weighted Distribution). The first three columns of data display the PPH estimates side-by-side. Column "Diff %, PRCS SOW" displays the difference between the PRCS SOW PPH estimate with that of the Census Interpolated estimate. For example, the value for the Puerto Rico 2009 estimate demonstrates a 22.6 % overestimate of PPH using the SOW method,

Table 10.2 Persons per household and differences from census interpolated: by source, year, and geography

Year	Geography	Persons per household					
		Census interpolated	PRCS SOW	Weighted distribution	Diff %, PRCS SOW	Diff %, Weighted distribution	PRCS sample size
2005	Puerto Rico	2.83	3.08	2.76	8.9	−2.3	12,860
2006	Puerto Rico	2.80	3.13	2.75	11.7	−1.9	12,726
2007	Puerto Rico	2.77	3.24	2.71	16.9	−2.3	12,409
2008	Puerto Rico	2.74	3.29	2.68	20.2	−2.1	12,213
2009	Puerto Rico	2.71	3.32	2.66	22.6	−1.8	12,170
2005	San Juan Municipio	2.47	2.73	2.52	10.5	1.9	1565
2006	San Juan Municipio	2.45	2.73	2.47	11.4	0.9	1606
2007	San Juan Municipio	2.42	2.83	2.45	16.9	1.2	1576
2008	San Juan Municipio	2.40	2.79	2.41	16.6	0.5	1560
2009	San Juan Municipio	2.37	2.77	2.41	16.8	1.7	1588
2005	Bayamón Municipio	2.81	2.94	2.76	4.9	−1.8	847
2006	Bayamón Municipio	2.77	2.96	2.81	6.7	1.3	823
2007	Bayamón Municipio	2.74	3.00	2.68	9.4	−2.2	828
2008	Bayamón Municipio	2.71	3.04	2.69	12.1	−0.8	788
2009	Bayamón Municipio	2.68	3.09	2.63	15.4	−1.6	793
2005	Carolina Municipio	2.77	2.98	2.74	7.6 %	−0.9	630
2006	Carolina Municipio	2.74	2.94	2.70	7.4	−1.5	652
2007	Carolina Municipio	2.71	3.02	2.63	11.5	−2.9	643
2008	Carolina Municipio	2.68	3.11	2.61	16.2	−2.5	620
2009	Carolina Municipio	2.65	3.14	2.57	18.5	−3.0	637
2005	Ponce Municipio	2.86	3.02	2.73	5.3	−4.6	615

(continued)

Table 10.2 (continued)

Year	Geography	Persons per household					
		Census interpolated	PRCS SOW	Weighted distribution	Diff %, PRCS SOW	Diff %, Weighted distribution	PRCS sample size
2006	Ponce Municipio	2.83	3.12	2.86	10.2	1.0	600
2007	Ponce Municipio	2.79	3.13	2.78	12.0	−0.6	575
2008	Ponce Municipio	2.76	3.12	2.63	13.2	−4.8	598
2009	Ponce Municipio	2.72	3.20	2.82	17.5	3.5	551
2005	Caguas Municipio	2.82	2.97	2.82	5.3	0.1	488
2006	Caguas Municipio	2.79	2.90	2.78	4.0	−0.3	510
2007	Caguas Municipio	2.76	3.00	2.69	8.8	−2.4	494
2008	Caguas Municipio	2.73	3.01	2.75	10.4	1.1	494
2009	Caguas Municipio	2.69	3.09	2.75	14.7	2.1	494
2005	Guaynabo Municipio	2.71	2.85	2.64	5.4	−2.4	374
2006	Guaynabo Municipio	2.68	2.95	2.59	10.3	−3.3	363
2007	Guaynabo Municipio	2.65	3.08	2.68	16.3	0.9	365
2008	Guaynabo Municipio	2.62	2.91	2.51	10.8	−4.2	338
2009	Guaynabo Municipio	2.59	3.08	2.50	18.6	−3.6	344
2005	Arecibo Municipio	2.73	2.95	2.61	7.7	−4.5	328
2006	Arecibo Municipio	2.71	3.06	2.73	13.2	0.9	303
2007	Arecibo Municipio	2.68	3.41	2.57	27.1	−3.9	287
2008	Arecibo Municipio	2.65	3.41	2.64	28.6	−0.4	284
2009	Arecibo Municipio	2.62	3.32	2.59	26.5	−1.3	304

assuming that the Census Interpolated estimate is accurate. Column "Diff %, Weighted Distribution" displays the corresponding estimate using the Weighted Distribution method and demonstrates a 1.8 % underestimate of PPH vis-à-vis the Census Interpolated estimate. The latter estimate is substantially closer to the

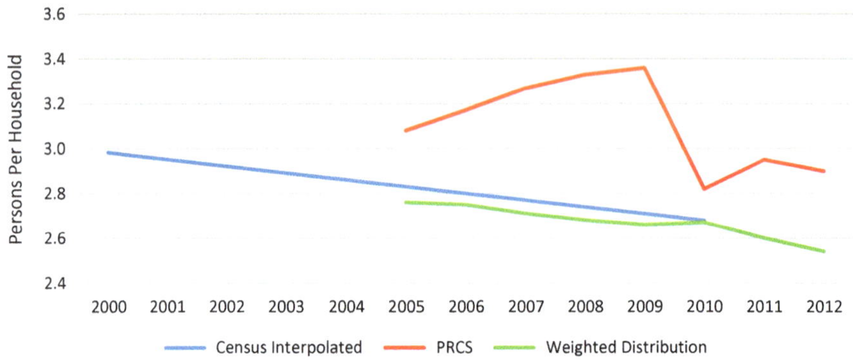

Fig. 10.4 Persons per household estimates for Puerto Rico: census interpolated, PRCS, and weighted distribution

Census Interpolated PPH estimate and thus suggests that the Weighted Distribution method produces far superior PPH estimates that what one would obtain using raw weights (as is currently produced in American Factfinder outputs).

Figure 10.4 graphically displays the yearly estimates of PPH using the three methods for all of Puerto Rico. The graph illustrates that the Weighted Distribution estimate is consistently much closer to the Census Interpolated estimate than is the PRCS estimate. Additionally, the Weighted Distribution estimate of PPH also trends downward alongside the Census Interpolated method. This suggests that the Weighted Distribution method is able to detect subtle changes over time. As seen on Table 10.2, these findings are not unique to the Puerto Rico aggregate estimates. Even as sample sizes get smaller for municipio-level estimates, one still finds that the Weighted Distribution method consistently produces superior results that generally trend in the direction of the Census Interpolated estimates.

Figure 10.5 displays the results of the method employed for Arecibo Municipio which has the smallest annual sample sizes of the municipios included in this study (between 284 and 328). Even with these small sample sizes, one finds that the resulting Weighted Distribution PPH estimates are relatively close to the Census Interpolated estimate. In this extreme case, one visualizes how divergent the PPH estimates can be as the 2008 PRCS estimate of PPH is 28.6 % higher than the Census Interpolated estimate whereas the Weighted Distribution is .4 % lower than that of the Census Interpolated estimate.

Figure 10.6 summarizes the relationship between the sample size of the PUMS 1-year data and the consistency between PPH estimates at the municipio level. Each point plots a % Difference between an ACS-based (PRCS SOW, or Weighted Distribution) PPH estimate and the Census Interpolated estimate by the respective logged 1-Year PRCS sample size for one municipio; the % Difference is on the y-axis and the logged sample size is on the x-axis. Points that are closer to 0 on the y-axis represent PPH estimates that are consistent with the Census Interpolated Estimate. As displayed on the graph, the Diff % for the Weighted Distribution (plotted in green) consistently tends to lie closer to 0 than the PRCS figures (red).

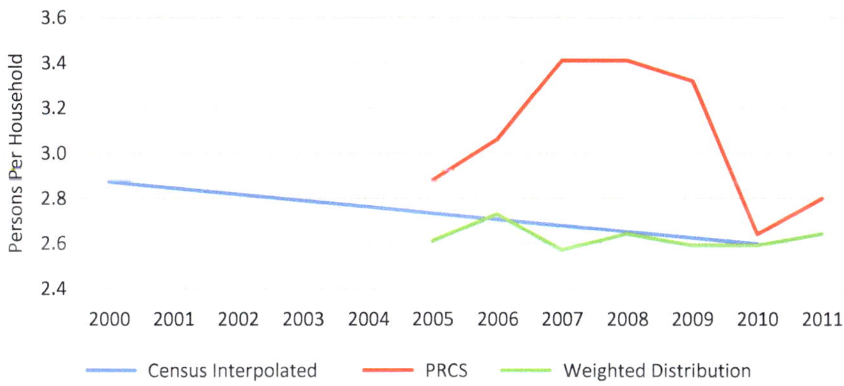

Fig. 10.5 Persons per household estimates for Arecibo municipio: census interpolated, PRCS, and weighted distribution

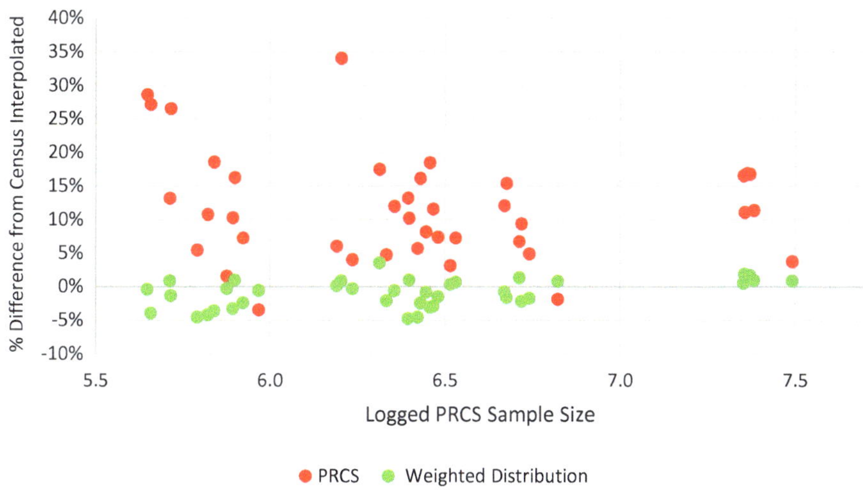

Fig. 10.6 Percent difference between PPH estimates and census interpolated PPH for municipios: by logged annual PRCS sample size and method

Whereas the % Difference for the PRCS method tends to improve as sample sizes increase, the Weighted Distribution method also performs well for smaller (as well as larger) sample sizes.

10.6 One Step Further: General Usage of the Weighted Distribution Method

The promising results of the Weighted Distribution method estimates of PPH prompt the question: "how well does the Weighted Distribution method perform when estimating other household distributions?" Toward this aim, this research compares the performance of the Weighted Distribution method against the Census Interpolated estimates for two other variables: Household Size and Age of Householder. In sum, this research suggests that the performance of the Weighted Distribution method is sensitive to the biases that are inherent in the ACS/PRCS samples vis-à-vis the Decennial Census enumeration process.

The bias in the ACS/PRCS that is explored in this research is tied to the differences in the types of persons who qualify for Decennial Census enumeration versus inclusion in the ACS/PRCS. The Decennial Census short form asks for information on "people… who live and sleep here *most of the time*", whereas the ACS/PRCS prompts for the inclusion of persons who are "living or staying here for *more than 2 months*" (emphases added). As such, the ACS/PRCS is less likely to include highly mobile populations in comparison with the Decennial Census such as persons who may have been working abroad (e.g. the continental United States) for the months prior to the PRCS.

As displayed in Fig. 10.7, the Weighted Distribution proportions of Households of Size 1 lie close to the Census Interpolated estimates. Conceptually, persons in Household Size 1 would be less likely to work seasonally in the Continental U.S. than those in larger-sized households since those in larger-sized households would have more persons who are at risk of working abroad (and thus be absent by the time of the PRCS). As such, shares of Household Size 1 are expected to be much more consistent than Households with 4+ persons across the Census and PRCS.

Figure 10.8 displays the Weighted Distribution proportion of households that are of size 4+ and, as expected, these Weighted Distribution estimates are lower than those of the Census Interpolated estimates. The same pattern is found when comparing shares of households by age of householder, as younger householders are conceptually more likely to work abroad and thus be absent by the time of the PRCS (see Appendix Figs. 10.9 and 10.10). These results suggest that the universe of persons included in the PRCS is slightly different from those enumerated in the Census as the PRCS is less likely to capture mobile populations. Thus, although the Weighted Distribution method performed well in determining PPH in comparison with the Census Interpolated method, the generalized use of the Weighted Distribution method must be done cautiously if one wishes to use the PRCS as a direct surrogate for the Decennial Census long form.

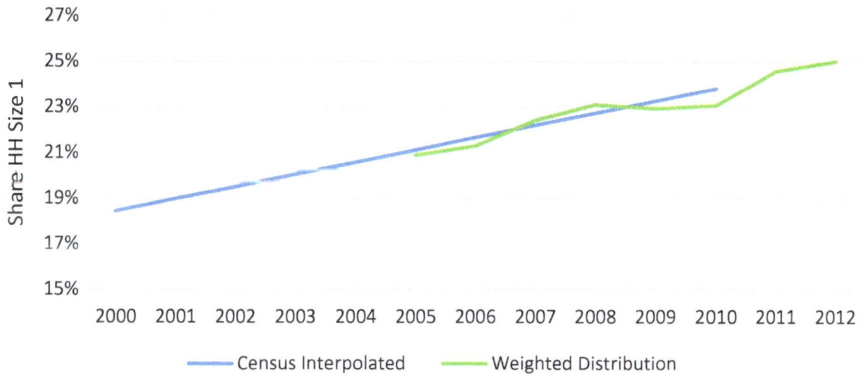

Fig. 10.7 Share household size 1 for Puerto Rico: census interpolated and weighted distribution

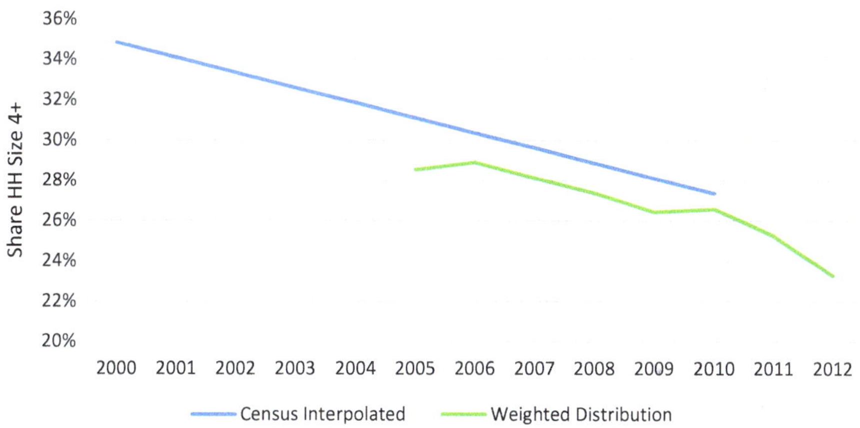

Fig. 10.8 Share household size 4+ for Puerto Rico: census interpolated and weighted distribution

10.7 Conclusion

This research demonstrates that the PRCS can be used to produce usable Persons-Per-Household estimates. Specifically, when treating the PRCS as simply a weighted household survey, PPH can be estimated as a weighted average of household size. This method works effectively at the Puerto Rico level as well as for smaller municipios. Despite the difficulties in determining PPH using Household and Person estimates and sums of weights (as has been a criticism of the ACS, see Swanson and Hough 2012), this method provides a work-around to some of the difficulties that are confronted in ACS weighting for smaller geographies. This case-study of Puerto Rico is particularly illuminating due to the large discrepancies between the Household and Person Estimates that were used to weigh the PRCS. Namely, the Census Bureau's Household estimates were declining while the Person estimates were increasing, although the opposite was the case. This resulted in a "worst-case-scenario" of PPH estimation as the numerators and denominators were going in the opposite direction of "reality," producing dramatically inflated PPH estimates. However, despite these difficulties, the Weighted Distribution method produced usable PPH estimates down to the municipio level.

Despite the efficacious use of the Weighted Distribution method to determine PPH, this research suggests that PRCS/ACS users should be careful when using the Weighted Distribution method for determining general household characteristics. Specifically, the PRCS/ACS user should understand the differences between Census enumeration and ACS inclusion as the ACS is less likely to include highly mobile populations. Evidence of this is found in the PRCS distributions of Household Size 1 (assumed to be less sensitive to mobile populations) vis-à-vis Household Size 4+ (assumed to have more persons at risk of being absent at the time of the PRCS); Household Size 1 estimates were highly consistent with Census Interpolated estimates, whereas Household Size 4+ was consistently below the Census Interpolated estimate. The same result was found for Age of Householder where younger householders were less likely to be included in the PRCS (versus Census) in comparison with older householders. This suggests that the use of the ACS/PRCS cannot serve as a direct surrogate for the "long form" of the Decennial Census as there are slight differences in the universes captured.

This research on the PRCS further highlights the importance of the Master Address File. As the MAF for Puerto Rico was not updated to the same extent as in the continental U.S., the PRCS may have encountered biases in the household sample as well as the household weighting algorithm. Further research may further explore some of the biases inherent in the PRCS and/or ways to use the PRCS to inform the MAF or weighting methodology.

Aside from the implications that this research has on general PRCS/ACS users, this research also has implications for the operations of the Census Bureau in an age of dramatic budget cuts. The Population Estimates of Puerto Rico in the 2000s can arguably be seen as a case study of a "worst-case scenario" for Population Estimates of the continental U.S. In a hypothetical world of dramatic budget cuts that negatively affect the Master Address File and the resources invested in producing Population Estimates, the resulting estimates (and ACS results) may face the same issues that were faced by Puerto Rico in the 2000s. The population and household estimates of Puerto Rico in the 2000–2009 decade were arguably of questionable quality. However, the household distributions of the PRCS were still usable. This suggests that the PRCS/ACS can open up further methodological opportunities for the Census Bureau, such as in PPH and thereby Household estimates. Future research for applied demographers and the Census Bureau may explore the potential that the ACS/PRCS has in producing estimates or simply in enhancing day-to-day activities (e.g. determining indicators of dynamic change in households to inform where/how sampling frames can be modified).

Appendix

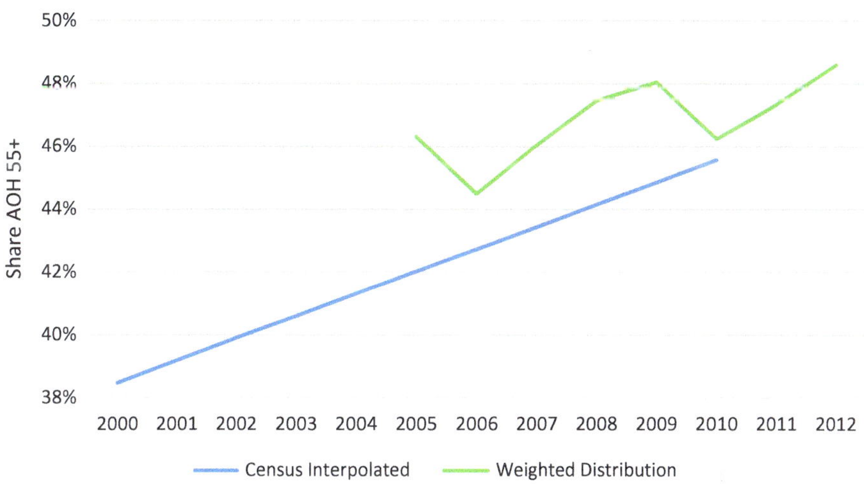

Fig. 10.9 Share age of householder 55+: census interpolated and weighted distribution

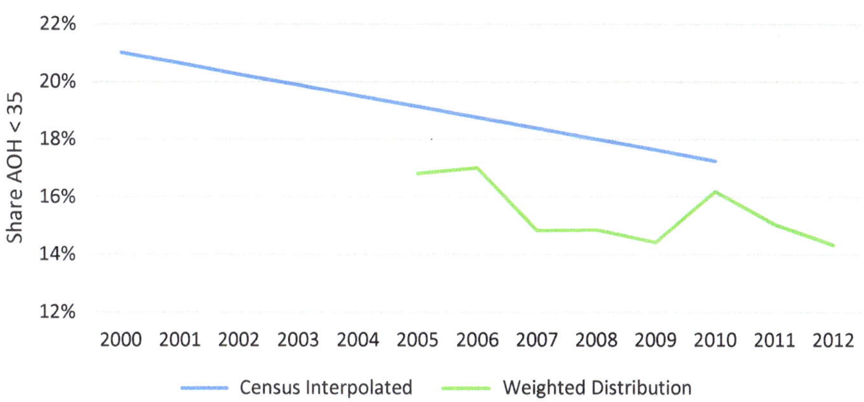

Fig. 10.10 Share age of householder <35: census interpolated and weighted distribution

References

Swanson, D. A., & Hough, G. C. J. (2012). An evaluation of persons per household (PPH) estimates generated by the American community survey: A demographic perspective. *Population Research and Policy Review, 31*, 235–266.

U.S. Census Bureau. (2009a). *Design and methodology: American community survey*. Washington, DC: U.S. Government Printing Office.

U.S. Census Bureau. (2009b). *Methodology for the Puerto Rico commonwealth and municipio population estimates by age and sex (Vintage 2009): April 1, 2000 to July 1, 2009*. Washington, DC: U.S. Census Bureau.

U.S. Census Bureau. (2011a). *Methodology for the Puerto Rico commonwealth and its municipios population estimates by age and sex (Vintage 2011): April 1, 2010 to July 1, 2011*. Washington, DC: U.S. Census Bureau.

U.S. Census Bureau. (2011b). *Puerto Rico accuracy document accuracy of the data (2011)*. Washington, DC: U.S. Census Bureau.

U.S. Census Bureau. (2012). *Methodology for the intercensal population and housing unit estimates: 2000 to 2010*. Washington, DC: U.S. Census Bureau.

Chapter 11
State Longitudinal Data Systems: Applications to Applied Demography

George C. Hough Jr. and Melissa M. Beard

Abstract Although originally built to support metrics for educational accountability under the No Child Left Behind Act of 2001 supporting elementary and secondary schools, state longitudinal data systems (SLDS) developed to incorporate state administrative records from pre-school all the way into the workforce (early learning, Kindergarten-12, higher education and workforce sectors). As it stands now, all 50 states have received funding from the federal government to build a K-12 SLDS and 43 states have received funding to build a P-20/Workforce SLDS (P20W). Washington has a comprehensive P20W data system because of its breadth of data sources and depth of data shared by state agencies. In addition, Washington has an office dedicated to P20W work. This is unique as most states' P20W office is within the K-12 system where resources are split between K-12 reporting and P-20 work. Washington state and the Education Research and Data Center (ERDC) has been at the forefront of developing and utilizing their SLDS to follow cohorts both over time and across sectors. Examples for Washington State provided herein demonstrate the usefulness of the cohort approach as well as provide direction for future developments both in education-related research and applied demography.

Keywords State longitudinal data systems • Cohort • P20 • Data warehouse • Identity matching

11.1 Introduction

Over the past decade, all 50 states and outlying territories have developed longitudinal data systems. These State Longitudinal Data Systems (SLDS) are essentially administrative records databases, i.e., partial population registers. The federal

G.C. Hough Jr. (✉) • M.M. Beard
Education Research & Data Center, Office of Financial Management, 43124, 98504-3124 Olympia, WA, USA
e-mail: George.Hough@ofm.wa.gov; Melissa.Beard@ofm.wa.gov

D.A. Swanson (ed.), *The Frontiers of Applied Demography*, Applied Demography Series 9, DOI 10.1007/978-3-319-43329-5_11

government has financially supported states in their development of these partial population registers, whose application to demography has been noted long before (Poulain and Herm 2013 and Siegel and Swanson 2004, pp. 31–35).

After a brief history of the evolution of SLDS and federal investments in making SLDS a reality for most states, this paper will provide a critical analysis of the strengths and weaknesses of using this newly created SLDS database for research. The following sections will then outline the Washington state data warehouse/data store elements, proceed to provide a few concrete examples already being displayed on the ERDC website and research briefs, and finally extensions to future research that could be performed once the data warehouse is fully operational.

11.2 Evolution of Federal Policies that Have Developed and Supported Longitudinal Data Systems

No Child Left Behind (NLCB) can be credited with changing the way educators across the nation looked at school performance (Snow-Renner and Torrence 2002) because of its accountability requirements, although some states had experimented with a version of accountability-focused standards-based reform in the decade. However, key provisions of NCLB—assessment requirements, adoption of standards, and the concept of adequate yearly progress (AYP)—were all introduced in the 1994 version of the Elementary and Secondary Education Act (ESEA), known as the Improving America's Schools Act of 1994. For those states that had not already started defining content standards or administering statewide assessments, this reauthorization of ESEA required all states to engage in this work (Improving America's Schools Act 1994).

11.2.1 Initial Impetus and Support for Longitudinal Data Systems

While the 1994 reauthorization kicked off statewide standards and assessments, No Child Left Behind (NCLB), the 2001 reauthorization of ESEA, assigned a timeline to AYP. It required that schools make AYP towards a goal of 100 % of students achieving academic success by the 2013–2014 school year. Progress would be measured by looking at students disaggregated so achievement gaps between ethnic or socio-economic groups could be more easily observed. In many cases, state education agencies collected and stored demographic, assessment and graduation data in separate systems, often called "data silos." In addition, systems were not built to follow students longitudinally, across time or even district lines. Acknowledging the difficulty of doing this work with data silos, NCLB also permitted states to create a longitudinal data system where assessment, enrollment and graduation records for a student could be linked together to meet reporting requirements (No Child Left Behind 2001; Snow-Renner and Torrence 2002; Wayman 2007).

In recognition of the increased data demands due to NCLB, in 2005, the Institute of Education Sciences (IES) released a request for proposals (RFP) to provide funding to state education agencies to build K-12 statewide longitudinal data systems (SLDS). The RFP provided a glimpse of how the federal government expected states to use longitudinal data. IES requested states build SLDS that could be used to study the academic achievement growth of individual students and have this data linked to teachers, programs and interventions. IES believed the education community needed this individual-level data in order to conduct the evaluations necessary to learn what actions led to improved student learning (US Department of Education 2005).

Next, the federal government began encouraging P-16 (preschool through baccalaureate degree) systems in the America Creating Opportunities to Meaningfully Promote Excellence in Technology, Education and Science (COMPETES) Act in 2007. While the majority of this Act focused on science, technology, engineering and mathematics innovation and competiveness, it included a section on the alignment of K-12 education programs with higher education and workforce demands. Within this section, 12 required elements of a statewide P-16 education data system were listed and included elements such as a unique statewide student identifier, linkages with higher education data systems, and student-level transcript information. Grants were offered to K-12 state education agencies to build or expand their K-12 SLDS to include preschool, higher education and workforce data. The Act also required states to build the following functionality into their statewide P-16 education data system:

(i) identify factors that correlate to students' ability to successfully engage in and complete postsecondary-level general education coursework without the needs for prior developmental coursework;
(ii) identify factors to increase the percentage of low-income and minority students who are academically prepared to enter and successfully complete postsecondary-level general education coursework; and
(iii) use the data in the system to otherwise inform education policy and practice in order to better align State academic content standards, and curricula, with the demands of postsecondary education, the twenty-first century workforce, and the Armed Forces (America COMPETES Act 2007, p. 102).

During this activity around K-12 student data systems, the National Center for Education Statistics (NCES) within IES released a report in 2005 describing the feasibility of creating a national individual-level data system for higher education. At the time, NCES administered the Integrated Postsecondary Education Data System (IPEDS) that integrated school-level information about enrollment, completions, and finance from the majority of public and private higher education institutions in the United States. The study described how IPEDS could be changed to allow for this expansion, shifting from the collection of school-level data to individual-level data, and recommended that legislative authorization and funding be provided in the reauthorization of the Higher Education Act (Cunningham and Milam 2005).

Unfortunately, the Higher Education Opportunity Act of 2008 did not include funding for the expansion of the IPEDS system. In fact, it specifically forbade the US Department of Education to build a national database of higher education activity because Congress did not want a federal database of individual-level higher education student data (Higher Education Opportunity Act 2008). Because of this prohibition, any work linking higher education data with K-12 data would need to be completed through state level systems.

11.2.2 Further Federal Support for Longitudinal Data Systems

Other federal policies, less directly focused on K-12 accountability followed suit. And taken together they have added to the momentum and the means for developing and mounting longitudinal data systems at the state level.

American Recovery and Reinvestment Act of 2009 (ARRA) On the heels of the reauthorization of the Higher Education Act, the economy plunged to levels rivaling the Great Depression. In an effort to provide funding relief to states, the federal government created the American Recovery and Reinvestment Act (ARRA) of 2009, which included the State Fiscal Stabilization Fund (SFSF). Over $5 billion was provided to states to restore state funding for education. State applications for the monetary assistance had to include assurances that the state was improving data collection related to the data element requirements in the America COMPETES Act and required the reporting of postsecondary enrollment rates for high school graduates.

Another section of ARRA included $250 million for K-12 state education authorities to create

> . . .comprehensive P-20 systems that permit the generation and use of accurate and timely data, support analysis and informed decision-making at all levels of the education system,. . .support education accountability systems, and simplify the processes used by State educational agencies to make education data transparent through Federal and public reporting (Education Research and Data Center 2009, p. 3).

While the previous SLDS grant RFP in 2009 requested the K-12 state education agency build a P-16 system, this RFP requested that states expand beyond baccalaureate work into graduate enrollment, the workforce and other areas where linkages would support better decision-making. Another difference was that the ARRA P20W (preschool through workforce) SLDS RFP required that states build systems that met the data system capabilities and data element requirements outlined in the America COMPETES Act. The RFP also dedicated a number of pages to describing the potential uses and benefits of a longitudinal P-20 data system. Examples included identifying preschool programs correlated with high numbers of students ready for kindergarten, understanding what is necessary to insure what is needed for all students to succeed after high school, and "determining priorities for allocating resources" (Education Research and Data Center 2009, p. 3).

A final piece of ARRA, as related to education data systems, was the Race to the Top (RTTT) grants that offered $5 billion to states that were creative and innovative in solving perennial education issues: graduation rates below 100 %, gaps in achievement between students who differed in race or socio-economic status, and preparing students for postsecondary enrollment AND the workforce. The RTTT grant RFP included building data systems that tracked student achievement and provided educators with information to improve practice as one of the four core education reform areas.

Understanding that a likely outcome of all the states building their own P20W data systems was to have 50 data silos that could not talk to each other and working under the limitation that a national data system was prohibited by the Higher Education Opportunity Act, an effort to create a common data language across the states and education sectors began in 2010 within the Department of Education. The effort is called Common Education Data Standards (CEDS) and is a way to link the numerous data systems across education sectors and state lines. CEDS provides a list of common data elements in early learning, K-12, higher education and workforce; defines them and valid values associated with the data elements. Each system or state can then compare their list of data elements to the CEDS list and determine that the elements, while maybe named differently, may in fact, be the same thing. This allows all systems to compare to CEDS and create crosswalks but does not require all systems to use the same data element names, valid values, etc. The ability to have a common language assists in the communication between education sectors where there are a number of common words. For example, "retention" in one system is a negative outcome while it can be a positive outcome in another sector. Additionally, "program" in K-12 refers to some sort of assistance in addition to basic education, such as the free and reduced price lunch program or the special education program while program in higher education refers to area of study.

Workforce Data Quality Initiative (WDQI) Also in 2010 (and again in 2012), the Department of Labor (DOL) announced the Workforce Data Quality Initiative (WDQI) grant program that would award $12 million to states that would build a workforce database linked to education data. DOL incentivized states to build systems where workforce program participants could be followed longitudinally through education and the workforce (US Department of Labor 2010). The linkage of this data was valuable because the educational outcomes funded by labor programs could not be shared with the labor agencies due to education privacy laws.

Family Educational Rights and Privacy Act (FERPA) Many of the grant RFPs during this time period included language related to the Family Educational Rights and Privacy Act (FERPA) (US Department of Education 2005, 2008, 2011). IES required that States ensure student and staff privacy and student personally-identifiable data be kept confidential. These new data systems were pushing the limits of a privacy act written in 1974 when student education records were kept in a filing cabinet. K-12 state education agencies were eager to begin receiving data from postsecondary institutions but some states believed FERPA restricted this sharing of data "backwards" through the system. In addition, there was confusion about how data could be shared with employment agencies that needed student social security numbers in order to follow a student into the workforce.

Some states that had received SLDS grant funding to build a P-20 data system were facing resistance from people in various education sectors who believed FERPA restricted the very sharing states were being funded to build systems to support. The Department of Education published a notice of proposed rulemaking in April 2011. The notice requested comments on proposed rule changes that would eliminate confusion on sharing data with employment agencies, define terms such as authorized representative and education program, and make it clear that data could be shared across states lines, among other clarifications. The Department took comments on the proposed rule changes and finalized these changes in January 2012 (US Department of Education 2011).

Because none of the language in the Act itself changed, these rule changes represented a shift in interpretation of FERPA, not a change in the privacy law itself. These changes were important because the comments allowed the federal government to provide examples of the types of sharing that were legal under FERPA. Therefore, what changed was the interpretation of FERPA by the Department of Education, not the law.

Finally, Congress and federal officials were discussing the next reauthorization of the Elementary and Secondary Education Act. Most discussions included the reporting of longitudinal outcomes (Alliance for Excellent Education 2011). Taken together, these various policy actions at the federal level provided a compelling and supportive context for state level activity in this realm of data system development, and especially so in Washington State.

11.3 Washington State Context for P-20 Longitudinal Data Use

Education reform conversations began in Washington State prior to the 1994 Improving America's School Act. In fact, the 1993 Washington State Legislature began reforming education with the passage of House Bill 1209, a law that expected all students to meet statewide learning targets (Washington State House of

Representatives 1993). With passage of this bill, the state's K-12 agency, the Office of Superintendent of Public Instruction (OSPI), began defining statewide standards and creating a statewide assessment for students in the fourth, eighth and tenth grades. When No Child Left Behind was enacted, the state also created statewide assessments for third, fifth, sixth, and seventh grades.

Regarding data at OSPI, each program had its own reporting system and the students in each program could not be easily linked together at the state level to determine the various services students were receiving. In addition, it was difficult to link student data from 1 year to the next. While this process was sufficient for appropriating money to school districts, it did not support any actions related to data use and understanding what services or programs were helping students achieve on statewide assessments. For example, a district would report their special education students to the Special Education office at OSPI, their bilingual students to the Migrant/Bilingual office at OSPI and the total number of students served through the Learning Assistance Program. Meanwhile, the Assessment office at OSPI held the statewide assessment results. To study the assessment results by program participation was difficult because it required the matching of various data sets by program coordinators who did not have the expertise to do this work.

In 2003, the Legislature began appropriating money to OSPI, to build a longitudinal data system to link these various programs. The first step in a statewide longitudinal data system was the assigning of a unique student ID, called the statewide student identifier (SSID). This allowed students to be linked across programs and across districts, which was important when students moved to a new district, however, the program information was not consolidated into a centralized data warehouse until the state received $6 million in funding from the third round of SLDS grants in 2009. This K-12 SLDS grant, completed in October 2014, created CEDARS, the Comprehensive Education Data and Research System, and is the data system used to appropriate state funds and to study the longitudinal outcomes of students within the public K-12 system.

In 2007, the Education Research and Data Center (ERDC) was created as Washington's P-20 Workforce (P20W) office, as a result of Governor Gregoire's study of education called Washington Learns. The Governor and others were frustrated by the limitations of the data systems within each education sector because there were no linkages across the early learning, K-12, higher education and workforce silos. By statute, ERDC is tasked with compiling data and creating a data system for longitudinal analyses (Revised Code of Washington 43.41.400).

With the advent of the Great Recession, ARRA funding was made available to states through the Department of Education in the form of SFSF, SLDS grants and RTTT grants. For the benefit of school districts, OSPI received $820 million in SFSF funding and was not eligible to receive RTTT funding because of the prohibition of charter schools in the state.

In 2010, ERDC received $17 million in funding from the ARRA SLDS grant program to build a P20W data system that would span early learning, K-12, higher education and the workforce. As part of its application, Education Research and

Data Center (2009) requested funding to create a high school feedback report that would link K-12 graduate data with post-secondary enrollment data. The P-20 Reports on Washington Public High School Graduates (Education Research and Data Center 2012) were created to serve the data needs of district and high school administrators, went into production in 2012 and have been updated often based on the needs of users. This report was especially important to OSPI because it met the reporting requirements of SFSF.

Washington also applied for and received a $1 million WDQI grant in 2012. While the SLDS grant funded the incorporation of employment data for K-12 and post-secondary students in Washington, the WDQI grant funded the incorporation of employment data for all covered employees in Washington, along with Department of Labor workforce training program participants. The linking of this data will allow researchers to study the education outcomes and effectiveness of worker retraining programs.

11.4 Exploring Effects that State Longitudinal Data May Have on Data Needs and Research

As it stands now, all 50 states have received funding from the federal government to build a K-12 SLDS and 43 states have received funding to build a P-20/Workforce SLDS. With the millions of dollars that have been spent, it is important to understand the expectations of these data systems, by the policymakers who support these systems, and the educators and researchers expected to use the systems.

11.4.1 Strengths and Limitations of Statewide Longitudinal Data Systems

To understand how, and how well, P-20 longitudinal data systems will help educators and researchers achieve these hoped-for effects, one needs to recognize both their strengths and their limitations. The numerous policies and the millions of dollars from the federal government are evidence of their belief in the power a SLDS can have in reforming education. There are many advantages of a statewide P20W system but can they overcome the limitations?

Strengths A huge strength of these systems is the connection of education data, within and across education sectors. Prior to the building of SLDS, state education agencies consisted of a number of data collection systems specific to the program it was created to administer. There were financial, bilingual, special education, enrollment, assessment, and certification systems that may overlap in the data elements collected but were not connected in any way. Connecting the bilingual program data with a student's assessment and enrollment data was a project that took a lot of time, if completed at all. Data that needed to be combined required manual merging or keying data from paper (Data Quality Campaign 2007). Now apply this to the numerous systems in early learning, K-12, higher education and workforce. Having a P20W SLDS becomes a powerful resource in understanding program outcomes (Petrides 2003).

SLDS are also being built to be flexible, not transactional. Transactional systems typically only collect the data needed to allocate dollars or resources and are specific to the program. They do not collect all information about students, interventions or outcomes. SLDS, on the other hand, link data from a variety of systems in a way that can be easily accessed and related to other records. Mapping data to a flexible system means this only has to be done once, rather than project by project (Palaich et al. 2004). This flexible, longitudinal system means educators can look at outcomes beyond the current school year and account for student mobility (Wayman 2007).

Another strength of SLDS is that student data are matched and linked the same way every time, which also means this laborious work does not need to be repeated project by project. The state benefits from this efficiency and consistency because it maximizes the work completed and decreases the costs and burden to state agencies, especially when responding to data requests from researchers (Culhane et al. 2010; Data Quality Campaign 2011a, b, 2012).

Finally, these systems allow the data to flow back and forth between the education sectors and levels where previous processes included only a flow from K-12 to post-secondary or district to state agency. This flow of data encourages more cross-sector conversations, which leads to common language and potentially more effective solutions to problems (Palaich et al. 2004).

Limitations A major limitation in understanding student outcomes using a SLDS is that it only includes data from the state in which it is located and in most cases, includes only public schools' student data. While most K-12 systems lose only a portion of its activity by not including private school data, post-secondary loses half of its enrollment by not having private institutions' data. In addition, with our highly mobile society, it could prove ineffective to make decisions based on the outcomes of students or employees who stayed in the state.

These systems are also limited in how they can quantify student success. For example, the student success metrics include assessment scores, high school graduation, post-secondary enrollment and completion and wages. While this may resonate with state-level policy makers who fund colleges and universities to increase higher education attainment so students can get higher paying jobs and pay more in taxes, other stakeholders such as parents, may not be interested in these metrics. They may be interested in job satisfaction, exposure to new ideas, integrity and increased confidence.

Yet another limitation is based on the funding process for these systems. Because the funding has flowed through the K-12 state education agencies, many P20W systems are outgrowths of the K-12 SLDS and this may not be helpful in post-secondary or workforce decision-making (Ewell 2009).

Many of the articles point to the ironic disconnect between the assumptions that data-based decision making is apolitical and person-neutral and the fact that people influence all aspects of data use, from determining what data to use to transforming it into evidence to describing the theory of the problem and solution (Gamson 2007; Honig and Coburn 2008; Knapp, Copland, and Swinnerton 2007; Ikemoto and Marsh 2007; Phillips 2007; Spillane and Miele 2007). As Phillips (2007) points out, it is unrealistic to believe that decisions will be made on data or evidence only. Politics, knowledge, ethics, goals, budgets and more all play a role in identifying the problem and the solution. Indeed, many policies at the federal and state levels take into account a variety of factors, not just the data or evidence. It is naïve for policymakers to believe that districts and schools do not also operate within a context and are immune from these same influences. In addition to contextual issues, there may be issues with the data itself. For example, the data may be insufficient, missing or inaccurate, meaning some decisions should not be made using the data.

11.4.2 Focus of Inquiry and Research Questions

Washington has a comprehensive P-20 data system because of its breadth of data sources and depth of data shared by state agencies. In addition, Washington has an office dedicated to P-20 work. This is unique as most states' P-20 office is within the K-12 system where resources are split between K-12 reporting and P-20 work. In addition, Washington has focused on sharing data with educators and researchers, in the form of data sets or aggregated reports, such as the high school feedback report.

Research is needed that explores the process that turns these kinds of data into action, and more specifically, into particular actions that serve the needs of users who are positioned differently in the system. Mandinach (2012) maintains that data needs to be turned into information and then information is turned into knowledge, which leads to action. The existing research does little to describe how knowledge is turned into action.

The following sections will outline the data warehouse/data store elements, then proceed to provide a few concrete examples already being displayed on the ERDC website and research briefs, and finally extensions to future research that could be performed once the data warehouse is fully operational.

11.5 The Washington State P20W Data Warehouse/ Operational Data Store (ODS)

The Washington state P20W data warehouse is the storage facility for the various state longitudinal data systems. This section will detail some of the elements available in many of the various data systems in the P20W data warehouse. First, however, a discussion of the identity-matching process need be presented. Absent the identity-matching process of record linkage, there would be no P20W data warehouse, just data silos.

11.5.1 Record Linkage

Record linkage is the identification of records belonging to the same entity (e.g., a person, household, or housing unit) either within a single data set or across two data sets (Steffey and Bradburn 1994: 36). For Washington state, the utility of the entire SLDS rests on being able to link individuals within a sector (pre-school, K-12, postsecondary, and workforce) and the various programs within each sector, as well as link individuals across sectors (Popoff and Judson 2004: 720). Many P20W operations and research questions involve linking one list of administrative records to another.

According to Sabel (2013), there are three main components to identity matching at ERDC: (1) Identities come in from more than two sources; (2) Matching builds across time, not a one-time process; and (3) Each source of data has their own set of matching rules due to differences in data elements.

The identity matching process at ERDC involves numerous steps (Sabel 2013): (1) Data moves from source data to a staging database. Here identifiers are cleaned and standardized. Also, identity token IDs are assigned to each record. Token IDs are based on sets of identifiers that are concatenated together. Thus defined, each token ID is understood to be unique to a person. Whereas any individual might have more than one token ID, any single Token ID is never shared between individuals. The sets of identifiers that define a token ID vary by source and sector. In step (2), data move from the staging database to a master data management (MDM) hub in the database. In the MDM hub, new data (new token IDs) are merged with existing data. If a person exists already in the data warehouse, their new data (token IDs) are assigned their existing P20ID. New people get new P20IDs, and their new data (token IDs) are assigned to their data (token IDs). Now that token and P20IDs are assigned, matching/record linkages can occur. Exhibit 11.1 displays the source systems used for identity matching.

Exhibit 11.1 Washington P20W Data Sources

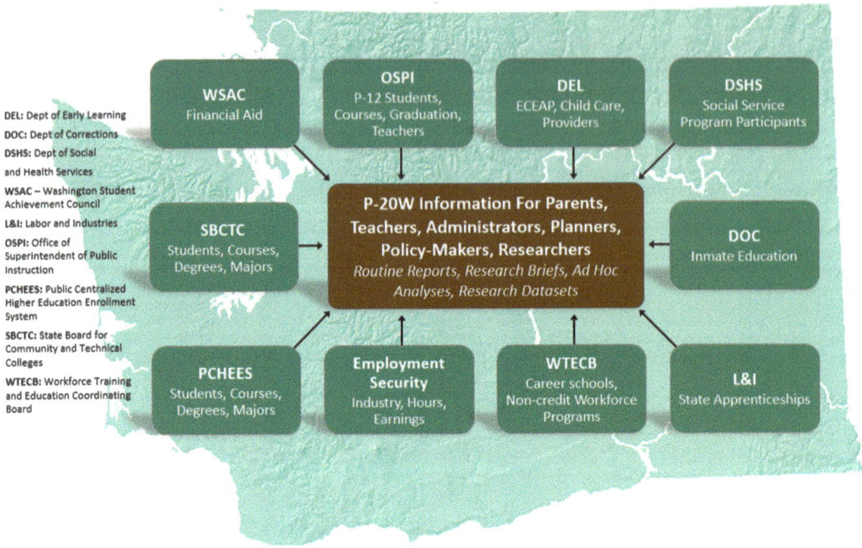

The matching process involves two steps: (1) Data is first merged using "Automerge" rules. Automerge rules are conservative rules for merging identities, merging P20IDs. As ERDC has implemented them, they are largely deterministic, similar to using SQL to merge data. For this step false negative rates are not too important. Rather, Automerge rules are designed to ensure extremely low false positive rates (Sabel 2013). And (2) Data are next subjected to a manual merge rule set. Manual merge rules are looser. They are designed so that the false negative rate is low. False positives are not much of a concern (since potential matches will be manually reviewed). Manual merge rule sets create a manual review table of potential pairs of identities. This table is brought into a spreadsheet, and a human evaluates each potential pair of identities to determine if there is a match or not, via an ERDC custom process. Matches are uploaded to the MDM hub, and results merged. Now that the P20W system has uniquely identified all persons in the system, researchers can begin to link and analyze the data from the various sources.

11.5.2 Washington State P20W PRO Model

The P20W PRO (Person-Role-Organization) model is based upon a **P**erson's **R**ole within an **O**rganization. Examples of Roles are student, teacher, staff, or employee. Organizations are represented by high school building, school district, college campus, institution, or employer. For each Role and Organization combination, a person receives a token ID. For example, a person who attended a certain school as a student, then taught there as a teacher will appear in the PRO table twice.

Exhibit 11.2 displays the inputs into the P20W Data Warehouse and the relationship of the original input data to the P20W Operational Data Store (ODS) with all personally identifiable data removed.

Exhibit 11.2 Washington State PRO Model and Operational Data Store (ODS)

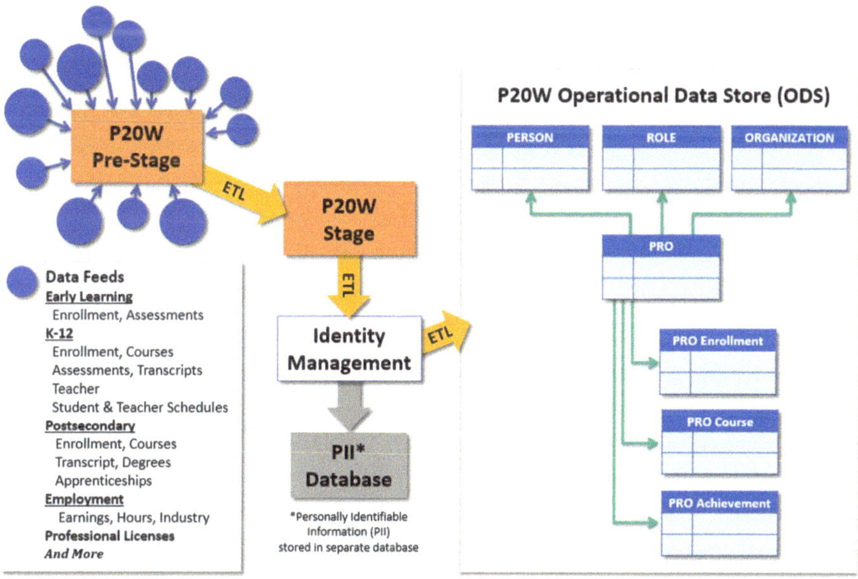

The elements of the P20W data warehouse may be expressed either as subject areas in the PRO Model (Enrollment, Achievement, GPA, Workforce) or as elements in data matrices from each of the contributing agencies. I will present the data by agency and PRO subject area/PRO table relevant to demographic research. Exhibit 11.3 provides a list of the PRO model subject areas common across the various agency data feeds.

Exhibit 11.3 Washington State PRO Model Subjects

PROxxx	
Achievement	Enrollment
Admission	GPA
Assessment	Program
Characteristic	Wage
CourseSection	Staff (Special)
Finance (Special)	IPEDS (Special)

To summarize, the purpose of linking all the Washington state data sources into a PRO Model is so that research can be conducted within and across the various sectors of education and work force participation, that is, a life course representing education and career. Exhibit 11.4 summarizes the four main sectors (Early Learning, K-12, Postsecondary, and Career) as well as the most commonly used PRO Model subject tables. Before identifying the elements in each of the subject area PRO tables, however, a discussion of two underlying elements of all the data must be presented: time and geography.

Exhibit 11.4 Longitudinal Data Spans the Sectors

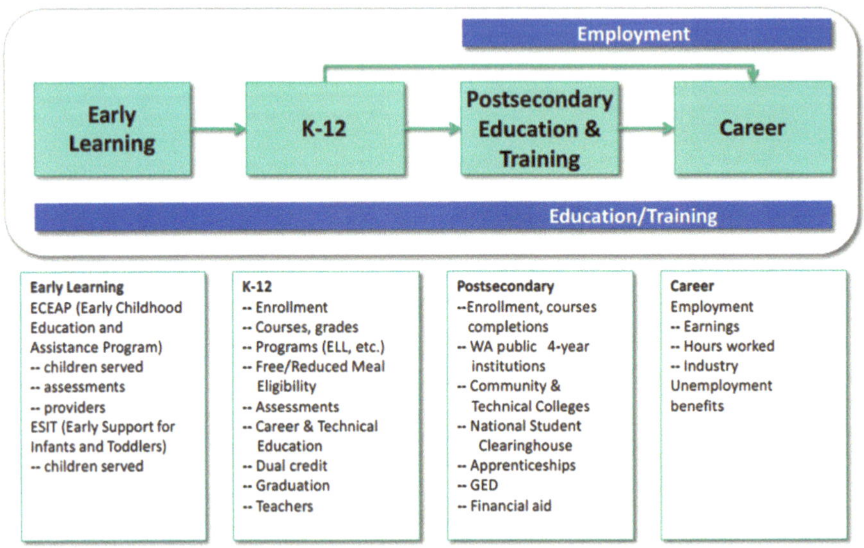

11.5.3 Time Dimension

Time is counted in numerous ways within the Washington State P20W data warehouse. Even the concept of a year has multiple definitions: Fiscal (July 1, Year (t) to June 30, Year (t + 1)), K-12 and Washington State University School Year (August 16, Year t to August 15, Year (t + 1)), all other Community and Technical Colleges and 4-year Universities/Colleges (September 16, Year t to September 15, Year (t + 1)), and workforce/calendar (January 1, Year (t) to December 31, Year (t)). As a result of these multiple calendars, ERDC adopted the concept of Org-Year-Term to cover all instances; i.e., each member is represented in time by the organization in which they are a member according to the year and term (annual, quarter, semester) the organization follows. Aligning these numerous calendars presents challenges for each research question raised, as well as to mathematical models employed to represent time (e.g., event history analysis).

11.5.4 Geographic Information

One of the largest shortcomings of the Washington state P20W data is the lack of geographic identifiers. In addition to aiding identity matching, "the more intensive use of administrative records centered on individual records with the potential for matching, merging, linking, and geographic coding... have the greatest impact on current and future programs of population and other demographic estimates" (Steffey and Bradburn 1994: 164). Although devoid of full addresses, there are, however, some geographic identifiers in the data, mostly linked to Organizations: school building address, school district address, and employer zip code. School districts and zip code tabulation areas have U.S. Census Bureau shape files for thematic mapping purposes. These lower level geographies can also be aggregated to county totals or other geographic/administrative areas of interest.

11.5.5 Data in the ODS by PRO Model Subject Areas

To review, all data are defined by your role within an organization and the time frame in which the relationship occurred. Exhibit 11.5 provides a matrix of Agency by available subject area data (not all subject areas from Exhibit 11.3 are listed in Exhibit 11.5). Characteristics data consist of: Date of Birth, Gender, Race, and Ethnicity, disability status. These data are available from most agencies, although their definitions may not always be the same (e.g., some agencies combine race/ethnicity into a single variable).

Exhibit 11.5 Matrix of Source Agency by PRO Model Subject Area and Years Available

	PRO Model Subject Area					
Agency	Characteristics	Enrollment	Achievement	Program	Wage/ Payment	Years Available
Department of Early Learning (DEL)[a]						
The Early Support for Infants and Toddlers (ESIT)	X	X				2013–2015
Early Childhood Education and Assistance Program (ECEAP)	X	X				2013–2015
Office of Superintendent of Public Instruction (OSPI) K-12	X	X	X	X	X	2005–2013
State Board for Community and Technical Colleges (SBCTC)	X	X	X			2005–2015

(continued)

Agency	PRO Model Subject Area					
	Characteristics	Enrollment	Achievement	Program	Wage/ Payment	Years Available
Public Centralized Higher Education Enrollment System (PCHEES)	X	X	X			2005–2015
Employment Security Department (ESD)						
Unemployment Insurance Wages					X	2002–2015
Unemployment Insurance Payments					X	2002–2015

[a]Note: Earlier years of data may be available for some sources, but with fewer data elements

Enrollment data refer to both enrollment type (full-time, part-time) as well as event type (e.g., dropout, transfer). Included in these data are also time dimensions defined on effective begin dates and effective end dates which allow the construction of spells or spans, and permit the study of stop-out students, seasonal students, or persistent students. They also allow for risk sets to be developed using the enrollment type as denominators and events to be used to calculate transition rates as the numerators.

Achievement data refer to accomplishments in education named Achievement Type (e.g., high school diploma, GED, baccalaureate degree). There may be multiple achievements by one individual within an organization as well as across a number of organizations. Included in these data are also time dimensions defined on effective begin dates and effective end dates which allow the construction of spells or spans between achievements as well as the sequencing of achievements.

The Program data file contains information for students participating in/receiving services from specific programs, eligibility for Free/Reduced Meal participation and individual student attributes. For example, Title I Migrant, Section 504, Limited English Proficiency, and Special Education all represent programs in the OSPI K-12 system. Attachment to these programs may be acute or chronic, constant or changing, hence students may have more than one record per year. These data also contain effective begin and end dates which allow for policy interventions and permit the effects of treatments to be studied over time spans.

Wage data refer to hours worked and wages earned by persons employed in the labor force. Industry of employer is also provided for the organization at the establishment level (headquarters, not site specific), reported in quarterly records. Unemployment benefit payments data are also included providing unemployment compensation amounts, reported on a weekly basis.

11.5.6 Population Universes for Source Agencies

The early learning data are subdivided by programs serving children of various pre-Kindergarten and early kindergarten ages. The Early Support for Infants and Toddlers (ESIT) program covers ages Birth to 3. The Early Childhood Education and Assistance Program (ECEAP) covers children ages 3 to Kindergarten in a state program, similar to the federal Head Start Program. These programs are undergoing expansion to providers along with enrollment.

The universe for OSPI, or pre-kindergarten through 12th grade, are persons aged 0–21, although most of the students are concentrated in ages 6–17. Table 11.1 presents the comparison of OSPI age data on April 1, 2010 versus the U.S. Census Bureau count for that date. Washington public students represent 56 % of persons 21 and under, and 91 % of all those aged 6–17.

The State Board for Community and Technical Colleges (SBCTC) serves students ranging in age from 13 to 85 and over. Community and technical colleges serve two types of students in Washington state: (1) degree-seeking students and (2) basic skills and lifelong learners. Table 11.2 presents the age group distributions for all students by degree-seeking status. Although most of the community colleges offer the traditional 2-year degree plus a number of specialized certificates, a few are now offering a 4-year applied baccalaureate. Overall, some 63 % of students are degree-seeking, with the maximum being represented by 86 % of teens and reaching a 50/50 split somewhere between the ages of 40–44. Lastly, almost 38 % of those 40 and over are still listed as degree-seeking, and still almost 32 % for those ages 55–59.

The Public Centralized Higher Education Enrollment System (PCHEES) data covers students enrolled in Washington state public 4 year institutions of higher education. Table 11.3 displays the age distribution for students in public 4-year institutions in Washington state. Less than 20 % of the students are ages 30 and over, whereas for the community and technical colleges, 43 % of the students are ages 30 and over. This should be no surprise as almost all students in the public 4 year institutions are degree seeking.

Finally, the Washington state Employment Security Department (ESD) is somewhat isolated in the data it provides in terms of employment (hours and wages) and unemployment (compensation). There are no characteristics data available from administrative wage/unemployment forms. Wages, hours and compensation are used as outcome variables in ERDC research (e.g., returns on investments to education), hence population characteristics data linked to ESD data must originate in the K-12 or higher education data. This limitation leads us to now discuss the cohort approach taken in most ERDC research studies.

Table 11.1 U.S. Census for Washington versus OSPI Enrollment by Age on April 1, 2010

Age	Washington 2010 Census	OSPI Enrollment	OSPI as % of Census
Under 1 year	87,016	641	0.7 %
1 year	87,607	1810	2.1 %
2 years	89,399	3346	3.7 %
3 years	89,097	5815	6.5 %
4 years	86,538	10,618	12.3 %
5 years	86,550	39,840	46.0 %
6 years	85,890	77,521	90.3 %
7 years	84,916	78,046	91.9 %
8 years	85,058	78,466	92.2 %
9 years	87,463	80,155	91.6 %
10 years	88,129	79,440	90.1 %
11 years	87,626	79,186	90.4 %
12 years	87,497	79,572	90.9 %
13 years	87,399	79,177	90.6 %
14 years	87,582	79,574	90.9 %
15 years	89,006	82,011	92.1 %
16 years	90,813	83,642	92.1 %
17 years	93,768	83,714	89.3 %
18 years	94,467	55,855	59.1 %
19 years	94,074	11,612	12.3 %
20 years	94,086	4119	4.4 %
21 years	91,883	1090	1.2 %
Total	**1,955,864**	**1,095,250**	**56.0 %**
Total 6–17	**1,055,147**	**960,504**	**91.0 %**

11.6 Using the Washington State P20W ODS for Demographic Research

Administrative records have become the mainstay of federal, state and local government. They play a major role in various demographic programs and their outputs of these systems can be seen in a variety of Washington state publications, be it the open data site at https://data.wa.gov/ or the *Washington State Data Book* (http://www.ofm.wa.gov/databook/default.asp). Usually, the statistical information found in these reports reflect aggregated data from administrative records managed at the state or local level, and usually within one organization. However, with the P20W ODS it is now possible to study individuals across organizations and across sectors (education, career, corrections). To properly study these new data, it is useful to follow a consistent group of persons over time and experiences.

Table 11.2 Age distribution for students enrolled at SBCTC campuses by degree-seeking status, 2009–2010 school year

Age Group	All Students	Degree-Seeking	Non-Degree-Seeking	Percent Degree-Seeking	Percent Non-Degree-Seeking
13–19	92,418	79,567	12,851	86.1 %	13.9 %
20–24	107,241	80,890	26,351	75.4 %	24.6 %
25–29	68,523	45,449	23,074	66.3 %	33.7 %
30–34	47,917	27,316	20,601	57.0 %	43.0 %
35–39	37,815	19,318	18,497	51.1 %	48.9 %
40–44	30,327	15,049	15,278	49.6 %	50.4 %
45–49	26,080	12,597	13,483	48.3 %	51.7 %
50–54	21,445	8,832	12,613	41.2 %	58.8 %
55–59	16,262	5,173	11,089	31.8 %	68.2 %
60–64	10,785	2,022	8,763	18.7 %	81.3 %
65+	12,805	909	11,896	7.1 %	92.9 %
Total	**471,618**	**297,122**	**174,496**	**63.0 %**	**37.0 %**
Under 40	**353,914**	**252,540**	**101,374**	**71.4 %**	**28.6 %**
40 and over	**117,704**	**44,582**	**73,122**	**37.9 %**	**62.1 %**

Table 11.3 Age distribution for students enrolled at PCHEES campuses, Spring term 2010

Age group	All students	Percent of total
Under 20	15,584	17.8 %
20–24	42,356	48.4 %
25–29	12,784	14.6 %
30–34	6,297	7.2 %
35–39	3,664	4.2 %
40–44	2,857	3.3 %
45–49	2,247	2.6 %
50–54	1,165	1.3 %
55–59	310	0.4 %
60–64	116	0.1 %
65+	211	0.2 %
Total	**87,591**	**100.0 %**
Under 30	**70,724**	**80.7 %**

11.6.1 Cohort as the Unit of Analysis

The utility of a cohort as a unit of analysis first came to light in an article by Norman Ryder (1965). The cohort concept is a prime example of linking the micro (i.e., individual) to the macro (i.e., societal) levels of human behavior. For most cohort research to date, the defining event has been birth. However, cohort differentiation is not confined to characteristics fixed at birth; changes in formal education, technological change, and idiosyncratic historical experience as well as other factors play a role in structural transformation (Ryder 1965). The necessary condition for utilizing the cohort design is to follow a fixed groups of individuals over time, initially defined by some calendar dates or by some significant event. The following three examples will briefly describe some cohort based studies completed to date.

11.6.2 Examples of Current Washington State Studies using the Cohort Approach

In parallel with the development of Washington State P20W ODS, ERDC constructed externally matched databases so research could advance before the completion of the data warehouse. Linking and matching algorithms were developed that allowed ERDC and external researchers to explore the data and provide public views for parents, teachers, administrators, legislators, researchers and the general public. Public views of downloadable data and research reports were released on the ERDC website (http://erdc.wa.gov/), and requests for more reports and data followed. A few examples should suffice to understand the approach ERDC has taken to follow cohorts both within a particular sector, as well as across sectors for longitudinal analyses.

11.6.3 High School Feedback Report (OSPI)

One of the first reports developed for public display was the High School Feedback Report (HSFB) (http://www.erdcdata.wa.gov/hsfb.aspx). These reports follow a cohort of Washington state public school graduates (the defining cohort event) over their first year post-graduation. Exhibit 11.6 displays Table 1 from the HSFB report. It displays the percent of graduates attending post-secondary institutions anytime during the year following graduation. It also displays the percent of enrollments (students may enroll in more than one post-secondary institution in a year) by state of post-secondary attendance (in or out of Washington State), public or private institution, and 2- or 4-year participation. Other tables in the HSFB report display similar information: Table 2 – characteristics (race/ethnicity, gender); Table 3 programs (free/reduced meals, bilingual, etc.), Grade Point Average (GPA) and assessment scores; and Table 4 (limited to Washington public post-secondary institutions) college participation in remedial course-taking, persistence, and full- or part-time status.

Exhibit 11.6 High School Feedback Report: College Going and Enrollment in Post-Secondary Education.

What percentage of high school graduates enrolled in postsecondary education?

Table 1. Student enrollment by type of institution Enrolled in Postsecondary Education

Percent of Enrollments	2010	2011	2012	2013
Washington	33 %	33 %	33 %	32 %
Public 4-year	28 %	29 %	31 %	31 %
Private 4-year	5 %	5 %	5 %	5 %
Public 2-year	50 %	48 %	47 %	45 %
Private 2-year	0–1 %	0–1 %	0–1 %	0–1 %
Out of State	17 %	17 %	17 %	18 %
Public 4-year	6 %	6 %	7 %	7 %
Private 4-year	8 %	8 %	8 %	8 %
Public 2-year	2 %	3 %	2 %	3 %
Private 2-year	0–1 %	0–1 %	0–1 %	0–1 %
Total High School Graduates	65,706	66,350	66,241	66,103
% Going to College	62 %	60 %	60 %	62 %

11.6.4 Community and Technical College (CTC) Feedback Report

For the SBCTC feedback report, the cohort approach was defined for all first-time enrolled students within a given school year, 2011–2012. Members of this 2011–2012 cohort were then traced back in time to link to other records from the Washington public K-12 system (http://www.erdcdata.wa.gov/ctc.aspx). Exhibit 11.7 provides data from Table 1 of the CTC report. The tables also provide data on characteristics from the OSPI K-12 SLDS for the almost one-half of new CTC enrollees who are or did attend Washington public school. A third table provides this cohort data for CTC characteristics (age, gender, race/ethnicity). A fourth table provides information on the CTC 2011–12 cohort by intent/purpose of enrollment, full- and part-time status, and number of terms enrolled. Finally, a fifth table presents cohort data on remedial course-taking, credits, and CTC GPA.

Exhibit 11.7 CTC Feedback Report: 2011–2012 First-time Enrolled CTC Students and Prior Washington Public K12 Enrollment.

WA Public Involvement		Counts	Percent
WA K12 Public HS Grads	Graduated from HS in 2012 (within current year of HS Graduation)	920	3.9 %
	Graduated from HS in 2011 (within 1 year of HS Graduation)	14,297	61.1 %
	Graduated from HS in 2010 (within 2 years of HS Graduation)	3,530	15.1 %
	Graduated from HS in 2009 (within 3+ years of HS Graduation)	4,659	19.9 %
	Total HS Grads	23,406	28.6 %
Attended WA K12 Public but Did Not Graduate	Last Enrolled in HS in 2012	3,149	35.7 %
	Last Enrolled in HS in 2011	1,538	17.4 %
	Last Enrolled in HS in 2010	898	10.2 %
	Last Enrolled in HS in 2009 or earlier	3,233	36.7 %
	Total enrolled, Did Not graduate	8,818	10.8 %
No WA Public K12 enrollment or graduation		49,713	60.7 %
Total		81,937	100.0 %

11.6.5 Earnings Information for Washington State Graduates Employed in Washington State

A 2014 budget proviso from the Legislature requested that ERDC "create a report of employment and earnings outcomes for degrees, apprenticeships and certificates earned at institutions of higher education" (ESSB 6002, Sec 129). Until this request, many institutions have relied on graduate surveys to get important feedback information for accreditation and program improvement purposes. While wage outcomes are not the only way to evaluate a program, linking completion and employment data and providing this to institutions, the public and the legislature can assist them in decision-making.

Exhibit 11.8 displays the earnings of students completing certificates and degrees from Washington's public schools and universities and for those completing apprenticeship programs in Washington. There are a number of caveats regarding these public-facing data (e.g., size of group must be greater than 30) and users are advised to consult the following link as well as additional links to frequently asked questions (FAQ) and the about the data section. (http://www.erdcdata.wa.gov/esm.aspx). It must also be noted that a number of these degree grantees and award recipients are only present starting with the higher education data, i.e., they have no link back to the public K-12 SLDS.

Exhibit 11.8 Earnings Information for Washington State Graduates with a Bachelor's Degree Employed in Washington State

Institution	Field of Study	Award Year	Information	2009	2010	2011	2012	2013
Bachelor's Degree								
All organizations								
	All Programs	2007-08	Wage records	8,928	9,211	9,558	9,720	9,895
		2007-08	Median Earnings	$39,200	$41,600	$44,500	$47,400	$50,900
		2008-09	Wage records		8,005	9,016	9,413	9,624
		2008-09	Median Earnings		$36,500	$39,900	$43,500	$47,200
		2009-10	Wage records			8,583	9,569	10,026
		2009-10	Median Earnings			$36,900	$40,800	$44,600

11.6.6 Public Use Microdata Sample (PUMS) Files

ERDC has developed PUMS type files for the annual High School Feedback Reports as well as augmented them for specific data requests (Education Research and Data Center 2014). The base high school feedback report files contain individual level data on each high school graduate, their high school demographic characteristics (gender, race/ethnicity), flags for program participation (free/reduced meal status, bilingual, special education, Section 504, Title I Migrant), Grade Point Average (GPA), math and reading assessment flags (pass/fail), links to college enrollment for the first year following graduation (in-state/out-of-state, public/private, and 2-year/4-year), and remedial course-taking, continuous enrollment and full-time/part-time course taking in college.

The preceding three examples of cohort approaches discussed ongoing and continuing research at ERDC. With the completion of most of the components of the Washington state P20W ODS, research opportunities will provided to move beyond these reports.

11.7 Future Research Using the Washington State P20W (ODS) for Demographic Research

Members of the Washington state ERDC have just begun to utilize the P20W ODS when it comes to research. External requests and current grants will extend this research to corrections data linked to education and the workforce, and other grant deliverables (e.g., linking education to corrections data and linking Early Learning to K-12). In addition, a number of demographic concepts currently employed in the field can be augmented by utilizing this new data source.

11.7.1 Impact of Education and Other Characteristics on Jail and Prison Admission

The Washington State Statistical Analysis Center (SAC), located in the Human Services Section of the Washington State Office of Financial Management (OFM) received a grant to perform studies on the connection between education and corrections. In conjunction with ERDC, one study will examine a series of cohorts of ninth graders. These students will be tracked through the Washington public K-12 system and linked with postsecondary education/training and employment. For each student, data will be linked on their characteristics (gender, race/ethnicity, and free/reduced meal status), their student assessments in math and reading, their participation in various programs (special education, English language learner status), their GPA, contextual characteristics of the school/school districts they attend (free/reduced meal status, size, urban and metro delineations), and final withdrawal status (graduated or completed, dropout, or unknown status). Both prison and jail data will be linked to the aforementioned data. This will allow the inclusion of information such as age at first jail admission during the follow-up period. Six ninth grade cohorts beginning with the 2005–2006 cohort will be followed through 2014–2015 for this exploratory study.

11.7.2 Impact of Early Learning on K-12 Outcomes

Washington State created the Department of Early Learning 10 years ago by combining functions and services in three different state agencies. In addition, the Legislature increased funding for state-funded preschool programs and full-day kindergarten programs in the K-12 system. With the consolidation of services, increased funding and a kindergarten readiness assessment, policy makers and educators are asking questions that only longitudinal data can begin to address. Are children in the state-funded preschool programs "ready" for kindergarten? For children in the state-funded preschool program, are there differences in outcomes between those who attended half-day versus full-day kindergarten? What are the relationships between the quality rating of an early learning provider and K-12 outcomes? To answer these questions, ERDC will link the birth-to-3, state-funded preschool, and K-12 student data to understand the early learning to K-12 pathways, transitions and outcomes.

11.7.3 Demographic Accounting

A major principle of demographic methodology is the population balancing equation. Land and McMillen (1981) expanded the basic mathematical identity of population stocks being balanced by births, deaths and migration flows to include

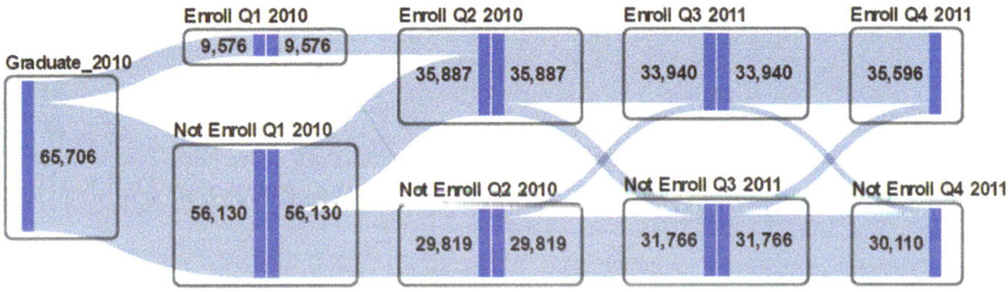

Fig. 11.1 Sankey chart for 2010 Washington Public HS Graduates and 2010–11 college participation

characteristics as well as state space. Figure 11.1 provides the stocks and flows of high school graduates from 2010 as they pass through their first four quarters of post-secondary enrollment. The boxes show the stocks at graduation as well as each quarter term (Q1 = Summer, Q2 = Fall, Q3 = Winter, and Q4 = Spring). The similarity of the stock numbers on either side of the bar in each box indicates that no censoring of the data exists; i.e., the ending stock for the fall quarter (Q2) is equal to the beginning stock for winter quarter (Q3). The curved lines (resembling sin and cosine graphs) represent the events (transfers or flows) between the two-state space of Enrolled (E) and Not Enrolled (NE). Given these Stocks and flows, probabilities can be also calculated between the states of E and NE. A discussion of types of multistate probabilities can be found in Schoen (1988, pp. 81–83). It is interesting to note that 40,978 2010 public high school graduates enrolled in college sometime during the summer of 2010 and spring 2011, however the largest stock of enrolled graduates occurs during Fall 2010 when 35,887 were enrolled.

As mentioned above, characteristics data are readily available from a number of P20W ODS source systems. And when linked with other systems, demographic accounting and models of social change can be examined for each of the cohorts. One could even model policy changes into these demographic accounts (e.g., increasing costs for college attendance leading to policies that encourage students in grades 11 and 12 to take more dual credit courses). Which populations/cohorts might benefit from such a policy shift? How do these changes then affect the composition of degree completion and educational attainment for the state and local areas?

11.7.4 Population Estimates and Projections

Using the Enrollment files from the OSPI K-12 SLDS, one can link student records at the County or sub-county (District or School). Transfers between districts across county borders would provide an estimate of migration from one school year to the next, or even smaller time intervals if the exit and entry enrollment dates are utilized. Identity matching from the P20W warehouse will be necessary as each combination of District and State Student Id will produce a Token ID and not an ID that can be linked from year to year if the student transfers between districts. County moves could also be modeled as the first 2 digits of the District ID identify a county code. An estimate of migration for the K-12 or grades 1–11 could be converted to age and then a scalar could be applied to this total for an estimate of migration for all ages in a county or school district, similar to the scalar relating employment growth to migration.

Following along these lines, in 2007, researchers at the Population Research Center (PRC) developed a proposal for a school district-wide enrollment forecast for the Portland, OR metropolitan area (Lycan et al. 2007). The PRC has a long history of contracting with school districts for school enrollment projections or forecasts. Largely, many of the models employed were limited to using grade progression ratios from one school year to the next (e.g., grade 2 in say 2011 divided by the grade 1 in 2010) for projections. School districts could only provide their District student ID and not the state ID and hence students could only be linked by grade if they stayed in the district. For example, the second grade class in a school district for 2011 is a result of the mathematical identity of students from the district from grade 1 in 2010 *minus* the number of those in first grade who could not be linked for grade 2 *plus/minus* some residual number that solves for the stock of students now attending second grade in 2011. Utilizing a statewide ID, and in particular, the P20ID (collecting all Token IDs) in Washington State allows for a more dynamic view of the changes that occur when moving from one grade to the next. For the Washington state example, the second grade class in a school district for 2011 is now a result of the mathematical identity of students from the district from grade 1 in 2010 *minus the outflows* (the number of students that transferred out of the district from grade 1 to grade 2 + the number of those in first grade who could not be linked) *plus* the number of students in grade 1 transferring into the school district for grade 2, *plus/minus* some residual number that solved for the stock of students now attending second grade in 2011. Hence, the transfers in and out can now be modeled and the size of the residual change has been minimized. A further extension would also model the drop-outs as exits to not enrolled, however, most do not begin until the ninth grade.

11.7.5 Tables of School Life

Similar to studies performed by Stockwell and Nam (1963) and Land and Hough (1989), the P20W ODS could be utilized to develop tables of school life; models more developed than the earlier research as now individual students would be linked across years and grades. If we utilize the same methodology as in the demographic accounting Sect. above (11.7.3) to incorporate transition rates between the enrolled and not enrolled states, we can improve on the Land and Hough (1989) model in a number of ways: (1) we will be able to use administrative data, not sample data, and so smoothing the rates should not have to be performed; (2) we will not be confined by prevalence rates at ages 3–13; (3) we can implement the two-state increment-decrement (Enrolled and Not Enrolled) rates and life table model beginning at age 3 and continuing throughout the years of available data; and (4) we are no longer limited to 1 year of transition rates; i.e., we can add students from succeeding years beyond year t + 1. In effect, one could begin constructing generational tables of school life.

11.8 Discussion

Over the past decade states, with financial assistance from the US government, have made notable strides in developing and linking partial population registers (administrative records systems) across a variety of sectors (early learning, K-12, higher education, and workforce). As it stands now, all 50 states have received funding from the federal government to build a K-12 SLDS and 43 states have received funding to build a P-20/Workforce SLDS (P20W). Washington has a comprehensive P20W data system because of its breadth of data sources and depth of data shared by state agencies. The fruitfulness of these data linkages are just now being realized.

The Washington state Education Research and Data Center has been developing cohort studies and sharing both public facing reports (see http://www.erdc.wa.gov/) as well as research data sets. As more data from relevant agencies (Early Learning, Corrections, Jail) are linked and added to the P20W warehouse, and additional data elements from existing K-16 SLDS are combined (K-12, SBCTC, PCHEES), research studies will begin to provide a more complex understanding of the process of the education and career life course.

In addition to contributing to life course research, even if only by linking two segments at a time, further advances in applied demography will be forthcoming. Population estimates, school enrollment forecasts/projections, demographic accounting and social change models, as well as tables of school and work life will all benefit from utilizing this rich P20W data warehouse.

Acknowledgements The authors are grateful to Marc Baldwin and Jim Schmidt for the encouragement and support for all projects ERDC. We are also grateful to our ERDC colleagues for comments and suggestions, including Tim Norris, Carol Jenner, Karen Pyle, John Sabel, Vivien Chen, Thomas Aldrich, Gary Benson, Liz Coker, Lynn Cole, Teresa Greene, Kent Meneghin, Toby Paterson, Katie Weaver-Randall and Greg Weeks.

References

Alliance for Excellent Education, Civic Enterprises, & Data Quality Campaign. (2011). *Education as a data-driven enterprise: A primer for leaders in business, philanthropy, and education.* Retrieved December 10, 2015, from http://all4ed.org/reports-factsheets/education-as-a-data-driven-enterprise-a-primer-for-leaders-in-business-philanthropy-and-education/

America COMPETES Act of 2007. (2007). Retrieved December 9, 2015, from http://www.govtrack.us/congress/bills/110/hr2272/text

Culhane, D. P., Fantuzzo, J., Rouse, H. L., Tam, V., & Lukens, J. (2010). Connecting the dots: The promise of integrated data systems for policy analysis and systems reform. *Intelligence for Social Policy, 1*(3).

Cunningham, A. F., & Milam, J. (2005). *Feasibility of a student unit record system within the Integrated Postsecondary Education Data System* (NCES 2005–160). U.S. Department of Education, National Center for Education Statistics. Washington, DC: U.S. Government Printing Office.

Data Quality Campaign. (2007). *The right data to the right people at the right time: How interoperability helps America's students succeed.* Retrieved August 21, 2013, from, http://www.dataqualitycampaign.org/files/events/resources/Meetings-DQC_Quarterly_Issue_Brief_061307.pdf

Data Quality Campaign. (2011a). *From compliance to service: Evolving the state role to support district data efforts to improve student achievement.* Retrieved August 21, 2013, from, http://dataqualitycampaign.org/files/1455_From%20Compliance%20to%20Service.pdf

Data Quality Campaign. (2011b). *Leveraging the power of state longitudinal data systems: Building capacity to turn data into useful information.* Retrieved August 21, 2013, from, http://www.dataqualitycampaign.org/files/1303_DQCResearch% 20capacity%20May17.pdf

Data Quality Campaign. (2012). *Investing in educator data literacy improves student achievement. Evidence of impact: The Oregon Data Project.* Retrieved August 21, 2013, from, http://www.dataqualitycampaign.org/find-resources/investing-in-educator-dataliteracy-improves-student-achievement/

Education Research and Data Center. (2009). *Washington ARRA SLDS grant application.* Retrieved September 8, 2013, from http://www.erdc.wa.gov/arraslds2009/application/

Education Research and Data Center. (2012). P-20 reports on Washington Public High School graduates. Retrieved February 9, 2016, from, http://www.erdcdata.wa.gov/

Education Research and Data Center. (2014). *Issue Trak Number 1025: Post-secondary attainment by students with disabilities*, Washington State Student Achievement Council. Retrieved February 22, 2016.

Ewell, P. T. (2009). "Shovel-ready" data: The stimulus package and state longitudinal data systems. *Assessment Update, 21*(5), 11–12.

Gamson, D. (2007). Historical perspectives on democratic decision making in education: Paradigms, paradoxes, and promises. *Yearbook of the National Society for the Study of Education, 106*(1), 15–45.

Higher Education Opportunity Act of 2008. (2007). Retrieved December 9, 2015, from, http://www.govtrack.us/congress/bills/110/hr41

Honig, M. I., & Coburn, C. (2008). Evidence-based decision making in school district central offices: Toward a policy and research agenda. *Educational Policy, 22*, 578–608.

Ikemoto, G. S., & Marsh, J. A. (2007). Cutting through the "data-driven" mantra: Different conceptions of data-driven decision making. *Yearbook of the National Society for the Study of Education, 106*(1), 105–131.

Improving America's Schools Act. (1994). Retrieved December 9, 2015, from, http://www2.ed.gov/legislation/ESEA/index.html

Knapp, M. S., Copland, M. A., & Swinnerton, J. A. (2007). Understanding the promise and dynamics of data-informed leadership. *Yearbook of the National Society for the Study of Education, 106*(1), 74–104.

Land, K., & Hough, G. (1989). On the combination of prevalence rate and increment-decrement methods for tables of school life, with applications to the 1969–70, 974–75, and 1979–80 school years. *Journal of the American Statistical Association, 84*(March), 63–75.

Land, K., & McMillen, M. (1981). Demographic accounts and the study of social change, with applications to the post-world war II United States. In *Social accounting systems: Essays on the state of the art* (pp. 241–306). New York: Academic.

Lycan, D., Hough, G., Rynerson, C., & Proehl, R. (2007). *A proposal for a metropolitan wide school enrollment forecasting system for the Portland, Oregon Metropolitan area*. Portland: The Population Research Center, Portland State University.

Mandinach, E. B. (2012). A perfect time for data use: Using data-driven decision making to inform practice. *Educational Psychologist, 47*(2), 71–85.

No Child Left Behind Act of 2001. (2001). Retrieved August 15, 2013, from, http://www2.ed.gov/policy/elsec/leg/esea02/index.html

Palaich, R., Good, D.G., & van der Ploeg, A. (2004). *State education data systems that increase learning and improve accountability*. Policy Issues, 16. Retrieved February 23, 2016, from, http://files.eric.ed.gov/fulltext/ED489522.pdf

Petrides, L. (2003). *Turning data into decisions*. Institute for the Study of Knowledge Management in Education, November, 2003, pp. 25–28. Retrieved February 23, 2016, from, http://www.issuelab.org/resource/turning_data_into_decisions

Phillips, D. C. (2007). Adding complexity: Philosophical perspectives on the relationship between evidence and policy. *Yearbook of the National Society for the Study of Education, 106*(1), 376–402.

Popoff, C., & Judson, D. (2004). Chapter 22: Some methods of estimation for statistically underdeveloped areas. In J. Siegel & D. Swanson (Eds.), *The methods and material of demography* (2nd ed.). San Diego: Elsevier Academic Press.

Poulain, M., & Herm, A., (2013). Le registre de population centralisé, source de statistiques démographiques en Europe , *Population* 2/2013 (Vol. 68), pp. 215–247, Translated to English by R. Depledge as Central population registers as a source of demographic statistics in Europe. www.cairn.info/revue-population-2013-2-page-215.htm

Revised Code of Washington 43.41.400. Office of Financial Management Education Data Center. Retrieved December 10, 2015, from, http://apps.leg.wa.gov/rcw/default.aspx?cite=43.41.400

Ryder, N. (1965). The cohort as a concept in the study of social change. *American Sociological Review, 30*, 843–861.

Sabel, J. (2013). *Identity matching at ERDC*. Presentation to the Research Coordination Committee September 6, 2013. Office of Financial Management, Education Research and Data Center. Retrieved December 29, 2015, from, http://qa-erdc.des.wa.gov/sites/default/files/publications/20130906_identity_matching.pdf

Schoen, R. (1988). *Modeling multigroup populations*. New York: Plenum Press.

Siegel, J., & Swanson, D. (2004). *The methods and material of demography* (2nd ed.). London: Elsevier Academic Press.

Snow-Renner, R., & Torrence, M. (2002). *No child left behind policy brief: State information systems*. Retrieved December 9, 2015, from, https://eric.ed.gov/?q=%22no+child+left+behind%22&ff1=subEducational+Policy&ff2=pubReports+–+Descriptive&ff3=autSnow-Renner%2c+Ravay&id=ED468187

Spillane, J. P., & Miele, D. B. (2007). Evidence in practice: A framing of the terrain. *Yearbook of the National Society for the Study of Education, 106*(1), 46–73.

Steffey, D., & Bradburn, M., (Eds.). (1994). *Counting people in the information age*. Panel to Evaluate Alternative Census Methods; Commission on Behavioral and Social Sciences and Education; Division of Behavioral and Social Sciences and Education; National Research Council.

Stockwell, E., & Nam, C. (1963). Illustrative tables of school life. *Journal of the American Statistical Association, 58*(304): 1113–1124.

U.S. Department of Education. (2005). *Statewide longitudinal data system grants request for applications*. Retrieved December 10, 2015, from, http://nces.ed.gov/Programs/SLDS/pdf/RFA_2005.pdf

U.S. Department of Education. (2008). *Statewide longitudinal data system grants request for applications*. Retrieved December 10, 2015, from, http://nces.ed.gov/Programs/SLDS/pdf/2009_84372A.pdf

U.S. Department of Education. (2011). *Family educational rights and privacy final regulations.* Retrieved December 10, 2015, from, http://www.gpo.gov/fdsys/pkg/FR-2011-12-02/pdf/2011-30683.pdf
U.S. Department of Labor. (2010). *Workforce data quality initiative solicitation for grant applications.* Retrieved August 12, 2013, from, http://www.doleta.gov/performance/workforcedatagrant09.cfm
Washington State House of Representatives. (1993). *House bill report, 1209.* Olympia: Washington State Legislature.
Wayman, J. C. (2007). Student data systems for school improvement: The state of the field. In Educational technology research symposium: Vol. 1 (p. 156–162). Lancaster, PA: ProActive.

Chapter 12
How Does Access to Primary Care Vary by Type of Insurance and by Rural and Urban Counties in Mississippi?

Ron Cossman

Abstract A common assumption is that access to primary care is less prevalent in rural areas. To measure actual access to primary health care in Mississippi we telephoned primary care physician practices (general practitioner, family practice, internal medicine, OB/GYN and pediatric) in Mississippi posing as a potential new patient. To measure access, we asked for a new patient appointment with a physician varying our insurance type such as Blue Cross & Blue Shield (BC&BS), Medicare, or Medicaid health insurance. Similar to other findings, only 50 % of practices were accepting new Medicaid patients. Likewise only 64 % were accepting new Medicare patients and 78 % were accepting new BC&BS health insurance patients. However, there were significant differences across the rural-urban continuum. Those practices in urban counties were significantly less likely to accept new Medicaid patients (<50 %) versus those in the most rural counties (up to 85 %). This trend was not evident for BC&BS or Medicare insurance patients. We speculate that lower operating costs in rural areas, plus smaller patient populations, may provide the financial incentives for physicians to accept Medicaid patients. As a result, access to health care for Medicaid-insured individuals is greater in rural counties than it is in urban counties in Mississippi.

Keywords Access to health care • Health insurance • Rural-urban continuum code • Medicaid • Mississippi

12.1 Introduction

Theoretically, higher levels of health insurance should lead to higher access rates for patients to the health care system. However, having health insurance does not guarantee that individuals will be seen by doctors. Thus, "discussions of health care quality are moot for Americans who *cannot get into* the health care system"

R. Cossman (✉)
Social Science Research Center, Mississippi State University, Starkville, MS, USA
e-mail: ronald.cossman@ssrc.msstate.edu

D.A. Swanson (ed.), *The Frontiers of Applied Demography*, Applied Demography Series 9, DOI 10.1007/978-3-319-43329-5_12

(Agency for Healthcare Research and Quality 2014, "National Healthcare Disparities Report 2013", p. H-10, emphasis added). Therefore, physical access to primary health care is the lynchpin of the entire U.S. health care system. It does not matter how many physicians are practicing or what diagnostic and treatment technologies exist if an individual with health insurance is unable to obtain an appointment with a primary health care practice.

Access is one measure of the availability of health care, along with supply and demand for health care. Access, or conversely, barriers, come in many forms. The measure examined in this paper is the ability to obtain a new patient appointment with a primary care physician. Barriers to access include the lack of health insurance, geographic distance, lack of transportation and lack of flexibility in one's schedule. These barriers are external to this investigation.

12.2 Current Research

Prior research shows that only 50 % of primary care physicians in Mississippi were accepting new Medicaid patients (Cossman et al. 2014). These findings have likewise been confirmed for metropolitan areas throughout the U.S. (Miller 2014). Thus, preliminary findings challenge the widespread assumption that having health insurance equates to having access to health care (Cossman et al. 2014). Barriers to primary health care and limitations come in many forms. Having health insurance does not guarantee access, which potentially emasculates current research on potential populations affected by the expansion of Medicaid, offerings through the health insurance exchange, and so forth. However, the rurality of one's residence was not found to be the barrier to access that one would suppose.

Supply and demand for health care and, separately, access to care, are measured in three ways. *Supply* metrics calculate the ratio of health care practitioners to the resident population in a county or some variation of that calculation. For example the Health Professional Shortage Area (HPSA) designation measures the supply of primary care physicians, dentists and mental health professionals in counties, calculates a ratio of health care professionals to the resident population and produces a HPSA score (U.S. Department of Health and Human Services 2015a). *Demand* indicators come from either survey questions, in which individuals were asked if they were able to obtain timely access to health care in the last year (National Center for Health Statistics 2015; Shi et al. 2010) or from a retrospective examination of physician reimbursement records (e.g., the National Ambulatory Medical Care Survey). In this study we focus on the third measure – *access* to health care. We measure this by calling primary care offices to determine whether potential new patients are accepted by the practice, and accepted or declined at different rates by health insurance type. A terminological nuance must be noted here. Physician practices may "decline to accept" a potential new patient, but does not "refuse" a new patient. Physician practices are similar to other professional association practices (e.g., attorney, accountant or architect), and the business is free to accept, or not accept, a new client.

We begin our literature review by looking at the *supply* of health care professionals. Two-fifths (40.2 %) of Mississippi's population resided in a primary care HPSA as of July 2014 (U.S. Department of Health and Human Services 2015b), just one measure of the *supply* of health care professionals. The same study estimated that the state would need to add 230 additional primary care practitioners to eliminate the shortage designation.

Several *demand* for health care studies measure the inability of individuals to see a physician or obtain a timely appointment. A 2011–2014 study found that nationally, 28–30 % of *privately insured* adults (age 50–64) reported difficulties in obtaining an appointment when seeking an initial appointment with a primary care physician for routine care (Medicare Payment Advisory Commission 2015a, b). Similarly, the study found that 23–26 % of *Medicare* insured adults (age 65 and older) reported some problem when seeking a routine care appointment. A 2011 study found that nationally, 20.4 % of office-based physicians (not restricted to primary care) were not accepting new *Medicaid* patients (Decker 2012). Finally, based on physician records, a study found that 33 % of primary care physicians did not accept new *Medicaid* patients in 2011–2012 (Decker 2013). Thus, inability to obtain an appointment with a primary care provider, depending on the type of health insurance, ranges from 20 to 33 %, based on either surveys of patients or analysis of payment records. The implication is that, regardless of demand or supply, there is a persistent unmet need for access to primary health care.

Looking at *acceptance* rates, a national survey in 2011 found that 68 % of *privately insured* adults (age 50–64) reported no problem when seeking an appointment with a new primary care physician, a result which is consistent with our finding that 78 % of Mississippi private insurance patients (all ages) were able to obtain an appointment with a core primary care physician (Medicare Payment Advisory Commission 2015b). Similarly, the national survey found that 65 % of *Medicare insured* adults (age 65 and older) reported no problem when seeking a new appointment, a result echoed by our finding of 64 % among Mississippi primary care physician offices. Nationally the estimate was 69.4 %, ranging from 40.4 to 99.3 % at the state level. In July 2014, the Centers for Medicare and Medicaid Services (CMS) announced future plans to conduct a large scale (1.5 million respondents) survey of *Medicaid* enrollees to measure their ability to obtain health care (Dickson 2014).

12.3 Methods

This study focused on the issue of access to primary health care in Mississippi, with the assumption that an individual had some type of health insurance. We used data from the Mississippi State Board of Medical Licensure (2010–2011) to obtain a list of licensed physicians. There were 5098 physicians licensed in the state. We used the HPSA definition of primary care to classify general practice, family practice, internal medicine, OB/GYN and pediatrics (U.S. Department of Health and Human Services 2015c). Next, we reduced the list to primary care physicians who were currently active in patient care in the state by eliminated those practicing outside of the state or retired from practice (N = 2138). We further excluded walk-in clinics because service is not dependent on type of insurance. We consolidated the list by combining physicians who shared a practice or office (N = 678). Physicians who share a practice will share business practices. That includes guidelines on acceptance of new patients. Thus, we focus on calling physician practices rather than individual doctors.

Approaching the telephone survey, in consultation with our Office of Compliance, it was felt that a telephone survey in which we self-identified as a university-based survey could potentially bias the results. For example, if physicians responded to a hypothetical survey instead, it is likely that physician offices would be less invested in the response (i.e., talking to a potential new patient) and would be more likely to respond with what they perceived as the "correct" response. It was important to avoid that potential bias in order to obtain a more realistic measure of access. The "secret shopper" or "mystery shopper" technique was used to measure access to a primary care physician. Posing as a new patient, we contacted each office by telephone using Gmail VOIP telephone service, thus masking our true location at Mississippi State University.

Offices were contacted in two separate waves. Wave 1 occurred from May to June 2012, prior to the Supreme Court's ruling on the Affordable Care Act. A total of 294 unique offices were contacted. Wave 2 occurred from November to December 2012, during which 384 unique offices were contacted. In each call, we (1) identified ourselves by a pseudonym (females called OB/GYNs, and for Medicare we called on behalf of our elderly mother); (2) volunteered information about our insurance type (private, e.g., Blue Cross & Blue Shield, or Medicare, or Medicaid), and (3) asked to schedule a new patient appointment with the physician for routine care. Of 678 individual and group practice offices that were identified, 580 were successfully contacted, for a completion rate of 86 %. We identified 489 *core* primary care offices (general practice, family practice, and internal medicine), 99 OB/GYN offices and 90 pediatric offices. The conservative margin of error for acceptance by all three kinds of insurance was +/− 1.6 %.

12.4 Results

Rates of new patient acceptance for all practices in the state varied widely by the type of health care insurance the new patient claimed. A new patient with private insurance (*BC&BS*) was accepted over the telephone by 78 % of all primary practices contacted (Fig. 12.1). Acceptance ranged from 91 % for OB/GYN to 75 % for core primary care practices. Outright decline-to-accept as a new patient ranged from 8 % for core primary care practices to 0 % for OB/GYN, averaging 7 %

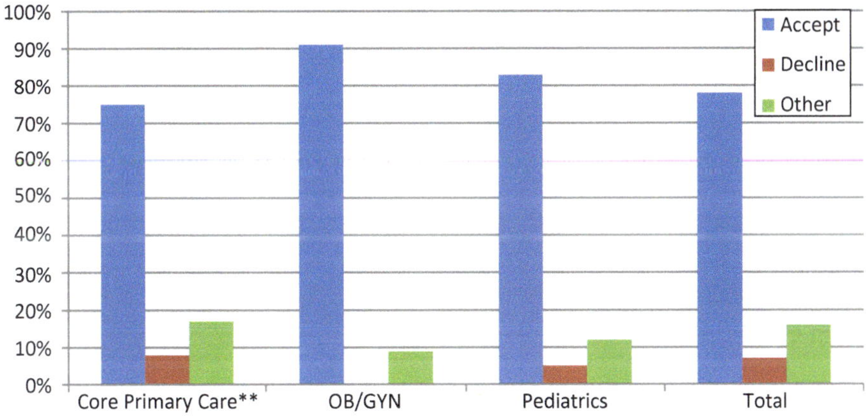

Fig. 12.1 Blue cross & blue shield health insurance acceptance/decline rates

for all categories. The unresolved category of "other" ranged from 17 % for core primary care practices to 9 % for OB/GYN, averaging 16 % for all categories. The "Other" category consisted of the calls that did not result in an appointment in the first telephone call because the practice required additional information (i.e., a valid health insurance number, valid social security number, valid date-of-birth, etc.), a requirement that the physician wanted to examine the individual before accepting them as a patient, or required a call-back to the patient prior to scheduling an appointment. Ultimately these practices may or may not have accepted the new patient.

When presented with a potential new patient with *Medicare*, 64 % of practices readily accepted and were willing to schedule an office visit (Fig. 12.2). Acceptance of new patients ranged from 76 % for OB/GYN to 63 % for core primary care practices (pediatrics were not surveyed). An average of 15 % of practices declined-to-accept the new patient. This category ranged from 8 % for OB/GYN to 16 % for core primary care. The "other" category was 16 % (OB/GYN) and 21 % (core primary).

Finally, when a potential patient with *Medicaid* health care insurance called a primary care practice, only 50 % of all practices would accept them as a new patient (Fig. 12.3). Acceptance ranged from a high of 61 % for pediatrics to a low of 47 % for core primary care. Outright decline-to-accept ranged from 43 % for primary care practices to 21 % for pediatrics, with an average for all groups of 38 %. The "other" category was lower than that for Medicare, with an average of 12 % for all groups, a low of 10 % for primary care and a high of 17 % for pediatrics.

It must be noted that these results potentially underestimate the lack of access to health care because of insurance status. Eventually all of the "other" responses (ranging from 9 to 21 % of practices) would have resolved into either accept or decline of the new patient.

Reported thus far are state-level aggregations of all practices contacted. Although acceptance rates differ by sub-specialty, they differ sometimes

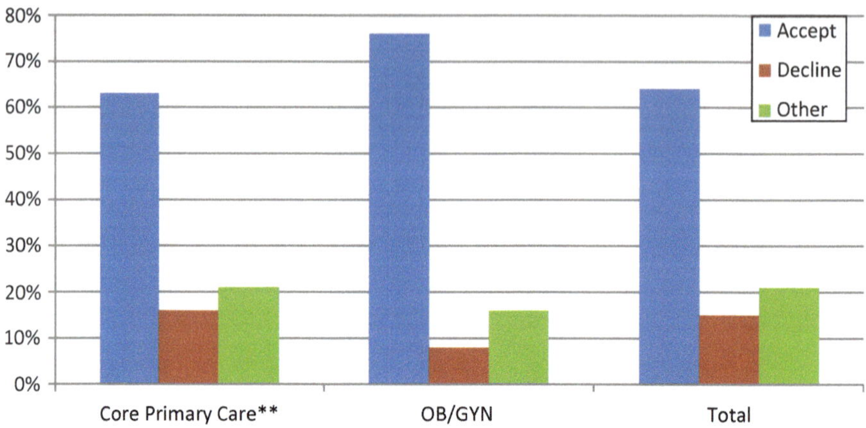

Fig. 12.2 Medicare health insurance

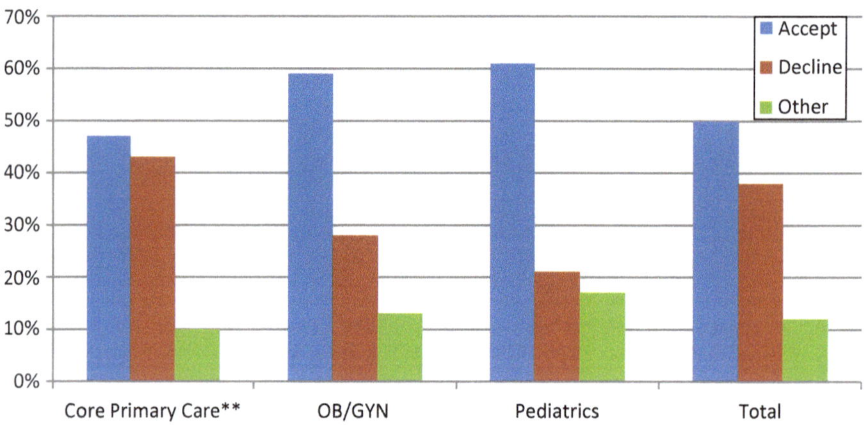

Fig. 12.3 Medicaid health insurance

dramatically across the rural-urban continuum (RUCC). The RUCC divides all counties into one of nine categories based on population and urbanicity or adjacency to an urban county (Economic Research Service 2015). For a listing of codes, see Table 12.1. Table 12.2 presents the number of Mississippi counties in each RUCC, the percent of the state's population and the percent of counties in each RUCC. Almost half (44.7 %) of Mississippi's population resides in counties that are part of a metropolitan area, yet that is only 14.7 % of all counties in the state. More than two-thirds (67.1 %) of Mississippi's counties are less than 20,000 in population, yet more than a third of the population (36.8 %) reside in those rural counties.

When we compare the distribution of the population by RUCC versus the distribution of primary care physician offices, we see that the two distributions

Table 12.1 Rural urban continuum codes

RUCC	Rural urban continuum code
1	Metro – Counties in metro areas of 1 million population or more
2	Metro – Counties in metro areas of 250,000–1 million population
3	Metro – Counties in metro areas of fewer than 250,000 population
4	Nonmetro – Urban population of 20,000 or more, adjacent to a metro area
5	Nonmetro – Urban population of 20,000 or more, not adjacent to a metro area
6	Nonmetro – Urban population of 2500–19,999, adjacent to a metro area
7	Nonmetro – Urban population of 2500–19,999, not adjacent to a metro area
8	Nonmetro – Completely rural or less than 2500 urban population, adjacent to a metro area
9	Nonmetro – Completely rural or less than 2500 urban population, not adjacent to a metro area

Source: "Rural-urban Continuum Codes", Economic Research Service, U.S. Department of Agriculture

Table 12.2 Rural urban continuum code distribution in Mississippi

RUCC	Population	% of population	# of counties	% of counties
1	246,789	8.3	5	6.1
2	937,824	31.6	9	11.0
3	142,842	4.8	3	3.7
4	163,885	5.5	3	3.7
5	386,372	13.0	7	8.5
6	391,061	13.2	15	18.3
7	444,225	15.0	19	23.2
8	142,129	4.8	12	14.6
9	112,170	3.8	9	11.0

Source: "Rural–urban Continuum Codes", Economic Research Service, U.S. Department of Agriculture

are similar (see Fig. 12.4). The greatest (and not unexpected) mismatch between population and health care practices is in the most rural counties (RUCC 9).

However, access afforded to those populations is a different story. Acceptance rates for both *private-pay* BC&BS and *Medicare*-insured new patients were high across the RUCC, ranging from 78 to 100 % (see Fig. 12.5). There was marginal difference between the two in their acceptance rates. However, for new *Medicaid* patients the acceptance rates varied widely. More importantly they increased from urban to rural areas. For example, in the most urban counties (metro areas of 1 million or more) acceptance rate were only 45 %, whereas in the most rural counties (less than 2500 residents), the acceptance rate was 85 %. Although the total number of respondents in each category is small and subject to instability, the overall trend appears to be that acceptance rates for *Medicaid* patients increase as the counties become more rural.

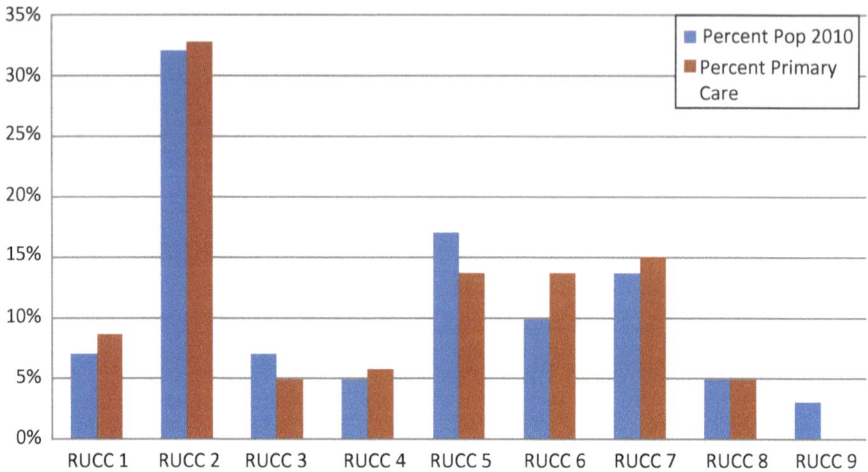

Fig. 12.4 Percent population and physicians Rural-urban continuum code

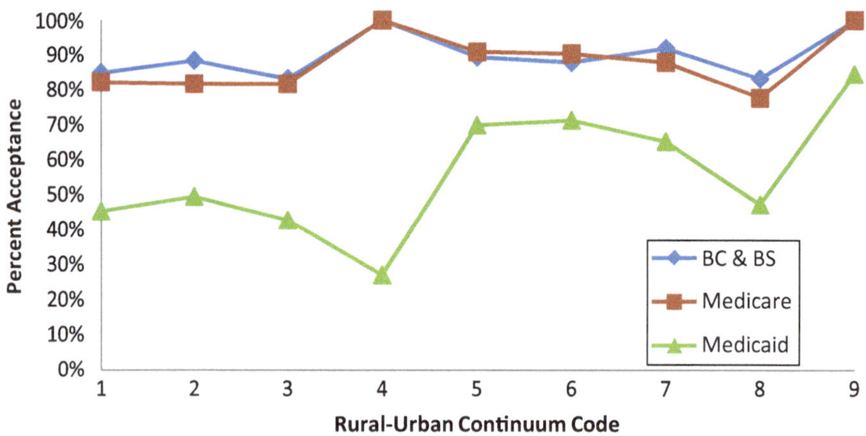

Fig. 12.5 Acceptance rate by type of health

12.5 Discussion

These findings have multiple implications. First, many states are implementing Medicaid expansion on the assumption that insurance equals access to health care (Reddy et al. 2012). Without analysis such as this, states are likely to overestimate the impact of the expansion on primary care offices and overestimate the reduction in demand at hospital emergency rooms. We are mindful that this was a measure at two points (waves of surveys) in time. Health care services are in flux and there are multiple reasons for a practice to decline a new patient. Nonetheless, on the day we called the practice, in up to 50 % of the calls, we were unable to secure an appointment to see a primary care physician. The second implication is the physician's business model. If new patient avoidance (especially Medicaid) is a financial decision, notwithstanding the ACA's incentives to primary care physicians to accept Medicaid patients, these research results have ramifications for the existing payment and delivery system.

The flip side to reimbursements is health care costs. Massachusetts, the model for the Affordable Care Act, has now targeted controlling health care costs as a primary goal (Steinbrook 2012). Third, the option of an appointment with physician extenders (e.g., nurse practitioners, physician assistants and registered nurses) was very low. Yet research has indicated that consumers would readily see a nurse practitioner or physician assistant (Dill et al. 2013) and other research indicates that nurse practitioners could be used to reduce the national shortage of primary care providers (Kuo et al. 2013). Finally, we have quantified the geographic differences in access to primary health care, specifically across the rural-urban continuum. We speculate that high acceptance in rural counties and, relatively, low acceptance rates in urban areas reflect two dynamics. First, practices situated in urban counties have access to a sufficiently large market such that they can afford to "cherry pick" only the most lucrative patients. A second dynamic is the cost of doing business. It is simply cheaper to own and operate a business in a rural county than it is in an urban county.

This study has demonstrated two important findings. First, having health insurance does not equate to access to primary care. The implication of that finding is that increased health insurance coverage will not necessarily increase access or reduce the burden on hospital emergency rooms. The second finding is that access to primary care, as measured in this study, is actually higher in rural places than it is in metropolitan places. This potentially means that metropolitan areas may experience the least benefits from health insurance expansion. Future studies should examine the physician practice business model to determine what factors would yield greater acceptance of newly insured patients.

References

Agency for Healthcare Research and Quality. (2014). *National healthcare disparities report 2013*. U.S. Department of Health and Human Resources. Available via: http://www.ahrq.gov/research/findings/nhqrdr/index.html. Cited 18 November 2015.

Cossman, R. E., Cossman, J. S., Rogers, S., McBride, D., Sutton, L., & Stubbs-Richardson, M. (2014). Access to primary care physicians differs by health insurance in Mississippi. *Southern Medical Journal, 107*(2), 87–90.

Decker, S. L. (2012). In 2011, nearly one-third of physicians said they would not – Accept new medicaid patients, but rising fees may help. *Health Affairs, 31*(8), 1673–1679.

Decker, S. L. (2013). Two-thirds of primary care physicians accepted new medicaid patients in 2011–12: A baseline to measure future acceptance rates. *Health Affairs, 32*(7), 1183–1187.

Dickson, V. (2014, July 8). Reform update: Can medicaid beneficiaries find a doctor? The CMS wants to know. *Modern Healthcare*. Available via http://www.modernhealthcare.com/article/20140708/NEWS/307089964. Cited 18 November 2015.

Dill, M. J., Pankow, S., Erikson, C., & Shipman, S. (2013). Survey shows consumers open to a greater role for physician assistants and nurse practitioners. *Health Affairs, 32*(6), 1135–1142.

Economic Research Service. (2015). *Rural-urban continuum codes*. United States Department of Agriculture. Available via http://www.ers.usda.gov/data-products/rural-urban-continuum-codes.aspx. Cited 18 November 2015.

Kuo, Y. F., Loresto, F., Jr., Rounds, L. R., & Goodwin, J. S. (2013). States with the least restrictive regulations experiences the largest increase in patients seen by nurse practitioners. *Health Affairs, 32*(7), 1236–1243.

Medicare Payment Advisory Commission. (2015a). *Chart 7–3: Medicare beneficiaries reported better ability to get timely appointments with physicians compared with privately insured individuals, 2011–2014*. In June 2015a data book, section 7: Ambulatory care. Available via http://www.medpac.gov/-documents-/data-book/. Cited 18 November 2015.

Medicare Payment Advisory Commission. (2015b). *Chart 7–5: Medicare and privately insured patients who are looking for a new physician report more difficulty finding one in primary care, 2008–2011* [Health care spending and the medicare program web site]. June, 2012b. Available at: http://www.medpac.gov/-documents-/data-book/. Cited 18 November 2015.

Miller, P. (2014). *Physician appointment wait times and medicaid and medicare acceptance rates: 2014 survey*. Merritt Hawkins: An AMN Healthcare Company. Available at: http://www.merritthawkins.com/2014-survey/patientwaittime.aspx. Cited 18 November 2015.

Mississippi State Board of Medical Licensure, Purchase Roster. (2011–2013). Available at: https://www.ms.gov/medical_licensure/renewal/rosterInstructions.jsp. Cited 18 November 2015.

National Center for Health Statistics. (2015). *Centers for disease control and prevention. National health interview survey*. Available via http://www.cdc.gov/nchs/nhis.htm. Cited 18 November 2015.

Reddy, S., Finley, M., Posey, D., et al. (2012). Expanding medicaid managed care: The right choice for Texas? *Southern Medical Journal, 105*(10), 545–550.

Shi, L., Lebrun, L. A., & Tsai, J. (2010). Access to medical care, dental care, and prescription drugs: The roles of race/ethnicity, health insurance, and income. *Southern Medical Journal, 103*(6), 509–516.

Steinbrook, R. (2012). Controlling health care costs in Massachusetts with a global spending target. *Journal of the American Medical Association, 308*(12), 1215–1216.

U.S. Department of Health and Human Services. (2015a). *Health resources and services administration. Health professional shortage area (HPSA) and medically underserved areas/populations (MUA/P)*. Available via http://datawarehouse.hrsa.gov/tools/analyzers/hpsafind.aspx. Cited 18 November 2015.

U.S. Department of Health and Human Services. (2015b). *Health resources and services administration. Quick access reports. State population and health professional shortage area designation population statistics*. Available via http://datawarehouse.hrsa.gov/topics/shortageAreas.aspx. Cited 18 November 2015.

U.S. Department of Health and Human Services. (2015c). *Health resources and services administration. Primary medical care HPSA designation criteria*. Available via http://bhpr.hrsa.gov/shortage/hpsas/designationcriteria/primarycarehpsacriteria.html. Cited 18 November 2015.

Chapter 13
Lifetime Migration in the United States as of 2006–2010: Measures, Patterns, and Applications

Albert I. Hermalin and Lisa J. Neidert

Abstract Though most US migration analyses in recent years have relied upon 1-year and 5-year residence information, analyses of lifetime migration may be more revealing of state-level trends in the relative ability to retain the native born and to attract in-migrants from other states and abroad, and of the effect of such exchanges on the composition of its population in terms of education and other characteristics. This chapter reviews a number of measures of native retention and migrant attraction, and examines the formal relationships among these measures; presents some state-specific lifetime migration measures as of 2006–2010, with special attention to education and the impact of immigration; analyzes the degree of change in these lifetime measures centering on 1990; and uses these measures to decompose a state's proportion of college graduates into elements that highlight the relative importance of retention and attraction and illustrates how these can contribute to appropriate policy formulation.

Keywords Lifetime migration • Retention • Attraction • In-migrants • Out-migrants • Immigrants

13.1 Introduction

The number of migrants is, along with fertility and mortality rates, a determinant of the size and distribution of the population; and the characteristics of the migrants are important factors in the social and economic development of the sending and receiving areas. Insofar as fertility and mortality levels do not vary widely across the areas of interest, migration levels will be the key factor in the distribution of the population and the changes that occur over a given period.

There are a large number of approaches to measuring migration and each has certain advantages and drawbacks. A basic distinction is between direct and indirect

A.I. Hermalin • L.J. Neidert (✉)
University of Michigan, Ann Arbor, MI 48109, USA
e-mail: alberth@umich.edu; lisan@umich.edu

D.A. Swanson (ed.), *The Frontiers of Applied Demography*, Applied Demography Series 9, DOI 10.1007/978-3-319-43329-5_13

measures. Direct measures utilize information obtained from individuals about their moves over some period either through questions in a survey or census, or through requirements for people to register their moves. Indirect measures rely on inferring a move, generally through a demographic analysis either by utilizing vital statistics and a bookkeeping equation or by contrasting actual and expected populations (the survival technique) (Bogue et al. 1982, Chapters 3 and 4; United Nations Manual VI, 1970, Chapter II; U.S. Bureau of the Census, 1973, Vol. 2, 625–637).

Among the direct measures are questions in censuses or surveys that ask where the respondent lived at some previous time, usually one or five years ago. Comparisons with current residence provide measures of mobility and migration. Using characteristics of the respondent, one can create age, education, sex, or race-specific rates. Another form of this approach is to compare current residence with place of birth. Place of birth is often collected as part of a census; in the United States it was included in every census from 1850 to 2000 and is also included in the replacement for the census long-form, the American Community Survey (ACS). Questions on previous residence were first introduced in 1940.[1] Techniques for utilizing the state of birth data to estimate interstate and interregional migration have been extensively developed, along with techniques for using successive censuses for estimating inter-censal migration (Lee et al. 1957; Eldridge and Kim 1968; United Nations 1970; Shyrock 1964).

A comparison of state of birth with state of current residence reveals whether a person has made at least one move over his or her lifetime. Measures derived from such data have several advantages as well as limitations. They can be used to study the level and direction of the flow of migrants over long periods; they reveal the degree to which a state is able to keep its native born population and the degree to which it can attract people from other states; and they provide measures of the degree to which states gain or lose population due to migration. In addition, characteristics like age, gender, education, race/ethnicity can be used to explore migration dynamics.

At the same time such data do not reveal when the move took place (or why), nor whether there were multiple moves. It is also not known if those currently residing in the state of their birth once were migrants and have returned after some time away.[2] Metaphorically, viewing migrants from this perspective is like viewing a

[1]In 1940, a question on residence 5 years ago was introduced; in 1950 it was changed to residence 1 year ago but then restored to the previous version in 1960 and a when did you move into your current dwelling was added. This latter question was limited to the householder in the 1980 and subsequent censuses. The 5-year ago question remained in the censuses from 1960 to 2000. The annual ACS has substituted a 1-year ago question for the decennial 5-year ago question.

With the advent of the 5-year questions from 1960 on it was possible to combine data on state of birth with that information to study progressive and return migration patterns. (See U.S. Bureau of the Census 1963, Table 1 for example).

[2]Some countries include a question in their census, which does make this distinction (see https://international.ipums.org/international-action/variables/MGCTRY1/#questionnaire_text_section) via Minnesota Population Center 2014.

game of musical chairs: you know who sits where at the start and when the music stops, but you do not know all the locations in between.

Questions that ask place of residence one or five years ago are often preferred because they delimit the period of a move (but also do not identify multiple moves within that period) and often collect more detailed geographic information about the earlier residence (e.g., not just state, but location with a state.)

Migration analyses that make use of one or five year information on prior residence are very useful for investigating the relationship of migration to current social and economic events, such as the deep recession starting in late 2007, or the impact of major changes within a state or region, such as the oil or natural gas boom in recent years in North Dakota, Oklahoma, and several other states. They are also useful for investigating special policies designed to deter, attract, or retain certain groups, such as changes in welfare policies, (Cebula 1979; Gelbach 2004; Moffit 1992) or the effect of taxation of pensions (Conway and Rork 2012; and Houtenville and Conway 2001) or to look at the impact of migration on the growth and educational composition of specific metropolitan areas (Frey 2004). When prior residence data are available over time, useful trend analyses can be developed, as in the paper on the migration of the younger, single and educated population between 1965 and 2000 (Gowоrowska and Gardner 2012; see also Liaw and Frey 2007).

Questions on place of birth are usually restricted to identifying the primary subdivision for native born (state of birth in U.S. censuses) and country of birth for immigrants. It is worth noting however that questions on residence 1 year or 5 years ago are often more useful in capturing mobility rather than major moves. For example, in the United States between 2005 and 2010, although 35 % of the population over 5 years of age moved, only 16 % of them were living in a different state than 5 years earlier (Ihrke and Faber 2012).

Though the preponderance of US migration literature in recent years has utilized the 1-year and 5-year residence information, analyses of lifetime migration may be very revealing of the longer-term trends in a state's relative ability to retain its native born, to attract in-migrants from other states and abroad, and the effect of such exchanges on the composition of its population in terms of education and other characteristics. They can also be informative as to policies and programs designed to maintain and enhance a state's skilled and educated work force and related population goals. The ability to examine lifetime migration at different time points can also reveal the extent to which trends are shifting and the magnitude of such changes.

This chapter has three main goals. It reviews and elaborates a number of measures of retention of natives and attraction of migrants and examines the formal relationships among these measures in Appendixes A13 and B13. These appendices present the basic algebra of the lifetime migration measures used throughout the text, and the decomposition of the college educated population, including their derivation and interrelationships. In the text therefore, we present only the key definitions and formulas. Secondly, it presents a number of state-specific lifetime migration measures as of 2006–2010 with special attention to education and the impact of immigration. Lastly, it utilizes the measures developed to present a decomposition of a state's proportion of college graduates into elements that highlight the relative importance of retention and attraction and illustrate how these can contribute to appropriate policy formulation (see Appendix B13).

13.2 Data

The source of data in this analysis is the American Community Survey (ACS) drawn from the Integrated Public Use Microdata Samples (IPUMS) (Ruggles et al. 2010). The ACS is the replacement for the census long-form, via an annual rolling census. An important distinction exists with regard to the explicit mobility question. Since the ACS are collected annually, the reference period for the where did you live question is 1-year rather than the 5-year reference period for the decennial census. Conway and Rork (2014) analyze the comparability issues between the ACS measure and the previous census long-form measure.

The ACS collects data on 250,000 households a month and cumulates the data into 1-year, 3-year, and 5-year files every year. We use the 5-year file covering the 2006–2010 period, which represents 6.9 million households and provides sufficient detail for each state.

We exclude Washington, DC from all the state rankings as it is more akin to a metropolitan area, which is unlike other states. But, we do include the data from Washington, DC in the national totals as well as totals for regions and census divisions.

13.3 Methods and Measures

The two key questions of place in birth and residence generate three independent measures for each state. ***B*** is the number born in a state; ***E*** is the number of native born who exit their state of birth; and ***D*** is the number of in-migrants to a state (domestic or foreign).

From these three counts a number of other measures can be readily derived:

$$\mathrm{C} = \mathrm{B} - \mathrm{E} \tag{13.1}$$

$$\mathrm{T} = \mathrm{C} + \mathrm{D} \tag{13.2}$$

In words, ***C*** is the number of current residents in a state who were born in the state in which they reside; and ***T*** is the total population.

If one wishes to study major streams of lifetime migration, one can determine for the out-migrants their current state of residence and for the in-migrants their state of birth. One can also identify the number of immigrants to a state from abroad, to study the impact of immigration on a particular state. For some states, the paucity of inter-state native migrants is masked by the influx of immigrants from abroad.

In addition to generating these basic dimensions for each state, one can aggregate data regionally and also determine the total pool of migrants for the country as a whole.

From the basic dimensions one can calculate the following measures of interest:

$$R = (B-E)/B \text{ or } C/B \tag{13.3}$$

$$A = D/T \tag{13.4}$$

$$N = D-E \tag{13.5}$$

$$G|L = (D-E)*100/T \tag{13.6}$$

The retention proportion, ***R*** is the proportion of current residents who were born in their state of residence. The attraction ratio, ***A*** is the proportion of current residents of state who are in-migrants to the state. The net gain or loss from migration, ***N*** is the difference between the number who migrated to a state, ***D*** minus those that exited the state, ***E***. Finally, the percent gain or loss from migration is the net gain or loss divided by the total population, ***T***.

Unlike the *retention* measure, the *attraction* measure is not a true probability, in that not all current residents are at risk of migrating to the state. To develop an alternative probability measure, ***P*** we determine for each state a denominator ***H***, which is the total number of lifetime migrants from each state (plus immigrants from abroad) less the state's number of out-migrants (since they are not at risk of being counted as in migrants). This yields a probability measure of:

$$P = D/H \tag{13.7}$$

where ***H*** is the total pool of in-migrants, i.e., all native born not living in their state of birth, less the migrants from the state in question, plus immigrants from abroad.

One drawback of the probability measure is that it is highly correlated with the size of the state, since the more populous states will generally have a larger absolute number of in-migrants, which will translate into higher probabilities. In fact, the correlation between population size and the probability of attraction is 0.95.

The influence of size of place on migration streams has long been recognized. Stouffer (1940) in his seminal article on intervening opportunities and Zipf's (1949) formulation of the amount of migration exchange as proportional to the product of the populations of the two areas and inversely proportional to the distance between them, have been very influential in generating so called gravity models that incorporate size and distance. See also Shyrock 1964, 265; Galle and Taeuber (1966); and Conway and Rork (2012) for additional discussion and illustrations. Peterson (1975, 314–327) and Bogue (1959, 486–509) also discuss other models of migration and the range of factors associated with migration.

In an attempt to adjust for the effect of size, we also generate an expected probability based on the state's share of total population as of 1990, part way through the period under study. The difference between the actual and expected probability yields a measure of relative attractiveness for the period, controlling for size. Appendix A13 defines the expected probability measure and related approaches.

As noted, in addition to identifying whether a person has migrated over his or her lifetime, the ACS data present a number of social and economic characteristics, which allow analysis of who did or did not move. In so far as the focus is on states, rather than individuals, it is cumbersome to employ too many of these characteristics in one analysis. In this chapter, we have chosen to look at two broad age groups, those aged 25–59 and those 60 and over, and their educational attainment (less than high school graduate; high school graduate and/or some college; college graduate or more) in examining the gains and losses and other measures for each state. While we use these two age groups and the population 25+, most of the analysis focuses on the age group in the prime working ages (25–59). In addition to age and education, we focus on the effect of immigration on a state's size and educational composition.

The younger age group captures those in prime work ages who have generally completed their education. Other studies have shown that for this group, migration tends to be highest in their 20s and early 30s and for the more educated. Accordingly, for those 25–59 in 2006–2010, it is likely that most of the interstate migration occurred since 1975 or 1980. It is important to note that lifetime migrants will also include current adults who move at an early age with their families to another state.

13.4 Basic Patterns

Table 13.1 presents basic dimensions of lifetime migration for the total population and for two age groups and three education levels as derived from the ACS data for 2006–2010. Of the total population of 200 million aged 25 and older at that time, half were residing in the state in which they were born, and half were either migrants from another state or immigrants from abroad. This even split holds for each of the two age groups, but among the younger group, immigrants from abroad are a larger proportion of all migrants (39 % vs 25 %).

The tendency for those with higher education to be more mobile shown in other studies also appears clearly in Table 13.1, with the proportion of each higher educational level who are domestic migrants increasing for both age groups. Among immigrants from abroad, educational attainment is lower than among domestic migrants, especially for the 60+ population. It is worth noting however, that among immigrants, the proportion with a college level education is somewhat higher than among non-migrants, in each age group.

It is also worth stressing, however, that despite the large numbers moving over their lifetime, the data also present a picture of substantial stability in residence, currently and in the past. Among the native population, as of 2006–2010, those residing in their state of birth was 62 % for the 20–59 age group and 56 % for those 60 years and older.

Table 13.1 Population of the United States according to lifetime migration status by age and education: ACS 2006–2010[a,b]

	Age 25–59				Age 60+				Age 25+			
Lifetime migration	<HS	HS	BA+	Tot	<HS	HS	BA+	Tot	<HS	HS	BA+	Tot
Living in state of birth	6.2	48.5	18.5	73.2	5.1	16.6	4.7	26.4	11.3	65.1	23.2	99.6
All migrants	9.6	37.7	25.2	72.4	5.2	15.0	7.6	27.8	14.8	52.7	32.8	100.2
Migrant: interstate	2.5	25.0	17.0	44.4	2.7	12.0	6.0	20.7	5.2	37.0	23.0	65.1
Migrant: from abroad	7.1	12.7	8.1	27.9	2.5	3.0	1.6	7.1	9.6	15.7	9.7	35.0
Total population	15.8	86.2	43.7	145.6	10.3	31.6	12.3	54.2	26.1	117.8	55.9	199.7
Interstate ÷ Native	27.4 %	34.0 %	47.9 %	37.8 %	34.6 %	41.7 %	56.0 %	43.9 %	31.5 %	36.2 %	49.8 %	39.5 %
Immigrants ÷ Total migrants	73.9 %	33.6 %	32.1 %	38.5 %	48.1%	20.0 %	21.1 %	25.5 %	64.9 %	29.8 %	29.6 %	34.9 %
All Migrants ÷ Total Pop	60.8 %	43.7 %	57.5 %	49.7 %	50.5 %	47.5 %	61.8 %	51.3 %	56.7 %	44.7 %	58.7 %	50.1 %

[a]Population in millions
[b]Source: American Community Survey, 2006–2010 from IPUMS-USA: https://usa.ipums.org/usa/

Table 13.2 Historical trend in lifetime migration measures: United States, 1850–2010

Year	Total US population	Native born	State of birth = State of residence	%
1850	19,984,671	17,718,556	13,121,831	74.1
1860	27,338,490	23,161,228	16,994,975	73.4
1870	38,410,618	32,844,758	24,827,899	75.6
1880	49,375,555	42,944,396	33,287,127	77.5
1890[a]	62,622,260	45,862,023	36,322,714	79.2
1900	76,028,172	65,532,243	52,005,561	79.4
1910	92,573,810	78,665,801	61,536,330	78.2
1920	106,010,594	91,532,213	71,239,030	77.8
1930	122,823,550	108,161,123	82,830,941	76.6
1940	130,343,838	117,605,281	91,034,543	77.4
1950	152,205,969	140,048,973	103,838,158	74.1
1960	179,292,732	163,974,037	119,387,705	72.8
1970	203,038,600	182,756,200	131,113,200	71.7
1980	226,862,400	210,632,200	145,167,260	68.9
1990	248,107,628	225,200,798	153,296,761	68.1
2000	281,421,906	246,765,636	167,838,872	68.0
2010	298,236,617	256,637,097	175,609,095	68.4

[a]Microdata for 1890 are not available because the records were burned in a fire. The 1890 figures were based on published table (Table 1) from a special report on State of Birth (U.S. Census Bureau 1953). The source for the rest of the years was produced with microdata from the IPUMS project (Ruggles et al. 2010: https://usa.ipums.org/usa/)

One advantage of the lifetime migration data is that it permits a long historical perspective on movements, as the state of birth question was added to the census in 1850. Table 13.2 shows that for the United States, for all age groups (including children) that about three quarters of the native population were residing in their state of birth between 1850 and 1870 and as westward migration slowed, that proportion increased to between 78 and 80 % in the censuses from 1880 through 1940. In 1950, the dramatic changes precipitated by World War II and its aftermath brought the proportion back down to three quarters, and it has declined quite steadily but slowly in the ensuing years so that by 1990, the proportion had dropped to about two-thirds and has remained there as of 2010.[3]

Despite this slight decline, it is an impressive socio-demographic aspect of our history that over 150 years of major social, economic, and cultural transformations,

[3]Earlier publications which trace the proportion residing in their state of birth include Lee et al. 1957, which presents for each state the current residence of those born in the state as well as the state of birth for current native residents, separately for whites and non-whites, between 1870 and 1950; and Eldridge and Thomas 1964, Table A1.38, which, gives for the US and for each state the place of birth of the urban and rural population between 1910 and 1940; and U. S. Bureau of the Census,1953, tables 1 and 6.

the proportion residing in their state of birth has remained in the narrow range of about two-thirds to three-quarters of the population for most of this period.

The relative stability in the measure of those residing in their state of birth for the total United States does not imply the absence of substantial changes for individual states. There can be substantial shifts in magnitude and ranking among a number of states without any changes in the US totals – analysis not shown.[4]

While the summary statistics of lifetime migration in Table 13.1 are useful, migration dynamics are much more apparent when examined across states. Table 13.3 and Table 13.4 show the magnitudes of lifetime migration across states and from abroad for the two age groups: 25–59 and 60+. The net effect of these migrations on current size is given as the percent gain or loss, measured as in-migrants minus out-migrants divided by number of current residents.

The actual magnitudes are very much associated with population size and the more refined measures of retention and attraction are presented in later tables, but from Table 13.3 it is worth noting that the 10 states with the largest number of out-migrants account for half of all the out-migrants for ages 25–59. Among those 60 and older shown in Table 13.4, the three states with more than one million out-migrants (New York, Pennsylvania, and Illinois) account for almost a quarter of all out-migrants.

The lifetime perspective tempers some of the buzz associated with current migration hot spots. North Dakota has the highest percent loss for either age group and yet it has had the highest current net domestic migration rate among states according to the components of change associated with Census Estimates for 2011–2013 (U.S. Census Bureau 2014). Even if we update the North Dakota lifetime profile to 2008–2012 to capture some of the shale driven boom, the results are very similar.

For the younger age group, in most states the number of those born in the state who currently reside there exceeds those who leave, as seen from columns 2 and 3 of the table, but in a number – Alaska, Idaho, Montana, Nevada, North Dakota, South Dakota and Wyoming – all wide open frontier-like states, the reverse is true.

Table 13.3 also shows that 15 states have a percentage loss from lifetime migration. The largest percentage losses occurred in North Dakota, South Dakota, West Virginia, and Iowa, which each experienced percentage losses between 25 and 50 %; Louisiana, Michigan, Mississippi, Nebraska, Ohio, and Pennsylvania each had percentage losses between 9 and 17 %.

Given the emphasis among states in developing a highly educated workforce, the degree of lifetime gain or loss in the college educated is of special interest. In every case where a state experienced an overall loss in the population aged 25–59, the percentage loss among the college educated was even greater. For example, for

---

[4]Gebeloff and Leonhardt (2014), in the NY Times Upshot series, discuss the political implications of the differential changes in the proportions residing in their state of birth that have been occurring. The interactive maps (Aisch et al. 2014) provided by the NY Times for these analyses together with the ability to manipulate the tables presented by us provide useful tools for continued exploration and research on these topics.

Table 13.3 Lifetime migration data for states for those 25–59: 2006–2010[a]

State	Total	Born in state		In-migrants		% Gain/ Loss
	Residents	Resident	Out-migrant	Domestic	Foreign	
Alabama	2,217,261	1,441,231	747,951	653,417	122,613	1.3
Alaska	347,600	85,794	241,343	221,104	40,702	5.9
Arizona	2,861,093	680,950	439,558	1,559,849	620,294	60.8
Arkansas	1,322,703	712,745	489,347	518,207	91,751	9.1
California	17,754,066	7,156,093	3,593,764	3,570,663	7,027,310	39.5
Colorado	2,442,576	738,993	615,066	1,344,353	359,230	44.6
Connecticut	1,714,442	819,759	644,505	524,215	370,468	14.6
Delaware	415,718	163,475	155,116	198,216	54,027	23.4
Florida	8,587,520	2,088,639	1,180,617	4,000,793	2,498,088	61.9
Georgia	4,632,452	2,075,837	863,762	1,867,070	689,545	36.5
Hawaii	639,942	293,596	256,401	182,858	163,488	14.1
Idaho	693,938	257,388	264,598	372,237	64,313	24.8
Illinois	6,168,638	3,649,024	2,566,635	1,273,984	1,245,630	−0.8
Indiana	3,029,757	1,944,675	1,156,164	881,266	203,816	−2.3
Iowa	1,386,553	946,574	787,426	351,636	88,343	−25.1
Kansas	1,297,585	659,700	649,440	504,289	133,596	−0.9
Kentucky	2,065,914	1,356,548	706,449	609,837	99,529	0.1
Louisiana	2,094,350	1,563,858	884,326	407,953	122,539	−16.9
Maine	644,778	372,909	305,179	239,598	32,271	−5.2
Maryland	2,819,143	1,076,634	754,847	1,189,855	552,654	35.0
Massachusetts	3,163,376	1,737,067	1,252,078	718,126	708,183	5.5
Michigan	4,724,818	3,500,812	1,764,827	834,988	389,018	−11.4
Minnesota	2,539,713	1,581,220	743,133	707,708	250,785	8.5
Mississippi	1,357,982	909,446	672,801	399,416	49,120	−16.5
Missouri	2,783,215	1,695,376	997,573	928,393	159,446	3.2
Montana	457,656	221,693	242,396	221,770	14,193	−1.4
Nebraska	833,465	492,348	445,611	265,325	75,792	−12.5
Nevada	1,288,645	130,130	141,210	789,614	368,901	78.9
New Hampshire	651,155	211,829	192,986	386,188	53,138	37.8
New Jersey	4,281,807	1,834,930	1,543,626	1,131,196	1,315,681	21.1
New Mexico	928,474	399,763	368,419	388,603	140,108	17.3
New York	9,364,951	5,210,819	4,229,145	1,206,793	2,947,339	−0.8
N. Carolina	4,457,234	2,228,308	858,825	1,718,675	510,251	30.7
N. Dakota	298,398	194,290	257,443	93,457	10,651	−51.4
Ohio	5,463,193	3,971,540	2,150,030	1,184,831	306,822	−12.1
Oklahoma	1,700,734	887,075	579,703	664,388	149,271	13.8
Oregon	1,821,421	668,490	481,945	895,530	257,401	36.8
Pennsylvania	5,948,447	4,199,377	2,255,083	1,203,424	545,646	−8.5
Rhode Island	502,492	261,160	239,912	143,755	97,577	0.3
S. Carolina	2,130,502	1,136,281	555,417	829,926	164,295	20.6

(continued)

Table 13.3 (continued)

	Total	Born in state		In-migrants		%
State	Residents	Resident	Out-migrant	Domestic	Foreign	Gain/Loss
S. Dakota	365,152	216,468	258,554	135,010	13,674	−30.1
Tennessee	2,996,053	1,615,466	763,749	1,171,137	209,450	20.6
Texas	11,577,167	5,694,759	1,820,204	3,036,583	2,845,825	35.1
Utah	1,171,747	611,625	329,867	401,460	158,662	19.7
Vermont	301,820	135,579	119,811	150,508	15,733	15.4
Virginia	3,869,935	1,552,600	1,004,814	1,660,214	657,121	33.9
Washington	3,230,958	1,210,694	689,824	1,424,144	596,120	41.2
West Virginia	878,816	596,296	552,694	264,489	18,031	−30.7
Wisconsin	2,699,669	1,816,884	819,775	703,581	179,204	2.3
Wyoming	261,765	90,869	157,325	158,105	12,791	5.2
United States	**145,491,530**	**73,195,256**	**44,438,402**	**44,438,402**	**27,857,872**	**19.1**

[a]Source: The source is microdata from the ACS, drawn from the IPUMS project (Ruggles et al. 2010): https://usa.ipums.org/usa/

North Dakota, South Dakota, West Virginia, and Iowa the percentage loss relative to the number of current college educated residents ranges from 49 to 76 %. Eleven additional states experienced losses for this population between 11 and 30 %, and seven states had losses less than 10 %.

By contrast, 11 states gained 40 % or more of their current college educated population through net migration, and in all cases but one, the percentage gain among the college educated exceeded their over-all percentage gain from migration. Some of these gains reflect immigration from abroad, and later we analyze the dynamics of college educated gains and losses.

Among the older population where the US total number of out-migrants approaches the number that remains, there are many more states where out-migrants exceed the native born who remain. We expect this, given the longer period of exposure to migration by the older population and the frequency of moves that take place in connection with retirement. It is worth emphasizing the dramatic numbers for North Dakota. In this age group, almost twice as many of those who were born in this state have left the state than remain (195,938 vs. 99,544) and only 27,488 native born moved in and, as above, the more recent 2008–2012 data do not alter this picture.

Among those 60 and older, over half the states show a percentage loss from lifetime migration with six states experiencing over a 50 % loss – North Dakota, West Virginia, South Dakota, Mississippi, Iowa and Nebraska. The next 14 states are evenly distributed between losses ranging from 10 to 40 %. Several states with relatively minor lifetime migration losses are those that the ACS "where did you live last year" question would describe as substantial net migration losers – Maine, Michigan, Indiana, Illinois, Missouri and Ohio (Census Bureau 2014). But as this

Table 13.4 Lifetime migration data for states for those 60+: 2006–2010[a]

State	Total Residents	Born in state Resident	Born in state Out-migrant	In-migrants Domestic	In-migrants Foreign	% Gain/ Loss
Alabama	890,184	635,970	504,673	236,899	17,315	−28.1
Alaska	81,547	13,106	48,499	58,947	9,494	24.5
Arizona	1,156,499	121,128	108,941	896,674	138,697	80.1
Arkansas	563,106	334,070	436,263	216,810	12,226	−36.8
California	5,734,852	1,519,287	738,638	2,430,593	1,784,972	60.6
Colorado	758,515	202,420	230,217	493,463	62,632	43.0
Connecticut	683,380	309,129	254,805	261,082	113,169	17.5
Delaware	172,304	48,778	38,296	110,879	12,647	49.5
Florida	4,197,952	426,960	234,889	2,824,896	946,096	84.2
Georgia	1,420,166	772,041	447,834	553,358	94,767	14.1
Hawaii	263,553	140,285	70,496	58,215	65,053	20.0
Idaho	258,663	87,770	130,232	159,139	11,754	15.7
Illinois	2,175,694	1,274,692	1,215,578	598,821	302,181	−14.5
Indiana	1,136,545	685,821	503,189	412,906	37,818	−4.6
Iowa	600,962	451,042	488,434	136,403	13,517	−56.3
Kansas	505,623	286,308	376,646	200,731	18,584	−31.1
Kentucky	789,141	584,865	505,267	189,200	15,076	−38.1
Louisiana	762,946	569,138	354,851	168,204	25,604	−21.1
Maine	285,018	172,531	136,703	98,459	14,028	−8.5
Maryland	970,630	362,623	238,146	490,040	117,967	38.1
Massachusetts	1,218,871	760,194	647,089	261,844	196,833	−15.5
Michigan	1,838,813	1,181,831	738,780	529,403	127,579	−4.4
Minnesota	913,400	618,758	431,731	251,065	43,577	−15.0
Mississippi	518,416	371,806	456,093	138,177	8,433	−59.7
Missouri	1,123,590	676,356	558,338	415,454	31,780	−9.9
Montana	196,921	89,940	117,338	101,257	5,724	−5.3
Nebraska	327,928	213,697	292,346	103,750	10,481	−54.3
Nevada	445,115	17,215	25,344	348,021	79,879	90.4
New Hampshire	244,845	78,195	85,480	151,427	15,223	33.2
New Jersey	1,605,898	642,220	634,025	594,928	368,750	20.5
New Mexico	367,986	128,689	134,326	206,246	33,051	28.5
New York	3,544,722	1,975,569	2,138,188	553,797	1,015,356	−16.1
N. Carolina	1,663,551	1,014,998	467,173	579,307	69,246	10.9
N. Dakota	129,102	99,544	195,938	27,488	2,070	−128.9
Ohio	2,191,353	1,384,413	945,453	705,911	101,029	−6.3
Oklahoma	680,391	406,847	498,117	253,563	19,981	−33.0
Oregon	721,341	200,060	161,295	465,568	55,713	49.9
Pennsylvania	2,610,631	1,977,035	1,427,000	473,637	159,959	−30.4
Rhode Island	205,707	122,494	114,905	55,501	27,712	−15.4
S. Carolina	851,228	474,251	313,907	347,174	29,803	7.4

(continued)

Table 13.4 (continued)

State	Total Residents	Born in state Resident	Born in state Out-migrant	In-migrants Domestic	In-migrants Foreign	% Gain/ Loss
S. Dakota	153,076	99,479	172,641	51,012	2,585	−77.8
Tennessee	1,159,287	687,379	433,914	442,760	29,148	3.3
Texas	3,533,326	1,880,352	663,583	1,150,273	502,701	28.0
Utah	332,720	176,342	108,397	132,559	23,819	14.4
Vermont	124,938	58,704	69,217	58,224	8,010	−2.4
Virginia	1,338,900	624,743	403,067	593,523	120,634	23.2
Washington	1,129,405	364,646	240,285	622,586	142,173	46.4
West Virginia	403,412	314,608	442,957	83,216	5,588	−87.8
Wisconsin	1,038,789	725,342	424,142	271,495	41,952	−10.7
Wyoming	96,653	28,203	69,041	65,406	3,044	−0.6
United States	**54,212,354**	**26,414,512**	**20,690,061**	**20,690,061**	**7,107,781**	**13.1**

[a]Source: The source is microdata from the ACS, drawn from the IPUMS project (Ruggles et al. 2010): https://usa.ipums.org/usa/

table reflects lifetime migration, it incorporates a time when these states had different migration profiles as well as recent dynamics.

13.5 Measures of Retention and Attraction

The measures of retention and attraction by state are presented in Table 13.5 for the age group 25–59. We break out the college educated in that age group, which are often of special interest to policymakers concerned with the state's economic development. In rank order from lowest to highest, the table presents the retention proportion, the attraction ratio, and the probability of attraction, defined above, as percentages.

There is considerable range in the effects of migration across the states. The retention proportion for the total population ranges from a low of 26 % for Alaska to a high of 76 % for Texas, and the range among the college-educated is similarly wide. A state's ranking in retaining its native-born college graduates is not far different from their overall ranking. The college graduates are, of course, one component of the total. In every state, the retention proportion among college graduates is lower than for the total population, testifying to the higher migration tendencies among the more educated. Note that in California, the two proportions are almost equal.

For the total native-born population of the United States aged 25–59, 62 % are living in the state of birth, and among those with a college education, that proportion is 52 %. Referring back to Table 13.1, for those with a high school education,

Table 13.5 State ranking on retention and attraction measures for the total and college educated populations: Age 25–59[a]

Retention %				Attraction %				Probability of attraction %			
ST	BA+	ST	Total	ST	BA+	ST	Total	ST	BA+	ST	Total
AK	16.21	AK	26.23	LA	34.00	LA	25.33	ND	0.15	ND	0.14
WY	25.39	WY	36.61	MI	34.97	MI	25.91	WY	0.18	SD	0.21
DE	34.31	ND	43.01	OH	37.16	OH	27.30	SD	0.19	VT	0.23
ND	34.42	SD	45.57	PA	38.23	PA	29.40	WV	0.28	WY	0.24
SD	35.01	MT	47.77	MS	38.43	IA	31.73	MT	0.32	MT	0.33
VT	35.18	NV	47.96	WI	38.94	WV	32.15	VT	0.32	RI	0.33
ID	37.67	ID	49.31	IA	39.44	WI	32.70	AK	0.32	DE	0.35
MT	37.85	KS	50.39	ND	39.52	MS	33.03	DE	0.35	AK	0.36
NM	38.09	DE	51.31	WV	41.03	KY	34.34	RI	0.36	ME	0.38
WV	38.30	WV	51.90	NE	44.27	ND	34.89	ME	0.41	WV	0.39
NV	39.10	NM	52.04	AL	44.84	AL	35.00	MS	0.43	NE	0.47
NH	39.67	RI	52.12	IN	45.07	IN	35.81	NE	0.46	HI	0.48
IA	40.52	NH	52.33	SD	45.48	MN	37.74	HI	0.46	ID	0.61
RI	42.49	NE	52.49	KY	46.15	MO	39.09	ID	0.49	NH	0.61
NE	44.07	VT	53.09	NY	46.69	SD	40.72	AR	0.55	IA	0.62
KS	44.07	HI	53.38	MN	46.88	IL	40.85	IA	0.62	MS	0.63
HI	45.52	NJ	54.31	MO	46.92	NE	40.93	LA	0.64	NM	0.74
CT	45.74	CO	54.58	IL	46.96	ME	42.16	NM	0.64	LA	0.74
ME	45.95	IA	54.59	AR	50.19	NY	44.36	NH	0.71	UT	0.78
NJ	46.81	ME	54.99	OK	50.40	MA	45.09	UT	0.75	AR	0.85
MD	46.82	NY	55.20	MA	52.51	TN	46.08	OK	0.83	KS	0.89
CO	46.86	CT	55.98	UT	53.21	AR	46.11	KY	0.86	KY	0.99
VA	47.28	MS	57.48	RI	54.45	SC	46.67	KS	0.92	AL	1.08
MS	47.43	OR	58.11	KS	54.60	UT	47.80	AL	0.94	OK	1.13
IN	47.65	MA	58.11	TX	57.07	OK	47.84	NV	1.03	WI	1.24

NY	47.86	IL	58.71	TN	57.40	RI	48.03	WI	1.21	CT	1.25
OK	48.47	MD	58.78	**US**	**57.48**	KS	49.16	SC	1.28	MN	1.34
AZ	48.89	AR	59.29	ME	57.96	**US**	**49.69**	IN	1.37	SC	1.39
OR	49.78	OK	60.48	MT	58.06	NC	50.01	MO	1.47	IN	1.53
AR	50.18	VA	60.71	HI	58.69	TX	50.81	OR	1.55	MO	1.53
MO	50.47	AZ	60.77	SC	59.41	MT	51.56	CT	1.59	OR	1.61
LA	51.13	**US**	**62.22**	CT	59.77	CT	52.19	MN	1.67	NV	1.61
OH	51.34	IN	62.71	NJ	62.52	HI	54.12	TN	1.72	MI	1.74
PA	51.63	MO	62.96	NC	62.85	VT	55.08	MI	1.86	TN	1.93
IL	51.91	WA	63.70	CA	62.92	GA	55.19	OH	2.24	MA	2.01
US	**52.11**	LA	63.88	NM	68.87	NM	56.94	AZ	2.54	OH	2.13
FL	52.87	FL	63.89	ID	69.14	NJ	57.15	PA	2.83	CO	2.38
MA	53.02	OH	64.88	GA	69.64	CA	59.69	CO	2.83	MD	2.44
KY	53.27	UT	64.96	WA	69.66	VA	59.88	MA	2.88	PA	2.50
WI	54.24	PA	65.06	WY	70.53	DE	60.68	WA	2.90	WA	2.82
AL	54.28	KY	65.76	OR	70.80	MD	61.81	NC	3.20	AZ	3.03
MI	54.33	AL	65.83	DE	71.43	WA	62.53	MD	3.26	NC	3.12
SC	54.65	MI	66.48	MD	75.30	ID	62.91	GA	3.79	VA	3.25
TN	55.16	CA	66.57	VA	75.71	OR	63.30	IL	4.03	NJ	3.46
WA	55.99	SC	67.17	VT	75.82	WY	65.29	NJ	4.21	GA	3.58
UT	56.29	TN	67.90	CO	77.20	NH	67.47	VA	4.33	IL	3.61
GA	58.53	MN	68.03	NH	77.69	CO	69.75	NY	6.65	NY	6.10
MN	58.80	WI	68.91	FL	80.91	AK	75.32	TX	7.25	TX	8.35
NC	61.25	GA	70.62	AZ	82.88	FL	75.68	FL	7.67	FL	9.14
CA	65.16	NC	72.18	AK	85.56	AZ	76.20	CA	14.43	CA	15.43
TX	69.37	TX	75.78	NV	91.00	NV	89.90	**US**	–	**US**	–

[a]Source: The source is microdata from the ACS, drawn from the IPUMS project (Ruggles et al. 2010): https://usa.ipums.org/usa/

the retention percentage is 64 % and for those with less than a high school degree, it rises to 73 %.

Table 13.5 also presents the two measures of attraction, described above. The first is the attraction ratio, which is simply the ratio of in-migrants (including those from abroad) to the total current population; the second more refined measure is the probability of the state attracting the migrants from another state, or from abroad. Like the retention percentages, the attraction ratios have a wide range across the states.

For the total population aged 25–59, the attraction measure ranges from 25 % in Louisiana to 90 % in Nevada, and for the college-educated, the same two states range from 34 % to 91 %. As before, a state's relative ranking on the total measure does not differ greatly from its ranking among the college educated. For the United States as a whole, domestic and foreign migrants represent 50 % of the total population, and college-educated migrants represent 58 % of all the higher educated, for those 25–59 years of age. Again, drawing from Table 13.1, for the native-born population, domestic migrants represent 38 % of the total and 48 % of the college educated.

The ranking of the states on the probability measure of attraction is quite different from the simpler attraction ratio. For the probability measure, the range for the total population is from 0.14 % for North Dakota to 15.4 % for California, and for the college-educated, the same two states bracket a similar range, 0.15 %–14.4 %. Only four states – New York, Texas, Florida, and California each attract more than 4 % of the relevant pool of all in-migrants; for the college-educated these four states are the only ones which attract more than 5 % of the pool.[5]

Note also that while Georgia and Michigan are roughly the same size in 2013, Georgia has a much higher attraction probability (3.58 vs. 1.74 %). This very much reflects the migration dynamics of the last 30 years. Georgia's population has grown by 50 % since 1990 while Michigan has only grown by 6 %. For perspective, the US population has grown by almost 25 %.

As Table 13.5 shows, 22 of the states each attracts less than 1 % of the total pool of migrants, and that is true for 24 of the states for the college-educated pool.

From Table 13.5 alone it is hard to discern whether there are regional patterns and the degree of relationship among the measures of attraction and retention. Several states with high retention percentages have low attraction ratios, but others have high rankings on both.

Some states like Michigan, which retain a high proportion of their native born, attract relatively few in-migrants, while others like Georgia with high retention also attract a high ratio of in-migrants. Across the 50 states the correlation between the overall retention proportion and the attraction ratio is −0.35 for ages 25–59 and −0.13 for ages 60 and older. The relationship is not linear, so the correlation

[5]The probabilities will not add to one because of the varying denominators. Out-migrants from a state are not eligible to be in-migrants to that state.

coefficients can be misleading and it is more informative to examine this as a table of quartiles of retention vs attraction.

Table 13.6 distributes the states by their location on the intersection of these two measures for the 25–59 age group. We divided the retention and attraction measures roughly into quarters, taking into account the breaking points noted on the table.

Table 13.6 shows that there are no states in the highest categories of the attraction ratio and retention proportion and only North Dakota is in the lowest category of both measures. By contrast, 11 states are at the extremes of the other diagonal – the lowest category for attraction and the highest for retention (7 states); and the highest for attraction and the lowest for retention (4 states). In the former group (low attraction, high retention) are several rust belt states, along with Kentucky, Alabama, and Minnesota. By contrast, those with high attraction and low retention include several western states.

States known for attracting migrants for climate or economic reasons, like California, Texas, Arizona, and Florida appear as states high in the attraction ratio and also high in retention in the lower right area. These are highest on one of the measures and third-highest on the other. They are joined by three states in the South Atlantic – Maryland, North Carolina, and Georgia and the other two continental states in the Pacific region, Oregon and Washington. Connecticut and Virginia round out the group of states with above average retention and attraction rates.

13.6 Decomposing the College-Educated Population for States Measures

In this section we take a closer look at the college-educated population age 25–59 as a great deal of current attention has focused on the degree to which states are able to develop and maintain an educated workforce. We proceed in two ways. First, we compare the actual lifetime migration of this population for a state with that expected as a function of its size. Secondly, we decompose the current proportion of a state's college educated population as a function of the relative importance of the production, retention, and attraction factors. In presenting these analyses, we focus on a dozen states which vary in terms of region, size, and rates of growth over the last 20–30 years, as well as their status as gateways of immigrants, rust belt classification, climate, and other factors.

For the first analysis we start by calculating an expected probability of college in-migration, based on the states' share of the US population and the total number of life-time college migrants (both domestic and foreign). Since this is a lifetime measure, we chose 1990 as the point to measure the states' share of the US population. We could have chosen other years as a baseline; the particular year chosen is not critical to the analysis. The elements of the calculation are as follows (see also Appendix A13):

Table 13.6 States classified by attraction and retention rate Quartiles for the population 25–59: ACS 2006–2010[a]

RETAIN		ATTRACT			
		1Q	2Q	3Q	4Q
		<40 %	40–50 %	51–60 %	>=61 %
	1Q <=50 %	**ND**	SD,KS	MT	ID,WY,AK,NV
	2Q 51–55 %	IA, WV	NE,ME, NY,RI	HI,VT,NM, NJ	DE,NH,CO
	3Q 56–64 %	LA,MS, IN,MO	IL,MA,AL, OK	CT,VA	MD,WA,OR, FL,AZ
	4Q >=65 %	MI,OH,PA,WI,KY, AL,MN	TN,SC,UT	NC,TX,GA, CA	**No States**

[a]Source: Based on the attraction and retention columns from Table 13.5

$$U = S \times H_G \quad (13.8)$$

where $\boldsymbol{U}$ is the expected number of college in-migrants to a state; $\boldsymbol{S}$ is the states' share of the US population in 1990; and $\boldsymbol{H_G}$ is the pool of college graduates at risk of moving to state i. More specifically, H_G is the potential number of domestic and foreign migrants to a state with a college degree minus the out-migrants from state i with a college degree.

Table 13.7 shows measures of expected number of college-educated in-migrants sorted by the relative gain in the state's college educated population (Column 5). The second column shows the actual number of college in-migrants to each state. The expected number is calculated using the terms described above.

For instance, for California, the value for relative size of the state to the US population in 1990 is.1197. The at-risk pool is the number of college educated domestic migrants (17,009,032) + foreign migrants (8,009,081) minus the share of college educated migrants who left California (3,593,764). The product for California is 2,864,454, which is smaller than the actual number of college educated in-migrants to the state. California has approximately a half million more college in-migrants than expected (Column 4). This is 2 percent of the 1990 population (588,074/29,760,021). But, it is almost 11 % of the current number of college graduates in the state (588,074/5,586,733).

From column 3 it will be noted that the states that did less well than expected in the number of college-level migrants – New York, Illinois, Ohio, Michigan, Iowa, and West Virginia – are states that have been slow-growing or losing population (Michigan) in recent years. They were also in the lowest two quarters of the attraction ratio for 25–59 year olds in Table 13.6.

However, the percentage loss for New York and Illinois were small compared to the other states with losses on this measure (columns 4 and 5). The deficiency for West Virginia in the number of college-level migrants received as against the number expected on the basis of its size, approached the deficiencies experienced

Table 13.7 Measures of actual and expected college in-migration for states, for the population 25–59: ACS 2006–2010[a]

State	College in-migrants		Actual – Expected	+/− in college graduates		Actual ÷ Expected
	Actual	Expected		% 1990	% 2006–2010	
NV	256,808	119,897	136,911	11.4	48.5	2.14
AZ	631,440	365,761	265,679	7.2	34.9	1.73
FL	1,888,371	1,280,296	608,075	4.7	26.1	1.47
GA	937,377	642,943	294,434	4.5	21.9	1.46
CA	**3,452,528**	**2,864,455**	**588,073**	**2.0**	**10.7**	**1.21**
TX	1,770,316	1,668,569	101,747	0.6	3.3	1.06
NY	1,537,076	1,670,588	−133,512	−0.7	−4.1	0.92
IL	966,656	1,104,303	−137,647	−1.2	−6.7	0.88
IA	152,444	276,354	−123,910	−4.5	−32.1	0.55
OH	541,820	1,052,932	−511,112	−4.7	−35.1	0.51
MI	451,979	909,255	−457,276	−4.9	−35.4	0.50
WV	68,467	178,989	−110,522	−6.2	−66.2	0.38

[a]Source: The source is microdata from the ACS, drawn from the IPUMS project (Ruggles et al. 2010: https://usa.ipums.org/usa/)

by the much larger states of New York and Illinois; this deficiency represents 66% of its college level population as of 2006–2010. For Ohio, Michigan, and Iowa, the losses in college-level in-migration represents about a third of their current college level populations in 2006–2010 (column 5).

The other six states presented in Table 13.7 all experienced gains in the actual minus expected measure, and these also appeared in the third-highest or highest quarter of over-all attractiveness in Table 13.6. Somewhat surprisingly, the gain for Texas in college-level migrants was rather minimal given its general rapid rate of growth. This indicates that in spite of Austin's reputation as a hotbed for college-educated migrants this might not apply to the entire state.

California did somewhat better and its gain over expected represented 10.7 % of its 2006–2010 college-level population. However, Texas and California were outpaced by Florida, Georgia, Arizona, and Nevada, where the gains in college-level migrants over the expected number ranged from a fifth to nearly a half of their college level populations.

Column 6 of Table 13.7 presents the ratio of the actual probability of attracting migrants to that expected and confirms the previous analysis. The six states with losses have ratios less than one; for West Virginia the actual probability is less than 40 % of the expected. The six states with gains ranged from little more than one (Texas) to almost twice the expected probability in the case of Arizona and Nevada.

Analyses that help reveal why a state gains or loses college-educated adults can be helpful for developing programs and policies. To this end, the next section presents a partitioning of the percent of the college-age population residing in the same 12 states into elements that focus on the production of the college-educated native born and their retention, and the level to which migrants with a college education are attracted from other states and from abroad. The key formula of the decomposition, which is derived and discussed at greater length in Appendix B13 may be expressed as follows:

$$\mathrm{T_G} = [\mathrm{B} \times \mathrm{B_G}/\mathrm{B} \times \mathrm{C_G}/\mathrm{B_G}] + [\mathrm{H_{N.G}} \times \mathrm{P_{N.G}}] + [\mathrm{H_{I.G}} \times \mathrm{P_{I.G}}] \quad (13.9)$$

The first term on the right gives the number of college graduate residents in the state that arises from the production of college graduates from the state's native born population and the level of retention of these graduates. It is made up of the product of the total number born in the state (which is the size potential for native born production of college grads), the rate at which these native born achieve college graduate status, and the rate at which these graduates remain in the state (or leave and return). The second term reflects the in-migration of native born college grads from other states, and the third term the immigration of college graduates from abroad.[6]

Tables 13.8 and 13.9 give the major components of this decomposition. In Table 13.8 we show in column one, in rank order, the percentage of a state's 25–59 year old population which has a college level education or higher as of 2006–2010. The succeeding columns show the percentage distribution of this level due to the production and retention of college level natives (column 2) and the percentage due to in-migrants, in total, and further subdivided by domestic in-migrants and immigrants from abroad.

Three of the states, California, New York and Illinois, have attainment levels of 30 % or more and present some interesting contrasts. New York and Illinois are similar in showing that about half of that attainment level is due to their native population, but for New York the proportion due to in-migration is heavily weighted to immigrants while for Illinois, domestic in-migrants predominate. By contrast, for California, only 37 % of its higher educational attainment is accounted for by native-born production and retention, and of the remainder, it is almost equally divided between domestic and foreign in-migrants.

Seven of the remaining states have a college level of attainment between 26 and 30 % but they show a wide variation in their component elements. In Ohio, Michigan, and Iowa 60 % or more of their attainment level results from production and retention of college-level natives, and of the in-migrants, foreign immigrants play a relatively small part. By contrast, in the faster growing states of Texas, Florida, Georgia, and Arizona the proportion due to the born in state population is less than half in all cases, and for Florida, it is less than 20 %. In addition, Texas and

[6]Dividing each term by the total population of the state expresses the decomposition in terms of a state's proportion of college graduates. See Appendix B13 for a fuller explication.

Table 13.8 Components of the percentage of residents who are college graduates for selected states: Ages 25–59[a]

State	% BA+ Total Pop	Production & retention of BA+ Natives	Due to in-migration of BA+ Total	Domestic	Foreign
NY	35.2	53.3	46.7	19.9	26.8
IL	33.4	53.0	47.0	29.0	18.0
CA	30.9	37.1	62.9	29.1	33.8
GA	29.1	30.4	69.6	53.8	15.8
IA	27.9	60.6	39.4	32.3	7.1
MI	27.4	65.0	35.0	22.6	12.4
FL	27.2	19.1	80.9	52.2	28.7
TX	26.8	42.9	57.1	37.0	20.1
OH	26.7	62.8	37.2	28.3	8.9
AZ	26.6	17.1	82.9	66.7	16.1
NV	21.9	9.0	91.0	66.7	24.3
WV	19.0	59.0	41.0	36.2	4.8

[a]Source: The source is microdata from the ACS, drawn from the IPUMS project (Ruggles et al. 2010: https://usa.ipums.org/usa/)

Florida show relatively high contributions of college-level immigrants from abroad compared to Georgia and Arizona.

The two remaining states in Table 13.8, West Virginia and Nevada, which have the lowest college-level attainment of the 12, offer a sharp contrast in their components. In West Virginia, 59 % is due to production and retention of natives. And of the in-migrant portion little is due to foreign immigrants. In Nevada, only 9 % of their higher education level is due to their native born (the lowest portion among the 12 states) and of the in-migrants, a significant share is due to foreign immigrants.

Table 13.9 provides the data for the three elements of the production and retention of college-level natives, the first term of the decomposition, sorted by college graduation percentage from high to low. As noted, the first column is primarily the size potential factor representing the total native born as a portion of the total population, and as such is affected by both retention and attraction factors. The second two factors are more readily interpretable and subject to policy and program actions. Among the 12 states, New York, Illinois, and Iowa are highest in the proportion of their native born who attain a college level education, each well above 30 %, and California Ohio and Michigan are very close to the 30 % mark. The remaining six states show a range of college level production among natives from 22 to 26 %, with West Virginia at the low-end and Florida and Texas at the higher end.

The third column measures the retention level of college-level natives. California and Texas are at the high-end with well over 60 %, but is worth noting that the three rust belt states – Illinois, Ohio, and Michigan – which each displayed a high

Table 13.9 Decomposition of attainment and retention proportions for the population 25–59[a]

	Native born BA+					
	Production potential	Attainment	Retention	Production & retention	%BA	Relative share
State	(1)	(2)	(3)	(1) × (2) × (3)	C(g)/C	P & R / %B
NY	1.008	0.388	0.479	18.7	35.2	0.533
IL	1.008	0.338	0.519	17.7	33.4	0.530
IA	1.251	0.333	0.405	16.9	27.9	0.606
MI	1.114	0.294	0.543	17.8	27.4	0.650
OH	1.121	0.292	0.513	16.8	26.7	0.628
CA	0.605	0.290	0.652	11.4	30.9	0.371
FL	0.381	0.258	0.529	5.2	27.2	0.191
TX	0.649	0.255	0.694	11.5	26.8	0.429
NV	0.211	0.239	0.391	2.0	21.9	0.090
GA	0.635	0.238	0.585	8.8	29.1	0.304
AZ	0.392	0.238	0.489	4.6	26.6	0.171
WV	1.307	0.224	0.383	11.2	19.0	0.590

[a]Source: The source is microdata from the ACS, drawn from the IPUMS project (Ruggles et al. 2010: https://usa.ipums.org/usa/)

level of college-level production, also show retention levels of over 50 % and are at the high-end of this measure.

As we illustrate in the summary, this decomposition can be one tool for policymakers to examine the relative importance of developing programs for increasing the production and retention of college grads among their native born versus programs that might increase the flow of educated migrants from other states or abroad.

13.7 State-to-State Migration Exchanges

These considerations lead us to conclude these analyses with two issues. First, we will look briefly at some of the actual exchanges between states. For the states that have net losses due to migration, where do their out-migrants move to? And, for states that have had substantial gains, which states have been the largest suppliers? An analysis of these exchanges can provide clues to the features and facilities that might assist in retaining a state's native population and in attracting others. Secondly, we will examine the role international migration plays in shoring up the population of states, especially when they are losing in state-to-state migration exchanges.

Table 13.10 looks at the exchanges among the 12 states we have been examining in some detail. The top portion examines the five states that have gained from net

Table 13.10 Lifetime domestic migration exchanges across 12 states for the population 25–59, 2006–2010[a]

Gainers			
State	**Net domestic gain**	**Receiving from: Top 5**	**Sending to: Top 5**
Arizona	1,120,291	CA(306,880);IL(124,930);NY (105,573); OH(78,584);MI (78,144)	CA(95,565);TX(40,132);CO (23,290); WA(22,902);NM (20,449)
Florida	2,820,176	NY(747,124);OH(273,021);PA (255,838); NJ(237,887);IL (209,706)	GA(191,902); TX(86,832);NC (85,917); CA(73,175); AL (63,658)
Georgia	1,003,308	FL(191,902);NY(178,781); AL (130,518);TN(102,721);OH (97,607)	FL(172,622);AL(78,418); SC (71,202);NC(60,076);TN (54,529)
Nevada	684,404	CA(225,228);NY(46,623);IL (37,573); TX(29,128);OH (25,475)	CA(32,607);UT(10,793); WA (9,802);AZ(8,862); TX(8,768)
Texas	1,216,379	CA(324,389);LA(280,044);IL (179,620); OK(166,767);NY (155,262)	CA(265,923);OK(121,766); FL (105,905);CO(86,731);AZ (76,352)
Losers			
State	**Net domestic loss**	**Sending to: Top 5**	**Receiving from: Top 5**
California	−23,101	WA(330,996);TX(324,389); AZ (306,880);OR(305,081); NV (255,228)	NY(348,453);TX(265,923);IL (250,765); OH(170,559);MI (168,468)
Illinois	−129,651	CA(250,765);FL(209,706); WI (185,176);AZ(179,620); IN (174,936)	IN(107,793);MO(106,204);MI (90,368); WI(86,852);IA (72,169)
Iowa	−435,790	IL(72,169);MN(69,407);CA (57,191); MO(55,320);TX (51,513)	IL(58,239);NE(35,527);MN (31,000); CA(26,478);MO (23,171)
Michigan	−929,839	FL(200,791);CA(168,468); TX (116,630);OH(92,949);IL (90,368)	OH(109,301);IL(84,939);IN (64,437); CA(46,711);NY (45,095)
New York	−3,022,352	FL(747,124);NJ(480,923); CA (348,453);PA(258,123); NC (229,435)	NJ(153,464);PA(133,013);CA (83,652); MA(77,260);CT (62,202)
Ohio	965,199	FL(273,021);CA(170,559); KY (129,691);TX(125,141);MI (109,301)	PA(132,920);WV(121,629);MI (92,949); KY(92,332);NY (77,162)
West Virginia	−288,205	OH(121,629);VA(64,037); NC (49,327);FL(48,707); PA (28,435)	OH(52,104);MD(36,078);VA (33,642); PA(25,758);KY (10,526)

[a]Source: The source is microdata from the ACS, drawn from the IPUMS project (Ruggles et al. 2010): https://usa.ipums.org/usa/

interstate migration of the native population. It presents the total gain, and then the five states that have contributed the most migrants (in descending order); also shown for each are the top five destinations of the out-migrants from these states. For the five states that have gained, the leading sending states reflect size, geographic proximity, and economic conditions. Arizona, for example, received the most migrants from California, a large neighboring state, but also gained substantially from Illinois, New York, Ohio, and Michigan. New York appears as a leading sending state to all five, Illinois and Ohio as leading sources for four of them, and California is the dominant sending state for three of the gainers.

The out-migrants from the states that have gained also show the influence of size and geography. California is a destination from four of the states, Texas for three (out of the four possible). The leading destinations for Georgia's out-migrants are all nearby southern states.

The bottom portion of Table 13.10 shows similar data for the seven states that have lost native population to lifetime migration, but the emphasis in column 2 is on the top five destinations of these losing states and secondarily, in column 3, the states that have sent them out-migrants. Geography plays a large role in the destination of these losing states, as many of the top destinations are to states adjoining or relatively nearby. The fast-growing states of Florida, Texas, and Arizona are also prominent destinations, as is California.

The leading sources of in-migrants to these losing states are also often nearby geographically, so that exchanges between leading destinations and sources of migrants appear more often among these losing states than for the gaining states. Iowa, as an example, has major exchanges with Illinois, Minnesota, and Missouri, in each case sending more out-migrants than it receives from those states. Michigan by contrast gains more from its exchanges with Ohio than it loses, and it is almost on a par with Illinois, but these exchanges do not compensate for the relatively large net losses to Florida, California, and Texas.

Table 13.11 shows, of the 12 states, the five leading states in terms of immigrants from abroad, which again illustrates the substantial impact immigration makes on the size of many states. Three of the leading states, California, New York, and Illinois all lost population based on exchanges of native born, but in the case of California, a small loss was swamped by the large number of immigrants. For New York and Illinois, the large number of immigrants almost matched their losses from the net migration of native born. In the case of Texas and Florida, the number of immigrants added substantially to their gains from native exchanges.

13.8 Summary and Conclusions

Analysis of lifetime migration rests on two pieces of information about respondents in a survey or census – their place of birth and current residence – that are almost always collected in a census. From these two building blocks a number of measures can be developed, such as the probability of a state retaining its native born or the

Table 13.11 Lifetime international migration sources across five states for the population 25–59, 2006–2010[a]

State	Immigrants	Receiving: continent > = 100,000[b]	Receiving: country > =100,000
California	7,027,310	C. Amer (3,748,844); Asia (2,281,651); Eur(459.320); S. Amer(158,498)	MX(3,089,144); PH(517,416); VT(330,361);SV(315,220); CN (237,242); IN(225,095); KR (214,431);GT(182,812)
New York	2,947,339	C. Amer(1,020,327); Asia (729,771); Eur(459,023); S. Amer (399,192); US Outlying(198,530)	DR(274,171);CN(193,511); MX(174,362); PR(154,568); JM(148,223); EC(123,065); GY (106,129), IN(100,981)
Texas	2,845,825	C. Amer(1,936,007); Asia (494,942); Eur(157,179)	MX(1,657,502); SV(125,910); VT(99,658); IN(96,824)
Florida	2,498,088	C. Amer(1,258,025); S. Amer (403,702); US Outlying (251,240); Asia(241,575); Eur (211,141);	CU(413,621); MX(200,261); PR(190,492); HT(168,524); CO (141,686); JM(113,821)
Illinois	1,245,630	C. Amer(569,719); S. Amer (303,467); Eur(252,045)	MX(516,706); PL(103,760)

[a]Source: The source is microdata from the ACS, drawn from the IPUMS project (Ruggles et al. 2010: https://usa.ipums.org/usa/)
[b]Below in the first column are the country codes and country names abbreviated in the tables above. In the next two columns are the range of IPUMS place of birth codes (BPL) included in the various continent designations

Code	Country	Continent	Place of birth codes – BPL (range)
CN	China	Outlying (US Outlying & North America)	100–199
CO	Colombia	C.Amer (Central America & Caribbean)	200–299
CU	Cuba	S.Amer (South America)	300–399
DO	Dominican Republic	Europe	400–499
EC	Ecuador	Asia	500–599
SV	El Salvador		
GT	Guatemala		
GY	Guyana		
HT	Haiti		
IN	India		
IR	Iran		
JM	Jamaica		
KR	South Korea		
MX	Mexico		
PH	Philippines		
PL	Poland		
PR	Puerto Rico		
TW	Taiwan		
VT	Vietnam		

degree to which it attracts domestic or foreign in-migrants, and a formal algebra developed which connects and decomposes these various measures. In addition, these measures can be made specific to gender, race, educational level, age, and other characteristics collected at the same time. The measures developed can also be employed in multivariate analysis in which they are combined with social, economic, geographic, and other aggregate characteristics of the states.

The analyses presented in this chapter examines lifetime migration as of 2006–2010. Several of the key findings are as follows:

Half the population of the United States aged 25 or older was residing in their state of birth. One third of all migrants in this age group are immigrants from abroad. Of the native born, 60 % are residing in their state of birth; those with higher education are more likely to be interstate migrants. As much as the popular press describes the US as a nation on the move, for the total US population from 1850 to 2010, between two thirds and three quarters was living in their state of birth at each measurement point.

For the age group 25–59, the 10 states with the largest number of out-migrants accounted for half of all out migrants. For those 60 years or older, the three states with one million or more out-migrants accounted for almost a quarter of all migrants. For the older age group, one half the states showed a percentage loss due to lifetime migration.

The probability of a state retaining its native born varied widely across the states from 26 to 76 %. The attraction ratio also varied widely across states – from 25 to 90 %. The more refined probability measure of attraction showed that 22 states each attracted less than 1 % of the pool of migrants, while only four attracted 4 % or more.

An analysis of the interstate exchanges for 12 select states, shows that many of the exchanges are between contiguous or nearby states, along with movements to the fast-growing states of Arizona, Florida, and Texas. California, which had a small net loss of lifetime domestic migration, was both a large sending and receiving state.

The analysis of lifetime migration can be a useful adjunct to other migration studies, which often focus on residence changes over one or five years, by providing a longer-term perspective on the interstate moves that have taken place and the consequences of these changes on the states' size and composition. As such it serves to counteract over-interpretation of short-term fluctuations. Examining lifetime migration at different points of can help reveal when major shifts occurred and their impact.

For example, attention of late has been paid to the relative growth and economic status of Georgia vs. Michigan, as the former has overtaken Michigan in size as of 2012. As Table 13.3 shows, Georgia has a somewhat higher retention probability and a much higher attraction ratio than Michigan for the 25–59 population as of 2006–2010. But has this always been true?

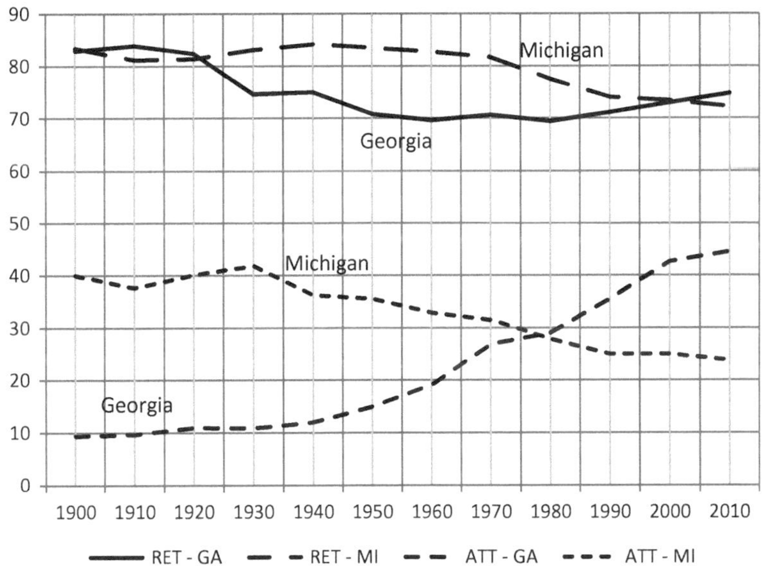

Fig. 13.1 Attraction and retention measures from 1900–2010 for Michigan and Georgia[a,b] Source: [a]The universe for the figure is the total population – all ages. [b]The source is microdata from the ACS, drawn from the IPUMS project (Ruggles et al. 2010: https://usa.ipums.org/usa/).

Figure 13.1, shows the retention and attraction ratios for Georgia and Michigan from the twentieth century on, for the total population.[7]

Both states have had above average retention rates over this period, but Georgia's diverged from Michigan's starting in 1930 as its residents moved out of Georgia – and the South in general. The difference in retention rates peaked in 1960 and both are quite similar by the 1990s with a slight advantage to Georgia by 2010. But, the big story is in attraction rates. Georgia and much of the South had low attraction rates until the 1970s. Georgia's attraction rate overtook Michigan's in 1980 (both close to 30 % at that time). They are now separated by 20 percentage points (44 vs. 24) in favor of Georgia. At the beginning of the twentieth century, they were separated by 30 percentage points (10 vs. 40) in favor of Michigan.

Although this broad historical sweep requires additional analyses (for example by race, education, destinations, etc.) to provide a detailed picture of the factors at work, it illustrates how these data can provide useful information for a state's policy and program planning in showing how its population evolved vis a vis key other states.

A more direct use of the migration and related data is illustrated by the decomposition of the college level population developed above. By identifying how well a

[7]We chose 1900 as the starting point because of the absence of microdata for 1890 and to allow both states to be established. By definition, new/frontier states will have very high attraction rates. We also used the total population for this graph, although the same conclusions are true for the population 25–59.

state is holding on to its talented people and how well it is attracting talent from other states or abroad can impart insights into policy and program planning. As the decomposition of the proportion of college educated in each state reveals, states vary widely in the extent to which this proportion arises from the production and retention of college educated native born versus the impact of in-migrants. Understanding the magnitude of these factors can help lead to appropriate policies, which may differ according to the primary goal – retention or attraction.

We illustrate this potential application by calculating certain hypotheticals for Michigan. From Tables 13.8 and 13.9 we can ask how much Michigan would increase its college-level proportion if it achieved the attainment level of Illinois (.338) and the retention level of Georgia (.585). Those changes, given the same production potential of column 1, would increase the proportion of college graduates for Michigan from 27 % to 31 %. Of course, this is simplistic because changing these terms would also change the production potential term.

Alternatively, we can keep constant the terms in Table 13.8 and ask how much difference increasing the probability of attracting college-level in-migrants would make to Michigan's higher educational percentage. As of 2006–2010, Michigan's probability of attracting domestic in-migrants was 1.8 %; an increase in this probability to 3.1 % would add 4 % to Michigan's higher education percentage and place it ahead of California and Georgia. It should be noted that even with a probability of 3.1 %, Michigan would still be attracting migrants at a lower rate than Georgia (4.3 %) or Illinois (3.7 %).

Of course, wishing does not make it so. The hypothetical calculations can point to some possible paths for improvement but the many policy and program steps needed to achieve the desired gains need to be addressed. In order to increase the production and retention of college graduates one must look to the adequacy of funding for education at the state level (both K-12 and higher education) for producing sufficient numbers of college ready students and for meeting the financial and capacity dimensions for the increased demands on higher education. Likewise, there needs to be the availability of appropriate jobs and quality-of-life considerations for college graduates to remain in-state. Many of the same considerations come into play in a program seeking to boost the number of college-educated in-migrants, especially those aspects dealing with job availability and the ability of a state's cities and towns to have the housing supply and quality-of-life amenities that will attract educated newcomers.

Although no single analysis of lifetime migration data can by itself generate effective and efficient programs, judicious scrutiny of this material can sharpen the focus of deliberations about the best policies to adopt. For example, Michigan has recognized the value of having more educated immigrants (Stateside Staff – Michigan Public Radio 2014), and the importance of a more highly educated population. But, as states reduce financial support for higher education it will likely reduce the production and retention of native born college graduates – sometimes in states where this is their major source of college graduates.

The tendency in a number of states to cut funding for higher education and pass along the resulting higher tuition (See Dynarski 2014), has been referred to as a college user tax (French 2012). In so doing, these states run the risk that eligible students will start to go elsewhere for their degrees, or will leave the state soon after graduation for jobs enabling them to pay off their high student debt. Even those that remain will have on average less discretionary income due to their debts and thus contribute less to the state economy. And the higher tuition structure may discourage talented in-migrants who will be facing college costs for their children.

These considerations point to the importance of adequate state support for both K–12 education and to higher education. The former will have the benefit of making more in-state students college-ready as well as attracting migrants who value high-quality schools for their children. The support for higher education will allow more local residents to study in-state and also serve as a draw for foreign-born students, who often remain in the local area after graduation as pointed out in a recent Brookings study (Ruiz 2014).[8]

The basic picture to be garnered from lifetime migration is where everyone was seated when the musical game of life started and where they are now when we stop the music, at least temporarily, and take stock of their locations. Although this picture is incomplete – it does not identify when the move took place or any of the intermediate moves that may have taken place prior to current residence – the analysis of lifetime migration is a useful complement to more detailed, shorter-term techniques by providing a long-term perspective on how states developed their mix of population from different geographic locations. The ability to analyze this evolution for key groups of the current population – by age, gender, race/ethnicity, education – provides insights to a state's population history and can provide useful input into policy and program planning.

Appendices

Appendix A13: Basic Algebra of Lifetime Migration Measures

As noted in the text, the two questions on place of birth and current residence yield three independent measures for each state: ***B*** is the total number born in a state, ***E*** is the number of native born leaving the state, and ***D*** is the number of in-migrants to a state. These building blocks can be combined in various ways to yield a number of measures, which capture several facets of the lifetime migration process. These measures are interrelated since they utilize one or more of the same elements. This appendix reviews a number of these measures and their interrelationships.

[8]The report includes an interactive table, which allows one to compare cities across the US in the volume of foreign students and the retention of them in the local labor market: http://www.brookings.edu/research/interactives/2014/geography-of-foreign-students#/M10420

A13.1 Basic Notation

The three independent quantities observed are ***B***, ***E***, and ***D***. $\mathbf{B}_i$ is the total number born in state *i*, who are alive and living somewhere in the enumerated area, not necessarily the state of birth. $\boldsymbol{E}_i$ is the number of native born who have left state *i* at some point prior to enumeration. $\boldsymbol{D}_i$ is the number of in-migrants to state *i* – those enumerated in state *i* but born elsewhere, including another country. To simplify the notation, *i* will be dropped from the equations.

As lifetime migration analysis often focuses on inter-state exchanges, it is often restricted to the native born, since immigrants from abroad cannot have a domestic place of birth. In this analysis, however, we often include both native-born migrants and immigrants from abroad to show the total impact of migration; and this also serves to show the importance of immigration from abroad. Immigrants from abroad are in effect treated as coming from a (mythical) place with no in-migrants and only out-migrants.

When necessary we use the notation D_N for native in-migrants and D_I for immigrants from abroad to distinguish these two sources.

A major focus of the analysis is on the effect of education on migration and to distinguish the three educational levels employed we use the letters P, S, G to distinguish those among three different educational categories.

P indicates those who had a primary school education, but less than a high school diploma. S represent those who have graduated secondary school or have some college. G represents those who have graduated college with a BA degree or higher. Thus, B_G would indicate the count of those who were born in state *i* and attained a college degree. Likewise, $D_{I.G}$ represents in-migrants from abroad with a college degree.

Simple addition and subtraction produce two additional quantities of importance:

$$C = B - E \tag{A13.1}$$

$$T = B - E + D \tag{A13.2}$$

Equation A13.1 represents the number of those born in a state who are still residing in the state at the time of enumeration. Equation (A13.2) is the total population of the state at the time of enumeration as a sum of the exits, E and the entrants D. Note that Eq. (A13.2) can be simplified by using C instead of the sum of B and E.

A13.2 Basic Measures

A prime use of lifetime migration data in past studies has been to study exchanges between states or the degree of gain or loss due to migration (Shyrock 1964; Eldridge and Kim 1968; Lee et al. 1957; and Siegel and Swanson 2004).

The key measure employed is the **Net Migration Rate for a State**, which is equivalent to the **Percent Gain or Loss due to migration**:

$$G = [D_N - E] \times 100/T \qquad (A13.3)$$

where the numerator measures the net gain or loss in native migrants and the denominator the total native population residing in state *i*, regardless of state of birth. The numerator, which is the numerical amount of gain or loss due to migration is sometimes referred to as the Birth Residence Index (Siegel and Swanson 2004, pp 511–512; Shyrock 1964, pp 19–20).

For the country as a whole, the net gains and losses of the native population across the states balance out to zero. The Interstate Migration Rate for the country is defined as the number of native born not living in their state of birth, divided by the total native population. As shown in Table 13.1, for the US as a whole in 2006–2010, 38% of the native population 25 years of age and over was not living in their state of birth.

In addition to the gain and loss measures, other key measures include those reflecting **retention** and **attraction** measures of a state.

$$R_N = C_N/B_N = (B_N - E_N)/B_N = 1 - (E_N - B_N) \qquad (A13.4)$$

The **Retention proportion** is the probability that a native person born in a state resides in the state at the time of enumeration. This is of course equal to 1 minus its complement – the proportion of native born in a state who emigrate from the state. For the country as a whole, the complement of the retention proportion is the **Interstate Migration Rate**, as it represents the proportion of the native population not living in their state of birth.

These measures may be made specific for race, gender, age, or education as well as other measured characteristics.

The **Attraction proportion** is defined as the proportion of the current population who are native or foreign in-migrants.

$$A = D/T = D/(B - E + D) \qquad (A13.5)$$

Note that while R is a true probability in that all members of the denominator are at risk of emigrating that is not true for the attraction measure. A portion of the denominator – those born in the state and still residing there – are not at risk of migrating there.

To construct a true probability of attraction, we first define for each state the pool of possible in-migrants (H), which is the sum of in-migrants from all states minus the out-migrants from that state, since they are not at risk of being in-migrants to their state of birth.

The pool is defined as:

$$H = \Sigma(D - E) \tag{A13.6}$$

so that

$$P = D/H \tag{A13.7}$$

P is the **probability of a state receiving a migrant** from among all those migrating.

For the reasons touched on in the text, larger states will most often attract a larger absolute number of migrants than smaller ones. As discussed there, this observation has been incorporated into a number of models of migration. As a result, the probability measure is highly correlated with population size.

In the current analysis, the correlation between the probability of a state receiving a lifetime migrant aged 25–59 as of 2006–2010 and the size of the state is 0.95. In a statistical analysis, it would be possible to control for size, but it is also desirable to develop a measure of expected probability for each state so that its actual probability can be contrasted with its expectation. To analyze the success of areas in attracting migrants over a period, independent of size, we have developed a measure of expected probability of in-migration and contrasted that with the actual.

This follows the precedent of Bachi (1957), who developed a Migration Preference Index based on the ratio of actual to expected number. As Bachi's interest was in analyzing specific migration streams his measure of expected is based on the population of both the place of origin and of destination. A detailed explanation of Bachi's index and a sample calculation is presented in U.S. Census Bureau 1973, v.2, pp 656–657 and Shryock 1964, pp 267–275.

As our goal is for a more general measure of expected to compare with actual, we use the simpler method of employing a state's proportion of the total population as of 1990 – part way through the relevant time period – as an estimate of the expected probability to compare with the actual probability.

Thus, our measure may be defined as:

$$X = T_{1990}/\Sigma T_{1990} \tag{A13.8}$$

Where T_{1990} is the total population in 1990 for a state divided by the sum of the population across all states in 1990. Thus, X is nothing more than the relative size of a state in 1990 compared to all other states.

$$M_G = X \times H_G \tag{A13.9}$$

The expected number of lifetime college graduate migrants, M_G, is the product of X by the pool of migrants with college degrees, H_G.

To illustrate for Florida, its expected probability is .0520 based on its population size in 1990 relative to the US total. The expected number of lifetime college grads is thus.052 multiplied by the pool of eligible domestic and foreign college grads,

(24,621,086) to yield the expected number of in-migrant lifetime college graduates of 1,280,296 as shown in Table 13.7.

A13.3 Interrelationships Among Measures

As noted in the text, since the basic measures utilize one or more of the independent factors, they are interrelated in various ways. This section illustrates a number of these interrelationships that are relevant to the analyses presented.

$$\text{Since } A = D/T \text{ and } P = D/H \text{ then } P = A \times (T/H) = A/(H/T)$$

so that a state's probability of receiving an in-migrant, ***P*** is a function of its attraction ratio (A), its current size (T), and the total pool of in-migrants for that state (H). Since the total pool will vary less from state to state than state size, the formula shows that for a given attraction ratio, larger states will generally have a higher probability than smaller ones, as the denominator in the formula will be smaller.

This relationship is illustrated in Table A13.1 for ages 25–59 with four states selected from Table 13.5 (for measures) and Table 13.3 for population size.

Although the four states have almost identical attraction ratios, as shown in column 1, they vary a great deal in population size (column 2), and consequently they vary widely in the probability of gaining in-migrants, with those states with greater populations showing higher probabilities.

By the same token, for a given probability, a larger state will have a smaller attraction ratio than a smaller one, as illustrated in Table A13.2.

As shown in Table A13.2, although Colorado, Maryland, and Pennsylvania have almost identical attraction probabilities of attraction, the attraction ratio decreases as the population size increases across the three states.

The attraction ratio gives a current snapshot for a state of the relative prominence of in-migrants in their total population as a function of the gains and losses that have occurred over the years. This can be seen more clearly if the attraction ratio is rewritten as:

$$A = D/T = D/(C + D) = 1/[C/D) + 1] \qquad \text{(A13.10)}$$

Table A13.1 The relationship between attraction probability and the attraction ratio[a]

State	A	Population (1000)	P
Illinois	40.8	6,169	0.036
Missouri	39.1	2,782	0.015
Nebraska	40.9	833	0.005
South Dakota	40.7	365	0.002

[a]The input for this table is drawn from Tables 13.3 and 13.5

Table A13.2 The relationship between the attraction ratio and population size[a]

State	P	Population (1000)	A
Colorado	0.024	2,443	69.8
Maryland	0.024	2,819	61.8
Pennsylvania	0.025	5,948	29.4

[a]The input for this table is drawn from Tables 13.3 and 13.5

Table A13.3 The relationship between the attraction ratio and the relative share of the migrant population[a]

	Population (1000)			
State	A	C	D	C/D
Nevada	89.9	130	1,158	0.112
Florida	75.7	2,088	6,498	0.321
Pennsylvania	29.4	4,199	1,749	2.404

[a]The input for this table is drawn from Table 13.3

As a reminder, ***D*** indicates the number of in-migrants to a state and ***C*** is the number of those born in a state who are still living in the state of birth. As the equation makes clear, the larger the ratio of native born to in-migrants, the larger the denominator and therefore the lower the attraction ratio, and vice versa.

Table A13.3 shows the attraction ratio and its two elements, C and D, for Nevada and Florida – two states with high attraction ratios and for Pennsylvania, a state with a low attraction ratio.

The high ratio of native born still residing state of birth (C) in Pennsylvania relative to the number of in-migrants (D), leads to the low attraction ratio, in comparison to Florida and Nevada, where the heavy influx of in-migrants leads to a low ratio of C/D and hence to a high attraction ratio.

On the other hand, the probability measure reflects a state's success, relative to other states, in attracting in-migrants from the pool of those moving over the years. It is not dependent on the size of the resident native population. It should be noted from Table A13.3 that Pennsylvania, despite its low attraction ratio has experienced many more in-migrants in absolute terms than Nevada and as a result also has a higher probability of in-migration (.025), as defined above, than Nevada (.016). This illustrates further the point that the probability measure is highly correlated with population size.

There is also a close relationship between the retention proportion and the attraction ratio.

Since: $(C + D)/T = 1$, then $C/T + D/T = 1$, which may be written as

$$(B/T) \times (C/B) + D/T = 1, \text{ so that } R \times B/T + A = 1 \text{ and}$$
$$A = 1 - (R \times B/T), \text{ and}$$
$$R = (1 - A) \times T/B$$

The retention probability is a function of the attraction ratio and the ratio of those born in the state to the total population, and conversely.

Appendix B13: Decomposition of the Number of College Graduates in a State and the Percentage of Residents Who Are College Graduates

A useful analytic device is to decompose a measure of interest into components that can point to key factors that affect that measure. Given the strong interest that most states have in increasing their college-educated population to enhance economic growth and development, the following section presents a decomposition of this number and its share of the total population into elements. The elements are the state's production of college graduates from their native born, the ability to retain these graduates, and the level of success in attracting college-educated in-migrants from within the country and from abroad.

Equation (B13.1) simply presents the number of college grads from each key source: college graduates born in the state and living there; native college graduates from other states who migrate in; and college-educated immigrants who move to the state.

$$T_G = C_G + D_{N.G} + D_{I.G} \qquad \text{(B13.1)}$$

Equation (B13.2) decomposes each element into factors of interest, so that:

$$T_G = B \times [(B_G/B) \times (C_G/B_G)] + [H_{N.G} \times P_{N.G}] + [H_{I.G} \times P_{I.G}] \qquad \text{(B13.2)}$$

The first term on the right represents the number of native born with a college degree residing in the state, arising from the total number born there, the proportion obtaining a college degree (which can occur out of state) and the retention rate of these college graduates. The second term on the right represents the number of college graduates residing in the state who were born in another state expressed in terms of the probability of attracting such a person and the total pool of these college graduates. The third term is similar but refers to the probability of attracting a college graduate from abroad and the total pool of college graduated immigrants.

The proportion of a state's population who are college graduates is simply the number in Eq. (B13.2), divided by the state's total population, as shown in Eq. (B13.3):

$$T_G/T = B/T \times B_G/B \times C_G/B_G + P_{N.G}/T \times H_{N.G} + P_{I.G}/T \times H_{I.G} \qquad \text{(B13.3)}$$

Table 13.8 in the text uses Eq. (B13.1) to show the proportion of college graduates coming from each major component and Table 13.9 uses Eq. (B13.3) to further subdivide the college graduates of each state into the production and retention elements.

The elements of Eq. (B13.3) are given below for three very different states – California, Michigan, and West Virginia – to show how the decomposition can reveal important dynamics that can prove useful in policy and program decisions.

Table B13.1 Decomposition of the sources of the college population for selected states, population 25–59: 2006–2010[a]

Decomposition elements		CA	MI	WV
Production & retention				
(1)	Size potential	0.605	1.114	1.307
(2)	Production of native college graduates	0.290	0.294	0.224
(3)	Retention of native college graduates	0.652	0.543	0.383
In-migration of college graduates				
(4)	Pool of native graduates (thousands)	15,921	16,303	16,851
(5)	Probability of attracting	0.1004	0.0178	0.0036
(6)	Pool of immigrant graduates (thousands)	8,009	8,009	8,009
(7)	Probability of attracting	0.2314	0.0199	0.0010
Percent of college graduates from				
(8)	Production & retention = (1) × (2) × (3)	11.4	17.8	11.2
(9)	In-migration of native grads = [(4) × (5)]/T	9.0	6.2	6.3
(10)	In-migration of foreign grads = [(6) × (7)]/T	10.4	3.4	0.9
(11)	Total percent college grads = (8) + (9) + (10)	30.9	27.4	19.0
Relative distribution of college graduates				
(12)	Production & retention = (8)/(11)	0.371	0.650	0.590
(13)	In-migration of native graduates = (0)/(11)	0.291	0.226	0.362
(14)	In-migration of foreign graduates = (10)/(11)	0.338	0.124	0.048

[a]Source: The source for cells (1) – (3) are drawn from Table 13.9; cells (4) – (7) are the authors own calculations from a table of lifetime migration counts - http://bit.ly/2bTwTeJ - produced from microdata based on the ACS, drawn from the IPUMS project (Ruggles et al. 2010: https://usa.ipums.org/usa/). All other cells are products from cells (1) – (7)

The percentage distribution of the components of the total proportion of college graduates in a state, as shown in Table 13.9, is obtained simply by dividing the total percentage by the percentage from each source. For example, the percentage of California graduates due to production and the retention of native born is 11.4/30.9 or 37.1 as shown in Table 13.9.

References

Aisch, G., Gebeloff, R., & Quealy, K. (2014, August 19). Where we came from and where we went, state by state [data visualization]. *The New York Times*. http://nyti.ms/1oLsIgy. Accessed 16 Oct 2014.

Bachi, R. (1957). Statistical analysis of geographic series. *Bulletin de l'Institut International de Statistique, Stockholm, 36*(2), 234–235.

Bogue, D. J. (1959). Internal migration. In P. M. Hauser & O. D. Duncan (Eds.), *The study of population: An inventory and appraisal* (pp. 486–509). Chicago: The University of Chicago Press.

Bogue, D. J., Hinze, K., & White, M. (1982). *Techniques of estimating net migration*. Chicago: Community and Family Study Center at the University of Chicago.

Cebula, R. J. (1979). A survey of the literature on the migration-impact of state and local government policies. *Public Finance, 34*(1), 69–84.

Conway, K. S., & Rork, J. (2012). No country for old men (or women) – Do state tax policies drive away the elderly? *National Tax Journal, 65*(2), 313–356.

Conway, K. S., & Rork, J. (2014). *How has elderly migration changed in the 21st century? What the data can – and can't – Tell us*. Paper presented at the Population Association of America 2014 annual meeting, Boston, Massachusetts, 1–3 May 2014.

Dynarski, S. (2014, September 21). Where college ratings hit the wall. *The New York Times*.

Eldridge, H. T., & Kim, Y. (1968). *The estimation of intercensal migration from birth-residence statistics: A study of data for the United States, 1950 and 1960* (Analytical and technical report no. 7). Philadelphia: University of Pennsylvania, Population Studies Center.

Eldridge, H. T., & Thomas, D. S. (1964). *Population redistribution and economic growth: United States, 1870–1950*. Philadelphia: The American Philosophical Society.

French, Ron. (2012). 'College tax' burdens students, state. *Bridge Magazine*. http://www.lssu.edu/ir/documents/bridgemi-com-College_tax_burdens_students_state_000.pdf. Accessed 16 Oct 2014.

Frey, W. H. (2004). Brain gains/brain drains. *American Demographics, 26*(5), 19–23.

Galle, O. R., & Tauber, K. E. (1966). Migration and intervening opportunities. *American Sociological Review, 31*(1), 5–13.

Gebeloff, R., & Leonhardt, D. (2014, August 24). The growing blue-state diaspora. *The New York Times*.

Gelbach, J. B. (2004). Migration, the life cycle, and state benefits: How low is the bottom? *Journal of Political Economy, 112*(5), 1091–1130.

Goworowska, J., & Gardner, T. K. (2012). *Historical migration of the young, single, and college educated: 1965 to 2000* (Population Division Working Paper No. 94). Washington, DC: U.S. Census Bureau.

Houtenville, A. J., & Conway, K. S. (2001). Elderly migration and state fiscal policy: Evidence from the 1990 census migration flows. *National Tax Journal, 54*(1), 103–124.

Ihrke, D. K., & Faber, C. S. (2012). Geographical mobility: 2005 to 2010 (*Current population reports*, pp. 20–567). Washington, DC: U.S. Census Bureau.

Lee, E. S., Miller, A. R., Brainerd, C. P., & Easterlin, R. A. (Prepared under the direction of S. Kuznets and D. S. Thomas.). (1957). *Methodological considerations and reference tables* (Population redistribution and economic growth, United States, 1870–1950, Vol. 1). Independence Square, Philadelphia: The American Philosophical Society.

Liaw, K.-L., & Frey, W. H. (2007). Multivariate explanation of the 1985–1990 and 1995–2000 destination choices of newly arrived immigrants in the United States: The beginning of a new trend. *Population, Space and Place, 13*(5), 377–399.

Minnesota Population Center. (2014), *Integrated public use microdata series, International: Version 6.3* [Machine-readable database]. Minneapolis: University of Minnesota.

Moffit, R. (1992). Incentive effects of the U.S. welfare system: A review. *Journal of Economic Literature, 30*(March), 1–61.

Petersen, W. (1975). *Population* (3rd ed.). New York: Macmillan Publishing Co., Inc.

Ruggles, S. J., Alexander, T., Genadek, K., Goeken, R., Schroeder, M. B., & Sobek, M. (2010). *Integrated public use microdata series: Version 5.0 [Machine-readable database]*. Minneapolis: University of Minnesota.

Ruiz, N. G. (2014). *The geography of foreign students in U.S. higher education: Origins and destinations*. Washington, DC: Brookings Institution. http://www.brookings.edu/~/media/research/files/reports/2014/08/foreign%20students/foreign_students_final.pdf. Accessed 14 Oct 2014.

Shyrock, H. S., Jr. (1964). *Population mobility within the United States*. Chicago: Community and Family Study Center, University of Chicago.

Siegel, J. S., & Swanson, D. A. (Eds.). (2004). *The methods and materials of demography* (2nd ed.). San Diego: Elsevier Academic Press.

Stateside Staff. (2014, May 4). EB-5 will bring investment, immigrants to Michigan. *Michigan Public Radio.* http://michiganradio.org/post/eb-5-will-bring-investment-immigrants-michigan. Accessed 14 Oct 2014.

Stouffer, S. (1940). Intervening opportunities. *American Sociological Review, 5*(December), 845–867.

U.S. Census Bureau. (1953). *United States census of population: 1950*. Volume 4. Special Reports, Part 4, Chapter A, State of Birth. Washington, DC: Government Printing Office.

U.S. Census Bureau. (1963). *United States census of population: 1960. Subject Reports, Lifetime and recent migration* (Final report PC(2)-2D). Washington, DC: GPO.

U.S. Census Bureau. (1973). *The methods and materials of demography*. By Henry S. Shryock, Jacob S. Siegel, and Associates. Second Printing. Washington, DC: U.S Government Printing Office.

U.S. Census Bureau. (2014). *National, state, and Puerto Rico commonwealth Totals datasets: Population change: April 1, 2010 to July 1, 2013*. Washington, DC: Population Division.

United Nations. (1970). *Manuals on methods of estimating population. Manual VI: Methods of measuring internal migration.* Population studies, no. 47. ST/SOA/Series A/47. Department of Economic and Social Affairs. New York: United Nations. Zipf, G. 1949. *Human Behaviour and the Principle of Least Effor*t. Cambridge, MA: Addison-Wesley.

United Nations. (1970). *Manuals on methods of estimating population. Manual VI: Methods of measuring internal migration.* Population studies, no. 47. ST/SOA/Series A/47. Department of Economic and Social Affairs. New York: United Nations.

Zipf, G. (1949). *Human behaviour and the principle of least effort*. Cambridge, MA: Addison-Wesley.

Chapter 14
Census Costs: Rationale for Re-Designing Traditional Census Data Collection Methodology with the Census-Enhanced Master Address File

Alison M. Yacyshyn and David A. Swanson

Abstract Debates in countries around the world focus on the need for accurate census data, the overall costs, response rates, confidentiality and legislated changes. With technological advancements, one would expect population counts to be calculated with more ease and fewer associated costs. In this paper, the costs of recent censuses per person, using overall budgetary costs of censuses in Canada and the United States, the costs per housing units, and the costs per person are analyzed. Overall census costs per household have gone up over recent years (and a slightly greater extent in the US than in Canada). Given that costs have risen; We recommend that the "Census-Enhanced Master Address File" (CEMAF) approach be considered in the re-designing of census methodology.

Keywords Census • Census data collection methodology • Housing units • Census-Enhanced Master Address File

14.1 Introduction

A discussion about the census is timely, given that the United States conducts one every 10 years and Canada every five. In Canada, participation of the census is mandatory according to the Statistics Act (1985) and penalties can be imposed for not completing the census. Similarly, in the United States, the United States Code requires that people fill out their U.S. census forms and a \$100 penalty for not answering the census and up to \$500 for giving false information is stipulated (United States Code 2010). Not completing a census has legal ramifications in both

A.M. Yacyshyn (✉)
University of Alberta, Edmonton, AB, Canada
e-mail: Alison.Yacyshyn@ualberta.ca

D.A. Swanson
University of California Riverside, Riverside, CA 92521, USA
e-mail: David.swanson@ucr.edu

D.A. Swanson (ed.), *The Frontiers of Applied Demography*, Applied Demography Series 9, DOI 10.1007/978-3-319-43329-5_14

Canada and the United States. Both countries statistical agencies encourage co-operation and explain the importance of participation gathered in their censuses and surveys.

The 2011 Canadian census was unique as elected politicians of the Canadian government, in June 2010, made a surprising move by eliminating the mandatory long questionnaire for the 2011 Canadian census. Historically, and up until 2006 census, questionnaires included both long- and short-forms in Canada. In 2015, shortly after changes in the federal government (the Liberal Party of Canada won a majority in the federal election that year), the Government of Canada restored the 2016 mandatory long form census (as per the political party's platform). The questions in both the short and long form remain stable to those asked in 2006 and 2011, allowing historical comparability. Where the 2011 Census had a response rate of 97.1 %, the unweighted response rate of the 2011 National Household Survey was 68.6 %.

Unlike Canada, which uses a mandatory long form census, the United States transitioned "the decennial long form to the American Community Survey (ACS), leaving the entire decennial survey with only 10 short-form questions designed for easier response" (U.S. Department of Commerce 2011: 1). The ACS, which began collecting annual data since 2005, is a mandatory survey and respondents are legally obligated to answer the questions in it. Implementing the ACS to collect annual data was "viewed by many as the single most important change in the way detailed decennial census information is collected since 1940, when the Census Bureau introduced statistical sampling as a way to collect 'long-form' data from a sample of households" (U.S. Census Bureau 2009: iii). As the ACS collects demographic, social, economic, and housing data every year, the reliance on (long-form) decennial census data has been eliminated in the United States.

To address the cost issues surrounding the census, this paper looks at past and current enumeration, and the decreased reliance on the enumerator. In order to assess current expenses associated with the census, the overall costs are analyzed for Canada and the United States for recent censuses. The effects of the Internet methodology used in conducting a census are discussed and how the digital divide will affect future censuses is addressed. Finally, limitations of the data used in the paper are noted.

14.2 Background: Population Counts from the First Censuses to Now

Although Canada's first census took place in New France in 1666 (with only 3215 inhabitants officially counted), the first national census of Canada was conducted some 205 years later, in 1871 (Statistics Canada 2006a) totaling 3,689,000 persons (Statistics Canada 2010b). The first census of the United States was conducted in 1790 and totaled 3,231,533 individuals, excluding counts related to slaves (Bureau of the Census 1916). The traditional means of conducting a census was a complex and massive undertaking (United Nations 2006). Historical accounts of census enumerators, including the famous Dr. William Mayo (who was named deputy by the sheriff in the Minnesota hinterland to take census counts) (Clapesattle 1941), noted that the accuracy of the census lists were troubling. The days, weeks, and even month's enumerators faced when dealing with challenging geographical terrain and conditions were often described as being difficult. Over the years, the census has been conducted with the same constitutional mandate; times have changed, and census coverage is still of great concern (Martin and Dillman 2008).

During the twentieth century, it is estimated that over 190 countries in the world conducted a population census and the majority used a traditional approach for enumeration (United Nations 2006). Since countries do not conduct a census every year, most governments use estimates. Recent population numbers approximate that there are 34,108,752 Canadians (Statistics Canada 2006b) and 310,472,998 Americans (U.S. Census Bureau 2010) residing as current residents in 2010. These population estimates are deemed to be accurate, timely population figures in their own right, which take into account net census under-coverage. Conducting a census is an obligation of the (federal) government to enumerate the population, apportioning the Congress, and it is part of the constitution (National Research Council 1995). The census is important for providing a framework for the electoral system, for allocating central government funds to local administrations, and for public planning.

Because a census is the primary source of data about the size and characteristics of the population, it provides a demographic profile of a country and often serves as the basis for developing area sampling frames for use in surveys. In a speech by James Garfield (Ridpath 1881: 219) in 1867:

> The chief instrument of American statistics is the census, which should accomplish a two-fold object. It should serve the country by making a full and accurate exhibit of the elements of national life and strength, and it should serve the science of statistics by so exhibiting general results that they may be compared with similar data obtained by other nations.

Still today, traditional type censuses are one of the largest and most costly statistical activities that governments and/or their national statistics office undertake, and costs are immense.[1] Thousands of temporary workers are hired every census year to assist in population enumeration (Lu 2009). As a result of monetary costs, countries have been forced to delay or even cancel a census due to funding constraints (United Nations 2006). For example, India, in 2011 pushed back a census "due to the unavailability of resources" (Shah 2012). Similarly, the 2001 census in South Africa "began very late because of an adequate budget allocation" (Fanoe 2011: 90).

---

[1] Direct costs for register based and partially register based censuses are much lower; however, the traditional type of census is focused upon here.

14.3 Eliminating the "Middleman" Enumerator

Countries around the world use various techniques for collecting population based statistics. Approaches to census design include: the classic approach (which is currently used by: Canada, and Colombia, for example), the register-based approach (e.g., Norway), combination of classic and register-based approaches (e.g., Spain), register-based censuses with sample surveys (e.g. Israel, and the Netherlands), the rolling census approach (e.g. France), and traditional enumeration with yearly updates of characteristics (e.g. United States and Peru) (United Nations 2010).

Of the techniques to collect population data by census enumeration (i.e. the classic approach, the combination of classic and register-based approaches, and the traditional enumeration with yearly updates of characteristics), two types of methods have been used: a direct interview (also known as the canvasser method), or self-enumeration (or the householder method) (Shryock, Siegel et al. 1976). Self-enumeration is a major innovation in twentieth and twenty-first century. Self-enumeration occurred for the first time in the 1960 United States census and in Canada's 2006 census about 98 % of households were enumerated using this method. In comparing these two methods, a substantial portion of costs associated with taking a census is due to field operations (Statistics Canada 2003) and the direct interview method is more costly.

While being interviewed at the Royal Statistical Society's Annual conference, Robert Groves (head of the U.S. Census Bureau, at the time) noted: "the major cost is not printing the forms, or distributing them; it is in physically chasing up the households who do not fill in or send them back" (Champkin 2010, p. 174–175). "Experiments in self-enumeration led to its successful use in the 1960 (U.S.) census, when householders in urban areas were asked to complete and mail back questionnaires containing the sample items" (U.S. Census Bureau 2000: 3). With technological advancements, the use of electronic means to conduct census counts would reduce the overall costs associated with employing enumerators (because fewer enumerators would be hired). So, statistical agencies in Australia, Canada, New Zealand, the United Kingdom, and the United States have embarked on using the Internet as a means to count their respective populations (U.S. Department of Commerce 2011).

As Rosemary Bender, the Director General, Social and Demographic Statistics Branch of Statistics Canada noted in a 2009 report: "using a number of pilot tests and phasing changes over the 2001, 2006 and 2011 Census cycles, the Agency (Statistics Canada) moves from a decentralized, manually intensive collection and data entry operation to a more centralized and automated approach. This in particular addresses key concerns regarding confidentiality and security of personal census data" (Bender 2009). So, with reductions in enumerator costs, it is assumed that as Internet participation rises, cost reductions should materialize (Statistics Canada 2003). Similarly, the United States, Commerce Secretary, Gary Locke has commented in 2010 that "the next decennial headcount (in 2020) will have to be done differently" and that "obviously, we need to look at the use of technology and the use of the Internet to make it easier for people to respond and to avoid mailing back the questionnaire" (Reilly 2010).

Given that there are risks in sensitive statistical records held by federal agencies (Cecil 1993), the U.S. Census Bureau cancelled all plans to use the Internet for data collection in the 2010 population census (Castro 2008). The U.S. Census Bureau had already developed computer screenshots for the 2005 National Census Test Internet application (U.S. Census Bureau 2005). It was predicted that the U.S. Census Bureau would continue to incorporate technological advancements in future census (Bounpane 1986).

In terms of cost-effectiveness, the aim for any country is to plan and carry out a census as inexpensively as possible, in a manner consistent with the content and quality requirements (United Nations 2006). The United States Department of Commerce (2011, p. 2) explicitly states that the "Census must make fundamental changes to the design, implementation, and management of the decennial census to obtain a quality count for a reasonable cost." The research focus of this paper is to provide background information regarding the overall costs of the census and rationale for methodological issues surrounding the census.

14.4 Methodology

As Canada and the United States incorporate technological advancements in to census methodologies, the overall costs are analyzed for the two countries last four official censuses. To determine the cost of the census per person, the entire budgetary costs of the census for Canada and the United States, and the number of housing units, are used to calculate the census cost per person. To convert the census costs from housing units in to census costs per person, the entire population counts are used in the calculations.

In order to properly compare the overall census costs of 1 year to another, dollars must be adjusted (to be consistent with a particular year). Economists often use more than one indicator to determine the relative value of money from 1 year to another (Williamson 2010) depending on the context. In this research, the purchasing power parities (PPPs), which reflect the rates of currency conversion, are used to equalize the purchasing power of different countries by eliminating differences in price levels between countries (OECD Statistics 2009). The costs per dwelling are also adjusted using the Consumer Price Index (OECD Statistics 2009) to be in constant Canadian dollars as represented in January, 2010.

Data used to evaluate census costs are obtained from Canada's national statistical agency, Statistics Canada (2010d), and the "congressional watchdog," the United States Government Accountability Office (United States General Accounting Office 2001), although the costs pertain to the US Census Bureau (who actually conduct the US census). Data from the two national statistical agencies will allow appropriate comparisons.

14.5 Results

Data from only the last five official censuses are obtained for analysis of this paper, as Canada conducts quinquennial (i.e. 5-year) censuses, the data are for years: 2011, 2006, 2001, 1996, and 1991.[2] As the United States conducts decennial censuses the data representing the years 2010, 2000, 1990, 1980, and 1970 are obtained. Since the years of the Canadian and the United States censuses do not match, the data are presented separately.

For Canada, the census data are for the 1991 census through 2011 census and spans 20 years. Table 14.1 indicates that the census cost per housing unit has increased by about 6 dollars and although the overall costs associated with the 1996 census decreased from the 1991 census, a larger increase occurred between the 2001 census and the 2006 census (an increase of about 5 dollars per housing unit, or roughly two-and-a-half dollars per person). The increase in the 2006 cost per unit was due in part "to a one time investment made to implement the Internet solution and a number of other changes" (Statistics Canada 2010d). Even with investments to census Internet applications, the costs did not increase by a gross amount. With population increase and more housing units, the overall census costs are reflected in the increases in associated costs per person and costs per housing unit. Naturally, with a larger population, more individuals need to be accounted for, and associated costs of conducting the Canadian census are reflected in the data of Table 14.1. Costs per person are rising faster than costs per households because households are getting smaller. The costs per household are important in carrying out a traditional type census, so if the population is spread out over more households this will contribute to increased overall costs. The costs per person between the 2006 and 2011 census were similar, demonstrating that the investment of the Internet solution and other changes to the Canadian census may have stabilized the overall costs.

Table 14.2 shows that from 1970 when the U.S. census cost per person was approximately five dollars, the increase to roughly 23 dollars in 2000 and about 43 dollars in 2010 seems large. However, from the 1970 census to the 1980 census costs only increased by 5 dollars per person. Similarly, between the 1980 and 1990 censuses there was an increase of about 5 dollars. It is between the years 1990 and 2010, that the largest costs per person for the census occurred (roughly a 20 dollar increase) resulting in the census cost per person to be approximately 43 dollars. Although the cost per person may increase slightly over time, analyzing the costs per housing unit is informative, especially as single dwellings households increase in number and more household need to be accounted. In 40 years (from 1970 to 2010) the United States population has increased by more than 100 million people and the number of housing units has increased by 59 million. So, not only does the

[2] Given that the data are from an unpublished data request from Statistics Canada and not available in the public realm, costs for census in Canada conducted prior to 1991 reduce the comparability to censuses conducted in the United States in the 1980s and 1970s, for example.

Table 14.1 Census costs for Canada

Year	Census costs (in dollars)	Housing units	Cost per housing unit (in dollars)	Population (persons)	Cost per person (in Canadian 2010 dollars)
1991	432,688,952	10,018,267	43.19	27,296,859	15.85
1996	459,094,722	10,820,050	42.43	28,846,760	15.91
2001	511,892,903	11,562,975	44.27	30,007,095	17.06
2006	611,330,163	12,435,520	49.16	31,612,895	19.34
2011	652,000,000	13,349,445	48.84	33,476,690	19.48

Sources:
Statistics Canada. (2010). Special Tabulation (unpublished data)
Statistics Canada. (2013). "Type of dwelling and population by type of dwelling (1961 to 2011 Censuses)"
http://www.statcan.gc.ca/tables-tableaux/sum-som/l01/cst01/famil66-eng.htm
http://www.macleans.ca/politics/ottawa/the-cost-of-scrapping-the-long-form-census/
Note: The costs are per dwelling, all data are adjusted to be in constant Canadian dollars as of January 2010, using the Consumer Price Index to make the adjustments. Private household refers to a person or a group of persons (other than foreign residents) who occupy a private dwelling and do not have a usual place of residence elsewhere in Canada. The housing units and population are adjusted for non-response (see reference: Statistics Canada, 2008, for population data)

increase in population increase total census costs, but as more independent housing units exist (which may be a house, apartment, a mobile home, etc.), the related costs to conduct a census by household also increases.

As seen in Fig. 14.1, the overall census costs have increased over time for both Canada and the United States; however, the data indicate a greater stability in census costs for Canada as compared to the significantly increasing census costs the United States has experienced. One could say that the census cost increases reflect the incorporation of technology; however, as Tables 14.1 and 14.2 have demonstrated, the increase in population size is a factor in the increases of overall costs. One example might be changes in the rate of cases requiring non-response follow-up, which may in turn have been caused by changes in the composition, attitudes and values of the population. Other census costs include: data processing, analysis, evaluation and dissemination. In 2010, the United States census costs per person superseded any previous U.S. census costs and any Canadian census costs to date. Statistics Canada officials expected "the 2011 Cost per Unit to go back down to a comparable level to previous censuses because of the efficiencies realized with the Internet approach" (Statistics Canada 2010d). The trajectory of census costs (using historical census cost data) suggests that costs will continue to increase until the Internet adoption is completely accepted and utilized methodology in conducting the census, or other census data collection methodology is used when censuses are conducted.

The results demonstrate that as census data collection methodology changes, with less reliance on an enumerator and incorporation of the Internet, costs are still associated with technological advancements in conducting a national census online.

Table 14.2 Census costs for the United States

Year	Census costs (in dollars)	Housing units	Cost per housing unit (in dollars)	Population (persons)	Cost per person (in Canadian 2010 dollars)
1970	920,000,000	70,700,000	13.01	203,302,031	4.53
1980	2,159,000,000	90,100,000	23.96	226,545,805	9.53
1990	3,275,000,000	104,000,000	31.49	248,709,873	13.17
2000	6,553,000,000	117,300,000	55.87	281,421,906	23.29
2010	13,000,000,000	130,038,080	99.97	303,965,272	42.77

References:

United States General Accounting Office. (2001). Significant Increase in Cost Per Housing Unit Compared to 1990 Census

Report to Congressional Requesters. GAO-02-31

Census Scope. (2010). Social Science Data Analysis Network. Retrieved from http://censusscope.org/

U.S. Government Accountability Office (2011). 2010 Census: Preliminary Lessons Learned Highlight the Need for Fundamental Reforms (April 6, 2011). Report number GAO-11-496 T, http://www.gao.gov/products/GAO-11-496T

Note: A housing unit may be a house, an apartment, a mobile home, a group of rooms, or a single room that is occupied (or, if vacant, is intended for occupancy) as separate living quarters. The costs are per dwelling, all data are adjusted to be in constant Canadian dollars as of January, 2010, using the Consumer Price Index to make the adjustments

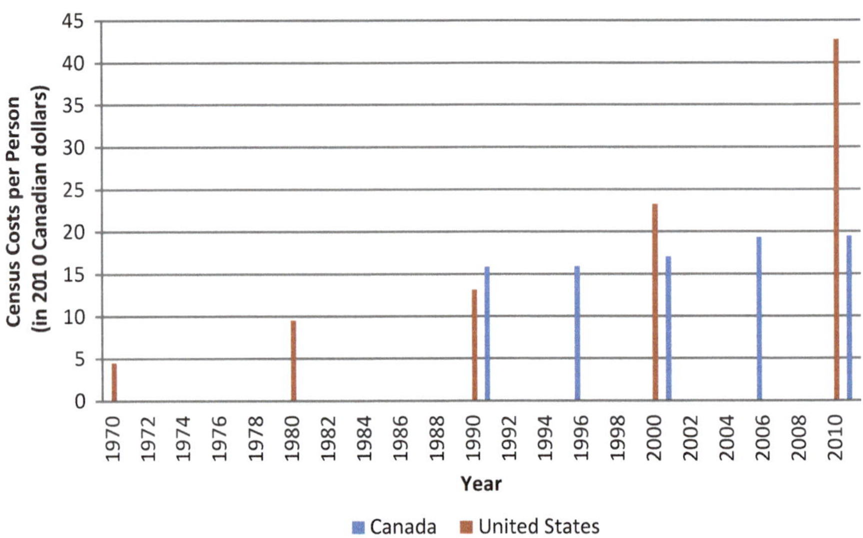

Fig. 14.1 Canada and United States census costs per person (in 2010 Canadian dollars)

Historically, neither Canada nor the United States have experienced a reduction in census costs.

14.6 Discussion: Completing the Census Online = Future Internet Use

According to the International Telecommunications Union (2010) of the United Nations, approximately 77.7 % of Canadian and 77.3 % of the United States populations are online (Internet World Stats 2010). From the "Canadian Internet Use Survey," conducted by Statistics Canada (2010b), the percentage of Internet use (by individuals aged 16 years old and over who responded to have used the Internet for personal non-business purposes in the past 12 months from any location) was approximated to be 80.3 % of the 2009 Canadian population being online. The 2009 data collected by Canadian Internet Use Survey suggests that there is no difference between the sexes in terms of Internet use (as there are 81.0 % male Internet users aged 18 years and over, and 79.7 % female Internet users aged 18 years and over) (Statistics Canada 2010b). There are fewer reported Internet users as age increases (with only 40.7 % of those 65 years of age and over using the Internet) (Statistics Canada 2010b). Similar age and sex differences in Internet use are found in data for the United States (Jones and Fox 2009). Research has demonstrated that "the Internet is still out of range for a lot of persons, especially low-income households, big families and many elderly persons.

Access to the Internet and the ability to use it (not only for surfing but for complex transactions) are limitation factors for the successful promotion of online surveys, not only in developing but also in developed countries" (Haug 2001, p. 8). However, as the numbers of respondents are increasingly able to respond to census questionnaires online, the Internet will affect a country's census data collection methodology. The Internet is an important driver for future change in census costs.

14.7 The Digital Divide

In the October, 1999, "Speech from the Throne," the Government of Canada declared that it would become "known around the world as the government most connected to its citizens, with Canadians able to access all government information and services on-line at the time and place of their choosing" (Government of Canada Privy Council Office 1999). The Government of Canada saw the Government On-Line Initiative (GOL) as: (1) connecting to Canadians (2) improving efficiencies and generating savings (3) assuring security and privacy and (4) developing a legacy: serving Canadians better (Public Works and Government Services Canada 2010). Introducing the Internet in the 2006 Canadian census would fulfill the commitment to make government services available to the public online. To add the Internet response channel to the 2006 Canadian census would also require additional investment. One of the more important benefits of Internet response is the improvement in quality one gets, as well as enhanced privacy; however, costs are accrued. In fact, "countries invest in technology and/or alternative methodologies in order to reduce costs, improve the quality and the timeliness of the dissemination of their census results" (United Nations Statistics Division 2010, p. 4). Maintenance and upgrading of the Internet census data collection methodology require a significant amount of money and it does take initiative and commitment by governments to adopt the Internet approach. Providing an Internet response channel requires a significant investment in development and infrastructure. So, before any money is saved, Internet response needs to exceed some minimal threshold. Careful analysis by the statistical agency (Statistics Canada) would be essential.

Just as Canada has invested in online channels, the United States has recognized the importance of including the Internet as part of census data collection methodology. Robert Groves (former Director of the U.S. Census Bureau) maintained that by "2020 (the census) will be multi-media. We shall be looking at using data already submitted- through tax returns and other official forms- but there are issues of intrusion here: will people regard that as more intrusive than the traditional census form, or less? That is something we have to explore" (Champkin 2010, p. 175). As both the Canadian and United States governments have begun to use the Internet with multi-methods census data collection methodology, incorporating a combination of methodologies seem to be the direction for current census undertakings. However, issues with the census that were once handled in one way, now need to be managed in other creative ways. Change is eminent, but how governments count their populations and incorporating technological advancements is critical to the overall census costs.

14.8 Census-Enhanced Master Address File

The Census-Enhanced Master Address File (CEMAF) is a census methodology which is not based direct interviews or self-enumeration, but rather on a combination of four elements: (1) administrative records; (2) the continuously updated Master Address File; (3) survey data; and (4) modeling techniques. The CEMAF methodology is built on the Master Area File (MAF) (Wang 1999) and Enhanced Master Area File (EMAF) (Swanson and McKibben 2010) methodologies. Where a Master Area File focuses on the estimates of housing, an Enhanced Master Area File also include estimates related to housing units, demographic and socio-economic characteristics for sub-areas and geographic areas as a whole.

> EMAF (i)s an integrated file that contains not only existing MAF variables (e.g., geocode, address, and structure type), but also information on the occupancy status of housing units and the people within these units and non-household living arrangements (group

> quarters)... generated using a combination of decennial census and ACS (American Community Survey) and administrative records data largely in conjunction with a combination of record matching, imputation and microsimulation methods. (Swanson and McKibben 2010, p. 809)

The term CEMAF is derived from "EMAF" (Enhanced Master Address File), a proposal by Swanson and McKibben (2010) for a re-designed population estimation system. CEMAF is aimed at curtailing both increasing non-response rates and increasing costs while maintaining reasonable levels of accuracy, functionality, and usability. As a hybrid approach, the CEMAF method for conducting the census would use administrative records like federal income tax returns as one of its sources for information. Similarly, CEMAF as a hybrid approach is also,

> an integrated file that contains not only existing MAF variables (e.g., geocode, address, and structure type), but also information on the occupancy status of housing units and the people within these units and non-household living arrangements (group quarters). Demographic and socio-economic characteristics would be generated using a combination of administrative records and survey data largely in conjunction with a combination of record matching, imputation and microsimulation methods (Swanson and Walashek 2011, p. 34).

The technical aspects of CEMAF use existing data and methods may seem complex, but in fact it is technically, administratively, and politically quite feasible. One of the keys to CEMAF identified by Swanson and Walashek (2011) is record matching, which in the United States depends on the virtually universal Social Security Number. Canada's equivalent identification is not as universal, but the Canada Revenue Agency does create an identification number for dependents, especially if they qualify for tax credits and benefits in order to track them over the years. In fact, the Canada Revenue Agency uses three identification numbers: the social insurance number (SIN), the temporary tax number (TTN) assigned to wage earners who do not have (or have not applied for) a SIN, and a number assigned to dependents of tax-filers (DIN) who do not have a SIN qualifying for government benefits or tax credits. With the use of identification numbers (such as SIN\TTN\DIN), the CEMAF idea for Canada is not out of the question as the conversion undertaken by Statistics Finland from traditional census enumeration to an administrative record system demonstrates, with the proviso that Finland uses a population registry system (Statistics Finland 2004).

Using this method would reduce costs in the long run and provide more detailed data than what the current census collects. A benefit of CEMAF is that it could largely negate and eliminate the need for many of the traditional demographic methods of population estimation and even reduce the number of sample surveys due to the diverse type of data collected. The US Census Bureau has in the past and is again exploring the uses of administrative records in regard to its census and other operations through the Center for Administrative Records Research and Applications. Examples of some of the current studies can be found online at https://www.census.gov/srd/carra/?cssp=SERP; earlier studies were done under the "StARS" (Statistical Administrative Records System) program.

14.9 Limitations

Certain limitations face the data used in this paper. Due to the inconsistent years of the Canadian and United States census, comparability between the two countries census costs is limited. Similarly, the accuracy of the costs of the censuses are entrusted to Statistics Canada and the General Accounting Office of the United States reporting and disclosure. In a symposium on the Global Review of (the) 2000 Round of Population and Housing Censuses, Richard Leete (2001) noted that:

> Rising costs of censuses, coupled with a lack of detailed data about census costs, led the United Nations in its recommendations for population and housing censuses, to emphasize the need for countries to keep account of the cost of each census activity. Summary cost indicators, such as total census cost per capita, are subject to limitations which make it difficult to say that one census is more or less expensive than another. They do not take account of variations in the quality, quantity and timeliness of census results.

Similarly, in a United Nations (2007, p. 6) workshop regarding Census Management and Planning, it was noted that:

> the cost of census-taking could vary widely not just between countries but also within-countries – for example, people living in tribal conditions in remote areas could drive up the cost by a factor of almost 20 in comparison to urban areas, in some cases. Therefore, it was suggested that the notion of a universal per unit costs was not particularly useful, and could in fact be misleading.

With secondary data analysis, the assumption that the data are accurate is entrusted. Even with some concern about the data, the information invokes discussion about the monetary costs of a national census and how the incorporation of technology has affected census data collection methodology and future possibilities.

14.10 Conclusions and Future Research

With technological advancements of conducting a census on the Internet, the once familiar census enumerator knocking on dwelling doors, seems to be part of many countries methodological past. How the costs associated by the incorporation of Internet methodology to the census, will only be assessed years after Internet implementation. We can look back to the days when enumerators, visited households in untamed wilderness, to compare today's methods and recommend that historical comparisons of costs per housing unit and costs per person be addressed in the future. Every census is unique due to the moment when it is conducted and the historical time it is important in counting the population. In comparing the overall costs Canada has incurred in conducting a census to the United States, the data suggest that the larger the population, the larger the total incurred census costs and costs per housing unit or per person are useful measures. It is important to note that there is no replacement for a census and incorporating technological advancements to census data collection methodology inherently affects the overall census costs during time of development and implementation.

Future research should include: analyzing the costs of census data for other countries, to explore a model which relates census costs and population (and assessing whether the relationship between census costs and a population is a quadratic), gaining access to additional years of census costs and this will assist in the development of a reasonable model.

Once a household opens its door to answer a census survey (or the technological door via the Internet), the number of questions should not make a significant difference to overall census costs. The role of the Census in collecting socio-economic information, and what the marginal cost increase over the basic census enumeration, needs to be analyzed in the future. This current research demonstrates that addressing the costs of a census is indicative of the importance governments place on census accounting. When a country conducts a census, attention may focus on various aspects. Recent census issues in Canada (the elimination of the census long-form in the 2011 census and its reinstatement for the 2016 census) and the United States (2010 census), has provided a spotlight on contemporary census issues.

Acknowledgements The authors would like to thank anonymous reviewers, along with Rosemary Bender, Marc Hammel, Don Royce, and André Cyr of Statistics Canada for valuable discussions and assistance. The authors are grateful to the late Wayne McVey, Jr. for providing comments on earlier drafts of this paper and referring the authors to Kalbach and McVey (1971) for inspiration. The views expressed in the paper are those of the authors. Any errors or omissions are the responsibility of the authors. Versions of the paper were presented: as a poster at the 21st Annual Warren E. Kalbach Population Conference, at the University of Alberta, Edmonton, Canada, on March 18th, 2011 and as a paper presentation at the 2012 Applied Demography Conference in San, Antonio, Texas, in January, where travel funding was generously provided by the Society of Edmonton Demographers.

References

Bender, R. (2009). *2011 Canadian census of population*. Available at http://www12.statcan.ca/census-recensement/2011/gen/2011-census-recencement-eng.cfm

Bounpane, P. (1986). How increased automation will improve the 1990 census of population and housing of the United States. *Journal of Official Statistics, 2*(4), 545–553.

Bureau of the Census. (1916). *The story of the census 1790–1916*. Washington, D.C.: Department of Commerce.

Castro, D. (2008). *e-Census unplugged: Why Americans should be able to complete the census online*. The Information Technology & Innovation Foundation, February.

Cecil, J. (1993). Confidentiality legislation and the United States federal statistical system. *Journal of Official Statistics, 9*(2), 519–535.

Champkin, J. (2010, December). US census: unsustainable?. *Significance, 7*, 174–175.

Clapesattle, H. (1941). *The Doctors Mayo*. Garden City: Garden City Publishing Co., Inc.

Fanoe, J. (2011). Lessons from census taking in South Africa: Budgeting and accounting experiences. *Journal Statistique Africain, 13*, 82–109.

Government of Canada Privy Council Office. (1999). *Speech from the throne to open the second session of the Thirty-Sixth Parliament of Canada*. Available at http://www.pcobcp.gc.ca/index.asp?lang=eng&page=information&sub=publications&doc=aarchives/sft-ddt/1999-eng.htm. Accessed Aug 2011.

Haug, W. (2001). *Population censuses on the Internet*. IUSSP General Population Conference, Session 06: The Collection of Demographic Statistics.

International Telecommunication Union. (2010). *About ITU*. External Affairs and Corporate Communication Division. Available at http://www.itu.int/net/about/index.aspx. Accessed Nov 2010.

Internet World Stats. (2010). *Top 20 countries with the highest number of Internet users. Internet world stats*. Usage and population statistics. Miniwatts Marketing Group. Available at http://www.internetworldstats.com/top20.htm. Accessed Nov 2010.

Jones, S., & Fox, S. (2009). *Generations online in 2009*. PEW Internet & American life project. Available at http://www.pewinternet.org/~/media//Files/Reports/2009/PIP_Generations_2009.pdf. Accessed Nov 2010.

Kalbach, W., & McVey, W. (1971). *The demographic bases of Canadian society*. Toronto: McGraw-Hill Series in Canadian Sociology.

Leete, R. (2001). *Population and housing censuses: A funding crisis? Symposium on global review of 2000 round of population and housing censuses: Mid-decade assessment and future prospects*. Statistics Division. Department of Economic and Social Affairs. United Nations Secretariat. New York, 7–10 August. Available at http://unstats.un.org/unsd/demographic/meetings/egm/symposium2001/docs/symposium_04.htm#_ftn2. Accessed Aug 2011.

Lu, X. (2009). *Need a job? Apply to become a Census enumerator*. Wise Bread: Living on a Small Budget. Available at http://www.wisebread.com/need-a-job-apply-to-become-a-census-enumerator. Accessed Jan 2011.

Martin, E., & Dillman, D. (2008). Does final coverage check identity and reduce census coverage errors? *Journal of Official Statistics, 24*(4), 571–589.

National Research Council. (1995). In B. Edmonston & C. Schultze (Eds.), *Modernizing the U.S. census*. Washington, DC: National Academy Press.

OECD Statistics. (2009). *PPPs and exchange rates*. Organisation for Economic Co-operation and Development. Available at http://stats.oecd.org/Index.aspx. Accessed Nov 2010.

Public Works and Government Services Canada. (2010). *Government On-Line (GOL)*. Available at http://www.tpsgc-pwgsc.gc.ca/apropos-about/fi-fs/ged-gol-eng.html. Accessed Aug 2011.

Reilly, S. (2010). 2010 census costs $13.1 billion, $1.6 billion less than expected. *Federal Times*. Available at http://www.federaltimes.com/article/20100810/AGENCY01/8100306/1055/AGENCY. Accessed Nov 2010.

Ridpath, J. C. (1881). *The life and work of James A. Garfield*. Available at http://www.todayinsci.com/QuotationsCategories/C_Cat/Census-Quotations.htm. Accessed Jan 2011.

Shah, R. (2012). Economic census delayed again due to lack of personnel. *The Economic Times*. Available at: http://articles.economictimes.indiatimes.com/2012-02-18/news/31075041_1_enterprises-and-people-bpl-census-economic-census. Accessed July 2012.

Shryock, H., Siegel, J., et al. (1976). *The methods and materials of demography* (Condensed Ed.). San Diego: Academic Press.

Statistics Canada. (2003). *Symposium 2003: Challenges in survey taking for the next decade*. In Statistics Canada international symposium series-proceedings. Catalogue no. 11-522-XIE.

Statistics Finland. (2004). *Use of register and administrative data sources for statistical purposes: Best practices of Statistics Finland* (Handbook no. 45) Helsinki: Statistics Finland.

Statistics Canada. (1985). *Statistics act*. R.S.C., c. S-19. Available at http://laws-lois.justice.gc.ca/eng/acts/s-19/fulltext.html. Accessed Aug 2016.

Statistics Canada. (2006b). *Canada's population estimates*. Available at http://www.statcan.gc.ca/daily-quotidien/100929/dq100929b-eng.htm. Accessed Nov 2010.

Statistics Canada. (2010b). *Canadian Internet use survey, Internet use, by location of access, Canada, provinces and selected census metropolitan areas (CMAs), every 2 years (percent)*.

Retrieved November 24, 2010, from CANSIM (database), Using E-STAT (distributor): http://estat.statcan.gc.ca.login.ezproxy.library.ualberta.ca/cgi-win/cnsmcgi.exe?Lang=E&EST-Fi=EStat/English/CII_1-eng.htm.

Statistics Canada. (2010c). *Table 075–0001 – Historical statistics, estimated population and immigrant arrivals, annual (persons)*. Retrieved November 23, 2010, from CANSIM (database), Using E-STAT (distributor): http://estat.statcan.gc.ca/cgi-win/cnsmcgi.exe?Lang=E&EST Fi=EStat/English/CII_1-eng.htm.

Statistics Canada. (2010d). *Special tabulation*. Unpublished data.

Statistics Canada. (2013). *Type of dwelling and population by type of dwelling (1961 to 2011 Censuses)*. Available at http://www.statcan.gc.ca/tables-tableaux/sum-som/l01/cst01/famil66-eng.htm. Accessed Sept 2016.

Swanson, D., & Walashek, P. (2011). *CEMAF as a census method: A proposal for a re-designed census and an independent U.S. census bureau*. New York: Springer.

U.S. Census Bureau. (2000). *Factfinder for the nation*. Available at www.census.gov/prod/2000pubs/cff-4.pdf. Accessed Nov 2010.

U.S. Census Bureau. (2005). *2005 National census test Internet application: Person-based screens*. Available at www.census.gov/procur/www/2010dris/omb-person-basedscreens.doc. Accessed Nov 2010.

U.S. Census Bureau. (2009). *Design and methodology: American community survey. U.S. department of commerce economics and statistics administration*. ACS-DM1. Available at http://www.census.gov/acs/www/methodology/methodology_main/. Accessed Aug 2011.

U.S. Census Bureau. (2010). *U.S. and world population clocks. Population division*. Available at http://www.census.gov/main/www/popclock.html. Accessed Nov 2010.

U.S. Department of Commerce. (2011). *Census 2010: Final report to Congress*. Washington, D.C: Office of Inspector General. Final Report No. OIG-11-030-I. June 27, 2011. Available at https://www.oig.doc.gov/OIGPublications/OIG-11-030-I.pdf. Accessed Sept 2016.

United Nations. (2006). Principles and recommendations for population and housing censuses. Revision 2, September, 20. Available at http://unstats.un.org/unsd/demographic/sources/census/docs/P&R_%20Rev2.pdf. Accessed Nov 2010.

United Nations. (2007). Report of the sub-regional workshop on census cartography and management. Department of Economic and Social Affairs. Statistics Division. ESA/STAT/AC.146. Port of Spain, Trinidad, October, 22–26.

United Nations. (2010). Population and housing censuses. Available at http://unstats.un.org/unsd/demographic/sources/census/alternativeCensusDesigns.htm. Accessed Aug 2011.

United Nations Statistics Division. (2010). Report on the results of a survey on census methods used by countries in the 2010 census round. Department of economic and social affairs. Working Paper: UNSD/DSSB/1.

United States Code. (2010). *13 USC Sec. 221. Title 13- Census. Chapter 7- Offenses and penalties*. Subchapter II- Other persons. Available at http://uscode.house.gov/download/pls/13C7.txt. Accessed July 2011.

United States General Accounting Office. (2001). *Significant increase in cost Per housing unit compared to 1990 Census Report to Congressional Requesters*, GAO-02-31.

Wang, C. (1999, June 8). *Development of national accounting of address and housing inventory: The baseline information for post-censal population estimates*. Available at http://www.census.gov/population/www/coop/popconf/paper.html. Accessed Mar 2011.

Williamson, S. H. (2010). *Seven ways to compute the relative value of a U.S. dollar amount, 1774 to present*. Measuring Worth. Available at http://www.measuringworth.com/uscompare/. Accessed Nov 2010.

Part III
Projection & Estimation Methods: Evaluations, Examples and Discussions

Chapter 15
How Accurate Are Japan's Official Subnational Projections? Comparative Analysis of Projections in Japan, English-Speaking Countries and the EU

Masakazu Yamauchi, Shiro Koike, and Kenji Kamata

Abstract With the country's total population having declined since its peak in 2008, future subnational population trends are a major public concern in Japan. The authors have been involved in official subnational population projections for Japan conducted at the National Institute of Population and Social Security Research (IPSS). This paper examines the accuracy of these subnational population projections. For the prefecture-level projections, median absolute percent error (MedAPE) increase with longer lengths of horizon, ranging from 0.5 to 1.4 % for 5-year projections, 1.3 to 2.1 % for 10-year projections, and 2.5 to 3.1 % for 15-year projections. For the municipality-level projections, MedAPE increase with longer horizon lengths, ranging from 1.3 to 1.7 % for 5-year projections, and at 3.5 % for the 10-year projection. Municipalities with small populations tend to have larger percent errors than those with large populations. In addition, we compare accuracy of IPSS subnational population projections with that of official subnational projections conducted by official agencies in English-speaking countries and by the EU. Using several measures, we find that the results of IPSS subnational population projections are more accurate than those conducted by official agencies abroad, although the IPSS projection model uses the cohort component method based on net migration rates, which is less sophisticated than the multiregional projection models used by other official agencies. The accuracy of the IPSS projections can be attributed to the relative stability of population changes in a country with an aging population and a lower level of international migration and number of foreign residents.

Keywords Official subnational population projections • Accuracy • Japan

This work was partially supported by a Scientific Grant from the Ministry of Health, Labour and Welfare of Japan (H26-Seisaku-Ippan-004), and JSPS KAKENHI Grant Number 15H01783.

M. Yamauchi (✉) • S. Koike • K. Kamata
National Institute of Population and Social Security Research, 2-2-3 Uchisaiwaicho, Chiyoda-ku, Tokyo, Japan
e-mail: yamauchi-masakazu@ipss.go.jp; koike-shiro@ipss.go.jp; kamata-kenji@ipss.go.jp

D.A. Swanson (ed.), *The Frontiers of Applied Demography*, Applied Demography Series 9, DOI 10.1007/978-3-319-43329-5_15

15.1 Introduction

Japan is one of the countries dealing with an aging and declining population. Faced with concerns about the repercussions of the anticipated decrease of the working population and the increase in elderly population in the future to the Japanese economy and society, the government of Japan has announced its "Regional Empowerment for Japan's Growth—Overcoming population decline and revitalizing local economies: Japan's long-term vision and comprehensive strategy"[1] in December 2014.

One of the basic materials for formulating government policies such as the above strategy is a population projection. In Japan, the government agency responsible for preparing population projections is the National Institute of Population and Social Security Research (IPSS).[2] In addition to national projections, IPSS prepares subnational projections by prefecture and municipality (described below). These subnational population projections of IPSS are used widely not only by the Japanese government, but also by subnational governments and private companies.

Some researchers, however, made critical assessments of these subnational population projections. For example, Abe (2006) pointed out major differences in projected versus actual populations in some prefectures. Also, the Japan Policy Council Commission on Declining Population Problems (2014) has proposed strategies for overcoming population decline and shrinking local economy based on subnational population projections that they calculated using assumptions that are different from those published by IPSS. This leads to the question, therefore, about how accurate Japan's subnational population projections are.

Oe (2011) and Esaki et al. (2013) discussed the projection errors of IPSS population projections, pointing out differences between the projected and actual populations. The former studied prefectural projection errors, while the latter studied municipal projection errors, and both authors found that the projection errors were relatively small from an overall perspective. Both studies, however, analyzed only a part of past IPSS subnational population projections[3]. Also, they did not examine whether the projection errors were significant or not in comparison to those of other countries.

In this study, therefore, we sought to evaluate the projection accuracy of the prefectural and municipal projections of IPSS. In addition, we compared the accuracy of IPSS subnational population projections with that of official

---

[1] http://www.kantei.go.jp/jp/singi/sousei/info/pdf/panf_eng.pdf (last accessed 30 October 2015).

[2] The former "Institute of Population Problem" had been responsible for preparing population projections. It was merged with the Social Development Research Institute in December 1996 to become the National Institute of Population and Social Security Research.

[3] Oe (2011) analyzed P3_1995, P4_2000, and P5_2005 projections mentioned in Table 15.1, while Esaki et al. (2013) analyzed M2_2005 projection, also mentioned in Table 15.1.

subnational projections conducted by official agencies in English-speaking countries and by the EU.

In this paper, we define "population projection" as having the same meaning as "population forecast" and use the term "population projection" throughout the paper. Population projections are calculations that show the future development of a population when certain assumptions are made about future population trends (Preston et al. 2001). Since they make no predictions as to whether those assumptions would be true, they are always correct barring a mathematical error in their calculation (Smith et al. 2013). Given the widespread use of official population projections as forecasts, it is valuable to assess the errors of such projections. Previous studies have been conducted regarding errors of population projections prepared by official agencies (Keilman 1997, 1998, 2008; Stoto 1983; Keyfitz 1981; Ato and Ikenoue 1987). These studies have not only discussed projection errors, but also contributed to improving the usefulness of population projections by looking into the suitability of projection methods and assumptions and generating empirical prediction intervals.

In this paper we also distinguished the use of the terms "projection error," "projection accuracy," and "projection bias." Projection error is a general concept for the difference between projected and actual populations, and includes concepts of projection accuracy and projection bias. Projection accuracy is a concept that expresses only the magnitude, without the negative/positive sign information, of the errors. Projection bias, on the other hand, is a concept that expresses both the magnitude and the direction of errors.

This paper is organized as follows. Section 15.2 reviews previous studies regarding subnational population projections conducted in English-speaking countries. Section 15.3 explains the overview of IPSS subnational population projections and projection accuracy indices used in this paper. Section 15.4 analyzes the accuracy of IPSS subnational population projections, while Sect. 15.5 looks into the factors associated with the accuracy. Section 15.6 compares the accuracy of IPSS projections with that of official subnational projections conducted by official agencies in English-speaking countries and by the EU. Section 15.7 presents an overall summary and future issues to address.

15.2 Literature Review

Most of the studies regarding projection error of subnational projections deal with clarifying the factors affecting the error. Smith et al. (2013; Chapter 13) has reviewed literature dealing with this particular criteria, and reported that population size, population growth rate, length of horizon, and length of base period generally have consistent impact on projection accuracy.

Many studies have shown that projection errors are larger for smaller population sizes (Murdock et al. 1984; Smith 1987; Smith and Sincich 1988; Tayman and Swanson 1996). This relationship was demonstrated by comparing projection errors for different levels of geographical units. Smith and Tayman (2003) and Smith et al. (2013) showed that lower geographic levels have higher projection errors. This opposite correlation, however, is generally seen only for projection accuracy.

Population growth rate has a strong impact on projection errors. The relationship between projection errors and population growth rate can be represented by a U-shaped curve, with higher projection errors typically found for places that have greater absolute values for population growth rate (Smith 1987; Smith and Shahidullah 1995; Tayman et al. 2011). Since migration plays an important role in population growth rates at subnational levels, a place with rapidly growing or declining population is generally characterized by large in- or out-migrations.

In regard to length of horizon, since it is more difficult to forecast changes in society farther into the future, the longer the length of horizon, the higher is the error of projection. Many studies have demonstrated this linear relationship between projection accuracy and length of horizon (Smith 1987; Smith and Sincich 1991).

Likewise, a shorter length of base period leads to a higher projection error (Smith and Sincich 1991). Since population projections are made by looking into past population trends to make assumptions for projection, a shorter base period tends to result in observations of the past that are insufficient for making proper assumptions.

Aside from the above factors, Smith et al. (2013) also mentioned projection method and base year as factors that do not have a consistent impact on projection error.

Although there are a variety of projection methods used, ranging from complex ones to simple ones, simpler projection methods do not necessary result in higher projection errors than complex methods (Chi 2009; Chi et al. 2011; Smith 1987; Smith and Tayman 2003; Smith et al. 2013). Smith et al. (2013, pp.336) explained that there is a certain irreducible level of uncertainty regarding the future, and that no projection method can consistently improve forecast accuracy beyond that level.

As theoretical and empirical knowledge develop with advancements in research, it might be expected that projection errors are smaller for population projections with more recent base years than older ones. This possibility, however, has not been actually demonstrated. Many studies have shown that projection errors are larger for some base years than for others, but projection errors for population projections with more recent base years are not always smaller (Long 1995; Smith and Sincich 1988).

Most of the studies cited deal with each of the factors affecting projection error separately. In contrast, there are also studies that looked into the effects of multiple factors. Tayman et al. (2011) used a multivariate regression model to determine the effects of base-year population size, population growth rate, length of horizon, geographical location, and other factors on projection error.

Unlike the above theoretical studies on projection error, many studies on projection errors of regional projections by official agencies published successively from 2000 onwards focused on the level of projection error.

Rees et al. (2001) published a voluminous report aimed at improving the methods used by EU for projecting subnational populations. In Chap. 5 of their paper, they examined projection errors of NUTS2 population projections for 1980 and 1990 base years conducted by EUROSTAT. Aside from presenting projected and actual populations and algebraic percent error (ALPE)[4] for each of the NUTS2 areas, they also analyzed the index of dissimilarity[5], which expresses the level of overall error.

Wang (2002) analyzed the projection error for the state population projections for 1995–2025 conducted by the United States Census Bureau. In particular, the author looked into the projection errors for births, deaths, domestic migration, and international migration for the total population, and conducted multivariate regression analysis using the absolute percent error (APE)[6] of the total population as dependent variable to determine factors contributing to projection accuracy.

Statistics New Zealand (2008) examined projection errors for their subnational population projections for the regional council areas and territorial authority areas for 1991 onwards and projections for area units for 1996 onwards. Primarily, they looked into the distribution of ALPE according to base year and length of horizon.

Wilson (2012) conducted the following analyses regarding the state and territory population projections for 1978 onwards conducted by the Australian Bureau of Statistics. First, the author analyzed the APE for each of the states, territories, and projection durations, and looked into the effects of births, deaths, domestic and international migration, and base-year population size on APE. Next, the author also examined whether the projection error of population projections conducted by the Australian Bureau of Statistics is smaller than that of projections conducted using naive methods. Further, the author determined whether more recent projections have smaller projection errors than older ones. Finally, the author created empirical prediction intervals using past projection errors.

The Office for National Statistics (2015) analyzed projection errors of the subnational population projections conducted by the UK Office for National Statistics in England. They analyzed projections made using 2004, 2006, 2008, and 2010 as base years. They showed the root mean square error (RMSE)[7] by regions, counties, and local authorities, and also examined the ALPE for each region.

[4] ALPE of area i is defined as follows, $ALPE_i = ({}_eP_i - {}_aP_i)/{}_aP_i \times 100$ where P is population, with subscripts e representing the projected value, a the actual value, and the i area.

[5] Index of dissimilarity (D) is defined as follows, $D = \frac{1}{2} \times 100 \times \sum \left| \frac{{}_eP_i}{{}_eP_I} - \frac{{}_aP_i}{{}_aP_I} \right|$ where subscript I representing the entire area.

[6] APE for area i is defined as follows, $APE_i = \left|{}_eP_i - {}_aP_i\right| / {}_aP_i \times 100$.

[7] RMSE is defined as follows, $RMSE = \sqrt{\frac{1}{n} \sum_i^n \left\{ \frac{({}_eP_i - {}_aP_i)}{{}_aP_i} \right\}^2}$ where n represents the number of areas.

These studies, which are focused on the projection errors of the population projections made by different official agencies, provide a wide range of useful information. The indices for projection error used, however, differed among these studies, and no comparisons were made with the errors of the population projections conducted by the other agencies.

15.3 Data and Methods

15.3.1 IPSS Subnational Population Projections Data

IPSS has published population projections for the national, prefectural, and municipal levels based on the population census. The first prefectural population projection published by IPSS was based on 1985 census population, while the municipal projection was based on 2000 census population (Table 15.1). In this paper, we analyze total populations for five prefectural projections, namely P1_1985, P2_1990, P3_1995, P4_2000, and P5_2005, and two municipal projections, namely M1_2000 and M2_2005. Because we could not obtain actual population data for 2015 at the time of the writing of this paper, we did not include P6_2010 and M3_2010 in the analysis.

For this study we obtained results of total population for some of these subnational population projections from electronic data published in websites and published reports. Although P2_1990 and P3_1995 projections did not provide values lower than 1,000 persons, other prefectural and municipal projections published values down to single person counts.

These subnational projections include several sets of results. In this paper we analyzed the results from the most commonly used basic assumptions.

There were 47 prefectures included in the subnational projections conducted by IPSS, while municipal projections included 3,244[8] municipalities in M1_2000 and 1,805 in M2_2005. The large difference in the number of municipalities between M1_2000 and M2_2005 is due to the municipal mergers called "Heisei-no-daigappei (the big merger of "Heisei era")" carried out from 1999 to 2010. Table 15.2 shows the base-year population size of the prefectures and municipalities, while Fig. 15.1 shows the prefectures and municipalities covered in projections based on 2000. Due to the large number of municipalities, only the enlarged view of the Tokyo Metropolitan Area (TMA) is shown. As reference, there are 47 prefectures and 1,718 municipalities as of October 1, 2015.

[8] This number excludes Yamakoshi-mura (currently Nagaoka-shi), whose population significantly decreased as a result of the Mid Niigata Prefecture Earthquake in 2004.

Table 15.1 Official subnational population projections conducted by IPSS, Japan

ID	Level of geography	Number of areas	Base year	Final target year	Release date
P1_1985	Prefecture	47	1985	2025	January 1987
P2_1990	Prefecture	47	1990	2010	November 1992
P3_1995	Prefecture	47	1995	2025	May 1997
P4_2000	Prefecture	47	2000	2030	March 2002
P5_2005	Prefecture	47	2005	2035	May 2007
P6_2010	Prefecture	47	2010	2040	March 2013
M1_2000	Municipality	3,244	2000	2030	December 2003
M2_2005	Municipality	1,805	2005	2035	December 2008
M3_2010	Municipality	1,799	2010	2040	March 2013

Note: The number of areas in M1_2000 excludes Yamakoshi-mura (currently Nagaoka-shi), whose population significantly decreased as a result of the Mid Niigata Prefecture Earthquake in 2004

Table 15.2 Number of areas by population size in base year

	Population size in base year			
ID	All	<1 million	1 million–2 million	2 million+
P1_1985	47	7	22	18
P2_1990	47	7	22	18
P3_1995	47	7	21	19
P4_2000	47	7	20	20
P5_2005	47	7	20	20
	Population size in base year			
ID	All	<10,000	100,000–100,000	100,000+
M1_2000	3,244	1,553	1,443	248
M2_2005	1,805	482	1,039	284

15.3.2 IPSS Subnational Population Projection Method

The prefectural population projections assessed in this paper were all made using cohort component model based on net migration for populations disaggregated by sex and 5-year age groups. For detailed explanation of the projection method, refer to Nishioka et al. (2011) for P5_2005, which is basically the same method used for other prefectural projections based on other years. Assumptions were made for age-specific birth rate, sex- and age-specific survival rate, sex- and age-specific net-migration rate, and sex ratio at birth, basically by extending the values for the 5-year base period for each index into the future. The assumptions for the sex- and age-specific net-migration rate do not distinguish between domestic and international migration rates.

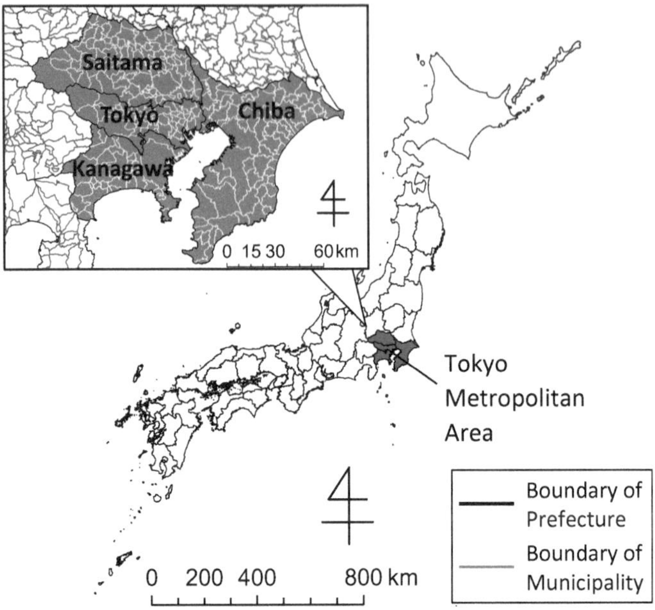

Fig. 15.1 Geographical distribution by Prefectures and Municipalities. **Grey* shaded area is Tokyo Metropolitan Area (TMA)

The municipal population projections assessed in this study were made using standard cohort component model using net migration for populations 5 years old and above to project future populations separately for males and females and for each 5-year age group. Sex-specific projections for the 0–4 population were made using the child-woman ratio (CWR). For detailed explanation of the projection method, refer to the materials summarized in the website[9] regarding M3_2010. M1_2000 and M2_2005 basically follow the same method explained in the website. Assumptions were made for CWR, sex- and age-specific survival rate, sex- and age-specific net-migration rate, and sex ratio of age 0–4 population, by extending the values for the 5-year base period for each index into the future. The assumptions for the sex- and age-specific net-migration rate do not distinguish between domestic and international migration rates.

15.3.3 Accuracy Measure

The main index used in the analysis of accuracy is APE, which is one among many indices used in assessing accuracy. Actual populations used for calculating APE

[9] http://www.ipss.go.jp/pp-shicyoson/e/shicyoson13/t-page.asp (last accessed 30 October 2015)

were based on the population data obtained through the population census. However, since P2_1990 and P3_1995 do not include population values less than 1,000, the actual population figures were rounded off before using them in the analysis. Due to changes in prefectural and municipal boundaries, population data used were those for the area representing the boundaries at the time the projections were made.

Rather than looking simply at the APE of each area, we focused on the overall distribution and relationship of the APEs of the different areas. Thus, in Sect. 15.4 and 15.5, for example, we used the median APE (MedAPE) and 90 percentile of APE (90%APE) in analyzing accuracy of IPSS subnational population projections. Also, in the comparative analyses of projections by different countries in Sect. 15.6, other than these two indices, we also used three other indices, namely, mean APE (MAPE), the proportion of the number of areas falling within the scope of a particular APE within the area covered in the analysis, and RMSE. These indices were used in order to make comparisons with results of existing studies.

15.3.4 Accuracy Data of Official Subnational Population Projections in the EU and English-Speaking Countries

Accuracy data regarding the subnational population projections conducted by official agencies in the EU and English-speaking countries used in this study were those from Rees et al. (2001), Wang (2002), Statistics New Zealand (2008), Wilson (2012), and Office for National Statistics (2015). Projected and actual subnational population data indicated in the papers, such as in Rees et al. (2001) and Wang (2002), were used in calculating the necessary accuracy measure.

It should be noted that the subnational population projections analyzed by these studies were based on different numbers of areas, population sizes, projection methods, and base years.

15.4 Accuracy of IPSS Subnational Population Projections

15.4.1 IPSS Prefectural Population Projections Accuracy

Table 15.3 shows the MedAPE and 90%APE for each base year and length of horizon. For prefectural projections, MedAPE increased with longer lengths of horizon. MedAPE for the 5-year projections ranged from 0.5 to 1.4 %, that of the 10-year projections from 1.3 to 2.1 %, and that of the 15-year projections from 2.5 to 3.1 %.

With longer lengths of horizon, the 90 % APEs also increased: ranging from 1.2 to 2.3 % for 5-year projections, from 3.0 to 4.5 % for 10-year projections, and from

Table 15.3 IPSS prefectural population projections MedAPEs and 90 %APEs by base year and length of horizon

	Median absolute percent errors (MedAPEs)			90 percentile of absolute percent errors (90 %APEs)		
	Length of horizon (Years)			Length of horizon (Years)		
ID	5	10	15	5	10	15
P1_1985	0.9	1.5	2.8	1.9	3.0	5.8
P2_1990	1.4	2.1	2.5	2.3	3.9	8.1
P3_1995	0.6	1.5	3.1	1.9	4.5	6.7
P4_2000	0.6	1.3		1.6	3.4	
P5_2005	0.5			1.2		

5.8 to 8.1 % for 15-year projections. Differences between 90%APE and MedAPE also increased with longer lengths of horizon: ranging from 0.7 to 1.3 % for 5-year projections, 1.5 to 3.0 % for 10-year projections, and 3.0 to 5.6 % for 15-year projections.

Figure 15.2 shows the geographic distribution of APE by prefecture for different base years for the 10-year projections. The map in the figure represents a cartogram created using the method proposed by Gastner and Newman (2004) to modify prefectural areas based on the total population in 2005. The APEs of each prefecture differed depending on base year, wherein no particular prefectures had consistently high APEs. It should be noted, however, that many prefectures in TMA have high APE. This is due to the fact that migration patterns in TMA are highly variable. For example, migration trends in TMA changed around the mid-1990s, wherein the outward flow of people from Tokyo to neighboring prefectures such as Saitama, Chiba, and Kanagawa was reversed starting in the mid-1990s (Shimizu 2004). These major changes in migration trends have led to increased APEs for IPSS population projections, which are based on assumptions that reflect past population trends.

15.4.2 IPSS Municipal Population Projections Accuracy

Table 15.4 shows the MedAPE and 90%APE by base year and length of horizon. For municipal projections, MedAPE increased with longer length of horizon, at 1.3–1.7 % for the 5-year projections and 3.5 % for the10-year projection. Likewise, 90%APE also increased with longer length of horizon, at 4.1–4.8 % for the 5-year projections and 9.4 % for the 10-year projection. The difference between MedAPE and 90%APE also increased with longer length of horizon, at 2.8–3.1 % for the 5-year projections, and 5.9 % for the 10-year projection. Also, MedAPE and 90% APE of municipal projections were larger than those of the prefectural projections for the same length of horizon.

A negative relationship between the base year population size and APE for municipal projections was found. Figure 15.3 shows the relationship between the

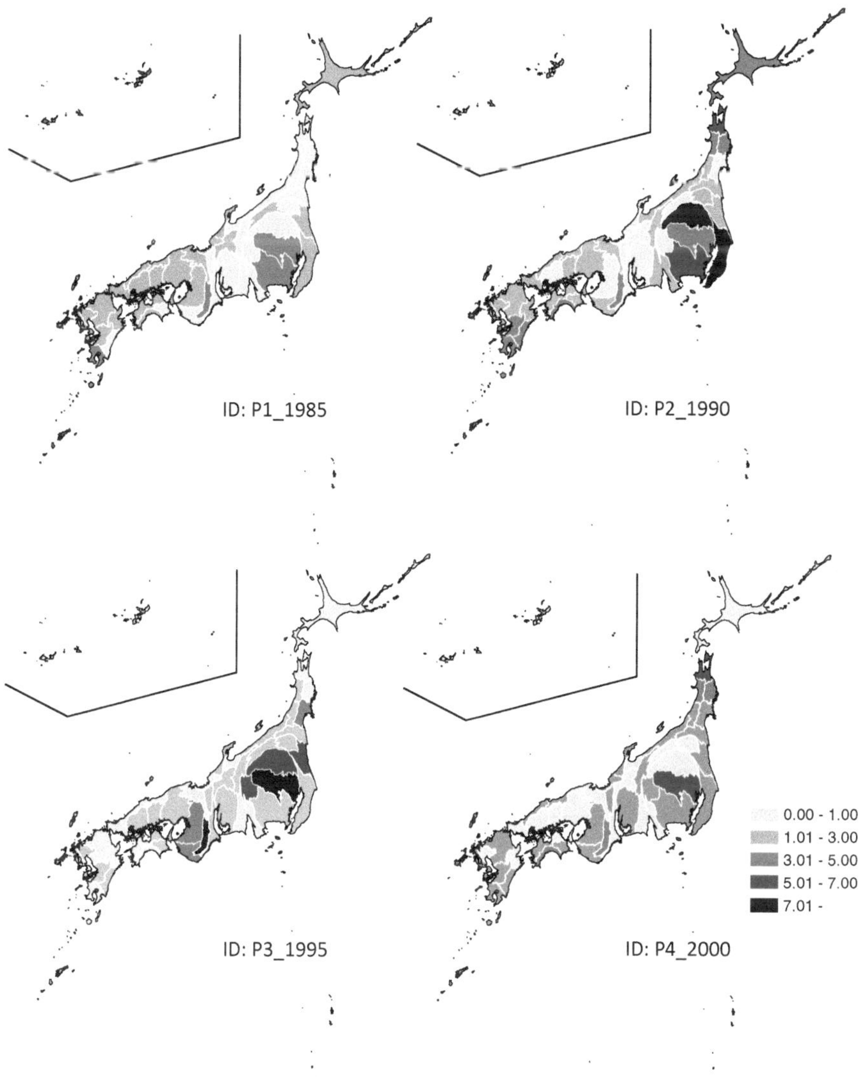

Fig. 15.2 Geographical distributions of APE by Prefecture, 10-year projection horizon. *The Cartogram (based on Total Population 2005) is created using the Gastner-Newman method in ArcGIS

APE and the base-year population size for the municipal projections. Many municipalities with smaller population sizes had high APEs, while few of those with larger population sizes had high APEs. As shown in Table 15.5, for the 5-year projections, MedAPE was 1.9–2.1 % and 90%APE was 5.5–6.7 % for municipalities with base-year populations at less than 10,000, in contrast to MedAPE of 1.0–1.3 % and 90%APE of 3.0–3.6 % for municipalities with base-year populations at 100,000 or higher.

Table 15.4 IPSS municipal population projections MedAPEs and 90 %APEs by base year and length of horizon

	Median absolute percent errors (MedAPEs)			90 percentile of absolute percent errors (90 %APEs)		
	Length of horizon (Years)			Length of horizon (Years)		
ID	5	10	15	5	10	15
M1_2000	1.7	3.5		4.8	9.4	
M2_2005	1.3			4.1		

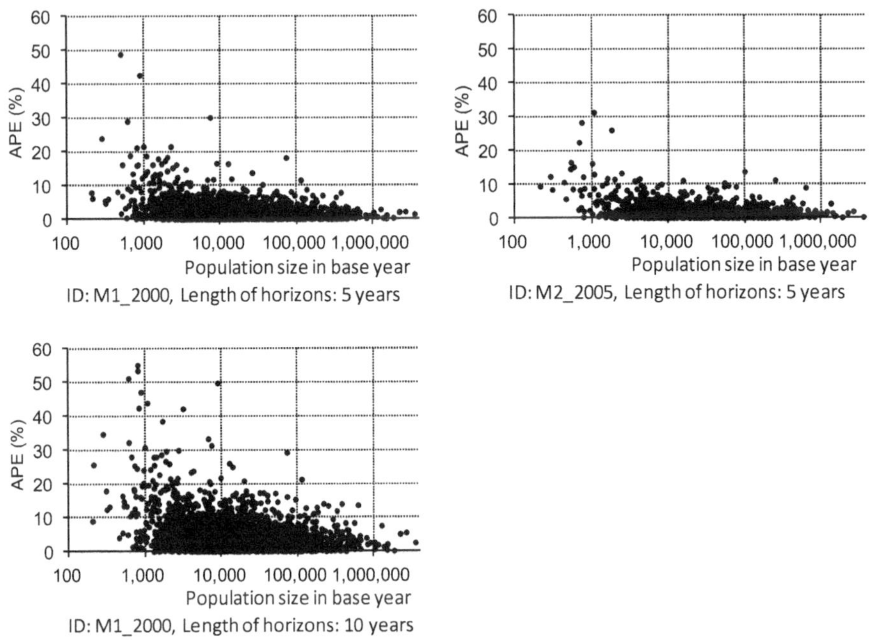

Fig. 15.3 IPSS municipal population projections APE by population size in base year, base year and length of horizon

15.5 Factors Affecting the Accuracy of IPSS Subnational Population Projections

We analyzed the factors associated with APE for prefectural and municipal projections using a multivariate regression model with base-year population size, base-period net migration rate, TMA dummy, base year, and length of horizon as covariates. Due to page width restrictions, we are only presenting here the results of the pooled model, which includes linear relationships of all variables.

The summary statistics of the dependent variables and covariates are shown in Table 15.6. Statistics for the base-year population represent natural log values, while absolute values were used for the base-period net migration. The TMA

Table 15.5 IPSS municipal population projections MedAPEs and 90 %APEs by base year, length of horizon, and population size in base year

		Median absolute percent errors (MedAPEs)			90 percentile of absolute percent errors (90 %APEs)		
		Population size in base year			Population size in base year		
ID	Length of projection horizon (Years)	<10,000	10,000–100,000	100,000+	<10,000	10,000–100,000	100,000+
M1_2000	5	1.9	1.5	1.3	5.5	4.2	3.6
M2_2005	5	2.1	1.2	1.0	6.7	3.3	3.0
M1_2000	10	4.0	3.1	2.7	10.9	8.1	7.4

Table 15.6 Summary statistics of variables for multivariate regressions

	Mean	S.D.	Min	Median	90 % tile	Max	N
Variables for prefectural pooled model							
APE	1.877	2.099	0.000	1.274	3.878	17.238	564
Population size in base year [log.]	14.492	0.721	13.316	14.397	15.656	16.347	564
Net migration rate in base period [Abs.]	1.174	1.088	0.007	0.833	2.762	5.873	564
TMA dummy	0.085	0.279	0.000	0.000	0.000	1.000	564
Base year							
Year 1985	0.250	0.433	0.000	0.000	1.000	1.000	564
Year 1990	0.250	0.433	0.000	0.000	1.000	1.000	564
Year 1995	0.250	0.433	0.000	0.000	1.000	1.000	564
Year 2000	0.167	0.373	0.000	0.000	1.000	1.000	564
Year 2005	0.083	0.277	0.000	0.000	0.000	1.000	564
Length of horizon							
5 years	0.417	0.493	0.000	0.000	1.000	1.000	564
10 years	0.333	0.472	0.000	0.000	1.000	1.000	564
15 years	0.250	0.433	0.000	0.000	1.000	1.000	564
Variables for municipal pooled model							
APE	3.118	3.795	0.000	2.027	6.977	55.035	8,292
Population size in base year [log.]	9.597	1.361	5.313	9.408	11.442	15.091	8,292
Net migration rate in base period [Abs.]	2.816	2.824	0.000	2.254	5.629	76.871	8,292
TMA dummy	0.091	0.288	0.000	0.000	0.000	1.000	8,292
Base year							
Year 2000	0.782	0.413	0.000	1.000	1.000	1.000	8,292
Year 2005	0.218	0.413	0.000	0.000	1.000	1.000	8,292
Length of horizon							
5 years	0.609	0.488	0.000	1.000	1.000	1.000	8,292
10 years	0.391	0.488	0.000	0.000	1.000	1.000	8,292

dummy represents a dummy variable using areas outside the TMA as reference category. This variable was used as an index to assess the effect to accuracy of the marked change in migration, i.e., reversal from outflow to inflow, observed for TMA. For the base year, the dummy variable for prefectural projections was based on 1985, while that for the municipal projections was based on 2000 as the reference category. In regard to the length of horizon, the dummy variable was based on 5 years as the reference category.

Based on the results of the prefecture pooled model (Table 15.7), the base-year population size was not related statistically to APE. However, net migration rate in base period and the TMA dummy were positively related to the APE. For the base-period net migration rate, this means that areas with high migration activity have

Table 15.7 Results from multivariate regressions on APE

Variables	Unstandardized Coefficients	Standardized Coefficients
Prefecture pooled model		
(Intercept)	−1.038	
Population size in base year [log.]	0.065	0.022
Net migration rate in base period [Abs.]	0.653 **	0.338
TMA dummy (ref. outside the TMA)	1.472 **	0.196
Base year (ref. Year 1985)		
Year 1990	0.064	0.013
Year 1995	0.449 *	0.093
Year 2000	0.338	0.060
Year 2005	0.003	0.000
Length of horizon (ref. 5 years)		
10 years	0.965 **	0.217
15 years	2.313 **	0.478
Number of data	564	
Adjusted R^2	0.412	
F statistics	44.8 **	
Municipality pooled model		
(Intercept)	8.439 **	
Population size in base year [log.]	−0.700 **	−0.251
Net migration rate in base period [Abs.]	0.125 **	0.093
TMA dummy (ref. outside the TMA)	1.229 **	0.093
Base year (ref. Year 2000)		
Year 2005	0.096	0.010
Length of horizon (ref. 5 years)		
10 years	2.340 **	0.301
Number of data	8292	
Adjusted R^2	0.175	
F statistics	353.5 **	

Significant level: 0.01 **, 0.05 *, 0.10 +

higher APEs. For the TMA dummy, it shows that the APE tends to increase when drastic changes in migration occur. For the base year, 1995 had a positive significant effect compared to 1985. The effect of the reversal in migration flow for TMA around the mid-1990s is reflected in the higher APE for 1995 than for the other base years. Over the length of the horizon, 10-year and 15-year projects were found to have higher APEs than 5-year projections.

Results for the municipality pooled model showed that the base-year population size is related statistically to APE, wherein larger base-year populations resulted in lower APEs. As was seen in the prefectural projections, no correlation with the APE was observed when population sizes were at least 500,000 or higher, but population size was associated with the APE for population projections of municipalities with a large variation in population size. Results for the base-period net migration rate and TMA dummy, as with the prefectural model, showed that they were positively related statistically to the APE. For the TMA dummy, this means that APE increases significantly when drastic changes in migration occur. In regard to the base year, 2005 did not have a statistically significant relationship with APE compared to 2000. Considering the horizon length, 10-year projections had higher APE than 5-year projections.

15.6 Comparative Analysis of Accuracy of Official Subnational Population Projections

Before comparing the accuracy of official subnational population projections, we sorted out the median population sizes of areal units for subnational population projections (Table 15.8). The highest value of median population size was observed for regions of England, followed by states of the U.S., states and territories of Australia, prefectures of Japan, and NUTS2 regions of EU. These values of median population size exceed one million, with the value of minimum population size at more than 100,000. On the other hand, the lowest value of median population size was that of unit areas of New Zealand, followed by municipalities of Japan, and territorial authority areas of New Zealand. These values of median population sizes are lower than 100,000, with the value of minimum population size at less than 1000 persons. Japan's municipalities, however, had the value of maximum population exceeding one million, unlike those of New Zealand's unit areas and territorial authorities, which had much lower values of maximum population.

Also, looking at the number of areas covered by the projections, we see that Japan's municipality and New Zealand's unit area projections cover a very large number of areas. Local authorities of England and NUTS2 regions of EU are more than 100, while the rest are less than 100.

Table 15.9 shows accuracy by horizon length. As a general trend, it was found that the projection accuracy decreases as the length of horizon increases and as the population size decreases. Also, although differences in projection accuracies among different agencies were found, the differences were not very large. The following discussion focuses on the accuracy of IPSS population projections.

For 5-year projections, RMSE ranged from 1.1 to 4.8 %, MedAPE from 0.5 to 2.3 %, and 90 %APE from 1.9 to 4.9 %, with the percentage of areas having APE under 5 % ranging from 60.3 to 100.0 %. The accuracies of projections for prefectures in Japan were high with almost indexes. Although accuracies of municipal projections were relatively lower, those for areas with small population size were high. For example, based on the percentage of areas with APE under 5 %, for M1_2000 the percentage is high at 91.0 %, against 60.3 % for unit areas and 85.0 % for territorial authority areas of New Zealand, and is almost at the same level as projections for EU NUTS2 regions based on 1990 population.

Table 15.8 Distribution of population size for subnational population projections

Country/ Region	Geographical area	Number of areas	Minimum (×1,000)	Median (×1,000)	Maximum (×1,000)	Source of population size data
England	Region	9	2,542.2	5,326.7	8,125.2	1
United States	State	51	493.8	4,012.0	33,871.6	2
Australia	State and Territory	8	210.6	1,813.7	6,816.1	3
Japan	Prefecture	47	588.7	1,706.2	13,159.4	4
EU	NUTS2(base year 1980)	68	114.5	1,519.4	8,941.7	5
EU	NUTS2(base year 1990)	165	115.3	1,508.5	10,649.6	5
England	County	41	287.3	681.7	7,389.1	1
New Zealand	Regional Council Area	16	31.3	148.1	1,303.1	6
England	Local Authority	352	24.7	114.7	995.5	1
New Zealand	Territorial Authority Area	73	0.6	32.4	404.7	6
Japan	Municipality (M2_2005)	1,805	0.2	25.1	3,579.6	7
Japan	Municipality (M1_2000)	3,244	0.2	10.7	3,426.7	8
New Zealand	Unit Area	1,633	0.1	2.2	9.5	9

The following are the sources of population size:
(1) Base year population derived from "Subnational population projections, 2004-based projections" conducted by the Office for National Statistics
(2) U.S. Census 2000 population
(3) Australia's 2006 Census population
(4) Japan's 2010 Census population
(5) Base year population derived from Rees et al.(2001)
(6) New Zealand's 2006 Census population
(7) Japan's 2005 Census population
(8) Japan's 2000 Census population
(9) Base year population derived from "2006-base area unit population projections" conducted by Office for Statistics New Zealand

For 10-year projections, RMSE ranged from 2.6 to 6.6 %, MedAPE from 1.6 to 3.5 %, and 90%APE from 3.8 to 9.4 %, with the percentage of areas having APE under 5 % ranging from 40.5 to 94.1 %. The accuracies of projections for prefectures in Japan were high regardless of index used. Although accuracies of municipal projections were relatively low compared to those of projections for states and territories of Australia, which have large population sizes, and to those of

Table 15.9 Official subnational population projection accuracy by countries/regions and length of horizon

Country/ Region	Geographical area	Median population size of geographical area (×1,000)	MedAPE (%)	90 %APE (%)	MAPE (%)	RMSE (%)	Percent of geographical area with APE (%)		
							Under 5 %	5-10 %	Over 10 %
5-year projection horizons									
England	Region	5,326.7				1.5			
United States	State	4,012.0	2.3	4.9	2.6	3.2	90.2	9.8	0.0
Australia	State and Territory	1,813.7	1.8	3.6					
Japan	Prefecture	1,706.2	0.7	1.9	0.9	1.2	100.0	0.0	0.0
EU	NUTS2(base year 1980)	1,519.4	0.5	2.1	0.8	1.1	100.0	0.0	0.0
EU	NUTS2(base year 1990)	1,508.5	1.2	4.3	2.0	3.1	92.0	5.6	2.5
England	County	681.7				2.0			
New Zealand	Regional Council Area	148.1					95.0	3.8	1.3
England	Local Authority	114.7				4.8			
New Zealand	Territorial Authority Area	32.4					85.0	12.0	3.0
Japan	Municipality (M2_2005)	25.1	1.3	4.1	1.9	3.0	93.2	5.4	1.3
Japan	Municipality (M1_2000)	10.7	1.7	4.8	2.3	3.5	91.0	7.2	1.7
New Zealand	Unit Area	2.2					60.3	24.5	15.2
10-year projection horizons									
Australia	State and Territory	1,813.7	3.2	6.5					
Japan	Prefecture	1,706.2	1.6	3.8	2.0	2.6	94.1	5.3	0.5
EU	NUTS2(base year 1980)	1,519.4	2.1	5.1	2.4	3.0	88.2	11.8	0.0

New Zealand	Regional Council Area	148.1					75.0	20.8	4.2
New Zealand	Territorial Authority Area	32.4					58.0	27.9	14.2
Japan	Municipality (M1_2000)	10.7	3.5	9.4	4.6	6.6	66.8	24.2	9.0
New Zealand	Unit Area	2.2					40.5	28.5	31.0
15-year projection horizons									
Australia	State and Territory	1,813.7	4.6	8.6					
Japan	Prefecture	1,706.2	2.7	6.7	3.4	4.5	82.3	14.2	3.5
EU	NUTS2(base year 1980)	1,519.4	5.7	10.5	5.8	6.8	45.6	38.2	16.2
New Zealand	Regional Council Area	148.1					62.5	18.8	18.8
New Zealand	Territorial Authority Area	32.4					35.6	24.7	39.7

Source: Wilson (2012), Statistics New Zealand (2008), Office for National Statistics Center for Demography(2015), Wang (2002) and Rees et al.(2001)
Note: The MedAPEs and 90 %APEs in Australia are average numbers of each State and Territory's MedAPE and 90 % APE derived from Wilson (2012)

projections for EU NUTS2 regions based on 1980, they were high compared to those of projections for unit areas and territorial authority areas of New Zealand.

For 15-year projections, RMSE ranged from 4.5 to 6.8 %, MedAPE from 2.7 to 5.7 %, and 90%APE from 6.7 to 10.5 %, with the percentage of areas having APE under 5 % ranging from 35.6 to 82.3 %. The accuracies of projections for prefectures in Japan were high regardless of index used.

15.7 Summary and Discussion

In this study, we analyzed the accuracy of prefectural and municipal projections conducted by IPSS. For prefectural projections, MedAPE increased with longer lengths of horizon, ranging from 0.5 to 1.4 % for 5-year projections, 1.3 to 2.1 % for 10-year projections, and 2.5 to 3.1 % for 15-year projections. For municipal projections, MedAPE increased with longer horizon lengths, ranging from 1.3 to 1.7 % for 5-year projections, and at 3.5 % for the 10-year projection.

We conducted a multivariate regression analysis to clarify the factors associated with the accuracy of prefectural and municipal population projections conducted by IPSS. The accuracy of prefectural projections was correlated to base-period net migration rate, TMA dummy, length of horizon, and base year. In particular, it was found that accuracy was adversely affected for areas with high absolute migration rates and those that went through drastic change in migration patterns, for projections with longer lengths of horizon and those wherein the base year coincided with the turning point for marked changes in migration pattern.

The accuracy of municipal projections was associated with base-year population size, base-period net migration rate, TMA dummy, and length of horizon. In particular, it was found that accuracy was adversely related for areas with small base-year population size, those with high absolute migration rates, and those that went through drastic change in migration patterns, and for projections with longer horizon lengths.

Among the factors associated with accuracy, those that had the same relationship with both the prefectural and the municipal projections were base-period net migration rate, TMA dummy, and length of horizon – a result that is consistent with findings in previous studies.

However, factors that varied between the prefectural and the municipal projections were the base-year population size and base year. This means that while accuracy was not correlated when there was a large base-year population size, as with the prefectural projections, it was correlated when there was a large variation in population size, as with municipal projections. Also, in regard to the base year, there is a possibility that no relationships were observed for municipal projections simply because of the small sample size – there were only two projections analyzed. It is possible that the base year would be associated with accuracy for future municipal projections to be conducted by IPSS.

To assess the accuracy of IPSS subnational population projections relative to those done in other countries, we compared them to subnational projections conducted by official agencies in English-speaking countries and in the EU. Specifically, we made comparisons with projections conducted by the Australian Bureau of Statistics, the UK Office for National Statistics, Statistics New Zealand, the United States Census Bureau, and the EU. Although variations in projection accuracy due to length of horizon and population size were found, there were no major differences as a whole. The accuracy of IPSS projections was found to be relatively high.

We suggest two reasons for the high accuracy of the IPSS subnational population projections. First, it is the case that the effect of foreigners on population changes in Japan is limited. Since the 1990s, as in other countries, there has been an increasing flow of foreigners into Japan, and in various aspects their activities have become more conspicuous (Ishikawa 2015). However, Japan's international migration rate remains very low (Table 15.10), wherein foreigners account for only 2 % of Japan's population. We believe that the limited effect of foreigners on changes in Japan's population (e.g. Yamauchi 2015) contributed to the high accuracy of IPSS subnational population projections.

The second reason is that Japan's population is aging. The low fertility levels found from the mid-1970s have aggravated the aging of the population, making Japan one of the countries experiencing extreme aging. An aging population means that the size of younger generations, which have a significant effect on migration and births, is also decreasing. As such, the effect on population change of migration and births, for which accurate assumptions are difficult to develop, becomes relatively smaller, leading to the high accuracy found in IPSS subnational population projections.

In regard to the relationship between projection accuracy and projection method, the approach used in the IPSS projections (the cohort component method using net migration) lacks the theoretical refinement[10] found in the methods used by other official agencies. However in comparing results of the cohort-component method combined with net migration in the IPSS projections (which was mainly due to data availability), to the more refined methods found in the projections of other official agencies, which are based on theoretically refined methods, did not reveal any major differences in projection accuracy.[11]

[10] For M2_2005, P6_2010 and M3_2010, we used cohort component method with alternative net migration rates which are calculated by dividing the net-migration by population changing by the sign of the net-migration. If the sign of net-migration in area j which constitute Japan is positive, the denominator population of alternative net migration is the difference between whole Japan and area j population. If the sign of net-migration is negative, then the denominator population is area j population. Alternative net migration rates model is better than normal net-migration rates model for calculating future population (Koike 2008).

[11] State population projections for 1995–2025 by the United States Census Bureau and the 1990-based NUTS2 population projections by EU used the multi-regional cohort component framework, which is theoretically refined projection method.

Table 15.10 Net migration rate by country

(per 1000 population)				
Country	1990–1995	1995–2000	2000–2005	2005–2010
Japan	0.73	0.03	0.99	0.70
Australia	3.99	4.06	5.84	10.68
Belgium	2.29	1.32	4.75	4.97
France	1.05	0.64	2.45	1.55
Germany	8.05	1.82	0.00	0.08
Greece	8.95	5.51	2.06	1.44
Italy	0.54	0.78	5.61	3.40
Netherlands	2.90	1.96	1.80	0.68
New Zealand	6.68	2.26	6.74	2.94
Spain	1.62	4.45	13.38	9.95
United Kingdom	0.71	1.71	3.25	4.96
United States of America	3.52	6.33	3.56	3.35

Data source: United Nations, Department of Economic and Social Affairs, Population Division (2015). World Population Prospects: The 2015 Revision
NUTS2 regions in 8 European countries in this table were evaluated by Rees et al. (2001)

Subnational population projections conducted by official agencies are widely used as forecasts of future populations. Recently, official agencies not only publish their subnational population projections, but also report the errors of their projections. Although these activities are very useful for the users of their projections, there have been few studies based on comparative analysis of these projections. In this study, we compare IPSS subnational population projection errors with those found in the subnational projections done by other official agencies ones. In the future, it will be necessary to conduct more rigid comparative analysis of projection errors of not only total population projections, but also of age-specific population projections. Further, it would also be necessary to compare and delve deeper into the methods used for projection and assumptions on which the projections are based. We believe that these comparative analyses will not only improve the reliability of official population projections, but also add to our knowledge of population dynamics and how best to model them.

References[12]

Abe, T. (2006). The role of population in regional planning. In Y. Miyagawa & J. Yamashita (Eds.), *Regional structure and regional planning* (pp. 118–133). Tokyo: Minerva shobo. (J).

[12] (J): written in Japanese.
(JE): written in Japanese with English abstract.

Ato, M. & Ikenoue, S. (1987). On the reliability of population projections. *The Journal of Population Problems, 182*, 56–61. (J). http://www.ipss.go.jp/syoushika/bunken/data/pdf/14167902.pdf. Last accessed 30 Oct 2015.

Chi, G. (2009). Can knowledge improve population forecasts at subcounty levels? *Demography, 46*, 405–427.

Chi, G., Zhou, X., & Voss, P. (2011). Small-area population forecasting in an urban setting: A spatial regression approach. *Journal of Population Research, 28*, 185–201.

Esaki, Y., Nishioka, H., Suzuki, T., Koike, S., Yamauchi, M., Suga, K., & Kishi. M. (2013). Population projection and its implications: The future image of municipalities in Japan. *E-journal GEO, 8*(2), 255–267. (JE). https://www.jstage.jst.go.jp/article/ejgeo/8/2/8_255/_pdf. Last accessed 30 Oct 2015.

Gastner, M. T., & Newman, M. E. J. (2004). Diffusion-based method for producing density-equalizing maps. *Proceedings of the National Academy of Sciences of the United States of America, 101*, 7499–7504.

Ishikawa, Y. (Ed.). (2015). *International migrants in Japan: Contribution in an era of population decline*. Melbourne: Trans Pacific Press.

Japan Policy Council Commission on Declining Population Problems. (2014). *The strategy for overcoming low fertility and regional empowerment*. (J). http://www.policycouncil.jp/pdf/prop03/prop03.pdf. Last accessed 30 Oct 2015.

Keilman, N. (1997). Ex-post errors in official population forecasts in industrialized countries. *Journal of Official Statistics, 13*(3), 245–277.

Keilman, N. (1998). How accurate are the United Nations World Population Projections? *Population and Development Review, 24* Supplement (Frontiers of Population Forecasting), 15–41.

Keilman, N. (2008). European demographic forecasts have not become more accurate over the past 25 years. *Population and Development Review, 34*(1), 137–153.

Keyfitz, N. (1981). The limits of population forecasting. *Population and Development Review, 7* (4), 579–593.

Koike, S (2008). Comparative research of the migration models in regional population projections. *The Journal of Population Problems, 64*(3), 87–111. (JE). http://www.ipss.go.jp/syoushika/bunken/data/pdf/18879106.pdf. Last accessed 30 Oct 2015.

Long, J. (1995). Complexity, accuracy, and utility of official population projections. *Mathematical Population Studies, 5*, 203–216.

Murdock, S. H., Leistritz, F. L., Hamm, R. R., Hwang, S., & Parpia, B. (1984). An assessment of the accuracy of a regional economic-demographic projection model. *Demography, 21*, 383–404.

Nishioka, H., Koike, S., Yamauchi, M., Suga, K., & Esaki, Y. (2011). Population projections by prefecture in Japan:2005–2035 Outline of results and methods. *The Japanese Journal of Population, 9*(1), 1–39. http://www.ipss.go.jp/webj-ad/WebJournal.files/Population/2011_Vol.9/Web%20Journal_Vol.9_01.pdf. Last accessed 30 Oct 2015.

Oe, M. (2011). On the reliability of prefectural population projections. *Tokei, 62*(12), 27–33. (J).

Office for National Statistics. (2015). *Subnational population projections accuracy report*. http://www.ons.gov.uk/ons/guide-method/method-quality/specific/population-and-migration/population-projections/index.html. Last accessed 30 Oct 2015.

Preston, S. H., Heuveline, P., & Guillot, M. (2001). *Demography: Measuring and modeling population processes*. Malden: Blackwell Publishing.

Rees, P., Kupiszewski, M., Eyre, H., Wilson, T., & Durham, H. (2001). *The evaluation of regional population projections for the European Union*. ERDF STUDY No.97/00/74/018, European Commission.

Shimizu, M. (2004). The age pattern of net migration rate in central Tokyo: The case of Chiyoda ward. *The Journal of Population Problems, 60*(4), 34–54. http://www.ipss.go.jp/syoushika/bunken/data/pdf/13879104.pdf. Last accessed 30 Oct 2015.

Smith, S. K. (1987). Tests of forecast accuracy and bias for county population projections. *Journal of the American Statistical Association, 82*(400), 991–1003.

Smith, S. K., & Shahidullah, M. (1995). An evaluation of population projection errors for census tracts. *Journal of the American Statistical Association, 90*(429), 64–71.
Smith, S. K., & Sincich, T. (1988). Stability over the distribution of population forecast errors. *Demography, 25*, 461–474.
Smith, S. K., & Sincich, T. (1991). An empirical analysis of the effect of length of forecast horizon on population forecast errors. *Demography, 28*, 261–274.
Smith, S. K., & Tayman, J. (2003). An evaluation of population projections by age. *Demography, 40*, 741–757.
Smith, S. K., Tayman, J., & Swanson, D. A. (2013). *A practitioner's guide to state and local population projections*. Dordrecht: Springer.
Statistics New Zealand. (2008). *How accurate are population projections? An evaluation of Statistics New Zealand population projections, 1991–2006*. Wellington: Statistics New Zealand. http://www.stats.govt.nz/browse_for_stats/population/estimates_and_projections/how-accurate-are-population-projections.aspx. Last accessed 30 Oct 2015.
Stoto, M. (1983). The accuracy of population projections. *Journal of American Statistical Association, 78*, 13–20.
Tayman, J., & Swanson, D. A. (1996). On the utility of population forecasts. *Demography, 33*, 523–528.
Tayman, J., Smith, S. K., & Rayer, S. (2011). Evaluating population forecast accuracy: A regression approach using county data. *Population Research and Policy Review, 30*, 235–262.
Wang, C. (2002). *Evaluation of Census Bureau's 1995–2025 state population projections* (Population Division Working Paper Series (US Census Bureau), No.67). Washington DC,: US Census Bureau, http://www.census.gov/library/working-papers/2002/demo/POP-twps0067.html. Last accessed 30 Oct 2015.
Wilson, T. (2012). Forecast accuracy and uncertainty of Australian Bureau of Statistics State and Territory Population Projections. *International Journal of Population Research, 2012*, 1–16. http://www.hindawi.com/journals/ijpr/2012/419824/. Last accessed 30 Oct 2015.
Yamauchi, M. (2015). The fertility contribution of foreign women to Japan. In Y. Ishikawa (Ed.), *International migrants in Japan: Contribution in an era of population decline* (pp. 23–43). Melbourne: Trans Pacific Press.

Chapter 16
Integrated Local Demographic Forecasts Constrained by the Supply of Housing or Jobs: Practice in the UK

Ludi Simpson

Abstract Demographic forecasting models that integrate population, housing and jobs aim to help planning in two ways. First, the models describe the future needs of the population, with a range of scenarios reflecting the continuation of recent experience and uncertainty about which assumptions best reflect that experience. These models apply the standard mathematics of demographic cohorts and their change through births, deaths and migration, and of derived forecasts which apply age-sex specific household headship rates and economic activity rates to the future population. Second, the models are extended to calculate the impact of planned developments that will change the population by attracting or deterring people at a different rate from recent experience. This chapter focuses on the need for both types of forecast scenario in the context of local development plans which are required throughout the UK. It provides the mathematics to calculate migration in a forecast which has imposed a constraint or target future number of jobs or housing units. These balancing models are known as dwelling-led or housing-led forecasts in the UK, and are commonly used in the planning industry though seldom documented. The chapter includes an example application and discusses further developments.

Keywords Planning • Forecasts • Population • Housing • Workforce • Targets • Constraints • POPGROUP

16.1 Introduction

This chapter describes the use of models that integrate demographic forecasts of local population, households and workforce. The models have become standard in the UK, where planning law requires each elected authority to publish for public challenge a local development plan based on a review of need including

L. Simpson (✉)
University of Manchester, Manchester, UK
e-mail: ludi.simpson@manchester.ac.uk

D.A. Swanson (ed.), *The Frontiers of Applied Demography*, Applied Demography Series 9, DOI 10.1007/978-3-319-43329-5_16

demographic forecasts. The plans attempt to foresee the demand for housing and employment embodied in the current population's growth and ageing, and the impact on that population of alternative plans for supply of housing or employment.

The chapter begins by reviewing the contexts of planning and data availability that have determined the particular approach described in the rest of the chapter. This approach is embodied in the software POPGROUP, a planning industry standard in the UK, but the mathematics can be applied more widely. Standard cohort component and derived forecasting methods form the framework for the demographic modelling. The technical core of the chapter is the derivation of future migration from the imbalance between the supply and demand of either housing or jobs. These are specific instances of a more general methodology to implement constraints within demographic forecasts.

Scenarios have two distinct roles in planning: they explore factors either outside or within the control of planners. The chapter's approach is applied in the example of East Cheshire district's Local Plan. A discussion of future developments and research conclude the chapter.

16.1.1 Contexts of Planning Regulations and Data Availability

The demographic models used in the UK are standard ones advised by the United Nations and used by the official statistics agencies of most countries for national planning. A projection of the future population is made by applying assumptions about fertility, mortality and migration to each age-sex cohort. To this future population are applied household headship rates and economic activity rates to derive household and labour force projections. In the UK, these same methods are applied at the sub-national scale for most local planning. These are not the only methods available for sub-national demographic projections, and practice can be varied even within one country (see for example Smith et al. 2013 for the case of the USA, and Wilson and Rees 2005 for a general review).

The sub-national forecasting practice described here has been shaped on the one hand by government planning regulations which insist that the designation of land for development be based on forecast population and households, and on the other hand by data availability including population estimates with age-sex composition for all sub-national areas. While these two contexts have favoured the demographic models described, other strategies exist that may be appropriate at different geographical scales. There is a great deal of room for evaluation of different methods.

16.1.2 The Planning Context

Any country that co-ordinates land use according to social needs must assess the composition of the future population in sub-national areas. Planning is about the future. The future population and its composition represent the social demand for housing and for jobs. Planning regulations in Britain are relatively strongly led by national government (Oxley et al. 2009; Monks et al. 2013; Cullingworth et al. 2015). A forecast of need for housing is required by government guidance in England (DCLG 2015a), and the minimum release of land for development must be calculated on the basis of satisfying that need. The guidance oversees the Local Plan that each district planning authority must make to guide its decisions about proposed housing and commercial developments. Local Plans can be challenged, and usually are, leading to Examinations with legal status where the plan is judged by an Inspector appointed by government. This chapter focuses on practice in England, with 55 m of the UK's 65 m population, where guidance since 2010 has shaped the actions of the 336 district planning authorities, and has prioritised growth through economic targets. The legal planning structure is similar in Wales, Scotland and Northern Ireland, although strategic policies for regions larger than planning authorities have more weight outside England (Inter-Parliamentary Research and Information Network 2013).

The guidance for Local Plans in England (DCLG 2015a) stipulates that the starting point to calculate an 'objective assessment of need for housing' must be the official government household projections, in which recent local demographic levels of fertility, mortality, migration and household formation are assumed to continue or to change in line with national forecasts of each of these factors. The Local Plan (which has in past years come under varied names of strategic, structure, spatial or development plan) must identify the land that will be available to developers to satisfy the forecast number of households. This predict-and-provide approach may be tempered by physical constraints, legal constraints on development in Green Belt areas around cities and National Parks, and policies to shift provision between neighbouring districts.

Figure 16.1 describes the three common questions which demographic forecasts seek to answer in the analyses that support Local Plans. First, what is the demand for housing and jobs if migration continues as recent trends suggest? This is the 'business as usual' forecast that the official projections provide an answer to. Forecasters may attempt to improve it by using more recent data or more sensitive analyses of migration trends. It is not 'policy off' or 'policy free' as sometimes claimed; rather it assumes that any impact of past policies will continue unchanged. In the second part of Fig. 16.1, a plan or target number of dwellings determines the future trajectory of households, which in turn determines the level of migration required to meet that target. The revised population forecast is then used to estimate the future need for jobs. Finally, if the plan is for jobs growth, it is this that affects future migration, which in turn determines housing need.

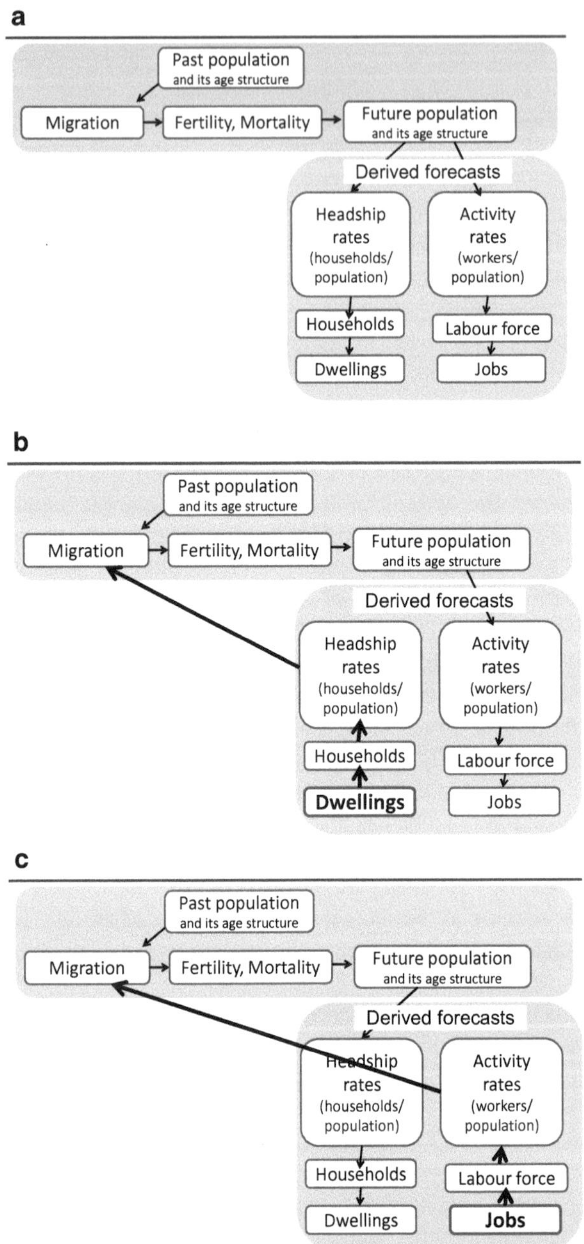

Fig. 16.1 Migration determines a population forecast, or is determined by constraints (**a**) Business as usual: migration, (**b**) A housing plan, (**c**) A jobs target

The Cameron government since 2010 has changed the guidance for Local Plans in England in a number of ways to encourage more land to be released than the forecast of housing need from demographic trends. The guidance has been translated into practical steps for implementation by the Planning Advisory Service

(2014). When draft plans are contested, alternative figures for future land need are proposed, not least by developers whose incentives include maximising the area identified as available for development, to reduce the price of housing land and to increase their choice of the most profitable sites. The new guidance allows no reductions from the demographic forecast unless balanced in other areas. On the other hand, the Local Plan must cater for more than the demographic demand to make up for any previous undersupply, and to meet local policies that must consider targets for future growth in the number of jobs (DCLG 2015a: paras 14–20).

Thus an aspirational target of jobs has become the 'new normal' and has created a focus on the future of economic activity. All other things being equal, a growth in jobs beyond that expected from demographic change would require more land for housing to accommodate the extra workers and their families. However, if economic activity were to continue to increase in the UK as in the 2000s, including the impact of a rising state pension age, then those extra jobs could be taken by the existing population without any requirement for extra housing. Thus aspirational jobs growth and future economic activity have become much more central to the examination of local plans than hitherto (for examples of the debate, see Peter Brett Associates 2014, especially Appendix D, Shropshire County Council 2014, and Simpson 2015). In the past, 'jobs-led scenarios' as illustrated by Fig. 16.1c, were only encountered where a demonstrable major jobs expansion or contraction was expected, such as North Sea oil expansion off Scotland, or de-industrialisation elsewhere, both in the 1970s and 1980s.

Housing-led scenarios have regularly been used to assess how a plan for future land release, expressed as numbers of housing units per annum, might impact on local population. Outside the preparation and debate of Local Plans, housing-led scenarios are common when assessing the impact of developers' proposals to use specific major land sites. The regulations covering permission to build allow *Planning Contributions* and a *Community Infrastructure Levy* from developers where a development impacts on school, transport or other services, stimulating the quantification of those impacts through the modelling implied by Fig. 16.1b (DCLG 2015b; North Devon Council 2015).

16.1.3 The Data Context

The UK system of official statistics has been fortunate that universal health services record the date of birth of every registered patient, which has led to the annual measurement of internal migration by age and sex since the 1970s (Scott and Kilbey 1999; ONS 2015a). Since 2001, registers from the health service have also been used to provide population estimates by single year of age for small areas within each country of the UK (NISRA 2011).

The decennial Census provides data for the household composition of sub-national areas. The Census also supplies three other key areas of information used in local planning: a sub-national series for economic activity; the local links between demographic demand for housing and its supply: vacant housing, shared housing, and second homes; and the local links between demographic demand for jobs and its supply: unemployment and commuting.

Tax and benefit systems also supply information on housing and unemployment. Together, vital statistics, universal health services, the Census, tax and benefit systems have been central in the development of the cohort component demographic forecasts and derived forecasts that are now essential ingredients of sub-national planning in the UK.

16.2 The Demographic Models

Official statistics agencies in the UK provide population forecasts by single year of age using cohort component methods for each local authority district (ONS 2015b). These districts are singly or sometimes in their aggregate the same as the planning authorities referred to above, with a range of about 50 thousand to one million population. The same cohort-component approach is favoured for many other planning purposes whether for districts or for smaller areas, taking advantage of the existence of data with detailed age-composition already referred to. The approach provides the future composition of the population which is directly useful to many services that address the needs of infants, young people, those of working age, elderly or other specific sections of the population. Detail of age and sex is also exploited for the derived household and labour force projections, since household composition and economic activity vary with age and sex in predictable patterns (Fig. 16.2).

In the official population projections, the future trajectory of each of fertility, mortality, international migration, and household rates is established by national analysis. The local characteristics of each of these components of change are established by an average of the recent past, usually 5 years, and its future trajectory assumed to be parallel to the national projection.

Official household projections in the UK use headship rates in a variety of ways (Welsh Government 2011). England and Scotland favour age-specific proportions of the household population (those not in communal establishments) that *represent* their household, classified into household types. Wales and Northern Ireland use age-specific proportions of the household population that are *members* of households of a particular type, then dividing the projected number of people in each type of household by its average household size. The latter approach avoids an algorithm to identify one person as the head or representative of a household while losing some rationale for being a stable indicator. Both approaches forecast the future household headship or membership rates by extrapolation from past census values, with some influence of recent national surveys.

There are no official projections by UK government departments of economic activity or the labour force, and those provided by the European Commission and the Office for Budget Responsibility are unsatisfactory in their treatment of young

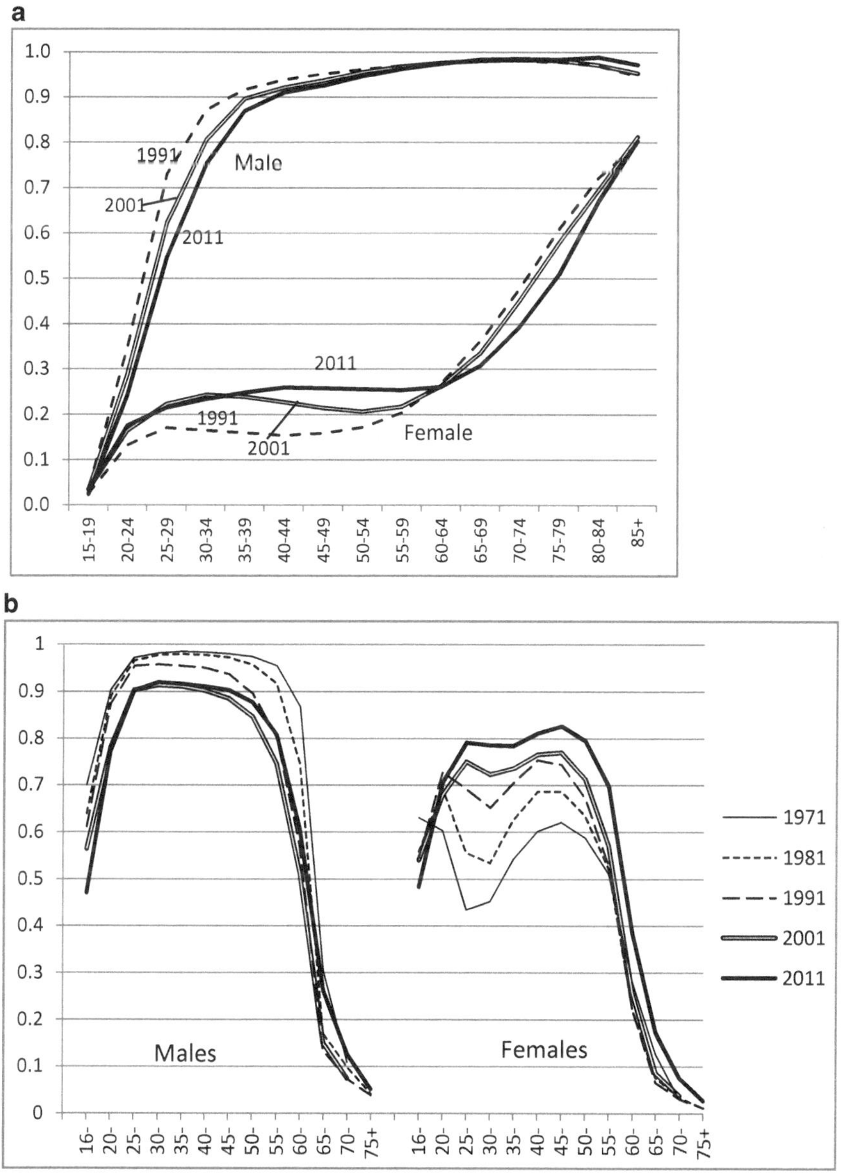

Fig. 16.2 Age-specific headship rates and economic activity rates (**a**) Household headship rates, England, (**b**) Economic activity, England and Wales (Sources: (a) DCLG (2015c). In this source, the 'head' is a statistical representative, calculated in a way which favours the male in a couple. (b) Population Censuses 1971–2011. '70–74' refers to 70 and older in 1971; 75+ was not recorded in 2001; school leaving age was raised from 15 to 16 in 1972)

adults and the elderly, leading local planners to provide their own assumptions (Simpson 2015).

These forecasts of future population, households and the labour force are projections under the assumption of 'business as usual' (Fig. 16.1a). However they do not give the demand for housing and jobs, which requires further adjustments which will now be described and have been part of the toolbox of strategic planning for many years (Breheny and Roberts 1980; Field and MacGregor 1987: chapter 8).

Some households share the same dwelling unit, reducing demand, while dwellings that are vacant or used as second homes or holiday homes increase the number of dwellings needed to house a forecast number of households. These relationships are used to compute the demand for dwellings from the forecast number of households by the following identity:

$$\begin{aligned} \text{Dwellings} &= \text{Households} * (1 - \text{sharing rate}) \\ &\quad /(1 - \text{vacancy rate} - \text{second homes rate} - \text{holiday homes rate}) \end{aligned}$$

Similarly the number of jobs demanded by a resident labour force is dependent on the future unemployment rate and the impact of commuting. Commuting is represented by the ratio of those residents in the area who are employed divided by the number of occupied jobs in the area. For example the commuting ratio is more than one when there is net out-commuting, because this decreases the number of jobs demanded by a projected resident labour force. The identity linking jobs to the labour force is:

$$\begin{aligned} \text{Jobs} &= \text{Labour force} * (1 - \text{unemployment rate})/(\text{commuting ratio}) \\ &= \text{Labour force} * (1 - [\text{unemployed living in area}]/[\text{employed} + \text{unemployed living in area}]) \\ &\quad /([\text{employed residents living in area}]/[\text{jobs in area}]) \end{aligned}$$

16.2.1 Software

Within the planning industry – spanning the public sector, agencies representing developers, and community groups – a variety of technical implementations are used to replicate the official demographic projections and to extend them with planning scenarios. POPGROUP is the only available software documented in the public domain and has become an industry standard in the UK (University of Manchester 2014; Edge Analytics Ltd 2013). It was developed by a consortium of local authorities in the late 1990s, and specified and developed by the author of this chapter. It is now owned by the Local Government Association and leased to Edge Analytics Ltd.

The POPGROUP software is a projection model-maker using the Microsoft Excel platform and VBA programs, free from data and appropriate for use with any geographical scale, so long as the user is prepared to locate and enter appropriate data. Up to 40 regions are forecast independently in the same model. POPGROUP implements the single year cohort component projection with gross flows of migration between each region and one or two external areas, chosen when setting up the model. These external areas are usually the rest of the UK and international, but may be specified as short distance and long-distance depending on the data available and the concerns of the forecaster. Each component of change may be represented by counts or rates, which in turn may be age-specific or general. Options allow controls to data for the aggregate of regions, which together with the use of a national population forecast for internal in-migration rates, introduces strong elements of bi-regional models. The minimum data entry comprises a base population and single-year demographic rates for fertility and mortality. Counts of events and population allow the model to be used to estimate past local demographic trends. Targets of housing or jobs form constraints which are discussed in the next section.

POPGROUP's sub-module Derived Forecasts (Edge Analytics Ltd 2010) applies rates to a forecast of the population. It is also a model-maker in that (a) the rates may be defined for the user's specified age-sex groups, aggregated from individual years of age, (b) the rates may refer to several categories such as household types, (c) adjustments before the application of rates cope with day-time, household or other population bases, (d) subsequent adjustments allow for household size or monetary or other factors applied to an initial forecast, and (e) the number of areas and their naming along with the naming of derived variables are under the control of the user when setting up the model. The flexibility has allowed users of Derived Forecasts to implement a great variety of models, including each of the UK official household projections, labour force forecasts and forecasts of the disabled population.

The integration of population and derived forecasts is key to successful modelling in the planning context. The Derived Forecasts software may take its population from POPGROUP outputs. One or two derived forecasts may be run in the background from POPGROUP, providing for example the consequences of a population forecast for the future number of households, housing units, the labour force, and jobs. When a plan of housing or jobs is used as a constraint on the population forecast, POPGROUP and Derived Forecasts are used together to determine the level of migration required so that the population will be consistent with the plan, as discussed in the next section.

16.3 Migration Determined by a Mismatch of Households and Housing, or a Mismatch of Labour Supply and Jobs

For a given supply of housing, how will the population adjust and what will be the consequent demand for jobs? Alternatively, for a given supply of jobs that may be an aspirational target, how will the population adjust if the target is achieved, and what will be the consequent demand for housing? To answer these questions, extensions are required to the 'business as usual' projection forward of recent experience, as described in Fig. 16.1a and the previous section. The extensions integrate estimates of future housing or jobs as described in Figs. 16.1b, c. This section briefly reviews approaches to making these extensions and specifies the common solution in UK modelling.

In the USA context, Smith et al. (2013: 222–227) describe econometric models linking population, jobs and migration, and balancing models in which migration is adjusted to match future population to the supply of jobs. In the UK, econometric modelling is represented by the work of commercial companies such as Oxford Economics and Cambridge Econometrics, and academic modellers Jeff Meen (2011) and Glen Bramley (Bramley et al. 2010). Econometric models can create a target of jobs within the context of past relationships between jobs and wage levels, house prices, migration and other aspects of the economy. In debates at Local Plan Examinations, target jobs growth is often set by econometric models embodying assumptions about demographic and economic trends. However, there is some circularity in the modelling if these targets based on assumptions about demographic change are then used to determine demographic indicators of housing need (Planning Advisory Service 2014: sections 6.10–6.13).

In contrast to econometric models, balancing models assume an exact equation between housing, jobs and population, as described above. With assumptions for future headship rates, economic activity, sharing households, unoccupied housing, unemployment and commuting already set by past experience in the 'business as usual' projection, or adjusted by policy, a new level of migration is calculated to balance the population with the target of housing or jobs.

Since the planning question asks for the demographic consequences of fixing the housing or jobs target, the balancing model has been accepted as an appropriate tool, if not the only analysis considered. Smith, Tayman and Swanson also observe its simplicity:

> Unlike an econometric forecasting model, a balancing model does not require formal statistical equations or time series data to project future levels of migration. In addition, it does not require implementing a large-scale model of the economy. Consequently, balancing models are less costly to implement and easier to use than econometric models, and are more accessible to a wider range of practitioners (Smith et al. 2013: 225).

If labour demand exceeds labour supply, it is projected that workers will move into the area. If labour supply exceeds labour demand, it is projected that workers will move out, and equivalently with the balance of housing demand and supply.

The constraint on the population forecast is either an estimate of future house-building, which may vary each year, or an annual growth rate in the number of jobs, also variable over time. A scenario that fixes the future level of both is not considered relevant unless operating at different geographical scales: for example a target of jobs growth for a region whose consequences for housing is satisfied with a plan for each neighbourhood.

The balancing model is technically described in the remainder of this section as implemented in the POPGROUP software. Its simplicity does not avoid making multiple assumptions and somewhat complex calculations, in order to respect what is known about the age-composition of migrants and the way in which households are formed. In what follows, a housing target is assumed and its impact on a population forecast is calculated; note is made where the operation of a jobs-led forecast would be different.

The impact of achieving a target level of house-building of say 10,000 dwellings (housing units) per annum may be varied. Vacancy levels, sharing households, numbers of second residences or holiday homes, may each change. Households may spread over a larger number of dwellings with higher headship rates. All those factors are 'levers' that the projectionist must consider and may change within the forecasting model, by making alternative assumptions about the future. It is possible that the provision of housing also affects fertility or mortality of the population. However, in this Chapter's treatment the challenge is to calculate how the population may change due to extra housing being filled by migrants into the local area.

When more housing is available, it is unreasonable to assume that it will be occupied directly by migrants from outside the area. In general, the extra housing allows movement from within the area by residents whose own housing then becomes available to others. Thus the *type* of new housing is not usually taken into account when calculating the migration attracted by extra housing stock. Instead new migrants will arrive in the general housing stock, or fewer migrants will leave the area. The specific flows of migrants that are affected – in- or out-flows, national or international, short-distance or long-distance, may be specified by the projectionist.

16.3.1 Calculations

In general, the derived forecast at time t is a linear function of the age-sex composition of forecast population. For households, the function involves deducting those in communal establishments, and then multiplying by a headship rate:

$$D_t = f\left(\underline{P}_t\right),$$

where $\underline{P}_t$ represents the age-sex set of population results projected for time t.

As mentioned, it is possible to alter the function f in order to meet a target population derivative, for example taking the target as evidence of a change in the assumed headship rates. The headship rates would then be scaled to meet the estimated housing without any change to the population projection. On the other hand, if the estimated future housing is considered as a cause of change in future population, then it is the population that should be adjusted to achieve the target. This latter option is considered here. The discussion assumes a housing target, household headship rates and a relationship between housing and households for simplicity of explanation. The same calculations are correct for a jobs target, economic activity rates and a relationship between jobs and workforce.

Let $D^1_{t+1} = f\left(\underline{P}^1_{t+1}\right)$ be the derived forecast at future time $t+1$ according to all the assumptions without a constraining target. The aim is to compute the population consistent with a target derived forecast number of houses, D^2_{t+1}. This target forecast is usually the plan for house-building or release of housing land expressed as a number of housing units. It is assumed that births and deaths are accurate, so the aim is to compute new flows of migration at each age and sex that are consistent with the housing target. The distance between the target and the unconstrained forecast of housing units is the difference between the linear function of the two population projections, and so it can be expressed as the same linear function applied to the difference in populations at each age and sex. With births and deaths fixed over a single time period, the difference is the same linear function of the difference in migration:

$$\begin{aligned} D^2_{t+1} - D^1_{t+1} &= f\left(\underline{P}^2_{t+1}\right) - f\left(\underline{P}^1_{t+1}\right) = f\left(\underline{P}^2_{t+1} - \underline{P}^1_{t+1}\right) \\ &= f\left(\sum_{i=1}^{M} \left(\underline{m}^{2i}_t - \underline{m}^{1i}_t\right)\right) \end{aligned} \tag{16.1}$$

where $\underline{m}_t$ represents a collection of age-sex-specific migration from time t to t + 1, and i = 1,. . .,M are the flows of migration to or from the area (out-flows being negative). A projection has been made without the constraint, so the left hand side and the initial unconstrained migration $\underline{m}^{1i}_t$ is known, as is the function that involves headship rates and the ratio of households to housing units. The aim is to find the adjustment to migration so that new migration flows $\underline{m}^{2i}_t$ may be substituted which will be consistent with the constraining target.

The solution is a proportional fitting procedure that respects the age-composition of migrants, which is strongly skewed towards young adults and whose local characteristics have already been estimated for the initial forecast. As there is more than one migration flow that may be adjusted, the solution requires a choice between them, which is kept flexible by allowing a weight w_i for each migrant flow i, adding to 1.

In practice there are four flows of migration in a POPGROUP model (M = 4) involving short-distance and long-distance in- and out-flows with the area's population. The weights are set by the forecaster, using knowledge of the housing and labour markets. A housing target has usually been considered as an attraction or deterrent to migration within the UK. It has been usual to set the weights for in- and out-flows equal to suggest that the impact may be equally via an attraction of in-migrants or a deterrence of out-migrants.

To solve Eq. (16.1), the impact of a single migrant on the derivative (housing) is first considered. The single migrant is expressed as a vector of all the age-sex population categories a and the M migrant flows, scaled to sum to 1. The age-sex 'composition' of this single migrant is based on the projected flows without the constraint, by dividing each m_{at}^{1i} by the total number of migrants in the flow across all age-sex categories, $m_{\bullet t}^{1i}$, and using the weights of each flow:

$$\sum_{i=1}^{M} w_i \underline{m}_t^{1i} / m_{\bullet t}^{1i} = \sum_{i=1}^{M} \sum_{a} w_i m_{at}^{1i} / m_{\bullet t}^{1i} = 1 \tag{16.2}$$

Equation (16.2) is simply an identity, adding the elements of the vector of migration flows $\underline{m}_t^{1i}$ after weighting and scaling. The sum of each flow's migrants by age and sex, divided by the flow total, is one. Since the weights w_i add to 1, the weighted total of all migrant flows expressed in this proportional way is also 1.

The impact on the population derivative of this single representative migrant, is $f\left(\sum_{i=1}^{M} w_i \underline{m}_t^{1i} / m_{\bullet t}^{1i}\right)$. It may indicate that the single representative migrant only fills half a dwelling, since many migrants are not household heads. In the case of jobs, not all migrants are workers, since workers have families. This impact of the representative migrant is then scaled up to calculate the adjustment to the total number of migrants required to meet the target population derivative:

$$\sum_{i=1}^{M} \left(m_{\bullet t}^{2i} - m_{\bullet t}^{1i}\right) = \left(D_{t+1}^{2} - D_{t+1}^{1}\right) / f\left(\sum_{i=1}^{M} w_i \underline{m}_t^{1i} / m_{\bullet t}^{1i}\right) \tag{16.3}$$

For example, if the representative migrant fills half a house, an extra 50 houses will require 100 extra migrants. Equation (16.3) provides the *total* adjustment to migrants required to meet the target population derivative. It is a single value and its division into flows and age-sex groups is not yet known. So far we have assumed that it is positive value, but the target D_{t+1}^{2} may be less than the demand implied without the constraint, D_{t+1}^{1}. For example an adjustment to migrants of -1000 would indicate that a *deduction* of 1000 migrants with the composition of the weighted initial flows will produce the target population derivative.

The separation of this total adjustment into the M flows and each age-sex category is achieved using the representative migrant vector again:

$$\underline{m}_t^{2i} - \underline{m}_t^{1i} = \left(\sum_{i=1}^{M} \left(m_{\bullet t}^{2i} - m_{\bullet t}^{1i} \right) \right) * \left(w_i \underline{m}_t^{1i} / m_{\bullet t}^{1i} \right) \tag{16.4}$$

The first term of the right hand side of Eq. (16.4) is the single value calculated from Eq. (16.3). For each flow i, this value multiplies the vector of age-sex-flow proportions that is the representative migrant.

Finally, the adjustments from Eq. 16.4 are made to each of the M initial flows, adding to in-flows and deducting from outflows, in order to meet the population constraint. Negative gross flows can result from an adjustment greater than the existing flow. These are avoided by adding the negative result to the opposite flow. Thus for example, an inflow of −100 is set to zero while 100 is added to the outflow from the same age-sex category.

To summarise: housing or employment constraints are achieved by scaling each age-sex migrant flow by the single factor that will reproduce the constraint, but allow for different weights for each flow of migration. The approach described above is a development of the 'plus-minus method' of Judson and Popoff (2004: 708–711) for adjusting gross flows to implement a constraint of a single net migration total. Alternatives could relax the requirement that the constraint be met exactly, or allow the migrants generated by a housing plan to have a different age-sex composition from that used in the initial projection.

This approach to constraints is extended in the POPGROUP software to allow constraints on each component of population change and on the population itself (Simpson 2006 provides the calculations and examples). The extensions are used to estimate missing demographic rates and migration age-sex profiles for small areas when a time series of vital events and population estimates are available (Simpson and Snowling 2011).

16.4 Scenarios for Uncertainty and Scenarios for Plans

Planners have to deal with uncertainty about the future. RJS Baker provided a framework for considering the relationship between forecasts and planning which remains useful today though seldom developed. He suggested four phases:

- Research and intelligence – inquiry into past and present events and situations…
- Forecasting future events and situations outside one's own immediate control.
- Planning action to be taken by oneself or under one's own control.
- Forecasting the effect of such action. (Baker 1972: 121–2)

In the current context of land use planning, the events and situations outside the planner's immediate control might include future fertility, mortality and the relationship between population and housing. The planner demands that the demographer forecasts these future events, and provides advice about the reliability of these forecasts. The demographer should provide a trend-based projection of population, with alternatives that are also indicated as equally likely due to past fluctuations in fertility, mortality and migration. Although difficult to make precisely, the likelihood of a range of scenarios should be provided to the planner. This completes the first two parts of Baker's framework.

Once the planner has determined possible actions to be taken, the demographer is then asked to forecast the effect of such action. This is achieved through further modelling such as the scenarios described in the previous section that test the impact on population of housing or job targets. The demographer should once again be clear about the assumptions made, and also describe the uncertainty in the estimates of the effect of the planner's proposed actions.

There are other approaches to scenarios which instead might intend to force planners' attention away from the past trends by imagining plausible different futures (Ramirez and Selin 2014).

16.4.1 Examples for East Cheshire

East Cheshire is one of the 336 planning authorities of England, including the towns of Nantwich, Crewe, Congleton and Macclesfield. It has no major institution of higher education, serves as a commuting belt for the conurbations of Manchester and Liverpool to its north, and receives older migrants from those cities. As such it is attractive to developers wishing to provide housing at the more expensive end of the market.

Figure 16.3 shows the growing population of East Cheshire 2001–2014. The thick continuous line shows the continued growth expected in the official projection by the Office for National Statistics (ONS), updated with the most recent population estimates. Two low scenarios showing steady or decreasing population from zero migration are not realistic but indicate the nature of migration to the area. Migration accounts for all of the district's expected population change – the 'Natural change' scenario shows a steady population. The 'Net Zero UK migration' scenario assumes a balance of gross flows of migration in total, but maintains the age structure of the district's recent flows to and from the rest of the UK. This scenario leads to a clearly decreasing population, showing that in net terms Cheshire East gains mostly older people who have already had children and are more likely to subsequently die than to give birth.

The other scenarios are intended to realistically show the impact of alternative population scenarios that are quite feasible without any intervention. They are alternative representations of 'business as usual'. The bounds of fertility labelled by ONS in their national projections as high and low are applied as proportional changes to East Cheshire's projected fertility. This range of fertility adds rather more uncertainty than the ONS national range of mortality. Two alternative assumptions about migration provide slightly more uncertainty. They are based on continuing flows of migration averaged from the 5 years prior to 2012 (mostly in a period of economic recession), or averaged from 10 years prior to 2012. The

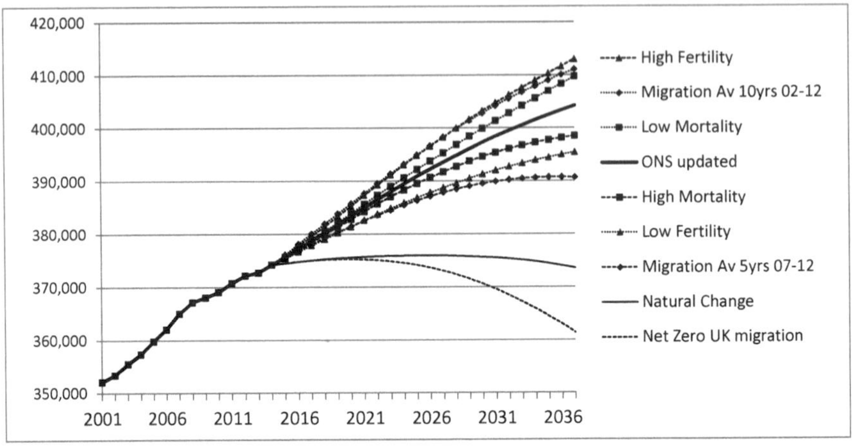

Fig. 16.3 Demographic scenarios testing the official population projection for East Cheshire
Notes: The scenarios are labelled in order of their projection at 2037 from highest to lowest. The scenario 'ONS updated' is the government 2012-based projection updated with 2013 and 2014 Mid-Year Estimates of population. Each other scenario applies alternative assumptions from mid-2014. Scenarios are as defined in the text

official projection was based on migration from the 5 years prior to 2012, but applied in a rates-based multi-regional model rather than constant flows at each age and sex.

Figure 16.3 shows that East Cheshire's future demographic growth cannot be easily disputed, but that by 2036 that growth may be between 20 and 40 thousand when compared to the population of 374 thousand in 2014. This range represents slight variations of the recent past. It shows uncertainty about the future even if the recent demographic experience continues.

Figure 16.4 translates these projections into need for housing by applying the officially projected age- and sex-specific headship rates for Cheshire East (DCLG 2015c) and household/dwelling ratios from the 2011 Census. The Figure includes two further scenarios based on modelling as follows:

- Housing required by a job growth of 0.2 % per annum, as proposed locally in 2014 (Cheshire East Council 2014), assuming that economic activity, commuting and unemployment remain the same as at the 2011 census.
- The same job growth of 0.2 % per annum but also a gradual increase of economic activity by 2 percentage points over the decade 2015–2025.

The two extra scenarios are examples of the implementation of Fig. 16.1c. In these scenarios, the jobs target is implemented as described earlier. Jobs are a derived forecast which acts as a constraint on migration, causing a change in population and its age structure. Figure 16.4 shows the impact on the second derived forecast, of housing.

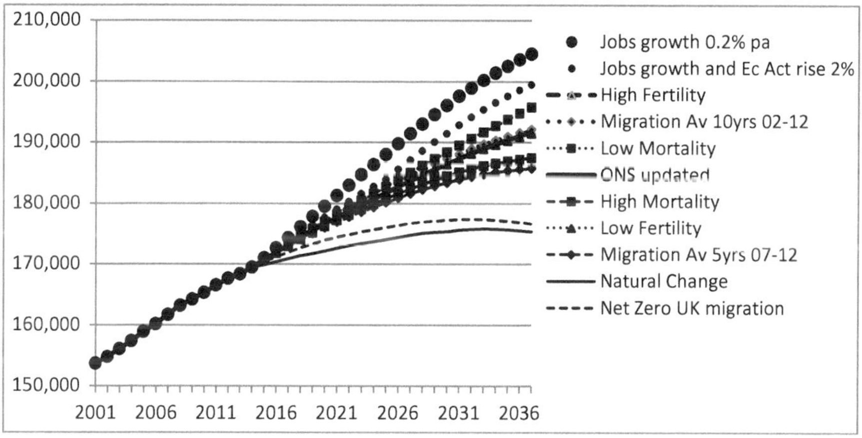

Fig. 16.4 Demographic scenarios and jobs-led scenarios: the impact on housing need: projections for East Cheshire. *Note*: Scenarios are as defined in the text

A comparison of Figs. 16.3 and 16.4 is instructive about the demographic processes modelled. While the range of fertility was a significant influence on the size of the population it has practically no impact on housing need in the 23 year period of the projection, as newly born take nearly a generation before they significantly contribute to the number of householders. On the other hand, the impact of Cheshire East's relatively older population among in-migrants is very significant on housing need. Although the population tends to reduce as older migrants die rather than produce children, the older population tends to live in households that are smaller than the average. Thus the impact of migration's age structure is to reduce population but to add to household need. This is seen by comparing the 'Net Zero UK migration' and the 'Natural Change' scenarios.

The population of Cheshire East is ageing as are all English Districts, mainly due to the growing old of the large cohorts born in the 1950s and 1960s and the continued lower fertility since then. As we have seen, in Cheshire East internal migration exacerbates ageing. Thus the demographic need for jobs in Cheshire East peaked in 2012 and is now reducing in spite of a growing population and growing need for housing (the figures for jobs are not in the graphs but are part of the output of the modelling). The Local Plan target of 0.2 % per annum growth in the number of jobs may therefore be rather optimistic, though developers have argued that it is insufficient. If achieved, Fig. 16.4 shows that the jobs target suggests a higher need for housing in the District than any of the 'business as usual' scenarios.

Some of that need to house new workers and their families would be eliminated if the economic activity of the residents of East Cheshire were to rise. Figure 16.4 shows one crude scenario in which all age groups' activity is increased by 2 percentage points over the 10 years 2015–2025. The housing need reduces by about 2 %, from 204.5 thousand to 199.4 thousand. The *extra* housing needed during the period 2014–2037 is reduced by 15 %, from 35.1 thousand to 30.0 thousand.

The modelling for Fig. 16.4 assumed that the migrants required to fill extra jobs will have the same age structure as those generally moving between Cheshire East and the rest of the UK, half of shortfall coming from deterred out-migration and half from increased in-migration. One might think that the jobs will result in younger population and this should be specified. In fact, meeting the constraint wholly by deterred out-migration which we know is younger than in-migration, makes little difference to the result, reducing housing need by less than 300 over the whole period. This is because both in and out-migration are largely composed of young adults. It is only when balanced in large gross flows that the net difference of older in-migration is revealed in the Net Zero Migration scenario that was mentioned above.

The planning legislation in the UK demands a single Plan releasing land to a single schedule allied to a single forecast. Which forecast should be chosen? This should depend on evidence about likely change due to trends expected to continue, and the feasibility of aspirational targets. The uncertainty in future development that the range of forecasts shows, even within a continuation of recent experience, suggests that planners should release land cautiously and review frequently. The cost of defending contested plans tends to mitigate against frequent reviews.

16.5 Discussion

Planning is about the future, and as such requires forecasts. In some countries including the UK there is a legal requirement that developments conform to a local plan based on an agreed set of demographic forecasts that may include planned target levels of development. This chapter has described how projections and targets are integrated to the benefit of strategic planning. Integrated population, housing and jobs forecasts involve describing not a single future but a variety of scenarios, which reflect on the one hand the uncertainty about the direction of recent trends, and on the other the impact of planners' and politicians' intention to change trends.

The first set of scenarios test alternative assumptions about fertility, mortality and migration, and lead to an understanding of the robustness of the 'trend-based' or 'business as usual' forecast, along with its implications for the demographic demand for housing and jobs. The second set of scenarios include alternative assumptions about the future supply of either housing or jobs, leading to a balancing adjustment of migration, a revised population forecast and a revised assessment of the need for other derived demands. The example explored a target growth of jobs, and its implications for population and the demand for housing. The modelling described in this chapter allows the exploration and comparison of both these types of scenarios. There is room for more clarity over the nature of scenarios, their calculation and their use in planning, ranging from the ways in which future migration is specified to the role of imaginative scenarios that address the way we live.

The main technical contribution of the chapter has been to describe how constraints of housing or jobs can be used to adjust population projections. A brief comment on two aspects indicates where technical development is in order. First, there may be multiple constraints available for the same projection year, which are not independent, and this commonly occurs when geographical sub areas or social divisions such as ethnic group are projected. Typically a prior population or economic projection for a larger area acts as a constraint for smaller areas that have observed or planned housing constraints. To achieve consistency the headship rates for each smaller area must also be adjusted.

Second, constraints have been accommodated above for a single time period. Constraints are often presented as estimates available at the end of longer intervals of say 2 or 5 years, or as a target 10–20 years ahead. In this case the question may be: which adjustment applied over n time periods, will produce a projected population consistent with the constraint? One practical solution is to impose a constraint for each time period by interpolation between the current population and the given constraint, to which the single period solutions that have been described here can be applied. While this solution will ensure neither exactly constant numbers of migrants nor constant migration rates, the aggregate migration over the longer period is of practical interpretation and use. A solution with a constant migration is only possible within an iterative framework, where the solution is amended after each interim projection with successive solutions providing closer consistency with the constraint. The iterations would continue until a pre-determined small convergence criterion is attained. The interaction of migration, births and deaths that occurs within one time period has been ignored in this paper, as is usual within projection models (Keilman 1985).

This paper has focussed on planning authorities in England, but the principles and modelling is applied to larger and to smaller areas. For smaller areas the availability of data is a constraining factor, but since 2001 the UK has been covered by annual small area population estimates which with vital statistics and socio-economic characteristics from the decennial census are sufficient to allow the cohort component and associated demographic forecasts to be estimated for many purposes (Simpson and Snowling 2011; Hampshire County Council 2015).

More generally, forecasters must warn planners that not only must demographers provide them with multiple scenarios to reflect uncertainty in the system and in the impact of their decisions, but also that our methods require improvement. There will be great value to be gained from evaluation of local demographic projections, including the assumed link between housing and population. There is a wealth of experience available from the analysis preceding local plans but precious little evaluation at a later date. By the time projections 10 or 20 years ahead can be compared with an outcome, those involved have long since gone to other projects and the previous data rarely remains in detail.

Long-term academic evaluations are in order, and would promise to guide improved practice and improved allocation of public and private investments. Post-hoc examination of demographic forecasts can establish confidence intervals to better judge future forecasts and to better choose the most appropriate methods to each context, as has been achieved in general terms by Smith et al. (2013) for the USA. While probabilistic forecasts should be considered, they are at present unusual in sub-national demography and may be more technically demanding than helpful in practice.

In parallel with a better understanding and practice of demographic forecasts, a dialogue between forecasters and planners is always fruitful. To what extent can different types of planners deal with uncertainty? How useful is the distinction between uncertainty in what is not under the planner's control, and the impact of the planner's intentions on demographic change?

These are some of the avenues for methodological research. It is however an equally pressing priority to document and preserve the current methods and practice of demography as applied in local public planning. The 40 % reduction of funding from national to local government in the UK since 2010 (Gainsbury and Neville 2015) has already seriously affected research and planning with the loss of skills and experience (Radical Statistics 2012). Support for strategic planning waxes and wanes in the UK, and is currently stronger in Wales and Scotland than in England.

References

Baker, R. J. S. (1972). *Administrative theory and public administration*. London: Hutchinson University Library.

Bramley, G., Pawson, H., White, M., et al. (2010). *Estimating housing need*. London: Department of Communities and Local Government.

Breheny, M. J., & Roberts, A. J. (1980). Forecasting methodologies in strategic planning: A review. *Papers in Regional Science, 44*(1), 75–89.

Cheshire East Council. (2014) *Local plan strategy background paper population projections and forecasts*, Document SD019. Crewe: Cheshire East Council.

Cullingworth, B., Nadin, V., Hart, T., Davoudi, S., Pendlebury, J., Vigar, G., Webb, D., & Townshend, T. (2015). *Town and country planning in the UK* (15th ed.). Abingdon: Routlege.

DCLG. (2015a, March 26). *Planning practice guidance: Housing and economic development needs assessments*. http://planningguidance.planningportal.gov.uk. London: Department of Communities and Local Government.

DCLG. (2015b). *Planning policy guidance: Planning obligations*. London: Department of Communities and Local Government. http://planningguidance.planningportal.gov.uk/blog/guidance/planning-obligations/planning-obligations-guidance/

DCLG. (2015c, March 10). *2012-based household projections: Detailed data for modelling and analytical purposes*. https://www.gov.uk/government/collections/household-projections#2012-based-projections

Edge Analytics Ltd. (2010). *Derived Forecast (DF) model reference manual Version 1.0*. London: Local Government Association. http://www.cmist.manchester.ac.uk/research/projects/past-projects/popgroup/

Edge Analytics Ltd. (2013). *POPGROUP v.4 population estimation and forecasting system reference manual*. London: Local Government Association. http://www.cmist.manchester.ac.uk/research/projects/past-projects/popgroup/

Field, B., & MacGregor, B. (1987). *Forecasting techniques for urban and regional planning*. London: Hutchinson.

Gainsbury, S.,& Neville, S. (2015, July 19) Austerity's 18bn impact on local services. *Financial Times*. http://www.ft.com/cms/s/2/5fcbd0c4-2948-11e5-8db8-c033edba8a6e.html#axzz3qRwnoLmk

Hampshire County Council. (2015). *Small Area Population Forecasts (SAPF) 2014-based*. Online: http://www3.hants.gov.uk/factsandfigures/population-statistics/pop-estimates/small-area-pop-stats.htm

Inter-Parliamentary Research and Information Network. (2013) *Comparison of the planning systems in the four UK countries SPICe Briefing 13/35*. Online: http://www.scottish.parliament.uk/ResearchBriefingsAndFactsheets/S4/SB_13-35.pdf

Judson, D. H., & Popoff, C. L. (2004). Selected general methods. In J. S. Siegel & D. A. Swanson (Eds.), *The methods and materials of demography* (pp. 677–732). San Diego: Elsevier Academic Press.

Keilman, N. (1985). Internal and external consistency in multidimensional population projection models. *Environ Plan A, 17*, 1473–1498.

Meen, G. (2011). A long-run model of housing affordability. *Housing Studies, 26*(7–8), 1081–1103.

Monks, S., Whitehead, C., Burgess, G., & Tang, C. (2013). *International review of land supply and planning systems*. York: Joseph Rowntree Foundation.

NISRA. (2011). *Small area population estimates across the UK*. Belfast: Northern Ireland Statistics and Research Agency. http://www.nisra.gov.uk/archive/demography/population/small_area/UK_SAPE_Dec_2011.pdf

North Devon Council. (2015). *Section 106, Community Infrastructure Levy (CIL) and other planning obligations*. Barnstaple: North Devon Council. http://www.northdevon.gov.uk/planning-and-building-control/section-106-community-infrastructure-levy-and-other-planning-obligations/

ONS. (2015a). *Statistical bulletin: Internal migration, England and Wales, year ending June 2014*. Tictchfield: Office for National Statistics. http://www.ons.gov.uk/ons/dcp171778_407692.pdf

ONS. (2015b). *Subnational population projections across the UK: A comparison of data sources and methods*. Titchfield: Office for National Statistics.

Oxley, M., Brown, T., Nadin, V., et al. (2009). *Review of European planning systems*. Leicester: De Montford Unviversity.

Peter Brett Associates. (2014). *Note outlining approach taken to job forecasts*, West Dorset, Weymouth and Portland Local Plan examination WDWP/Ex13, 12 December 2014. London: Peter Brett Associates.

Planning Advisory Service. (2014). *Objectively assessed needs and housing targets*. London: Local Government Association.

Radical Statistics. (2012). *The effect of the cuts on government statistics and their use, report of the Reduced Statistics working group*. London: Radical Statistics. http://www.statsusernet.org.uk/HigherLogic/System/DownloadDocumentFile.ashx?DocumentFileKey=4122199d-65df-4f2f-a90a-17240e458008&forceDialog=0

Ramirez, R., & Selin, C. (2014). Plausibility and probability in scenario planning. *Foresight, 16* (1), 54–74.

Scott, A., & Kilbey, T. (1999). Can patient registers give an improved measure of internal migration in England and Wales? *Population Trends, 96*, 44–56.

Shropshire County Council. (2014). *Labour force projections based on 2012-based sub-national population projections, in EV3c SHMA update addendum July 2014*. Shrewsbury: Shropshire County Council. http://shropshire.gov.uk/planning-policy/samdev-examination/evidence-base/

Simpson, L. (2006) *Integrating estimates and targets within a population projection*. Manchester: Research Methods Programme Working Paper, *18*, 89–104. http://www.ccsr.ac.uk/methods/publications/documents/WP18.pdf

Simpson, L. (2015). *National and local labour force projections for the UK* (CMIST working Paper). Manchester: Cathie Marsh Institute for Social Research. http://www.cmist.manchester.ac.uk/research/publications/working-papers

Simpson, L., & Snowling, H. (2011). Estimation of local demographic variation in a flexible framework for population projections. *Journal of Population Research, 28*(2), 109–127.

Smith, S. K., Tayman, J., & Swanson, D. A. (2013). *A practitioner's guide to state and local population projections*. New York: Springer.

University of Manchester. (2014). *Impact of demographic projections on sub-national planning* (Research Excellence Framework impact study). Manchester: University of Manchester. http://results.ref.ac.uk/DownloadFile/ImpactCaseStudy/pdf?caseStudyId=28160

Welsh Government. (2011). *Household projections across the United Kingdom, technical report*. Cardiff: Welsh Government.

Wilson, T., & Rees, P. (2005). Recent developments in population projections methodology: A review. *Population, Space Place, 11*(5), 337–360.

Chapter 17
Demographic Forecasting for Local Governments in Queensland, Australia – Difficult, But Effective

Kanan M. Saraiya

Abstract In Australia, demographic projections are available at national, state and local level through Census, Bureau of Statistics and variety of institutes, however, demographic projections for smaller areas of local government/County/Shire/Council/Municipality are often developed by the Local government authority.

In Queensland, Australia, basic demographic information on dwellings and population can easily be extracted from various Council resources. The availability of databases on Property tax rates and property development status as well as technological advances like Queensland Globe, Open-Data, Google Street-view, regular aerial view updates and the like have made validating demographic forecasts feasible. It has become an effective desktop exercise to model land use across the local government area. Broader statistics provide information on trends in dwellings and population at local government level. Physical constraints are mapped and shared by various government agencies. Land uses are guided by local governments via planning schemes and development codes. Through a "attractor and detractor" analysis, projections for next 15 years are provided in quinquennial cycles. Considering all of the above inputs, it is possible to project what is achievable for a given Council area down to the lot level. Basic forecasts on land use, dwellings & population can be developed efficiently for local governments in Queensland because of data availability and consistent structure and functioning of local governments. Forecasts can be regularly updated depending on the size of local government, although the task is relatively less complex and often more accurate for smaller Councils. This chapter describes forecast methods based on government agency data, methods that have been regularly applied over a number of years by various Councils.

Keywords Open-data • Land use • Attractor and detractor analysis • Demographic forecast

K.M. Saraiya (✉)
Scenic Rim Regional Council, Scenic Rim, QLD 4285, Australia
e-mail: Kanan.saraiya@gmail.com

D.A. Swanson (ed.), *The Frontiers of Applied Demography*, Applied Demography Series 9, DOI 10.1007/978-3-319-43329-5_17

17.1 Importance

Local government in Australia is the third tier of administration following, respectively the state and federal government levels. In Queensland, the state government defines the powers of local governments mainly through the Local Government Act 2009 (2015a) and other acts (e.g. city of Brisbane Act) whereby local government undertakes various functions including waste collection, public recreation facilities, infrastructure provision and town planning. It becomes very important for local government to accurately portray their demographic projections to achieve desirable long term objectives including sustainability, growth and prosperity, as well as justifying funding via loans and grants from the state and federal government.

Under the Queensland Sustainable Planning Act (SPA) 2009 (2015), it is a legislative requirement for the growing local governments (also referred to as Councils) to create and update demographic projections every 5 years. Councils that are unable to do so risk having their right to impose and collect infrastructure charges attributed for the development (the charges collected will only account for partial infrastructure funding but details of which would be covered elsewhere). This requirement fosters the financial sustainability and governance pressure necessary to regularly develop and revise demographic forecasts. Thus, demographic projections become a crucial element in strategic planning for local governments and the delivery of required services.

For the purpose of this chapter, "Demographic Projection" is defined as a projection of the number and characteristics of people such as, their age, sex, marital status, type of dwelling, and so forth. For a local government it is important to know these characteristics so that it can plan the future. Accordingly, a "Population projection" is distinguished from a "Demographic projection" in that the former only refers to the number of people while the latter involves both numbers of people and their characteristics. Continuing, local government demographic projections are theoretically classified as small area projections and they serve as base data for planning (Isserman 1984). These projections support a growth plan, which is used to influence decision making on infrastructure, funding, and grants. They are also used when undertaking land use planning, analysing development effects and preparing infrastructure plans. They also can be used as base data to undertake financial sustainability checks on local governments.

It is important to note that a projection is not a prediction but merely the result of entering hypothetical assumptions into a mechanistic quantitative procedure (Smith et al. 2013: 1–4). A forecast represents a best guess about the future, achieved by adding judgment about the most likely future rates of behaviour and other assumptions. Part of the judgment required for a forecast includes decision making about the quality of input data and what type of analytical model provides the most realistic results. Finally, a *plan* requires evaluation of the forecasted future for its level of desirability and potential alterability. Plans can be constructed to avoid undesirable futures, to make desired forecasts come true, or to create new, more desirable futures. This may seem difficult, but the good news is that demographic forecasts can be developed efficiently for local governments in Queensland, Australia. They can be derived accurately with the use of open-data and local government's in-house dataset using a Geographical Information System (GIS).

17.2 History

There has been strong emphasis on demographic projections in Queensland since 1997. The Integrated Planning Act (1997) (now superseded by SPA 2009) had a requirement of demographic projections for next 5 years which can benchmark development sequence. In addition, there is the IPA Infrastructure Guideline 1/04 (2004) by the state government, which has a requirement of demographic projections for the next 15 years, with planning projection cohorts at every 3 to 5 year interval. This guideline was revised in 2008, 2009, 2011 and 2014 along with the legislative changes in the Act. However, the same emphasis on the requirements of demographic projections remained. Over the years, policy for projections has been refined so that requirements can be consistently applied across different local government areas. The current requirement is to have population & dwelling projections at the lot level with designated type, scale and timing of development for the 15 years and the ultimate capacity at lot level for the life of the planning scheme. These assumptions are required to be reviewed every 5 years and need to be updated in response to the current trends. Finally, starting in 1824, the Government Statistician's office reports on demographic statistics, which has evolved over the years along with the technological advances. Currently, it publishes broad demographics at the local government level and update it every couple of years.

17.3 Data Revolution

National, state and local government projections are available through Census and Australian Bureau of Statistics; however, demographic projections for smaller areas of local government/County/Shire/Council/Municipality are developed by the local government authority.

Technological advances have created revolution in data availability and consistency, such as:

- #Open-Data (a platform for all data produced by various departments of federal, state and local government);
- Queensland Globe (a free plug-in to view a wide range of data spatially at a lot level);
- AURIN* (Australian Government initiative providing modelling & visualisation) – small area population projections

#Open-Data is defined as the data that can be freely used, re-used and redistributed by anyone – subject only, at most, to the requirement to attribute and "sharealike," as per Open Knowledge Foundation.[1] To implement Open-Data methodology, Australian government has created online source www.data.gov.au which provides an easy way to find, access and reuse public datasets from Government. It publishes data from three tiers of government – local, state and federal across all departments. The data.gov.au provides a central catalogue to discover public data. It also provides hosting for tabular, spatial and relational data with hosted APIs and the option for agencies to link data and services hosted by other government sources. Improving the quantity and quality of the government data will be an ongoing process. In addition to open datasets, the data.gov.au catalogue includes unpublished data and data available for purchase.

Queensland Globe is an interactive online tool as an extension to Google Earth™ application. Queensland Globe is used to view and explore Queensland spatial data and imagery (including up-to-date satellite images). The Queensland Globe is part of the Queensland Government's open-data initiative, where data.qld.gov.au is an online source of all Queensland Government's data at state and local level. A cadastral "SmartMap" can be downloaded or a current titles search can be purchased and downloaded. Information on Queensland Globe is updated instantaneously. For example, a new lot is created and registered with the state government; this change in cadastre is simultaneously (overnight) uploaded on Queensland Globe, so allows confirmation to customers/users in real time.

AURIN stands for Australian Urban Research Infrastructure Network, it is a national collaboration delivering platform to empower better decisions for Australia's urban settlements and their future development. Funded by the Australian Government, AURIN initiative is building the e-research infrastructure to enable better understanding of the current state of Australia's cities and towns and to meet the challenges they face. Led by The University of Melbourne, AURIN

---

[1] "Open Knowledge" is a worldwide non-profit network of people passionate about openness, using advocacy, technology and training to unlock information and enable people to work with it to create and share knowledge.

Open knowledge is what open-data becomes when it is useful, usable and used.

The key features of openness are:

- **Availability and access**: the data must be available as a whole and at no more than a reasonable reproduction cost, preferably by downloading over the internet. The data must also be available in a convenient and modifiable form.
- **Reuse and redistribution**: the data must be provided under terms that permit reuse and redistribution including the intermixing with other datasets. The data must be machine-readable.
- **Universal participation**: everyone must be able to use, reuse and redistribute — there should be no discrimination against fields of endeavour or against persons or groups. For example, 'non-commercial' restrictions that would prevent 'commercial' use, or restrictions of use for certain purposes (e.g. only in education), are not allowed.

collaborates with more than 60 institutions and data providers across Australia through contracted subprojects and Data Access Agreements. The AURIN[2] Workbench and its flagship application, the AURIN Portal, is delivering access to diverse data from multiple sources, and is facilitating data integration and data interrogation using open source e-research tools. This generates meaningful knowledge – urban intelligence – the evidence base for informed decisions for the smart growth and the sustainable development of Australia's cities and towns.

Other local government datasets used in this methodology include:

- Council's rates, bin and property development database (for existing demographics);
- Planning Scheme which provides lot level spatial information on possible land uses; and
- Street view, regular aerial view updates and information search (made validating demographics feasible).

Local governments in Queensland capture and retain a variety of data, information, documents, records and databases for their effective functioning as well as the provision of community services. The "Rates" database is heavily used in this exercise to determine land tenure, land use codes, emergency services category and improvements. All these characteristics are used to classify current land-use, which helps in modelling the number of dwellings, thereby setting a solid base upon which to forecast. Rates database gets a monthly update for land classification through digital cadastre database. Notices are sent to residents quarterly via the "Rates" system, which helps to capture any change of use and improvements as well as serve as a mechanism for customer feedback.

Another data source is Council's "bin" database; it keeps track of all the waste collection bins issued as new or as replacement. This database indirectly checks new lot creation. The "Property development" database is another active depository that holds all the information in regard to sub-dividing land over number of years; it

---

[2] AURIN:

- Brings together and streamlines access to more than 1,000 datasets previously difficult, time consuming or costly to obtain
- Provides the online capability to combine data at various levels of scale from multiple sources
- Delivers online access to open source e-research tools to interrogate, model and visualise data.
- Analytical capabilities include statistical and spatial modelling, planning and simulation tools, graphing and mapping routines (including 2D and 3D visualisation).
- Provides the network to facilitate collaboration, partnerships and knowledge sharing across academia, all levels of government, and the private sector.

Where needed, AURIN provides merit-based securitised access protocols to interrogate unit record data, maintaining privacy and confidentiality.

The AURIN Workbench is the collective term for the applications and services that the AURIN project has delivered. Most data and services are delivered through the 'Portal', a flagship application, but other tools and capability are made available through other websites; in many cases, these sites are open access and do not require the AAF login that the Portal does.

has links between land sub-division and change of use. This information is very critical in validating the "rates" database and helps in projecting growth for near future. "Development assessment" is one of the core functions of local governments and with a number of sophisticated interconnected systems, growth and future development can be easily monitored and reported.

The "planning scheme" is another enriched database which reflects strategic and statutory features of local governments. Through the planning scheme, an entire region can be modelled to achieve desired growth and shape of the region. Information on density, zones, precincts, constraints and future potential is specified in the planning scheme. This methodology is dependent on the planning scheme parameters to forecast future growth and exact location of the growth. Given the heavy dependency on parameters, projections are required to be updated with the planning scheme amendments, so that they remain relevant and can be implemented effectively.

Aerial photography and "streetview" provide a desktop experience of site visits and make validating data quicker and efficient. A number of providers are available who can under various licensing options (usually based on frequency of aerial photography updates) can update "streetview" images regularly. Google also updates its streetview images regularly. Local governments invest in regular aerial photography updates at a required resolution to a maintain the compliance function, development assessment, infrastructure planning, vegetation and environmental management, and asset management, to name a few. Given the manifold uses across Councils, these regularly updated images are readily available. Hence using them to validate existing land use patterns makes it very desirable. This dataset is useful in validating existing demographics, which in return, ensures that future demographics are based on solid information.

17.4 The Australian Statistical Geography Standard

In Australia, there is a defined hierarchical geographical classification called ASGS (the Australian Statistical Geography Standard). It is created by the Australian Bureau of Statistics, which is used in the collection and dissemination of official statistics. The ASGS provides a common framework of statistical geography and thereby enables the production of statistics which are comparable and can be spatially integrated. The ASGS provides a more comprehensive, flexible and consistent way of defining Australia's statistical geography (2011). Figure 17.1 explains the hierarchical categories where a lower category is a subset of a higher order category.

These hierarchical categories are important to create, analyse and use statistical data in Australia; they are the basics of the statistical analysis. In the remainder of this chapter, these statistical hierarchies will be used to explain projection methodology.

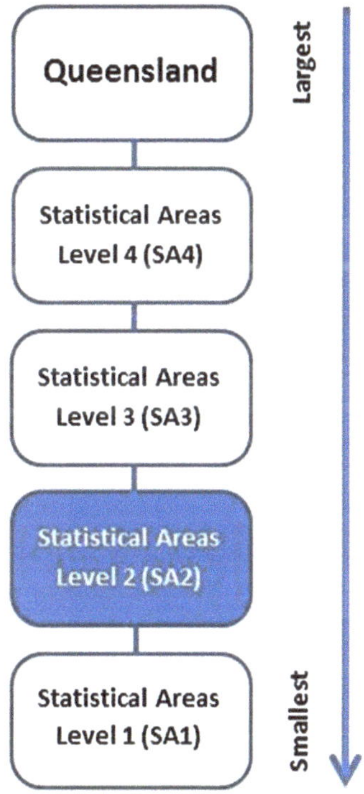

Fig. 17.1 Australian statistical hierarchy

In summary, data revolution by ease of access to a number of datasets due to an open-data approach has made projecting demographics feasible and less time consuming. Various datasets are available for cross-reference and validation. ***Hence, data revolution has made modelling of demographic forecast across the local government area an effective and quick process.***

17.5 Methodology

The author developed a demographic forecast for the Scenic Rim Regional Council in Queensland, Australia that serves as a working example of the process. The same methodology can be applied in principle to any local government in Queensland with similar datasets. Depending upon the variability of datasets, local calibration is advised if one wished to apply the example to other local governments in Queensland. Thus, the concepts and framework can be utilized for any local government demographic forecast with appropriate modification based on the availability/lack of dataset or parameters around how Councils are structured and functionally operated.

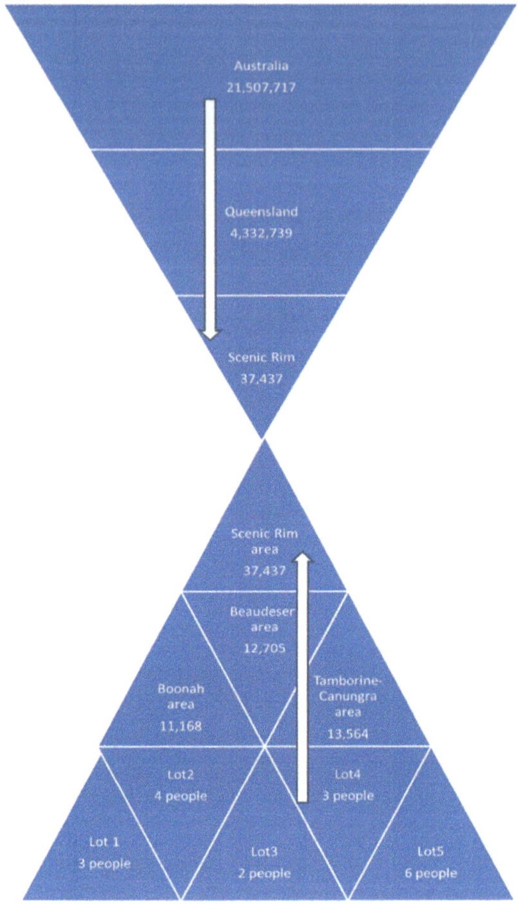

Fig. 17.2 Top-down bottom-up approach for population projection of Scenic Rim regional council

17.5.1 Overview

The demographic forecast is prepared using a *top-down bottom-up* approach. The top-down approach involves the forward projection of historical growth data to estimate future growth. The bottom-up approach involves limiting growth projections to the physical capacity available to accommodate growth in a locality. That is, development at a local level is projected to occur for each projection year until it reaches the adopted population and employment capacity (ultimate development) for a locality. This process can be briefly explained via infographic in Fig. 17.2.

This methodology can be easily replicated for variety of forecast including dwellings & population, non-residential gross floor area & jobs and household goods and consumers. Different assumptions and ratios apply for each type of projection. The methodology is indirectly used by number of planners over the years but not formally recognised. Brian Lister Planning[3] has used similar methodology for couple of local governments in Queensland. But author could not locate formalised publication on this methodology. As well as data revolution has made all these data openly available, which was not the case few years ago. This might be the reason for not publishing this methodology up till now. Below is the detailed method for demographic projection of dwellings and population for a local government.

17.5.2 The Top-Down Approach

Population is projected from National to State to Local Government (LG) level through Census. In Australia, Census occurs every 5 years which provides a snapshot of time. Intermediate studies and surveys are conducted by the Australian Bureau of Statistics (ABS) to validate the assumptions used in forecasting future demographics. Data from Census gives overall control totals to target the possible maximum growth for small area projections that is local government area. It also indicates trend in dwellings, population and employment at the local government level. LG population is available from Australian Bureau of Statistics (2009) for last 25 years which is revised 12 months later and rebased and final estimates after the following Census. This process provides flexibility to accommodate new technology, any statistical boundary changes and fix methodology issues identified over the period. This trend is accurate representation of history and is used for forecasting future demographics.

As shown in Fig. 17.2, *Top-down bottom-up approach for population projection of Scenic Rim Regional Council*, population of Australia as per Census in 2011 is 21,507,717, which filters down to Queensland (one of the state) population to 4,332,739 and further projects down to Scenic Rim Regional Council (one of the local government) population to 37,437. The local government population of 37,437 becomes control total for bottom-up distribution of existing population.

The Queensland Government Statistician's Office (QGSO) undertakes population projections for every Census year that is in five yearly cohorts. QGSO is the state department for statistics and refines ABS data for small area including local government area, suburbs and designated project areas. The work of QGSO is governed by a key piece of legislation that has a direct bearing on its statistical activities and functions — the *Statistical Returns Act 1896*. This Act facilitates the collection of official statistics by QGSO (2015).

[3] Brian Lister Planning approach

Brian Lister as a consultant worked with number of companies and local governments to develop a small area growth projection model for predicting population and employment at the lot level. This methodology has been automated by developing model suitable for individual local government. Over the years, methodology has evolved by resolving the shortcomings known through implementation of the model.

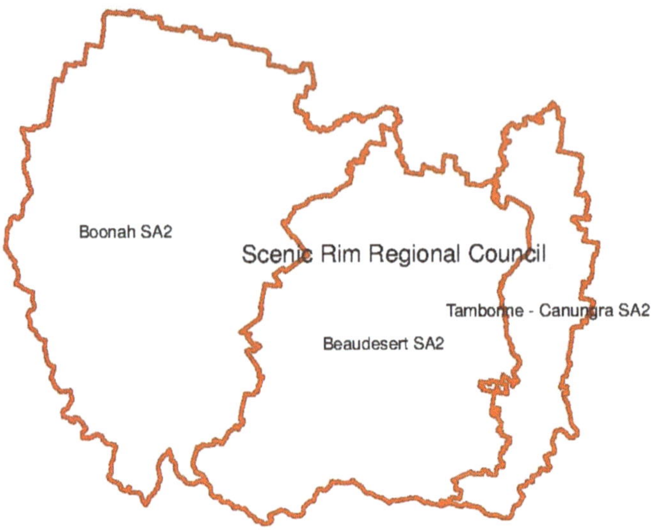

Fig. 17.3 Three SA2 forms local government area of Scenic Rim

QGSO forecast and publishes population projections for local government area and statistical areas 2 (SA2) level. For Scenic Rim Regional Council, local government boundary consists of three SA2, which makes analysis easier compared to where local government boundary does not match SA2 boundaries. The three SA2 are Beaudesert, Boonah and Tamborine-Canungra. Figure 17.3 shows their alignment to local government area.

The QGSO population forecast for local government areas include low, medium and high growth scenarios at 5 yearly intervals to align with Census years (e.g. 2011, 2016, 2021). The Information in Table 17.1 is taken from the QGSO publication, *Projected population, by local government area, Queensland, 2011 to 2036, 2013 edition.*

QGSO also forecast population for SA2 level for medium growth scenario at Census years. Information in Table 17.2 is sourced from QGSO publication on *Projected population (medium series), by statistical area level 2 (SA2), Queensland, 2011 to 2036, 2013 edition.*

The population projections from QGSO are used as control totals. "Scenic Rim" has the Bromelton State Development Area (SDA) declared for industrial growth along the freight rail corridor. The Bromelton SDA plays an important role in triggering growth for the region. Based on the current scenario and trends in industrial land demand, it is envisaged that Bromelton SDA will not commence until 2021. So, by applying the vital economic local factor, i.e. Bromelton SDA, to the population projections, the control totals are refined. Immediate future numbers are projected on low growth series till 2021. The medium series projections for

Table 17.1 Population projected for local government area

Period	Low series: persons	Medium series: persons	High series: persons
2011	37,437	37,437	37,437
2016	40,964	41,412	41,759
2021	45,704	47,407	48,999
2026	53,625	57,720	61,816
2031	62,003	68,917	75,989
2036	71,476	81,985	92,871

Table 17.2 Population projected for SA2 and local government area on medium series growth

SA2	2011	2016	2021	2026	2031	2036
Tamborine – Canungra	13,564	14,143	14,335	15,321	16,347	16,586
Boonah	11,168	12,483	13,918	15,317	16,704	18,067
Beaudesert	12,705	14,786	19,155	27,081	35,866	47,332
Scenic Rim	37,437	41,412	47,408	57,719	68,917	81,985

2031 are pushed out to 2036 and extrapolated accordingly from 2021 to 2036. The extrapolation considers a uniform growth rate from 2021 to 2036. This Growth rate is also consistent with the average growth rate over a decade for Scenic Rim, based on the estimated resident population (ERP) data from 1991 to 2014 published in Regional Population Growth publication (ABS 2015). Table 17.3 shows the modifications to control totals by applying the Bromelton SDA economic factor.

There are over-arching control totals at Local Government Area (LGA) level and at Statistical Area 2 (SA2) level. Based on the estimated resident population (ERP) for SA2, control totals are projected till 2036 to meet LGA totals. With the bottom-up approach, designing the growth at lot level, these control totals are adjusted at SA2 level to meet the LGA totals projected till 2041. I now turn to this approach.

17.5.3 The Bottom-Up Approach

To start, note that physical constraints such as flooding, bushfire hazard, slope, conservation areas, are spatially available and shared by various government agencies via #open-data. Land uses and future growth potential are guided by local government via planning schemes and development codes. Applying physical constraints to the lot excludes the undevelopable area. The remaining developable area is assigned respective land use information to generate ultimate development. Ultimate development is distributed throughout the region using attractors and detractors analysis.

Forecasts include the intensity and timing of development. This forecast is assigned to a 5 yearly cohort for the next 15–20 years. Grouped forecast should

Table 17.3 Population projected for local government area along with local factor

Period	Low series: persons total (Persons)	Medium series: persons total (Persons)	High series: persons total (Persons)	Bromelton SDA local factor
2011	37,437	37,437	37,437	37,437
2016	40,964	41,412	41,759	40,964
2021	45,704	47,407	48,999	45,704
2026	53,625	57,720	61,816	51,963
2031	62,003	68,917	75,989	59,656
2036	71,476	81,985	92,871	68,917

match the control totals from the top-down approach. Hence, being a two-way check of the forecast process.

17.5.3.1 Digital Cadastral Database

To create bottom-up population projections, dwelling projections and occupancy rates are used. In Queensland, once a month, Councils received updated Digital Cadastral Database (DCDB) from Department of Natural Resources and Mines (DNRM) (2015). The growth forecasting exercise for Scenic Rim Regional Council commenced in December 2014, hence version of DCDB for December 2014 is locked for this exercise; all other data will be surrounding this period. Data cleansing is required to remove easements and road segments from the DNRM database. There are slivers and splits found in the data, but as it is the core data from DNRM, local government is not in a position to amend it. "Slivers and splits" are created due to overlapping or gap between survey boundaries and is relatively minor compared to total geographical area. Slivers and splits can also be found when lot adjoins to road, river, dam, etc. Tenure is one of the important features of the cadastre; it shows the holding of the land. Tenure coding is used to sort, segregate and analyse the data. Exhibit 17.1 refers to the DCDB Tenure Codes and their description as published by DNRM.

This exercise is executed in GIS ESRI and MS Excel programs. Below I describe the manual edits that are undertaken to DCDB to refine the data, making it suitable for the demographic projections.

17.5.3.2 Manual Edits to the DCDB

1. all the lots with tenure covenants (CV), lands lease (LL), "profit a prendre" (PP) and railway (RY) are deleted; as LL & PP tenures are duplication over base lot while CV & RY are restricted and are for infrastructure purpose only;
2. developable area for the lots with tenure national park (NP), forest reserve (FR) and state forest (SF) is manually made zero; and

Department of Natural Resources and Mines
Specification – Queensland Interchange Format (QIF)
PBD/801/206 Issue: 3.31 Page 34 of 38

7.4 DCDB Tenure Codes

Code	Description
CA	COMMONWEALTH ACQUISITION - Land acquired by the Commonwealth of Australia and held prior to issue of a formal title. (Generally this land is used for military or government store purposes).
CI	CARBON ABATEMENT INTEREST - A registered interest in land for a term, generally of at least 100 years, to facilitate the sequestration of carbon stored in a carbon abatement product (for example the living biomass, or dead organic matter, or soil)
CV	COVENANT - A registered right or interest over a parcel of land used to restrict usage of that land. (Only those covenants that are shown on a survey plan depicting their extent and described in the 'Plan of' box are recorded in the DCDB).
EA	EASEMENT - A right or interest on a property that is registered against the title.
FD	BELOW THE DEPTH PLANS – A registered right or interest over a parcel of land whose location is defined as below a depth or to a depth below the surface of the earth (Underground Coal mines below Ipswich).
FE	FORESTRY ENTITLEMENT AREA - Land in a deed of grant or freeholding lease where the commercial timber and the land on which it stands is reserved by the State of Queensland
FH	FREEHOLD - Land held in Fee Simple (freehold title) which includes titles surrendered to the State of Queensland (or Crown) in terms of Section 358 of the Land Act 1994.
FR	FOREST RESERVE - Tenure of an interim nature for a maximum of 5 years and managed by the Department of National Parks, Recreation, Sport and Racing with associated conditions.
HM	BOAT HARBOURS - Land vested under the control of the Department of Transport and Main Roads (Maritime Division).
HL	HOUSING LAND - Land vested in the Department of Housing and Public Works.
ID	INDUSTRIAL ESTATES - Land vested under the control of the Department of State Development, Infrastructure and Planning for the development of State Government industrial estates.
LL	LANDS LEASE - Leasehold land administered by the Department of Natural Resources and Mines excluding Mining Homestead Tenement Leases.
MP	MARINE PARK - Land vested under the control of Marine Park Authorities.
MR	MAIN ROAD - Base Parcel – Land vested in the Department of Transportt and Main Roads prior to issue of title or road dedication. Strata Parcel – Declared Common Area under Transport Infrastructure Act (Main Road over Rail Corridor)
MT	MINES TENURE - Land leased as Mining Homestead Tenement Leases (eg. MHPL, MHL and SPMPL) originally issued by the Department of Energy. These leases are now administered by the Department of Natural Resources and Mines. This category does not include Mining Lease or Mining Application Areas, the exception being ML7024 at Weipa, being land set aside under the "Commonwealth Aluminium Corporation Pty Ltd Agreement Act of 1957" and SBML No.8 under the "Alcan Queensland Pty Limited Agreement Act of 1965" GG1965.2.1441.
NP	NATIONAL PARK - Land reserved by Department of National Parks, Recreation, Sport and Racing for a National Park and Regional Park. (Nature Conservation and Other Legislation Amendment Act [No. 2] 2013)
PH	PORT AND HARBOUR BOARDS - Land vested under the control of Port Authorities.
PP	PROFIT Á PRENDRE - A registered right or interest of use over the property of another that allows the holders to enter and take some natural produce (mineral deposits, timber).
RE	RESERVE - Land reserved by Department of Natural Resources and Mines for community or public purposes.
RY	RAILWAY - Land vested for railway purposes in Queensland Rail.
SF	STATE FOREST - Land reserved by the Department of National Parks, Recreation, Sport and Racing

Exhibit 17.1 DCDB tenure codes

Department of Natural Resources and Mines
Specification – Queensland Interchange Format (QIF)
PBD/801/206 Issue: 3.31 Page 35 of 38

	for State Forest purposes.
SL	STATE LAND - Land held by the State of Queensland as Unallocated State Land and other areas vested in the State (or Crown) but not held in Fee Simple or as a lease issued under the Lands Act 1994.
TP	TRANSFERRED PROPERTY - Land transferred to the Commonwealth of Australia on federation, usually for lighthouse or post and telegraph reserves.
TR	TIMBER RESERVE - Land reserved by Department of National Parks, Recreation, Sport and Racing for Timber Reserve purposes.
WR	WATER RESOURCE - Land vested in Sunwater (originally Water Resources Commission).

Exhibit 17.1 (continued)

3. The developable area is manually made zero for all Building Unit Plans (BUP) and Group Title Plans (GTP) lots, as they are the common areas required for individual development.

After these edits makes the DCDB ready for next step, which is the removal of constraints.

17.5.3.3 Constraints

Constraint analysis is undertaken for the draft Scenic Rim planning scheme in June 2014. The same information is used to build up the constraint layer for demographic projections as in December 2014 this constraint information is still relevant and the latest.

The following list provides the constraints used in determining developable areas:

1. Mining Development Licence
2. Mining Lease
3. Matters of State Environmental Significance (MSES) – regulated vegetation
4. MSES – protected areas
5. Declared catchment area (Dam)
6. Military base Beaudesert
7. Matters of National Environmental Significance (MNES) – wetlands
8. MNES – world heritage
9. Ipswich difficult topography (slope greater than 25%)
10. Beaudesert Landslide Hazard (slope greater than 25%)
11. Combined flood layer (Queensland Reconstruction Authority and Council studies)
12. Key Resource Area (KRA) resource process area
13. KRA separation area
14. KRA haul route
15. KRA transport route
16. KRA transport route separation

17. MSES – Wild Rivers
18. MSES – High ecological value waters
19. MSES – Wetlands
20. MSES – Wildlife habitat
21. Queensland Heritage Register
22. Commonwealth Heritage Register
23. State Development Area
24. Bushfire hazard area (medium to very high)
25. Ipswich slope 15%–25%
26. Beaudesert slope 15%–25% or high/med landslide hazard
27. Local Heritage Register of Scenic Rim (version 29 July 2014)

These constraints are removed in order to better inform a realistic yield. Certain modifications are done to slope, landslide and bushfire hazard constraints. For slope constraint, slope less than 20% can be mitigated through engineering solution and land can be developed. Therefore only Slope 20% or greater is removed from the projections as a constraint to development. Similarly through building design and construction solutions, medium landslide constraint and medium bushfire hazard can be mitigated. Therefore very high and high bushfire hazard and high landslide hazard is removed from the projections as a constraint to development. These variations are easily created in GIS environment.

On cadastre, all constraints are applied and developable area is calculated by removing area constraints from the shape area of the lot. The shape area of the lot can be different than the lot area which is registered on the title. This can be due to number of reasons, including but not limited to lot not being surveyed since decades, splits and slivers. This exercise is undertaken across the "Scenic Rim+. Output is the Developable Area (DA) for every lot. Figure 17.4 represents aerial photography showing existing site conditions and also portraits the Constraint layer on top showing flooding in the 100 year rainfall event. It is the worst scenario, so maximum constrained land is removed. This also helps in making informative decision on planning and constructing infrastructure by avoiding constrained land.

After removing constraints and calculating developable area, next step is to apply land development types and land uses feasible for every lot.

17.5.3.4 The Relationship Between Development Categories, Development Types and Planning Scheme Land Uses

The demographic forecast is prepared for a limited number of development types. Uses under the planning scheme, which are guided by Queensland Planning Provisions (QPP) draft version 4 (December 2014), are grouped into broader types of development that adequately reflect differences in infrastructure demand for various infrastructure networks. This creates consistency in land uses across Queensland for different local government areas. The standardisation of small components makes this methodology applicable across the state.

Fig. 17.4 Aerial photograph & constrained area

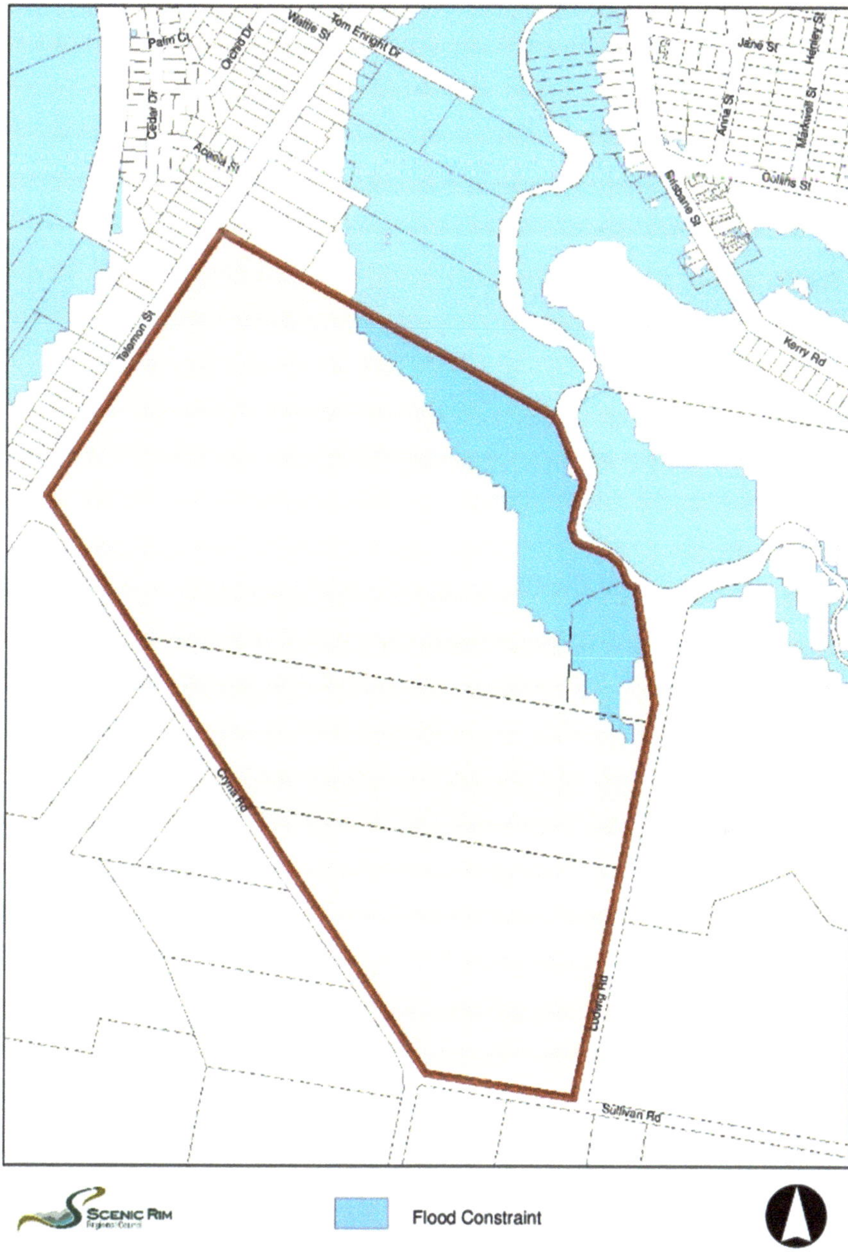

Fig. 17.4 (continued)

Development categories are mainly Residential and Non-residential. These categories are further distributed as per statutory guidelines (2014a, b) to sub-categories such as:

- detached dwellings;
- attached dwellings;
- other dwellings (tourist accommodation)
- retail;
- commercial;
- industrial; and
- community purposes.

The category, "Rural development type" is added to cater for the regional characteristics of the Scenic Rim. Extra category can be added because of flexibility in statutory guideline to satisfy individual geographical conditions. Table 17.4 shows the relationship between development categories, development types and planning scheme land uses.

17.5.3.5 Existing Land Use Classification & Existing Dwelling Forecast

Based on the categories in shown Table 17.4, all lots are classified using their existing land use. Three individual datasets are used to determine existing land use as follows:

1. Primary Use Land Use Codes Description – This dataset is available from Department of Natural Resources and Mines (DNRM), where each lot is classified in a category as per their primary use. Exhibit 17.2 represents the Land Use Codes from DNRM which are used in determining rates on each land parcel.
2. Emergency services land use codes – This dataset is collectively produced by Emergency services (fire & rescue and ambulance services) and local government in order to categorise emergency services levy.
3. Improvements values from Council's Rates Database – This dataset is an ongoing data where information is created from building applications which is again Council's development approval system.

All three datasets are compared with one another to check for alignment. Where three categories are inconsistent, validation is done via ground truth exercise, using Council's development assessment database, Nearmap (privately owned providing latest high resolution imaginary to Council) and Google street view. Nearmap is a service purchased by Council for its numerous functions where properties can be inspected from desktop. Through this exercise, it is evident that rates database is nearly accurate. Also, there is an opportunity to provide feedback to Rates section on errors and make Council's dataset precise. The Output is the number of existing dwellings, its type – attached or detached, occupied by resident or tourist. Non-residential categories are not analysed for this exercise, as concentration is on dwellings and population projections.

Table 17.4 Relationship between development categories, development types and uses

Development category	Development type	Planning scheme land uses
Residential development	Attached dwelling	Caretaker's accommodation
		Community residence
		Dual occupancy
		Dwelling unit
		Home based business
		Multiple dwelling
		Nature-based tourism
		Non-resident workforce accommodation
		Relocatable home park
		Resort complex
		Retirement facility
		Rooming accommodation
		Rural workers' accommodation
		Short-term accommodation
		Tourist park
	Detached dwelling	Dwelling house
		Sales office
Non-residential development	Commercial	Garden centre
		Hardware and trade supplies
		Outdoor sales
		Showroom
	Community purpose	Cemetery
		Club
		Community care centre
		Community use
		Crematorium
		Detention facility
		Emergency services
		Funeral parlour
		Hospital
		Outstation
		Place of worship
		Residential care facility
	Industry	Brothel
		Bulk landscape supplies
		Extractive industry
		Low impact industry
		High impact industry Medium impact industry
		Research and technology industry
		Special industry
		Transport Depot
		Warehouse
	Retail	Adult store
		Bar

(continued)

Table 17.4 (continued)

Development category	Development type	Planning scheme land uses
		Car wash
		Child care centre
		Educational establishment
		Food and drink outlet
		Function facility
		Health care services
		Hotel
		Indoor sport and recreation
		Major sport, recreation and entertainment facility
		Market
		Motor sport facility
		Nightclub entertainment facility
		Office
		Outdoor sport and recreation
		Parking station
		Service industry
		Service station
		Shop
		Shopping centre
		Theatre
		Tourist attraction
		Veterinary services
	Rural	Agricultural supplies store
		Animal husbandry
		Animal keeping
		Aquaculture
		Cropping
		Intensive animal industry
		Intensive horticulture
		Permanent plantation
		Roadside stall
		Rural industry
		Wholesale nursery
		Winery
	Other	Air services
		Environment facility
		Landing
		Major electricity infrastructure
		Park
		Renewable energy facility
		Substation
		Telecommunication facility
		Utility installation

Queensland Government
Natural Resources and Mines

Land Use Codes **ATTACHMENT A**

(This is a four-character code - Characters 1-2 represent the primary use and 3-4 the secondary land use flag.)

Primary Codes.

Urban Land use	**Transport & Storage**	**Sheep Grazing**	**General**
01 Vacant Urban Land	28 W'House & Bulk Stores	60 Sheep Grz- dry	96 Public hospital
Residential	29 Transport terminal	61 Sheep breeding	97 Welfare
02 Single unit dwelling	30 Service station	62 **Vacant**	home/institution
03 Multi unit dwelling	31 Oil depot & refinery	63 **Vacant**	98 **Vacant**
(Flats)	32 Wharves	**Cattle Grazing**	99 Community
04 Large Home site-vac	33 Builders yard,	64 Breeding	Protection Centre
05 Large Home site –Dwg	contractors	65 Breeding & Fattening	
06 Outbuildings	34 Cold stores- ice works	66 Fattening	
07 Guest house/private	**Industrial**	67 Goats	
hotel	35 General industry	**Dairy Cattle**	
08 Building units	36 Light industry	68 Milk- Quota	
Primary use only	37 Noxious / offensive	69 Milk- No quota	
09 Group Title	industry (include Abattoir)	70 Cream	
Primary use only	38 Advertising- Hoarding	**Agricultural**	
Retail Business/ Comm	39 Harbour Industries	71 Oil seed	
10 Comb. Multi Dwg &	40 Extractive	72 **Section 25 Valn**	
shops	**Special Uses**	73 Grains	
11 Shop single	41 Child care ex k/garten	74 Turf Farms	
12 Shops- group	42 Hotel/tavern	75 Sugar cane	
(More than 6 shops)	43 Motel	76 Tobacco	
13 Shopping group	44 Nurseries (Plants)	77 Cotton	
(2 to 6 shops)	45 Theatres cinemas	78 Rice	
14 Shops- main retail	46 Drive-in Theatre	79 Orchards	
(Central Business Dist)	47 Licensed club	80 Tropical fruits	
15 Shops- Second retail	48 Sports clubs/ facilities	81 Pineapples	
(Fringe central	49 Caravan parks	82 Vineyards	
business presence of	50 Other clubs (Non	83 Small Crops & fodder	
service ind)	business)	Irrig	
16 Drive in shopping	**Special Uses**	84 Small Crops & fodder	
centre	51 Religious	Non Irrig	
17 Restaurant	52 Cemeteries	**Other Rural Uses**	
18 Special tourist	(Include Crematoria)	85 Pigs	
attraction	53 **Vacant**	86 Horses	
19 Walkway	54 **Vacant**	87 Poultry	
20 Marina	55 Library	88 Forestry & Logs	
21 Residential Institution	56 S/Gnd, R/course,	89 Animal Special	
(Non-medical care)	Airfield	90 Stratum	
22 Car parks	57 Parks, gardens	91 Transformers	
23 Retail Warehouse	58 Educational include	92 Defence Force Estab	
24 Sales area outdoors	K/gtn	93 Peanuts	
(Dealers, boats,	59 **Vacant**	94 Vacant rural land	
cars,etc)		(Excl 01 & 04)	
25 Professional offices		95 Reservoir, dam,	
26 Funeral parlours		bores	
27 Hospitals, conv, homes			***February 2000***
(Medical care)(Private)			
1 Vacant urban			

Secondary Land Use Flag

Note: A Secondary Land Use Flag is Mandatory. Where no Secondary Land Use Flag is applicable adopt 00.

Primary Land Use Code	Secondary Land Use Flag
all	Flag substantive secondary use by using the primary code relevant for the secondary use where none of the following apply.
all	Land used by the Commonwealth Government use secondary land use code 53
all	Land used by the State Government use secondary land use code 54
all	Land used by the a local government use secondary land use code 59
all	Land with higher potential redevelopment use secondary land use code 72. (Except where code 98 Applies)
02, 03, 05, 08, 09, 44, 60-71, 73-88, 93.	Land valued under section 17 apply secondary land use flag of 98

Maintenance of Valuation Land Use Codes. Val-Not-ValOver-151 v3 Page 4

Exhibit 17.2 DNRM land use codes

17.5.3.6 Tourist Accommodation

Tourist accommodation is determined using the process just described. It involves comparisons of Primary Land Use Description, Emergency service category and Improvements values from rates database. Due to the limited number of tourist accommodation dwellings in the region, a separate exercise is undertaken to research number of accommodation units/rooms and to know caretaker's residential status. It is found that at majority of places, caretakers are residing on the property. This exercise gave accurate numbers of dwellings and resident population in them. Tourist accommodation dwellings are categorised as other dwellings and has different resident population rate compared to town dwellings. The tourist population is ignored for this exercise as the demand generation due to tourism is not significant compared to the total population of Scenic Rim. As well as complexities of vacancy rates and market conditions are required to be addressed to get tourist population number. Hence to avoid further complications, tourist population is ignored.

17.5.3.7 Ultimate Dwellings Forecast

Planned Density

Existing dwellings, types of dwellings and their occupancy are determined through existing dwelling forecast process. To project the ultimate dwellings, the assumed type and scale of development for a particular location is determined by applying a planned density to the developable area of the site. A planned density reflects the realistic level of development (ultimate development) that can be achieved for the premises. Considerations as per the Statutory Guideline 03/14 (2014a) for this include:

- the South East Queensland Regional Plan 2009–2031 framework for infrastructure planning;
- the strategic framework within the local government's planning scheme;
- zoning and development provisions within the planning scheme;
- other planning instruments such as State Development Area development schemes;
- approved plans for development; and
- current development trends in the area (or similar areas).

The planned densities, used to prepare the demographic projections, are clearly identified in terms of dwellings per developable hectare for residential development. While defining planned densities for each precinct/zone of the planning scheme, a broad assumption of 30% land removal for infrastructure purposes is made. This assumption particularly applies to the residential component of the precinct/zones, which includes township, low density residential and low-medium density residential. The removal of 30% of land area is for road, water, sewer, parks & stormwater infrastructure that is provided above or under the ground.

Applying planned density to the Developable Area (which recall is derived by removing constraints from the lot area), gives the ultimate dwellings that can be accommodated on the lot. Removing the existing dwellings from the ultimate dwellings gives maximum future growth in dwellings that can be accommodated on a particular lot. This growth forecast is the ideal situation, hence overwrites are applied to make forecast realistic and achievable.

Overwrites

A number of overwrites are applied to the ultimate dwellings on an individual lot basis, per the following:

1. All lots having value of ultimate dwellings lesser than value of existing dwellings are overwritten to equate to value of existing dwellings. This can be possible because development has nullified the constraints through engineering solution and there are existing dwellings.
2. All vacant lots having value of ultimate dwellings lesser than one are overwritten to zero to avoid development in constrained area.
3. Value of ultimate dwellings is round down to last integer, which reduced the total dwellings by 1100 dwellings.
4. For lots with existing development approval, value of ultimate dwellings is replaced with the approved number of lots on the plan. This overwrite is to consider the market conditions.
5. For lots with existing community use, for example – Cemeteries, Show Ground, Hangers, Reservoirs, Water & Sewer Treatment Plants, etc, value of ultimate dwellings is overwritten to zero.

This exercise has filtered data for any unrealistic growth and helps local government to adequately supply infrastructure. Output is the refined ultimate dwellings for each lot which can be achieved over the life of planning scheme (generally 25 years). Further this growth needs to be distributed for each Census cohort (every 5 years). Urban footprint and Priority Infrastructure Area serves as a guide in modelling appropriate distribution.

Urban Footprint and Priority Infrastructure Area

Both the "Urban Footprint" and the "Priority Infrastructure Area" (PIA) are used to determine the timing of the growth & development. Below is the significance of these geographical boundaries and how they are used to determine timing of the development. South East Queensland Regional Plan 2009–2031 (2014) governs the urban development across the region through Urban Footprint. It is the area marked for urban growth and for urban purposes around each town. Certain development activities are prohibited in urban footprint. It also preserves the rural area for agriculture, cattle breeding and rural living. Majority of the towns will have 90% growth inside the urban footprint.

Priority Infrastructure Area is a subset of urban footprint. The PIA is an area used, or approved for use, for non-rural purposes; and serviced, or intended to be serviced, with development infrastructure networks; and that will accommodate at least 10 (but no more than 15) years of growth for non-rural purposes as defined in Statutory Guideline 03/14 (2014b). It is very important to determine PIA accurately as it influences infrastructure requirements, timing of growth and financial sustainability of a local government. For this exercise, in addition to above criteria for PIA, following additional criteria are considered.

1. availability of existing water and sewer infrastructure network (as shown in Table 17.5); and
2. the local government must be able to fund and supply adequate trunk infrastructure to service the assumed urban development inside the PIA.

Considering the preceding criteria, the following urban centres are identified to accommodate growth for 10–15 years:

1. Beaudesert;
2. Boonah;
3. Canungra;
4. Kalbar; and
5. Kooralbyn.

The future timing of the assumed type and scale of development for a particular location is based on the population projections for that location. This involves making an assumption concerning the timing of development in a particular location; for example, inside PIA development occurs by 15 years while outside

Table 17.5 Infrastructure availability used to finalise priority infrastructure area

		Availability of infrastructure	
Town	2011 population	Water	Sewer
Aratula	500	Y	Y
Beaudesert/Gleneagle	6603	Y	Y
Boonah	2474	Y	Y
Canungra	746	Y	Y
Harrisville	430	Y	N
Kalbar	725	Y	Y
Kooralbyn	1372	Y	Y
Peak crossing	400	Y	N
Rathdowney	200	Y	N
Tamborine Mt	6813	N	N
Tamborine village	3500	N	N
Warrill view	0	Y	N

PIA development occurs after 15 years and before ultimate capacity of the planning scheme. Further attractors and detractors analysis is used to determine which lot gets developed first compared to its neighbouring lots. As the name suggest, it is analysis of various attractors – access to shopping centre, public transport, existing basic infrastructure for water, sewer, parks, etc. and detractors – transformer station, cemetery, sewer treatment plant, major highway, etc. All attractors and detractors are given a rating such as attractors are positive and detractors are negative. The lot with the maximum rating gets developed before the lot with lesser rating. This analysis can be quite intensive in larger local government areas, which requires further assumptions about individual sub-zones.

Outside urban footprint, rural subdivision occurs generally for the rural purposes and hence not much growth anticipated. For this exercise, priority in given to lots inside PIA, where development is serviced by basic infrastructure, and hence growth is financially sustainable for local government. In real world, there will be overlap and particular lots will not follow consistently and may develop earlier or later than predicted under market pressure. Hence it is a legislative requirement in Queensland to update demographic projections at least every 5 years. High growth local governments, who have developed GIS model, can update their demographics at every couple of years.

Population projections align with future ABS census years (e.g. 2016, 2021). It is therefore appropriate to prepare the planning assumptions for the base date i.e December 2014 and for each future ABS census date till 2031 (next 15 years) as well as the ultimate capacity of the planning scheme. Figure 17.5 provides an example of a development that shows the growth in number of dwellings spread across cohorts.

17.5.3.8 Occupancy Rates

The definition of an "Occupancy rate" is the number of households in a dwelling. They are calculated considering occupied and vacant private dwellings, where private dwellings include structural dwellings (e.g. houses, flats, townhouses) and temporary dwellings (e.g. tents, caravans). Non-private dwellings are excluded (e.g. hotels, hospitals, boarding schools, mining camps) for the calculation of occupancy rates.

The Queensland Government Statistician's Office (2013) has derived the occupancy rates by dividing the total population (including those in non-private dwellings) with the number of private dwellings (occupied and vacant). This rate is re-based on the Census information every 5 years. In Table 17.6, Occupancy rate for Scenic Rim Local Government Area, the 2011 figure is an estimate based on estimated resident population (ERP) from 2011 Census. 2011 ERPs for the local government area have been derived using published Statistical Area Level 2 and local government area ERP data. The projected occupancy rate at local government level is an average of occupancy rates for SA2 geographical area and different type of dwelling.

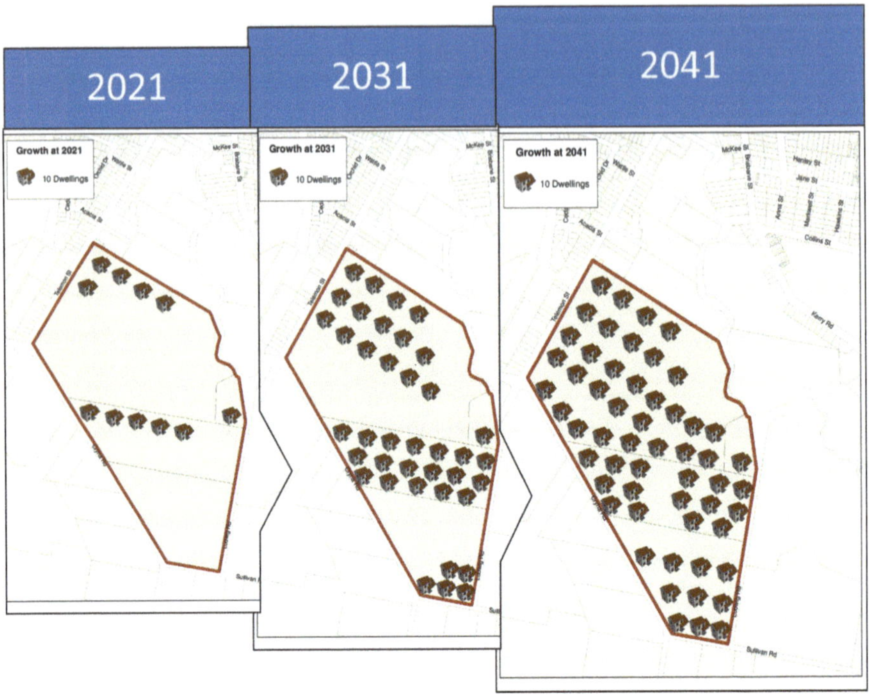

Fig. 17.5 Growth pattern across decades

Table 17.6 Occupancy rate for Scenic Rim local government area

Local government area	2011	2016	2021	2026	2031	2036	2041
Scenic Rim (R)	2.42	2.40	2.40	2.40	2.40	2.40	2.39

Occupancy rates for SA2 are derived for each individual geographical location and type of dwelling. This information is available in Census data from the household size and the number of households. Over the years, trend is portrayed from Census data where lifestyle choices influence the household size. Considering these variations, Table 17.7 represents various occupancy rates for year 2011 as used in this exercise. Tourist accommodation is standardised to account for care-taker population.

Similarly, based on trends and recent lifestyle options, occupancy rates for each 5 yearly cohort are projected for individual SA2's and type of dwelling.

Table 17.7 Year 2011 occupancy rates at SA2 level

	Type of dwelling		
Location/	Detached dwelling	Attached dwelling	Tourist accommodation
Beaudesert	2.56	1.32	1
Boonah	2.37	1.24	1
Tamborine – Canungra	2.45	1.28	1

17.5.3.9 The Ultimate and Existing Population Forecast

Forecasting population is relatively simple though highly dependent on dwelling forecast. Population is derived by multiplying number of dwellings with occupancy rates for individual SA2. These data is available at lot level, so it can be aggregated to any geographical boundary, for example, development project boundary, suburb, SA1 or local government area.

Population is generated using the occupancy rates previously described. The output is the existing population grouped for three SA2 making up local government area. The ame logic applies for future population projections at 5 yearly cohorts. These population totals are compared with the control totals from the top-down methodology, where population is forecasted at SA2 and local government area level. Where existing population from bottom-up methodology does not meet control totals from top-down methodology, occupancy rates are adjusted to match the numbers. As existing population i.e. for year 2011 is derived from latest Census and verified by Australian Bureau of Statistics with number of independent datasets, it is nearly accurate.

The Forecasted population for the nearest future cohort i.e. year 2016, for this exercise is required to be analysed thoroughly. So in order to update information from year 2011 to year 2015, migration & birth and death datasets are used. Migration considered for this exercise is of 3 types: International, Interstate and Intercity. International migration data is available from the arrival/departure information, which is published on quarterly basis by ABS. Interstate and intercity information is available from Medicare change of address dataset. Department of Health provides Medicare services to all residents in Australia. ABS publishes the interstate migration data every quarter. This dataset/model is reviewed and updated every 5 years using data from the latest Census information. Medicare information can be very useful in verifying this forecast. Similarly birth and death registers are used to verify the recent changes in population. Queensland Government Statistician's Office publishes estimated resident population at SA2 and local government area level intermittently using above mentioned datasets and ABS catalogue 3218.0, Regional Population Growth, Australia. The existing and forecasted population numbers are revised with Census information every 5 years and hence local governments undertake the population modelling exercise every 5 years to inform their city planning.

17.6 The "Consistent" Approach

Basic projections on land use, dwelling & population are easier to develop for local governments in Queensland, due to the data availability and consistent structure and functioning of local governments. Local governments in Queensland are governed under Local Government Act 2009 (LGA) (2015 b) which provides a system for local government. Hence all the Councils are required to function uniformly. Local government planning is governed by SPA, which requires Council to plan for sustainable growth, and provide infrastructure to facilitate the growth. Under these legislative requirements, Council need to develop & maintain various datasets to perform its functions. Rates database is developed and maintained under LGA with uniform land use categories generally standardised by State government. Similarly land use zoning and dwelling and population projections are standardised by SPA. Moreover, under this methodology there is substantial use of federal and state government information which is available for every local government area. Hence it becomes relatively easy to apply this methodology uniformly across the State. Localised or geographically factors will influence the outcome but the overall methodology remains the same.

Demographic projections can be regularly updated pending on the development activity in the local government area. It is relatively less complex and often more accurate for smaller Councils. In Fig. 17.6, a comparison pie chart shows the size

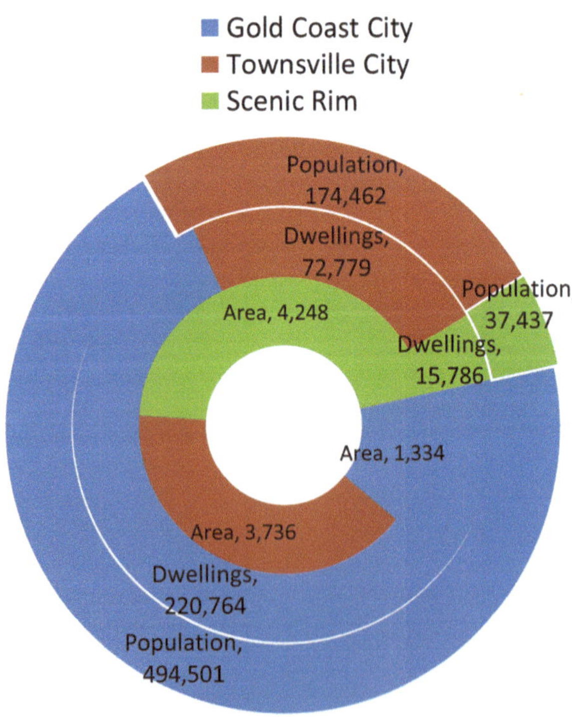

Fig. 17.6 Comparison of different local government using same methodology

and scale of various Councils where the author has consistently applied the methodology. Gold Coast City Council is densely populated and hence much more complicated and there are data integrity issues. It becomes harder to track growth and update projections regularly because of the quantum of development activity. While Scenic Rim Regional Council is sparsely populated, hence easily manageable data, which can be validated by ground truth exercise. Growth can also be monitored and projections can be updated on annual basis. Similarly Townsville City Council is intermediate in comparison Councils, relatively bigger compared to Scenic Rim but smaller than Gold Coast, so growth can still be monitored but may not be easy to validate data via ground-truth exercise. This shows that applying same methodology demographic projections can be developed with relative differences in outcome due to individual characteristics.

17.7 Application

Demographic projections can be used in numerous ways. Scenic Rim applies the methodology to model demographics of population and dwellings. These demographics are used to generate infrastructure demand on roads, transport, stormwater, parks and community facilities network at 5 yearly cohort intervals. It is further utilised to scope, design and deliver the infrastructure in timely manner by identifying construction dates for new infrastructure that is necessary to service development. For example, provision of park is required when certain population target is achieved for an individual small area catchment or with the appropriate planning land acquisition for new park can be undertaken in advance reducing the cost of park provision.

Demographic projections become crucial for local governments who have restricted access to finance and are liable for providing adequate infrastructure to the community. Under the SPA, Councils are required to prepare and adopt Local Government Infrastructure Plan (LGIP). LGIP is an infrastructure plan derived to accommodate projected growth for next 10 to 15 years in planning assumptions about population and employment growth; and the type, scale, location and timing of development. Together with the Desired Standard of Services they provide a logical and consistent basis for infrastructure planning. These demographic projections clearly identify a summary of the existing and future projected urban residential and non-residential development by development type for a small projection area in terms of:

- dwellings;
- population;
- non-residential gross floor area (GFA); and
- employment.

This information can be further utilised to consider land use supply and demand analysis. If inadequate availability of a particular land use, for example Industry then local government can rezone areas to provide enough options to the market and can drive to promote a particular industry in the area. In Scenic Rim, aged care industry has high potential to grow because of the characteristics of the town and still accessible to major medical and retail facilities. Council can rezone the parcels to provide ample of opportunities to market to develop aged care in the vicinity of existing infrastructure.

Demographic projections are modelled at a lot level which facilitates their use in the planning of each infrastructure network. That is, the demographic projections are allocated to the various service catchments of the infrastructure networks. For example, a small section of new foot path is to be constructed. These projections can be used to do the user analysis and the population it can serve in the catchment. It can also reflect on the land uses that will have benefit of the new footpath, by reflecting that it connects school to the new residential development or light industry to the retail precinct. Overall it means a more informative decision can be taken to provide right type of infrastructure at right location and timing. Further demographics can empower Councils to attract funding via state and federal grants, loan and public private partnerships.

The Queensland Department of Education uses population projections to undertake forward planning for new schools or expansion of existing ones. This projection enables state department to service the population growth and satisfy the catchment requirement to provide education to all students. It is a requirement in Queensland that catchment areas (2015) should ensure every Queensland student from Prep to Year 12 is able to enrol at their local state school. This rule makes population projections a very basic important requirement for state schools to function effectively. Schools are assessed for their capacity annually and compared with projections so that expansion or new schools are implemented before the shortage occurs.

The Queensland Department of Transport and "Main Roads Queensland" use the population and employment projections to plan the major roads, upgrades and analyse the shortfalls before congestion issues are created. Through the projections, the vehicle trips are forecasted and hence impact on the road network. In Queensland, local government and various state departments work together and co-ordinate the infrastructure planning at local and state level. Numerous applications of this methodology can be implemented by understanding the data correctly and acknowledging the assumptions used during modelling.

In summary, above stated applications reflect the importance of demographic forecasting for various levels of government. It is difficult to undertake the demographic forecasting as too many variables applicable for each locality, but modelling has become comparatively effective and quick. The principle behind demographic forecasting remains the same over number of years. However, it became efficient and sharp due to data revolution. Moreover, it became feasible to have multiple applications from a dataset which can be individually tailored to meet various requirements. Data revolution has made difficult demographic forecasting an effective approach.

Acknowledgements The author is grateful to a number of people for comments and suggestions, including David A. Swanson and Ashish G Shah.

References

AURIN – About AURIN, Australian Urban Research Intelligence Network, Victoria, Australia.

Australian Bureau of Statistics. (2009). 3228.0.55.001 – Population estimates: Concepts, sources and methods. Australian Bureau of Statistics, Australia.

Australian Bureau of Statistics. (2015). 3218.0 – Regional population growth, Australia, 2013–14. Australian Bureau of Statistics, Australia.

Australian Government. (2015). Population estimates and projections – Australian infrastructure audit background paper. Infrastructure Australia, Australian Government.

Australian Government data.gov.au – About, Background, Department of the Prime Minister and Cabinet ACT Australia.

Department of Education. (2015) – School catchment areas, Queensland State Government, Queensland, Australia.

Integrated Planning Act 1997. (1997). Chapter 5 Miscellaneous section 5.1.7 Fixing a charge for an item not included in plan, Queensland State Government, Queensland, Australia.

IPA Infrastructure Guideline 1/04. (2004). Chapter 4 Preparing the priority infrastructure plan Section 4.2.5 Mapping the PIA, Queensland state government, Queensland, Australia.

Isserman, A. (1984). Projection, forecast, and plan: On the future of population forecasting. *Journal of the American Planning Association, 50*(2), 208–221.

Local Government Act 2009. (2015a). Chapter 1 Preliminary Section 3 Purpose of this act, Queensland State Government, Queensland, Australia

Local Government Act 2009. (2015b). Chapter 4 Finances and accountability Section 96 Regulations for rates and charges, Queensland State Government, Queensland, Australia.

Open Knowledge Foundation, trading as open knowledge, a not-for-profit organisation, Cambridge, UK.

Queensland Globe, data.qld.gov.au – Department of Natural Resources and Mines, Queensland, Australia.

Queensland Government Business and industry portal. (2015). Digital Cadastral Data, Queensland State Government, Queensland, Australia.

Queensland Government household and dwelling projections. (2013) edition, Queensland Government Statistician's Office, Queensland Treasury, Queensland State Government, Queensland, Australia.

Queensland Government Statistician's Office. (2011). Queensland statistical areas, level 2 (SA2), Queensland Treasury, Queensland State Government, Queensland, Australia.

Queensland Government Statistician's Office. (2015). Overview of QGSO, Queensland Treasury, Queensland State Government, Queensland, Australia.

Queensland Planning Provisions version 4. (2014, December). Suite of use definitions, Queensland State Government, Queensland, Australia.

Smith, S., Tayman, J., & Swanson, D. (2013). *A practitioner's guide to state and local population projections*. Dordrecht: Springer.

South East Queensland Regional Plan 2009–2031. (2014). Division 1 Preliminary Section 1.5 When these State planning regulatory provisions do not apply, Queensland state government, Queensland, Australia.

SPA Statutory Guideline 03/14. (2014a). Local government infrastructure plans. Section 2.5.2 Planning assumptions & Section 2.5.2a Preparing the development projections, Queensland State Government, Queensland, Australia.

SPA Statutory Guideline 03/14. (2014b). Local government infrastructure plans. (Section 2.5.3 Priority Infrastructure Area, Queensland State Government, Queensland, Australia.

Sustainable Planning Act 2009. (2015). Chapter 3 Local planning instruments Section 94A Requirement to review LGIP, Queensland state government, Queensland, Australia

Chapter 18
Population Projections by Ethnicity: Challenges and Solutions for the United Kingdom

Philip H. Rees, Pia Wohland, Paul Norman, Nikolas Lomax, and Stephen D. Clark

Abstract This chapter describes the context, model, estimates and assumptions for projections of ethnic group populations in England at local authority scale, and in Wales, Scotland and Northern Ireland. A bi-regional cohort-component model is used; estimates of the component rates for ethnic groups are developed; assumptions are aligned to recent official projections with one exception. For international migration we assume higher immigration, emigration and net balances of 254 thousand in the long term compared with the most recent official assumption of 185 thousand. The projections show that the UK population grows significantly, from a population of 59.1 million in 2001 to 84.5 million in 2061. Black, Asian and other Minority ethnic groups expand their share of the UK population from 8 to 30 % in that period. This increasing diversity is greatest in the UK's largest cities but ethnic minority groups grow fastest outside those cities. We show through a comparison of 2001 based and 2011 based projections that there is considerable uncertainty both nationally and locally in future diversity although the direction of travel to a more diverse future is certain.

Keywords Population projections • Ethnicity • Local projections • Component estimates

---

P.H. Rees (✉) • P. Norman • N. Lomax • S.D. Clark
School of Geography, University of Leeds, Leeds LS2 9JT, UK
e-mail: p.h.rees@leeds.ac.uk; p.d.norman@leeds.ac.uk; n.m.lomax@leeds.ac.uk; s.d.clark@leeds.ac.uk

P. Wohland
Hull York Medical School, University of Hull, Hull HU6 7RX, UK
e-mail: Pia.Wohland@hyms.ac.uk

D.A. Swanson (ed.), *The Frontiers of Applied Demography*, Applied Demography Series 9, DOI 10.1007/978-3-319-43329-5_18

18.1 Introduction

Many developed countries are experiencing low or negative growth of their native born populations. They have completed the first demographic transition to low fertility and mortality and now experience the below-replacement fertility characteristic of the second demographic transition. The demand for labour is not met by the supply of native born young adults. In countries growing economically, high immigration of foreign origin populations occurs to meet labour demands. Family building in young immigrant groups occurs at first in the major centres of immigration and then more widely as socio-economic advancement and internal migration produce a diffusion of ethnic minorities. Bill Frey, in his book *The Diversity Explosion*, has documented these processes for the United States, using census data and projections (Frey 2015). Rees et al. (2011, 2012) have explored the population future of the UK through developing population projections by ethnicity at sub-national scale, which show the UK following the US in this transformation, 25 years in lag.

The extent to which such information is routinely produced varies widely between countries. At one extreme is France where enquiries into a person's race or ethnicity are forbidden under the constitution (Sabbagh 2008). At the other extreme the recording of race was embedded in the constitution of the United States in order to use counts of the "Negro" slave population in electoral apportionment. The US Bureau of the Census produces projections of the national and state populations by age, sex, four races, Hispanic origin and nativity (Colby and Ortman 2015). Statistics New Zealand (2015a) reports on the 2013 national ethnic population projections for four ethnic groups. The New Zealand census allows multi-ticking of ethnic categories. As a result projected group populations overlap. Statistics New Zealand also produces sub-national projections by ethnicity for territorial areas (Statistics New Zealand 2015b).

The aim of this chapter is to explain the challenges posed by extending conventional projections to produce forecasts of ethnic populations. A model is described for UK local authorities which finds solutions to the challenges of adding the ethnicity dimension to conventional projection models.

Why project ethnic populations? The fundamental reason is that it is important to understand the future mix of populations with different national and cultural backgrounds. If we do that then it is easier to work towards a harmonious society. Ethnic populations are groups with a common national origin, distinguished by birthplace, citizenship, migration status, race or language. Many groups face difficulties in entering or progressing in labour and housing markets and in education. Statistics on ethnic populations are needed for monitoring disadvantage and understanding discrimination. Reliance has been placed on census statistics for this background information but this quickly becomes dated. Population projections provide population numbers for short and medium term planning. Ethnic population projections also play a part in health studies, proving context population variables for conditions where ethnicity may be a factor affecting risk.

Table 18.1 Definitions of the harmonized ethnic groups used for projection

Abbreviation	Ethnic group description	Broad grouping	Figure 18.1 groups
WBI	White: British, Irish, Gypsy, Irish Traveller	White	WBR + WIR
WHO	White: Other White	White	WHO
MIX	Mixed or Multiple Ethnic Groups	Mixed	WBA + WBC + WAS + OMI
IND	Asian or Asian British: Indian	Asian	IND
PAK	Asian or Asian British: Pakistani	Asian	PAK
BAN	Asian or Asian British: Bangladeshi	Asian	BAN
CHI	Asian or Asian British: Chinese	Asian	CHI
OAS	Asian or Asian British: Other Asian	Asian	OAS
BLA	Black or Black British: African	Black	BLA
BLC	Black or Black British: Black Caribbean	Black	BLC
OBL	Black or Black British: Other	Black	OBL
OTH	Other Ethnic Group	Other	OAT

Notes: Minority groups = WHO to OTH. *BAME* Black and Asian Minority groups = MIX to OTH

In the UK ethnicity statistics depend on a self-identification question. Respondents to censuses or surveys are presented with lists of ethnic group titles and asked to tick one group. Prior to the UK censuses of 1991, 2001 and 2011, the UK National Statistical Agencies consulted with potential users about the ethnic question. As a result of consultation and because the UK population was becoming more ethnically diverse, the ethnic group classification changed over time and across the national territory. Nine groups were identified in the 1991 Censuses of England, Wales and Scotland; in 2011 18 groups were distinguished in England and Wales, 16 were used in Scotland and 11 were employed in Northern Ireland.

Table 18.1 presents the 12 groups used in the projections reported in this chapter, which have been harmonized across the last two censuses and the four home countries. For many uses, the 12 can be grouped into the five broader categories of White, Mixed, Asian or Asian British, Black or Black British and Other. The census classifications are also employed in official household surveys, in health studies, in administrative records and the National Pupil Census. However, they have not been used in the recording of births, deaths, internal or international migration. The methods for estimating demographic rates by ethnicity are described in a later section.

The outline of the chapter is as follows. The next section, Sect. 18.2, presents a checklist of design decisions that must be made when building a projection model for sub-national populations classified by ethnicity, and identifies the options used in the 2011 based projection described in this chapter. Section 18.3 lays out the sequence of equations implemented in our ethnic population projection model.

Section 18.4 summarises the methods used to estimate the ethnic components of change, presenting some new results on the ethnicity of international migration. Section 18.5 sets out the assumptions used in two rounds of projections, 2001 based and 2011 based. Section 18.6 compares the results. Significant changes in the ethnic composition of sub-national populations in Britain are forecast in both projections. Section 18.7 summarises and reflects on the findings. In the chapter we report on one projection, TRENDEF, taken from the 2001 based ETHPOP set and on one projection, LEEDS, prepared as part of the 2011 based NewETHPOP set.

18.2 Building Projections for Ethnic Groups

18.2.1 General Considerations

The model specifications for the 2011-based population projections by ethnicity are listed in Table 18.2.

The design of the projections was informed not only by recent practice and literature but also through face to face consultations with the UK National Statistical Agencies, the Welsh Government, a set of local authority districts (LADs) and research centres. Prior to formulating the projection assumptions a 2001–2011 time series of LAD components of change by ethnicity was estimated and constrained to the equivalent ONS all-person LAD components. An account of this reconciliation procedure is provided in Rees et al. (2017). The population and component estimates are heavily dependent on ethnic information from the UK censuses of 2001 and 2011. In the ETHPOP set of projections, reliance was placed on population estimates and components based on just the 2001 Census (Rees et al. 2011, 2012).

18.2.2 The Ethnic Classification

In general, ethnic groups are closed populations which do not exchange members. However, there are two exceptions: first, new-born infants may have a different ethnicity from their parents and second, all people have an opportunity to choose their group at the end of each time interval in the projection.

18.2.3 Geographical Coverage and Zones

The model covers the whole UK population, integrating data published by ONS, NRS (National Records of Scotland) and NISRA (Northern Ireland Statistics and Research Agency). In the 2011 based projections described in this chapter, the

Table 18.2 The 2011 based LEEDS model specifications

Features	Description
GENERAL AND DATA CONSIDERATIONS	
Purpose	To revise projections of ethnic group populations for local areas
Consultations	4 National Statistics Agencies, 8 LADs, 5 Other organizations
External constraints	Not yet implemented
Special populations	Armed Forces, Prisoners, not yet implemented
Evaluation/review/ reconciliation	Population and component estimates have been evaluated for 2001–2011 against ONS total populations and components and reconciled between "Census Based Book Ends"
Available data	2001 and 2011 Census population data, ONS mid-year population data, 2001 to 2011 component data, reconciled between mid-year 2001 and mid-year 2011
Making estimates	Ethnic component flows and rates must be indirectly estimated
GROUPS, REGIONS, GENDERS, AGES, TIMES	
Population groups	Projections in this chapter: 12 harmonized ethnic groups
Group transitions	Groups are independent, with two exceptions. New-born infants may be assigned an ethnicity different from their mothers. At the end of each time interval group members may transfer to another group
Geography: coverage	United Kingdom
Geography: regions	Projections in this chapter: 324 LADs in England plus Wales, Scotland and Northern Ireland (327 zones)
Gender detail	Males, Females. Female dominant fertility model
Age-detail	Single years of age: 0, . . ., 99, 100+
Age-time plan	Period-cohorts: Birth to age 0, age 0 to age 1, . . ., age 99 to age 100, ages 100+ to ages 101+
Time interval	One year, mid-year to mid-year (30 June/1 July)
Time horizon	Medium term, mid-year 2011 to mid-year 2061. Long term mid-year 2061 to mid-year 2111
HANDLING MIGRATION	
System representation	Bi-regional: 327 pairs of LADs & rest of the UK, + rest of world
Migration concept	Movement migration, derived from 2001 to 2014 NHS registers with 2001 and 2011 Census data used in estimation
Demographic accounts	Movement accounts and components of change
Population at risk	Average of start and final populations; computed iteratively
Internal migration model	Occurrence-exposure rates × Populations at risk (average in interval)
International migration model	Assumptions about emigration flows and immigration flows
PROJECTION CONSIDERATIONS	
Formulation of assumptions	Trajectories (short-term & long-term) of leading indicators. Constant distribution across LADs, ethnic groups, sexes and ages
Uncertainty	Deterministic projections plus scenarios
Projection outputs in online database	Populations by LAD, ethnicity, gender and single year of age; component totals

Notes: 1. *LADs* Local Authority Districts (lowest tier). 2. *ONS* Office for National Statistics (UK) geographical scale is Local Authority Districts (LADs) in England together with Wales, Scotland and Northern Ireland as aggregate areas.

18.2.4 The Number of Population Groups

The number of groups for which projections are reported in this chapter is 3924 (327 × 12). To project such a large number of population groups is challenging. However, there are good reasons for modelling populations in this detail. Projecting the populations of 327 LADs simultaneously ensures consistency and enables comparison of the results of one LAD with outcomes for the UK, home country, region or neighbouring LAD. When small numbers are converted into rates and applied to populations, the model provides an opportunity for small groups to become larger in future.

18.2.5 Genders, Ages and Time Intervals

Ethnic group populations and components of change are disaggregated by gender, with the exception that only women are at risk of giving birth. Single years of age are used from 0 to 99, with a final age of 100 and over. This matches the 1 year time interval employed in the model. All component information is organised for projection in period-cohort format, starting with the new born to age 0, followed by age 0 to age 1, age 1 to age 2 and so on to age 99 to age 100, with a final period-cohort being age 100+ to age 101+. There are therefore 101 population ages and 102 period-cohorts. Although estimates of the very oldest populations and period-cohorts are subject to some uncertainty, this age range is needed in anticipation of further population ageing in future.

18.2.6 Projection Horizons

The projections are designed to run for up to 100 years, because this is the full time span over which the current population could survive and the effects of current ethnic-age structures will be worked out. Local users are interested in the short term (the next 25 years) for planning purposes; national users are also interested in the medium term (up to 50 years ahead) for social security computations; infrastructure planners adopt a long term horizon out to 100 years. In Sect. 18.5, we report on results in the medium term, 40 and 50 years on from the jump off year of 2011.

18.2.7 How Migration Is Treated in the Projection Model

Sub-national projections include internal and international migration components but there many methods for handling these components. Table 18.2 lists five features associated with the way migration is handled.

System Representation refers to the way the regions in the projections are managed. The first option is to represent each region as a single isolated unit incorporating total migration as a net term (the uni-regional approach). The second option is to represent regions as full interacting sets within a country and to model internal migration as the product of an origin population multiplied by the rate of out-migration (the multi-regional approach). The third option is to represent regions as a set of pairs, the region itself and the rest of the country (the bi-regional approach). The fourth option is to represent regions in two layers in which the upper tier is modelled as a multi-regional system and the lower tier uses a simpler model with lower tier results constrained to upper level projections (the hierarchical approach). The NewETHPOP projection reported in this chapter uses the bi-regional approach.

The *migration concept* refers to the way migration is measured in the data. Two concepts can be employed: either the transition concept, which records a person's shift in region of residence between two fixed time points, or the movement concept, which records all changes of residence between fixed time points.

Every projection model is based on a set of *demographic accounts*. These ensure that in each time interval population inputs equal population outputs and that the final population is consistent with the start population and the components of change. Projection models can be built using either migration concept given suitable data. However, in practice it is difficult to handle the projection of births and deaths as transitions unless the time interval between censuses and the time interval of migration are coincident. Most national statistical agencies, including ONS, therefore use the movement concept (Raymer et al. 2015).

The final feature of a projection model is the *Population at risk*, which must match the demographic account type. For projections using the movement concept, a population-time exposure measure is needed. This is approximated as the average of start and final populations in a time interval. This poses the problem that the final population is unknown at the start of any time interval. The usual approach is to re-work the model equations to re-define the rates to eliminate the need for a final population (Rogers and Ledent 1976). The problem with this approach is that every time the model is changed, new equations for the rates are needed. In the NewETHPOP model we instead compute the population at risk iteratively. This method has the advantage of the population at risk equations remain the same irrespective of projection model.

18.2.8 The Modelling of Internal Migration

Internal Migration Model options are as follows. Migrations between regions can be projected either as (1) rates multiplied by populations at risk in the origin region or as (2) exogenous flows, which need to be projected using an independent model (e.g. a gravity model). The first option is used in the NewETHPOP model described here, using constant rates. There is a concern that this model, when applied using time-constant rates, leads to excessive drift towards a stable equilibrium. In reality migration trends lead to feedback effects which change the rates. ONS (2015) has introduced destination population controls to alleviate this drift in the 2014 based National Population Projections.

18.2.9 The Importance of International Migration

International Migration is projected to make a much bigger contribution to future UK population change than natural increase. Rees et al. (2013) computed, for the ETHPOP 2001-based projections, the effects of assumptions about the demographic components using a scheme of scenario projections designed by Bongaarts and Bulatao (1999). This analysis found that international migration (direct and indirect effects) accounted for 87 % of total UK population change over the 50 years from 2001. ONS (2015) computed a reference projection assuming no international migration for comparison with the Principal projection and estimated the direct and indirect contribution of international migration to population growth to be 67 % between 2014 and 2039. These impacts on the UK's future population mean that there are tensions in setting international migration assumptions between what the time series of flows tells us – international migration is substantially higher than recent official assumptions – and what government would like the figures to show – net international migration below 100,000 per annum.

18.2.10 The Modelling of International Migration

Little attention has been paid to the design of the *International migration model* used in population projections. Should future international migration be simply a set of judgements about net international migration flows? Or should they be framed separately as immigration and emigration assumptions? Or should immigration assumptions be set in absolute flow numbers and emigration as rates multiplied by a UK population at risk? Bijak (2012) recommended that ONS formulate assumptions using gross immigration and emigration flows which can be linked directly to determinants rather than using net international migration. Bijak proposed a model which adopted assumptions based on immigration flows and emigration rates, as used in Rees et al. (2011, 2012). ONS (2014) experimented with the Bijak proposal but found the results implausible. The UK Home Office's Migration Advisory Committee, which advises on official immigration policy, commissioned a report on how to best forecast international migration (Disney et al. 2015). The authors conclude that no single model could be recommended for forecasting immigration flows but that decomposition into streams to and from different parts of the world was useful.

18.2.11 Formulation of Assumptions, Uncertainty and Outputs

We took a pragmatic approach to the formulation of *assumptions* (see Sect. 18.5 for details). There is always huge *uncertainty* in any projection of the future population. This can be ascertained through variant projections (implemented by ONS) or through probabilistic projections (Statistics New Zealand 2015a, b). We focussed on the preparation of a deterministic projection coupled with variants (Rees et al. 2015). Each projection delivers a huge quantity of projected outputs, which are delivered via www.ethpop.org.

18.3 The NewETHPOP Model Equations

Table 18.3 sets out the notation used in the projection model. The variables and the equations are listed in Table 18.4. Cross-referencing these tables establishes the meaning of the variables. We use single letters for the demographic variables, lower case letters for rates or probabilities and upper case letters for counts of stocks and flows.

A key feature of the notation for the NewETHPOP model is that people stay in the same period-cohort during a time interval and change period-cohorts by ageing on at the end of the interval. Final populations in a period-cohort at end of one interval become start populations in the next period-cohort and time interval. This way of handling ageing was used by Stone (1971).

Projection computations begin with input of the jump-off populations (step 18.1) as start populations for the first time interval, mid-year 2001 LAD ethnic populations in the ETHPOP projections and mid-year 2011 populations in the NewETHPOP projections. After the first time interval, start populations are the transferred from final populations of the previous interval (step 18.24). Initial values are adopted for the populations at risk for birth, death and internal migration in step 18.2. The iterative loop for computing populations at risk starts here. Once all of the components have been projected a population at risk can be computed as the average of start and final populations in a time interval (step 18.22) and checked

Table 18.3 A notation for an ethnic population projection model

Variable	Description
Stocks	*Counts of people*
P^S	Start Population in a time interval (count)
P^F	Final Population in a time interval (count)
A	Armed Forces population
C	Prisoners
Flows	*Movements from one state to another*
B	Births
D	Deaths
E	Emigrations (international migration from UK to Rest of the World)
M	Migrations (internal to the country)
M^{ij}	Migration from LAD i (origin) to LAD j (destination)
M^{i+}	Total out-migrations from LAD $i = \Sigma_{j \neq i} M^{ij}$
R	Residual (balances)
M^{+i}	Total in-migrations to LAD $i = \Sigma_{j \neq i} M^{ji}$
I	Immigrations (international migration to the UK from the Rest of the World)
Intensities	*Either probabilities or occurrence-exposure rates*
f	Fertility rates (occurrence-exposure rates) for period-ages
d	Death rates (occurrence-exposure rates) for period-cohorts
m	Internal migration (transmission) rates
e^o	Emigration (transmission) rates
e^a	Emigration (admission) rates
i^o	Immigration (transmission) rates
i^a	Immigration (admission) rates
v	Sex proportion at birth
b	Mixing probabilities of the ethnicity of a new-born given the ethnicity of mother
y	Switching probabilities of a new ethnicity a new ethnicity given ethnicity at a prior census
s	Probability of survival
Indexes	*Subscripts or superscripts*
x	Age index (used for period-ages and period-cohorts)
g	Gender (or sex) index
e	Ethnic group
n	Nativity group (birth place)
i	Zone index for zone of interest (origin)
j	Zone index for zone of interest (destination)
z	Zone index for the last zone in the system
u(i)	Zone index for rest of the UK
w(u)	Zone index for rest of world or rest of world region
o	Transmission rate = migration/origin population
a	Admission rate = migration/destination population
t	Stocks: a point in time; Flows: an interval in time from t to t + 1
b	Coefficients in a migration model

Table 18.4 Projection model equations

Description	Variable or equation	Steps
Start populations	P^{S1}_{xgen}	(18.1)
Initial populations at risk	$P^{ARi}_{xgen} = P^{i}_{xgen}$	(18.2)
The fertility and nativity model for births	$B^{Si}_{bgen} = v^{i}_{g} \times \sum_{x=10}^{x-49} f^{i}_{xen} \times P^{ARi}_{xfen}$	(18.3)
Mixing: assigning ethnicity to the new-born	$B^{Si}_{bgfn} = \sum_{e} x^{i}_{ef} \times B^{Si}_{bgen}$	(18.4)
Mortality model: when x <90	$D^{i}_{xgen} = d^{i}_{xgen} \times P^{ARi}_{xgen}$	(18.5)
Mortality model: when x >= 90	$D^{i}_{xgen} = \left(1 - s^{i}_{xgen}\right) \times P^{Si}_{xgen}$	(18.6)
Subtraction of special population stocks (prisoners and armed forces)	$-C^{i+}_{xgen} - A^{i+}_{xgen}$	(18.7)
Emigration option (1) Exogenous projected emigration flows	E^{i}_{xgen}	(18.8)
Emigration option (2) Emigration (transmission) rates × Populations at Risk	$E^{i}_{xgen} = e^{ti}_{xgen} \times P^{ARi}_{xgen}$	(18.9)
Emigration option (3) Emigration (admission) rates × Populations at Risk in the Rest of the World	$E^{i}_{xgen} = e^{ai}_{xgen} \times P^{ARw(u)}_{xgen}$	(18.10)
Internal out-migration option (1) Multi-regional equation with constant or trended transition rates	$M^{ij}_{xgen}(t) = m^{ij}_{xgen} \times P^{ARi}_{xgen}\ (\mathrm{t})$	(18.11)
Internal out-migration option (2) Adjustment of migration flow to destination shares of populations	$\left(\frac{P^{Sj}_{xgen}(t)/\sum_{j} P^{Sj}_{xgen}(t)}{P^{Sj}_{xgen}(ref)/\sum_{j} P^{Sj}_{xgen}(ref)}\right) \times m^{ij}_{xgen}(ref) \times P^{ARi}_{xgen}$	(18.12)
Internal out-migration option (3) Gravity model based on origin, destination and impedance factors	$M^{ij}_{xgen}(t) = b_0 + \sum_{k} b_k X^{i}_{k} + \sum_{l} b_l Y^{j}_{i} + f(c^{ij})$	(18.13)
Total internal out-migrations are the sum of projected migration out-flows	$M^{i+}_{xgen} = \sum_{j \neq i} M^{ij}_{xgen}$	(18.14)
Residual balances	$R^{i}_{xgen} = P^{Si}_{xgen} - M^{i+}_{xgen} - E^{i}_{xgen} - C^{i+}_{xgen} - A^{i+}_{xgen} - D^{i}_{xgen}$	(18.15)
Total internal in-migrations are the sum of projected migration in-flows	$M^{+i}_{xgen} = \sum_{j \neq i} M^{ji}_{xgen}$	(18.16)

(continued)

Table 18.4 (continued)

Description	Variable or equation	Steps
Immigration option (1) Externally generated projected immigration flows	I^i_{xgen}	(18.17)
Immigration option (2) Immigration (transmission) rates × Population at Risk in Rest of the World	$I^i_{xgen} = i^{ti}_{xgen} \times P^{ARw(u)}_{xgen}$	(18.18)
Immigration option (3) Immigration (admission) rates × Population at Risk	$I^i_{xgen} = i^{ai}_{xgen} \times P^{ARi}_{xgen}$	(18.19)
Addition of prisoners and armed forces	$+C^{+i}_{xgen} + A^{+i}_{xgen}$	(18.20)
Final populations	$P^{Fi}_{xgen} = R^i_{xgen} + M^{+i}_{xgen} + I^i_{xgen} + C^{+i}_{xgen} + A^{+i}_{xgen}$	(18.21)
Populations at risk, convergence test	$P^{ARi}_{xgen} = 0.5 \times \left[P^{Si}_{xgen} + P^{Fi}_{xgen}\right]$	(18.22)
Ethnic switching	$P^{Fi}_{xgfn} = \sum_e y^i_{ef} \times P^{Fi}_{xgen}$	(18.23)
Ageing on	$P^{Si}_{x+1gen} = P^{Fi}_{xgen}$	(18.24)

Note: Steps 18.7 and 18.20, are not implemented in the interim projections reported in this Chapter

for convergence. Computations then return to step (18.3) if convergence has not been achieved.

At step (18.3) births by gender, ethnicity and nativity are computed by summing the products of fertility rates and female populations at risk, followed by application of probabilities of different ethnicity to the mother and sex proportions. Births are computed as "start populations" for the new-born period-cohort, so all period-cohorts can be computed together.

Step (18.5) computes the projected number of deaths by multiplying the population at risk by the period-cohort mortality rate. Period-cohort mortality rates are computed from life tables for each LAD-ethnic group population. A slightly different equation may be needed (step 18.6) at the oldest ages when conventional mortality rates may exceed unity because observed deaths and estimated populations come from different data sets in the UK rather than being part of one integrated population register. If this is the case then one minus the survivorship probability, a term which is always positive, can be substituted.

At step (18.7) prisoners (convicted persons) and armed forces populations for the projection year are subtracted from the start populations. New values of these special populations are added back in at step (18.19). The numbers are small, spread over about one third of LADs. In the NewETHPOP projection reported here, we do not include these numbers.

The next three steps in Table 18.6 concern options for projecting emigration. These can be introduced as exogenous flow assumptions (option 1, step 18.8), as exogenous (transmission) rate assumptions multiplied by LAD populations at risk (option 2, step 18.9) or as (admission) rate assumptions multiplied by the population at risk in the Rest of the World (option 3, step 18.10). Option 1 was used for both the 2001-based TRENDEF and 2011-based LEEDS projections. Option 2, the model recommended by Bijak (2012), was used for the 2001 based UPTAPER projections. As the populations of UK LADs grow, so does emigration. When a constant immigration flow long-term assumption is adopted, this leads to shrinking net international migration and slowing population growth. ONS regarded this scenario as implausible; we take the same view for our 2011-based projections. Projected emigration flows are subtracted from the start population.

Three options for internal migration are then set out. Option 1 (18.11) involves multiplication of out-migration rate assumptions by LAD populations at risk. This is used in our LEEDS interim projections in bi-regional form. Option 2 (18.12) is an adjustment to constant out-migration rates reflecting the influence of changing destination populations, introduced by Statistics Canada (Dion 2014) and adopted by ONS (2014) for modelling internal migration between home countries in NPP2014 (National Population Projections 2014-based). Option 3 (18.13) introduces exogenous results from a gravity model, specified here in general form. There is a very large body of work exploring the factors driving inter-regional migration but very little of this has been used in population projections. Projected out-migration flows (summed over all destinations) are subtracted from the start population.

Step (18.15) adds together the variables projected up to this point and computes the accounting residual balance. The remaining terms are then added to the residual balance to produce the end of interval final population. The first input term is sum of internal in-migrations to each LAD; the options for modelling have already been described earlier.

Immigration options are then set out in steps (18.17)–(18.18), which parallel those for emigration. In option (1) assumed immigration flows are input (used in the LEEDS projection); in option (2) assumed immigration rates are multiplied by populations at risk for the Rest of the World; for option (3) assumed immigration admission rates are multiplied by LAD populations at risk. The second option leads to rapid increases in immigration as the Rest of the World population grows and so is suitable only as a reference projection against which restrictive policies can be assessed for impact. Option (3) was considered in preparations for NPP2014 but thought implausible (ONS 2014).

Step (18.21) sums the population inputs including the residual balance to provide the final population at the end of the interval. Two final processes follow. In step (18.22) we allow ethnic groups to switch identities by multiplying the final populations by switching probabilities. These are based on an analysis of the Longitudinal Study which links the 2001 and 2011 Census for a sample of individuals (Simpson 2014). In effect, switching is envisaged as a repeated census question asked at the end of each projection time interval. However, in the projections reported here we did not implement this step because of concerns about the reliability of estimates of switching probabilities based on the 2001 and 2011 Censuses. The final step in the model, (18.24), is to age on the final population in a period-cohort to become the start population 1 year older in the next period-cohort in the next time interval.

18.4 Estimating the Ethnic Inputs

In theory, it should be easy to consult official demographic statistics to access ethnically classified components of change rates, as can researchers in the US and New Zealand. However, in the UK indirect methods must be used to estimate fertility, mortality, internal and international migration by ethnicity.

We have revised all of the demographic inputs by ethnicity in two ways: developing estimates based on the 2011 Census and by connecting those estimates to the 2001 Census estimates. The objective was to reconcile the 2001–2011 local ethnic group population estimates and components of change with the ethnic group populations from the 2001 and 2011 Censuses. This task turned out to require innovations such as the concept of the Census Based Book End (CBBE) and the development of algorithms to interpolate by age, period and cohort. Rees et al. (2017) provide an account of the reconciliation exercise.

18.4.1 A Triangular Approach to Ethnic Fertility Using Census Data, Vital Statistics and Survey Information

Norman et al. (2014) describe the methods and data sets used to develop fertility rates by ethnicity for 2001, employing Census data, births statistics and survey information. These ethnic fertility estimates have been updated to 2011 and extended to include a nativity classification, distinguishing native and foreign born potential mothers (Norman 2015). A time series of fertility by ethnicity has been developed for 2001–2011. Table 18.5 presents UK total fertility rates in 2011 for 12 harmonized ethnic groups. Overall TFR rose over the previous decade from 1.63 to 1.93 as a result of catch up (women having children in their 30s who had postponed children in their 20s). The range of TFRs is wide, with 6 below the average and 6 above the average, ranging from a low of 1.26 for the Black Other (OBL) group to a high of 3.47 for the Bangladeshi (BAN) group. An equivalent set of fertility rates have been estimated for LADs in England. For the projections reported in Sect. 18.6, we use the England estimates for ethnic groups in Wales, Scotland and Northern Ireland.

Table 18.5 Fertility estimates by ethnic group, England, 2011

Ethnic group	ASFRs						
	<20	20–24	25–29	30–34	35–39	40+	TFR
WBI	19	67	98	108	61	14	1.83
WHO	21	75	110	121	68	16	2.06
MIX	9	51	88	95	47	9	1.49
IND	20	84	129	130	64	12	2.20
PAK	65	150	171	160	83	11	3.20
BAN	71	162	185	174	90	12	3.47
CHI	8	43	74	80	40	7	1.26
OAS	13	71	123	133	66	12	2.09
BLA	45	115	140	138	73	17	2.64
BLC	30	76	93	91	49	11	1.75
OBL	8	42	73	78	39	7	1.23
OTH	11	60	104	113	56	11	1.77
Total	21	71	104	112	63	14	1.93

Notes: *ASFR* age specific fertility rate, births per 1000 women. *TFR* Total Fertility Rate, births per woman = sum of ASFRs/1000
Source: Author's computations from ONS Births, 2011 Census Data Tables & Samples of Anonymised Records and Annual Population Survey

18.4.2 A Geographic Distribution Approach to Ethnic Mortality

Revised ethnic mortality estimates have been made for 2011, using data on ethnic populations from the Census and LAD statistics on deaths. In Rees et al. (2009), two ways of estimating ethnic mortality were developed: (1) a method based on the relationship between limiting long-term illness (LLTI) and mortality and (2) a method that used LAD mortality rates weighted by the different geographical profiles of ethnic groups. Subsequent research has shown that LLTI and mortality are only partially correlated. The healthy immigrant effect was also important. So, we used a *Geographically Distributed Method* (Wohland and Rees 2015; Wohland 2015). Figure 18.1 presents results for 16 ethnic groups in 2001 and 2011. The difference between the 2011 (red plots) and 2001 (blue plots) shows the continuing improvement in life expectancy over the decade (between 1.5 and 2 years depending on group). The variation between ethnic groups is much smaller than between LADs. Deprivation, which varies radically between LADs, is more important in determining life expectancy than ethnicity. Two ethnic groups have markedly lower life expectancy than the White British majority: the Bangladeshi and Pakistani group, reflecting their lower economic status. However, over the decade Bangladeshi women improve their relative position compared with Pakistani women, because their concentration in London affords better chances of educational and occupational improvement than does the concentration of the Pakistani population in northern de-industrialized cities.

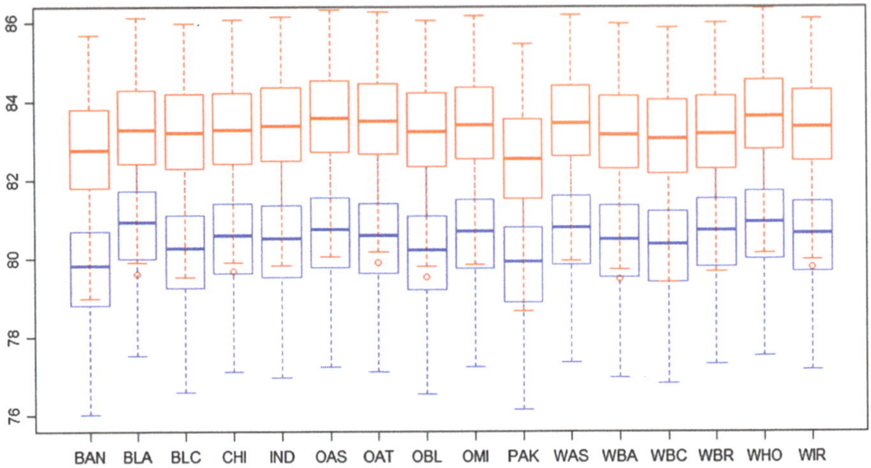

Fig. 18.1 Life expectancy at birth for women, estimated using the geographical distribution method, English LADs 2001 and 2011 (Notes: *Boxes* show the median LAD and the interquartile range (IQR). Whiskers show the minimum and maximum values with *dots* representing a few outliers. Red box and whisker plots refer to 2011. Blue box and whisker plots refer to 2001. See Table 18.1 for ethnic group definitions. Source: Wohland 2015b)

18.4.3 Using the Census to Estimate Ethnically Specific Internal Migration

New estimates of internal migration by ethnicity have been made using commissioned migration tables from the 2011 Census (Lomax 2015; Lomax and Rees 2015). These flow data show that ethnic minority groups are moving away from areas of highest concentration of the group and following the White British and Irish group in outward migration from metropolitan areas. Figure 18.2 reports on the variation across ethnic groups in the share of inter-LAD migration flows using a simple classification of LADs into metropolitan and non-metropolitan. For all minority groups the share of flows within metropolitan regions is much larger (37–54 %) than for the White British and Irish (WBI) (23 %). The shares of flows that are between metropolitan and non-metropolitan regions for minority groups (40–49 %) are closer to the WBI share (50 %). Where the balance is negative (e.g. the Chinese group), flows are concentrated in the 16–24 age group at which non-metropolitan teenagers leave for higher education in metropolitan areas. Flows within non-metropolitan areas are most important for the WBI group. Further analysis of the flows suggests that ethnic minority groups are spreading out spatially within the UK, though not yet as widely as the WBI majority.

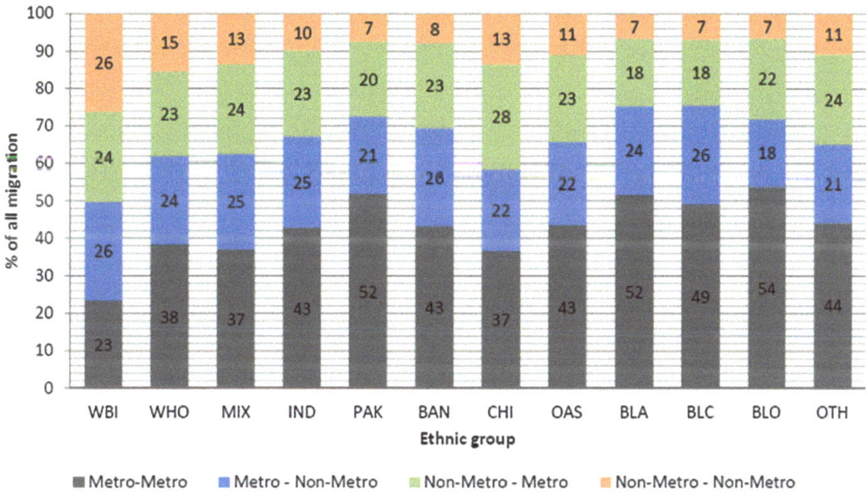

Fig. 18.2 Internal migration between LADs by ethnicity classified by metropolitan-non-metropolitan type of flow, 2011 Census, UK. (Notes: See Table 18.1 for the definitions of ethnic groups)

18.4.4 Using Census Data and Survey Data to Estimate International Migration

For international migration new estimates of immigration and emigration by ethnicity have been created using International Passenger Survey/Long Term International Migration (IPS/LTIM) tables published by ONS in combination with 2001 and 2011 Census tables of ethnicity by country of birth (Clark and Rees 2016). From published tables we reconstructed a five dimensional array of estimated flows. The dimensions included country of birth, country of next or last residence, region of last or next residence within the UK, broad age and gender. From the 2001 and 2011 Censuses we used tables that cross-classified ethnicity by country of birth to compute the probability that an immigrant or emigrant belonged to a particular ethnic group. The regional compositions by ethnicity were used to adjust LAD estimates of ethnicity produced through interpolation between CBBE estimates based on the 2001 and 2011 Censuses (Rees et al. 2017). The ethnic compositions of emigration and immigration streams are different. Over the period 2000–2014 the WBI group contributed 55 % of emigrants but only 33 % of immigrants. The ethnic minority which contributes most immigrants (24 %) and emigrants (20 %) is the White Other group. Individual Black and Asian Minority Ethnic (BAME) groups contribute smaller numbers but figure much more prominently in immigration than emigration. For example, the Indian group makes up 8 % of immigrants and 4 % of emigrants. The differences in ethnic composition between immigration and emigration streams contribute to the large differences in growth between the WBI group and ethnic minorities.

We illustrate our estimates of the ethnic composition of UK immigration and emigration by examining the inflows and outflows to London (Fig. 18.3), the most

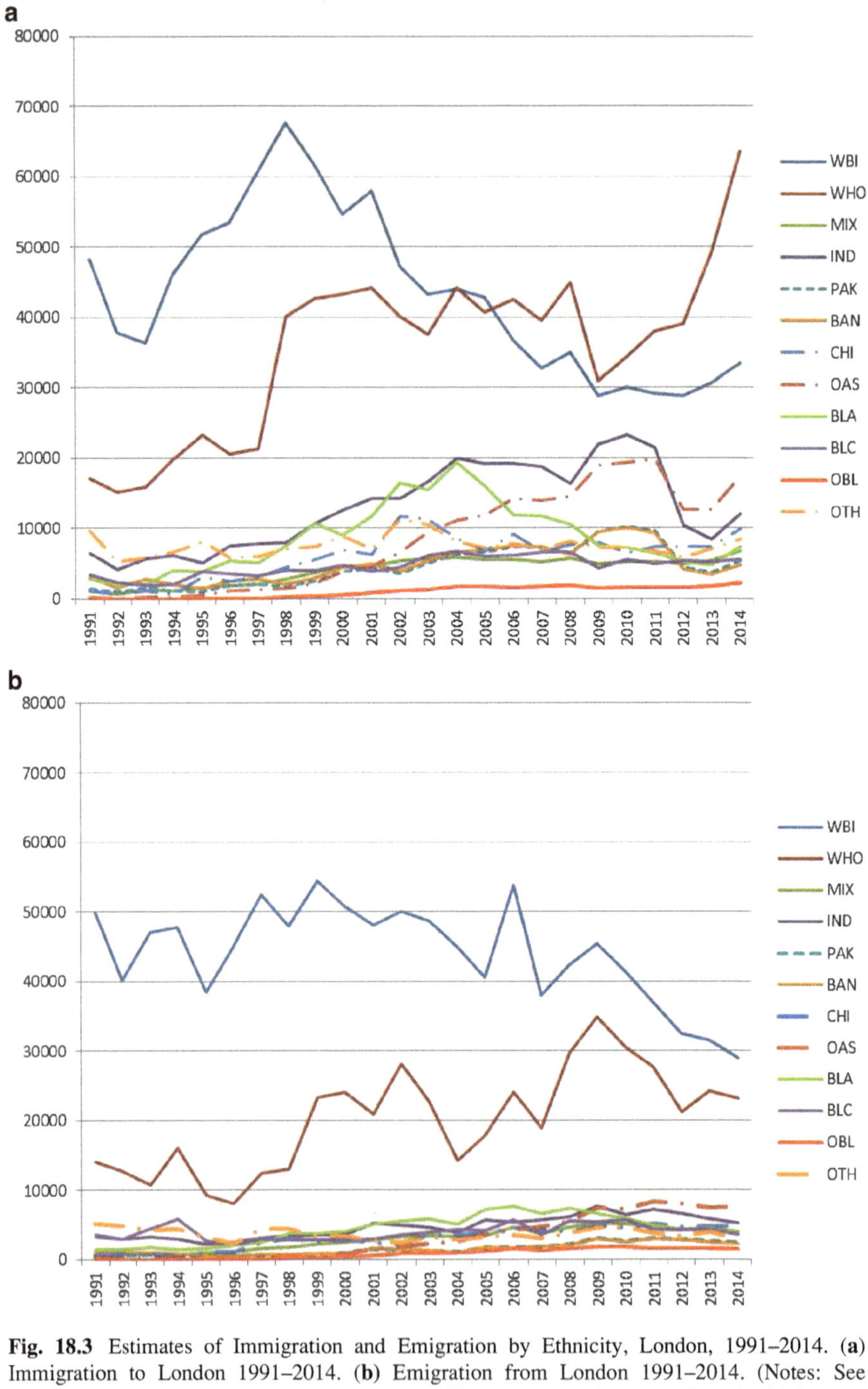

Fig. 18.3 Estimates of Immigration and Emigration by Ethnicity, London, 1991–2014. (**a**) Immigration to London 1991–2014. (**b**) Emigration from London 1991–2014. (Notes: See Table 18.1 for the definitions of ethnic groups. The London region comprises the 33 London Boroughs (LADs) which make up the Greater London Authority)

ethnically diverse UK region. The WBI and WHO groups make the largest contributions to immigration to London but the time trends are different. WBI immigration declines from a peak in the late 1990s to much lower levels after 2010. The WHO group sees higher levels from 1998 onwards, reflecting the addition of 10 countries to the EU in the 2000s and the impact of poor employment conditions in southern EU states since 2010. The BAME groups make smaller individual contributions to immigration than the two main groups but, because emigration is much lower, these immigration streams are more effective in adding to the population.

18.5 Assumptions for Projections

Table 18.6 lists the assumptions for the 2001-based and 2011-based projections described in Sect. 18.6. The method for assumption setting varies with the component. For fertility we assume a long term total fertility for the UK. The ratios of local, ethnic group TFRs based on the 2011 Census and associated data to the national TFR are applied to the national TFR to generate assumptions. The long term TFR was assumed to be 1.84 in the 2001 based projection and 1.89 in the 2011 based projection. No convergence of ethnic group fertility on the national average is assumed. There is minor short term downward adjustment of TFRs from the levels reported in Table 18.5.

Mortality assumptions draw on ONS practice in NPP2008 and NPP2014. An annual rate of decline constant across most ages is used: this was a 1 % decline in the TRENDEF projection and 1.2 % decline in the LEEDS projection. These decline rates are applied to all LAD-ethnic group mortality rates.

For internal migration we assumed constant application of the 2000–2001 based probabilities (ETHPOP) and the 2010–2011 rates (NewETHPOP). For international

Table 18.6 ETHPOP and NewETHPOP projection assumptions

Component	Assumptions
ETHPOP: TRENDEF projection	
Fertility	Long term TFR = 1.84
Mortality	Mortality decline rate = 1 % pa
Internal migration	Constant 2000–01 conditional probabilities
International migration	Long term immigration = 435 k, Long-term emigration = 293 k (net 142 k)
NewETHPOP: LEEDS (interim) projection	
Fertility	Long term TFR = 1.89
Mortality	Mortality decline rate = 1.2 % pa
Internal migration	Constant 2010–11 out-migration rates
International migration	Long term immigration = 592 k, Long term emigration = 338 k (net 254 k)

migration assumptions we specify long term levels of immigration and emigration as set out in Table 18.6. These totals are shared out across LADs, ethnic groups, genders and ages using 2001 and 2011 based shares. Note that the long term immigration assumptions rose more than the emigration assumptions, leading to a higher net inward balance in 2010–11. The LEEDS 2011 based assumption was set by fitting a logistic function to the IPS/LTIM time series discussed in Sect. 18.4. In the short term, 2014–2015 higher immigration and emigration levels are assumed to trend down to the long term assumptions.

18.6 Results

Populations projected using the LEEDS assumptions are listed in Table 18.7. The all-person population of the UK grows substantially from 2011 to 2061 by 34 %. Almost all this growth (80 %) occurs in the BAME groups while the most of the 20 % growth in the White grouping is due to the increase in the White Other population with the WBI population growing to 2021 and then decreasing. At the bottom of the table we make comparisons with other projections. In the 2001 based TRENDEF projection we over-projected the 2011 White population by 1.3 million. The difference in the White population decreases to 2051 as the WHO group grows. Our 2001 based projection of the BAME population under-shot by 1.2 million people. The difference between the 2001 and 2011 based projected BAME population increases to reach 4.5 million by 2051. Our 2001 based projections radically

Table 18.7 Projected ethnic group populations, UK, 2001-2061

Broad Grouping	MY2001	MY2011	MY2031	MY2051	MY2061
White	54,384	55,211	59,289	59,821	59,370
Mixed	687	1,260	2,297	3,543	4,183
Asian	2,627	4,333	8,209	12,697	15,080
Black	1,174	1,881	2,831	3,844	4,279
Other	238	592	1,038	1,480	1,670
All	59,111	63,278	73,664	81,386	84,582
BAME	4,726	8.066	14,375	21,564	25,212
White %	92.0 %	87.3 %	80.5 %	73.5 %	70.2 %
BAME %	8.0 %	12.7 %	19.5 %	26.5 %	29.8 %
LEEDS vs TRENDEF: White	–	−1,345	−502	−261	–
LEEDS vs TRENDEF: BAME	–	+1,249	+3,178	+4,456	–
LEEDS vs NPP2014: All	–	–	+1,957	+3,660	+4,333

Notes: 1. Broad groups are defined in Table 18.1. 2. TRENDEF = 2001 based, trended based projection using ETHPOP model. 3. NPP2014 = National Population Projections, 2014-based population (ONS 2015). 4. Populations are in 1000 s under-estimated the speed of the ethnic transition from a largely White population in 2001 to a White population making up only 7 in 10 of the population in 2061.

The final comparison in Table 18.7 is between the LEEDS 2011 based projection and the NPP2014 results. Our 2011 based projected population of the UK in 2061 is 4.3 million people higher than NPP2014. The main reason for this is our higher assumption for immigration. In net terms NPP2014 assumes a net 185,000 international migrants per year, while the LEEDS projection adds a net 254,000. Over 50 years to 2061 the difference cumulates to 3.5 million extra people, out of a total difference of 4.3 million.

Figure 18.4 graphs the trends, expressed as a percentage of the 2011 population, for the 12 harmonized groups used in the 2011 based projections. The most rapidly growing groups are the Chinese and White Other group in the 2011 based projection. The differences between the 2011 projection and the 2001 based are greatest for these groups. The Chinese and White Other groups are assumed to have high levels of immigration. The Pakistani group also grows strongly in future because of high fertility and a young age structure. For the Pakistani group we revised the long term fertility assumption upwards using 2011 information. Most BAME groups are expected to grow more in the 2011 based projections because the starting population is based on the 2011 Census. The only group which experiences a lowering of its growth path between the two projections rounds is the Black Caribbean group. This diminishes at younger ages because an increasing share of offspring of Black Caribbean parents are Mixed in ethnicity. At older ages the BLC group loses emigrants who return to the Caribbean at retirement ages.

So far, we have commented on national results. Our projections provide a wealth of projection detail for 324 LADs in England. Figure 18.5 presents one map set from a potential atlas of results. The variable plotted on each map is the Index of Diversity (see the note to Fig. 18.5). Diversity ranges from very low (0.05–0.15) in deep blue to very high (0.61–0.92) in deep red. The same classification scheme is used in each map so comparison can be made between two censuses, two projections and two times. Small extensions of high diversity occur between the two censuses (18.5a vs 18.5c). Over five decades (18.5a vs 18.5b) or four decades (18.5c vs 18.5d) high diversity spreads out from metropolitan cores to nearby non-metropolitan areas. This spread is slightly more extensive in the 2001 based projections (18.5b).

The degree of spread depends mainly on the structure of the internal migration matrix of out-migration rates. During favourable parts of the economic cycle, people are confident to move home outside of their origin location. During unfavourable economic phases the volume of internal migration declines and people are more conservative in their location choices. The spread in the 2001 based projections is driven by conditions in 2000–2001 (the interval measured by the census question) which was in the middle of a long boom; the spread in the 2011 based projections is affected by the financial crash of 2008–2009, which depressed economic activity in 2010–2011, particularly the housing market. This difference explains the diminished spread of diversity in the 2011 based projections.

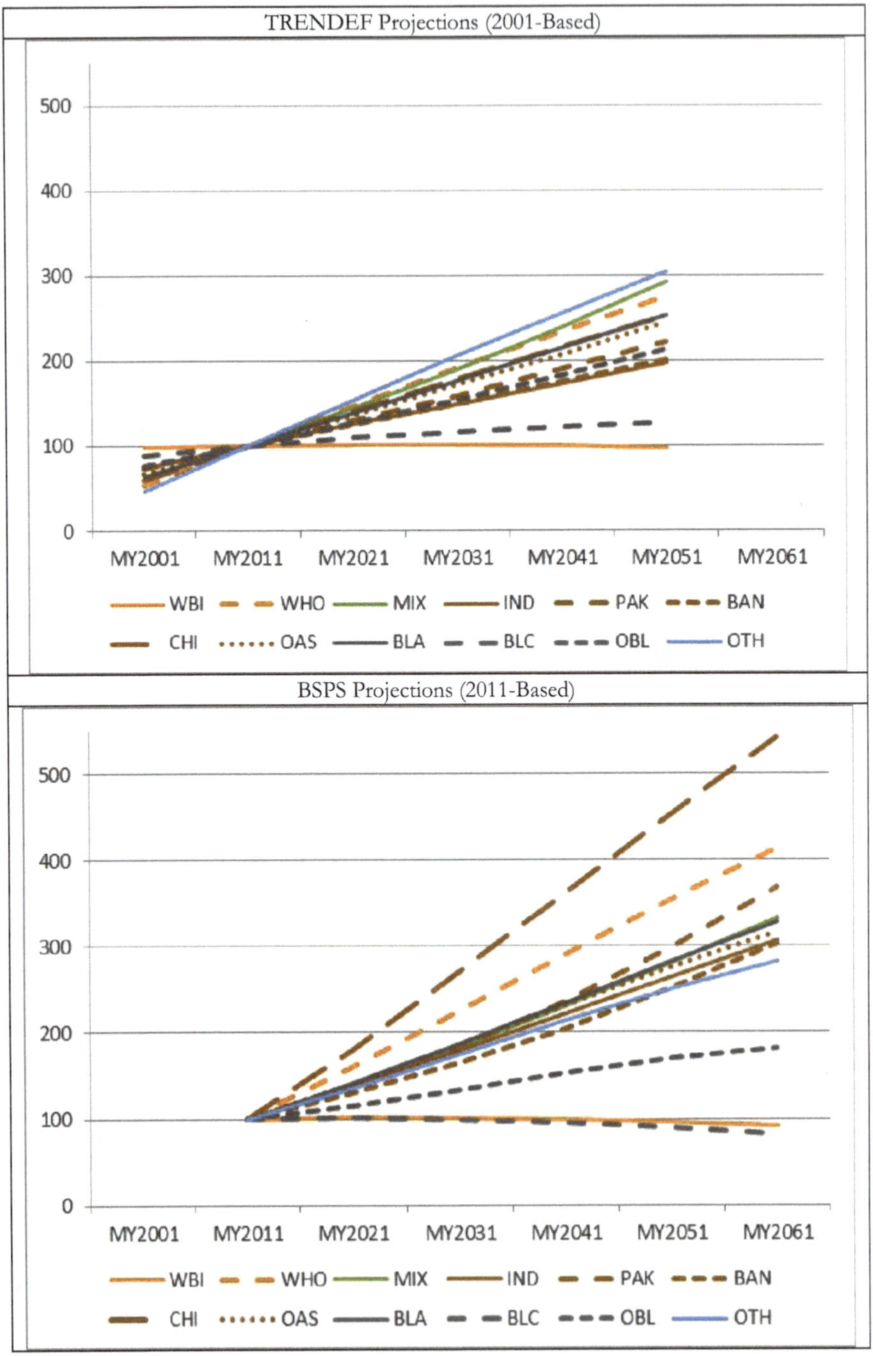

Fig. 18.4 Projected ethnic group trends: 2001-based and 2011-based (Notes: 1. See Table 18.2 for the definitions of the ethnic group abbreviations. 2. The vertical axis is a time series index set to 100 for MY2011. 3. MY = mid-year = 30 June/1 July. Sources: TRENDEF: Rees et al. 2011, 2012. LEEDS: NewETHPOP project)

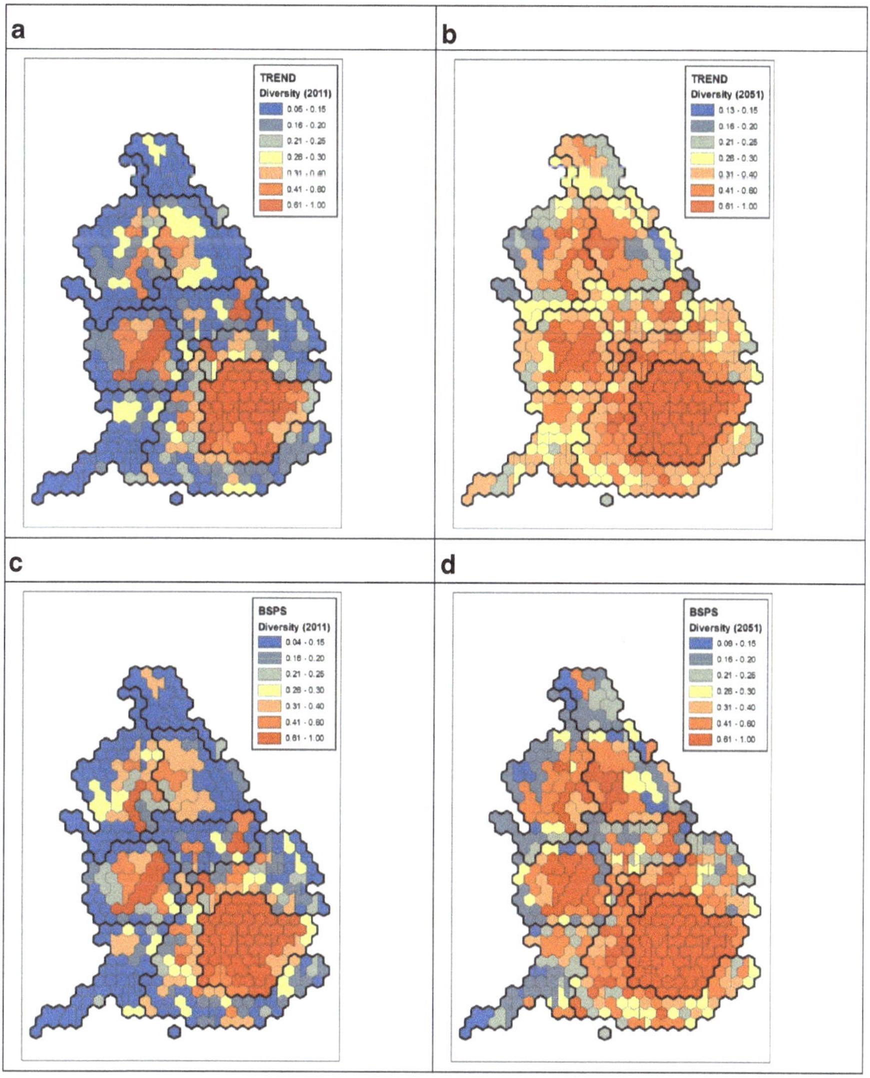

Fig. 18.5 The changing diversity of England's local populations. (**a**) TRENDEF Projection (2001-Based): Mid-Year 2011. (**b**) TRENDEF Projection (2001-Based): Mid-Year 2051. (**c**) Leeds Interim Projection (2011 Census based): Mid-Year 2011. (**d**) Leeds Interim Projection (2011-Based): Mid-Year 2051 (Notes: 1. Diversity = 1 minus the sum over all ethnic groups of the squares of the proportions of LAD populations in an ethnic group. Minimum diversity = 0, where the whole LAD population belongs to one ethnic groups. Maximum diversity is 0.917, where each of 12 groups has the same share (8.3 %). 2. The map base is a population cartogram from Dorling and Thomas (2004), adapted by Wohland and Clark. 3. Population cartograms assign areas to zones in proportion to population. LADs are composed of one or more hexagons which represent equal populations. The darker boundaries are for the 9 English regions)

18.7 Conclusion

This Chapter constitutes a case study of the demographic dynamics of a country in the midst of what has been termed the *third demographic transition* (Coleman 2006): when smaller birth cohorts reach the labour market, the demand for labour rises, and is filled by international migration. New ethnic communities are created with large demographic potentials because young age structures favour family formation. The *fourth demographic transition* (Frey 2015) involves the spatial re-distribution of ethnic minorities (of immigrant origin), with shifts from their initial places of settlement (mainly large cities) to other parts of the country. These two transitions can be termed collectively the *ethnic transition*.

Most researchers take a backwards look at such demographic processes. We have taken a forward look by building a population projection model which can detect the transitions. This is a challenging task in the UK because the ethnicity classification adopted in the population census and official surveys has not been ported over to registers and administrative databases that record demographic events. We have filled the void by using indirect, innovative methods to provide ethnic and local estimates of the necessary component rates. It has also been necessary to think hard about the additional processes that need to be added to a conventional population projection model, including the process when parents of different ethnicities have a child of mixed ethnicity and the process of changing identity.

Using the new estimations and projection model, we have carried out a second set of projections based on the 2011 Census and learnt valuable lessons from comparisons with previous projections and official projections. The United Kingdom can look forward to increasing ethnic diversity in future and this diversity will spread spatially.

Acknowledgements The research reported in this chapter was supported by ESRC award ES/L013878/1 for the project *Evaluation, Revision and Extension of Ethnic Population Projections – NewETHPOP* and by the University of Leeds Higher Education Innovation Fund. The datasets used in the analysis were sourced from the Office for National Statistics, National Records Scotland and the Northern Ireland Statistics & Research Agency under the terms of the UK's Open Government Licence and are subject to Crown Copyright.

References

Bijak, J. (2012). *Migration assumptions in the UK National Population Projections: Methodology review*. http://www.ons.gov.uk/ons/guide-method/method-quality/specific/population-and-migration/population-projections/npp-migration-assumptions-methodology-review/index.html

Bongaarts, J., & Bulatao, R. (1999). Completing the demographic transition. *Population and Development Review, 25*, 515–529.

Clark, S., & Rees, P. (2016). *International migration by ethnicity to UK regions: Results for 1991–2014 and implications for the future*. Paper presented at the first international conference on migration and mobility, July 18–20, Loughborough University.
Colby, S., & Ortman, J. (2015). Projections of the size and composition of the U.S. Population: 2014 to 2060. *Current Population Reports*, pp. 25-1143. https://www.census.gov/content/dam/Census/library/publications/2015/demo/p25-1143.pdf
Coleman, D. (2006). Immigration and ethnic change in low-fertility countries: A third demographic transition. *Population and Development Review, 32*(3), 401–446.
Dion, P. (2014). Projection of interprovincial migration, Chapter 8 in *Population projections for Canada (2013 to 2063), Provinces and Territories (2013 to 2038): Technical report on methodology and assumptions*, Statistics Canada Catalogue, 91-620-X. http://www.statcan.gc.ca/pub/91-620-x/2014001/chap08-eng.htm
Disney, G., Wiśniowski, A., Forster, J., Smith, P., & Bijak, J. (2015). *Evaluation of existing migration forecasting methods and models* (Report for the Migration Advisory Committee). https://www.gov.uk/government/uploads/system/uploads/attachment_data/file/467405/Migration_Forecasting_report.pdf
Dorling, D., & Thomas, B. (2004). *People and places: A 2001 census atlas of the UK*. Bristol: The Policy Press.
Frey, W. (2015). *Diversity explosion: How new racial demographics are remaking America*. Washington, DC: Brookings Institution Press.
Lomax, N. (2015, September 7–9). *Ethnic group migration patterns: A UK time series analysis*, British society for population studies, Annual Conference, Leeds. https://www.ethpop.org/Presentations/NewETHPOP/BSPS2015/NL%20BSPS2015%20Presentation.pdf
Lomax, N., & Rees, P. (2015, April 16). UK internal migration by ethnicity, presentation at the GISRUK2015 conference, University of Leeds.. http://www.ethpop.org/Presentations/NewETHPOP/NLGISRUK%20UK%20Internal%20Migration%20by%20Ethnicity%20NL.pdf
Norman, P. (2015). *We can't just 'Call the Midwife'! Are variations in fertility rates more about ethnicity, country of birth or local geography of residence?*, British Society for Population Studies, September 7–9, Annual conference, Leeds. http://www.ethpop.org/Presentations/NewETHPOP/BSPS2015/PN-ethnic-fertility-BSPS-2015.pdf
Norman, P., Rees, P., & Wohland, P. (2014). The use of a new indirect method to estimate ethnic-group fertility rates for subnational projections for England. *Population Studies, 68*(1), 43–64.
ONS. (2014). *Assessment of three potential methods for setting NPP migration assumptions as rates*. Internal Paper, Office for National Statistics.
ONS. (2015). *National population projections, 2014-based Statistical Bulletin*. Office for National Statistics. http://www.ons.gov.uk/peoplepopulationandcommunity/populationandmigration/populationprojections/bulletins/nationalpopulationprojections/2015-10-29
Raymer, J., Rees, P., & Blake, A. (2015). Frameworks for guiding the development and improvement of population statistics in the United Kingdom. *Journal of Official Statistics, 31*, 699–722.
Rees, P., Wohland, P., & Norman, P. (2009). The estimation of mortality for ethnic groups at local scale within the United Kingdom. *Social Science and Medicine, 69*, 1592–1607.
Rees, P., Wohland, P., Norman, P., & Boden, P. (2011). A local analysis of ethnic group population trends and projections for the UK. *Journal of Population Research, 28*(2–3), 149–184.
Rees, P., Wohland, P., Norman, P., & Boden, P. (2012). Ethnic population projections for the UK, 2001–2051. *Journal of Population Research, 29*(1), 45–89.
Rees, P., Wohland, P., & Norman, P. (2013). The demographic drivers of future ethnic group populations for UK local areas 2001–2051. *Geographical Journal, 179*(1), 44–60.
Rees, P., Clark, S., Wohland, P., Lomax, N., & Norman, P. (2015). *Ethnic population projections for the UK and local areas, 2011–2061: New results for the fourth demographic transition*. British Society for Population Studies, Annual Conference, Leeds, 7–9 September, Session, Demographic projections: ethnicity & projections. https://www.ethpop.org/Presentations/NewETHPOP/BSPS2015/BSPS%20Ethnic%20Population%20Projections%20Phil%20Rees.pdf

Rees, P., Clark, S., Wohland, P., Lomax, N., & Norman, P. (2017). Using the 2001 and 2011 censuses to reconcile ethnic group estimates and components for the intervening decade for English local authority districts. Chapter 21 in J. Stillwell, & O. Duke-Williams (Eds.), *Unlocking the UK 2011 census: Handbook of census resources, methods and applications*. Routledge: Oxford.

Rogers, A., & Ledent, J. (1976). Increment-decrement life tables: A comment. *Demography, 13*, 287–290.

Sabbagh, D. (2008). *The collection of ethnoracial statistics: Developments in the French controversy*. https://frenchamerican.org/sites/default/files/documents/media_policy_briefs/sabbagh_ethnicstatistics_brief_en.pdf

Simpson, L. (2014). How have people's ethnic identities changed in England and Wales? Briefing on the dynamics of diversity: Evidence from the 2011 Census. ESRC Centre on Dynamics of Ethnicity (CoDE). http://www.ethnicity.ac.uk/medialibrary/briefingsupdated/how-have-people's-ethnic-identities-changed-in-england-and-wales.pdf

Statistics New Zealand. (2015a). National ethnic population projections: 2013(base)–2038. http://www.stats.govt.nz/browse_for_stats/population/estimates_and_projections/NationalEthnicPopulationProjections_HOTP2013-38.aspx

Statistics New Zealand. (2015b). Subnational ethnic population projections: 2013(base)–2038. http://www.stats.govt.nz/browse_for_stats/population/estimates_and_projections/SubnationalEthnicPopulationProjections_HOTP2013base.aspx

Stone, R. (1971). *Demographic accounting and model building*. Paris: OECD.

Wohland, P. (2015). *Alternative methods to estimate mortality for UK ethnic groups*, British Society for Population Studies. Annual Conference, Leeds, 7–9 September 2015. http://www.ethpop.org/Presentations/NewETHPOP/BSPS2015/Alternative%20methods%20to%20estimates%20mortality%20for%20UK%20ethnic.pdf

Wohland, P., & Rees, P. (2015) *New estimates of ethnic mortality in the UK – Resolving the debate*. International conference on population geographies, University of Queensland, Brisbane, Australia, 30 June to 3 July. http://www.ethpop.org/Presentations/NewETHPOP/PWEthnicMortality1.pdf

Chapter 19
Revising Long-Established Population Estimates in Australia: Reasons, Methods and Implications

Andrew Howe

Abstract Official population estimates in Australia are derived from the five-yearly population census supplemented by undercount rates, which estimate the number of people missed in the census. Between censuses, population growth is modelled using component- and regression-based methods. After each census, these modelled estimates are superseded by the new census-based, or rebased, estimates. However, major differences between the 2011 modelled and census-based estimates, caused by a change in the way the census undercount rate was derived, led to a one-off 'recasting' revision to 20 years of historical population estimates. This extraordinary revision process led to a revisit of the approach to rebasing adopted in response to past censuses, where it was found that strict reliance on volatile census undercount rates, regardless of methodological changes, led to implausible population change between censuses. Discounting the modelled estimates or any other data source that indicates population change between censuses had a detrimental impact on the quality of the rebased estimates before recasting. Lessons learned from the recasting exercise could be used to improve the quality of Australia's population estimates from the 2016 census and beyond.

Keywords Undercount • Rebasing • Recasting • Intercensal difference

19.1 Introduction – Population Estimates in Australia

In Australia, the official measure of population is estimated resident population (ERP) which is produced by the Australian Bureau of Statistics (ABS). ERPs are the authoritative population estimates in Australia, and are used widely across government, private and academic sectors as the closest measure of truth for usual residence-based population levels, change and structure for Australia and its regions. Given their significant and diverse range of uses, population growth

A. Howe (✉)
Australian Bureau of Statistics, Adelaide, Australia
e-mail: andrew.howe@abs.gov.au

D.A. Swanson (ed.), *The Frontiers of Applied Demography*, Applied Demography Series 9, DOI 10.1007/978-3-319-43329-5_19

rates based on changes to ERP over time are of critical importance for evaluating policies and programs, and in measuring and assessing changing demands for and uses of services and infrastructure, especially at local area level.

At the national, and state and territory level,[1] Australia's population estimates are prepared every 3 months, as at 31 March, 30 June, 30 September and 31 December. Below the state/territory level,[2] referred to as sub-state level for convenience, estimates are prepared annually as at 30 June. Sub-state geographic units include parts of state (capital city/rest of state areas), Local Government Areas (or LGAs, for which local governing bodies have responsibility) and Statistical Areas Level 2 (or SA2s, Australia's base small to medium level geographic unit).

In addition to the quarterly (national, state and territory) and annual (sub-state) estimation cycles, Australia's population estimates also work around a five yearly census cycle. Since 1961, the Australian Census of Population and Housing (hereinafter the census) has been conducted in every year ending in 1 or 6.

Population estimates are calculated for 30 June of a census year based on census counts according to address of usual residence of the census respondent. Census counts are turned into population estimates by accounting for net census undercount, numbers of Australian residents temporarily overseas, and for any population change between the census date (early August) and the 30 June reference date.

Net census undercount is calculated based on the Post Enumeration Survey (PES), an intensive household survey conducted shortly after the Australian census to assess coverage of the census counts. To determine under- and over-count rates, person records from the PES are matched with census records. Matching records indicate the PES respondent was counted in the census, non-matching records indicate the respondent was not counted and from this (and other information) net undercount rates are calculated. Net undercount rates by age, sex, part of state and

[1] There are six states, two major territories (Northern Territory and Australian Capital Territory) and three external territories for which these quarterly estimates are prepared. The three external territories are referred to in aggregate as 'Other Territories'. The following abbreviations are used: NSW – New South Wales (state); Vic – Victoria (state); Qld – Queensland (state); SA – South Australia (state); WA – Western Australia (state); Tas – Tasmania (state); NT – Northern Territory; ACT – Australian Capital Territory; OT – Other Territories.

[2] At the sub-state level, population estimates are prepared for many different geographies. The main sub-state geographies are Statistical Areas Level 2 (SA2s) and Local Government Areas (LGAs). SA2s are medium-sized general purpose areas which aim to represent communities that interact together socially and economically, and are the lowest level of Australia's statistical geography for which ERP is made generally available. SA2s form or combine to form SA3s (which reflect widely recognised informal and administrative regions), SA4s (designed to reflect labour markets in urban areas, and aggregations of remaining areas) and Greater Capital City Statistical Areas or GCCSAs (which represent a broad socioeconomic definition of the capital city of each of the eight state and major territories), sometimes referred to as capital cities. 'Rest of state' areas are defined as those remaining areas of each state and the Northern Territory; the Australian Capital Territory consists wholly of the capital city GCCSA. LGAs are officially gazetted incorporated areas of Australia, which are legally designated regions for which incorporated local governing bodies have responsibility. In 2011 there were 2160 populated SA2s, 333 populated SA3s, 88 populated SA4s and 558 Local Government Areas.

Indigenous status are then applied to the census counts as part of the process of preparing the official population estimates.[3]

The 2011 PES involved a considerable degree of innovation, with the most important change being the introduction of Automated Data Linking (ADL). This new method used a range of personal and address characteristics to evaluate the likelihood that a PES and census record pertained to the same individual. ADL provided the opportunity to match respondents who would have been too difficult to match in earlier censuses, given the constraints of previous technology and processes. This method therefore resulted in better linking and matching of PES and census records, and a new measure of net undercount in 2011 (ABS 2012b).

Information on Australian residents temporarily overseas on census night (who were therefore not counted on a census form) for each area is obtained from overseas arrival and departure passenger card information, which is added to the census night population count for ERP purposes when the resident returns to Australia.

Population change between the census date (since 1991 Australian censuses have been held in early August) and 30 June reference date is calculated for each area using the component method, where the number of births, deaths and net migrants are backcasted from the census population date to reflect the 30 June population.

This process of turning census counts into ERP, referred to as 'rebasing', is applied for all geographic levels, from the national level down to SA2s.

ERPs are then prepared out from 30 June in a census year at the national, state and territory levels using component methods, where populations are updated iteratively from births, deaths and migration information using the component method (demographic accounting) equation. The component method equation simply adds births, subtracts deaths and adds net migration (which can be negative) to the base population estimate to arrive at an updated population estimate, which is then used as the base for the next estimation phase. The component method/ demographic accounting equation for calculating population change between time points t and t + 1 can be represented as:

$$\begin{aligned}\text{Population}_{t,t+1} = {} & \text{Population}_t + \text{Births}_{t,t+1} - \text{Deaths}_{t,t+1} \\ & + \text{Net Migration}_{t,t+1}\end{aligned} \qquad (19.1)$$

At the sub-state level, estimates are prepared out from the census year using regression methods, based on various administrative indicator data

[3] As the undercount rates are derived from a sample survey, the Post Enumeration Survey (PES), standard errors for estimates derived from the PES are also available (ABS 2012b). For Australia's 2011 net undercount rate of 374,500 people (or 1.7 %) for example, the standard error was 38,300. Under standard statistical survey treatment, there was 95 % confidence that the true 2011 Census net undercount rate was between 297,900 and 451,200, or 374,500 plus/minus two times the standard error. For rest of Victoria, the 95 per cent confidence bound ranged from −17,800 (ie. a net overcount) and 21,400.

sources (ABS 2013a). The regression method establishes relationships between changes in population and indicator data in the previous intercensal period (for groups of areas) and uses updated indicator data for each individual area to estimate population change from the census base for that area. Component methods have not been used to estimate sub-state populations due to the absence of adequate migration data at the sub-state level; the regression-modelled ERPs estimate the combined effect of adding births, subtracting deaths and adding/subtracting migration for sub-state areas. The sub-state population estimates are constrained to state/territory estimates, which are calculated first (ABS 2009).

To meet the competing demands for timeliness and accuracy, several versions of population estimates are prepared by the ABS. Preliminary estimates at the national and state/territory level are normally released within 6 months of the reference date, and within 9 months of the reference date for the sub-state estimates. The preliminary estimates are calculated up to and including the following census year; these are referred to as 'unrebased' estimates for the census year reference date, as they do not take into account information based on the census held that year. One or more revised versions of the preliminary estimates are later prepared, based on updated input data.

After the following census, all estimates calculated after the previous census are rebased. A first set of rebased estimates, referred to as preliminary rebased estimates, are released as early as possible after a census using first-release census data. A second or final set of rebased estimates are later prepared, based on final census and other input data (residents temporarily overseas, and the backcast census night to 30 June component data) used in the calculation of rebased 30 June ERP for a census year. For the 2011 rebasing process, an improvement to the distribution of Indigenous undercount below part of state level used in preliminary rebasing was also made for final rebasing.

The difference between unrebased and rebased ERP for the same 30 June census year reference date is referred to as intercensal difference. Rebasing encompasses revisions to all historical population estimates back to the previous census, by evenly distributing the intercensal difference to the unrebased ERPs over this 5 year period; quarterly for the national and state/territory (component-modelled) estimates, and annually for the sub-state (regression-modelled) estimates (ABS 2009).

Intercensal difference can be as a result of one or more of three factors: imprecision in the base population (the previous census-based estimate), imprecision in the rebased population (the latest census-based estimate), and/or imprecision in the component- or regression-modelled updating procedure, including input data not adequately reflecting population change. Intercensal difference has traditionally been assumed to reflect imprecision in the component- and regression-based population estimates, including error in the input data. Historically, the start- and end-point census-based estimates were assumed to be correct.

19.2 Consequences of 2011 Census Rebasing

Preliminary rebasing of ERP after the 2011 Census was conducted using the previously employed methodology, with the preliminary rebased estimates published in mid 2012. The preliminary intercensal difference for Australia in 2011 was +294,400, or 1.3 per cent; in other words, the unrebased ERP as at 30 June 2011 was 294,400 higher than the preliminary rebased estimate for the same date from the 2011 Census (ABS 2012c).

The preliminary intercensal difference for Australia for this 2006 to 2011 period was far greater than any other intercensal difference recorded in Australia. As the results of the 2011 preliminary rebasing were digested, it was found that the standard rebasing treatment could not credibly account for the large intercensal difference between 2006 and 2011, and the resulting rebased ERP series showed implausibly low growth for this period.

The relatively large preliminary intercensal difference in 2011 was regarded as being mostly due to the introduction of ADL in the PES, and its impact on the rebased 2011 population estimates relative to earlier censuses. The new ADL method improved the linking and matching of PES and census records, which is used to calculate net undercount. A statistical impact study conducted to determine the effect of using ADL found that the 2011 net undercount was approximately 40 % lower than it would have been if the previous PES methods had been used. The ADL method was calculated to account for around 246,900 or 84 % of the 2006–2011 preliminary intercensal difference of 294,400 (ABS 2013b).

The ABS estimated that the net undercount would have been substantially lower for all previous PESs had this methodology been available, and the large intercensal difference in 2011 was therefore likely to have been accumulated over a period greater than the usual 5 years.

The ABS then considered an extraordinary rebasing process for its historical population estimates extending back further than the standard 5 years of previous rebasing procedures. In September 2012 the ABS embarked on a consultation exercise with ABS and non-ABS stakeholders to determine how much further than 5 years, if at all, a new revised historical series should be prepared for (ABS 2012e).

Following this extensive consultation process, the ABS made the decision to revise ERP over a period of 20 years (ABS 2012f). Given the exceptional circumstance of this revision, where estimates previously regarded as final were being changed, this particular revision of 20 years of ERP from 2011 back to 1991 was referred to as 'recasting' the data, to distinguish it from the standard rebasing process.

The 20 year range, covering four intercensal periods, was decided on as the length of time that would result in estimates that reflected the growth observed in the historical data for population components (births, deaths and migration), which are the best data sources for measuring population change over time.

Recasting over the standard 5 years or over a 10 year period were both options that were considered, but eventually deemed sub-optimal, as the resulting ERP series would have low growth rates that would be more reflective of the change in methodology, rather than reflecting the best estimate of growth over the period. Recasting over a longer period, such as back to the beginning of the concept of usual residence-based ERP in Australia in 1971 (40 years), would have yielded very little statistical gain.

The processes and methods used to recast the data were developed and quality assured by a team of demographic and methodological specialists within the ABS. These methods were guided by a series of principles that were developed during the consultation process. These guiding principles established that:

- the credibility of population estimates, both level and growth, should be maintained for all spatial levels (ie. national, state/territory, and sub-state);
- the use of ADL in the PES has been a major improvement in how the ABS measures census coverage, and the 2011 net undercount should be used to inform historical understanding of census coverage;
- population growth for the 2006–2011 period should, as closely as possible, reflect the growth in the population components (ie. births, deaths and migration) or indicator data for all spatial levels (ie. national, state/territory, and sub-state);
- any assumptions should be based upon the best available data;
- any revision to the historical ERP series should maintain the demographically plausible relationships between the fundamental building blocks of the population series, for example age/sex profiles; and
- where revised data exist for population components data, they should be used regardless of whether they were available at the time of previous rebasing processes.

19.3 Method of Recasting – Overview

The usual rebasing treatment of intercensal difference distributes the difference evenly over the 5 year period, as there is no further data to inform the distribution of the difference within the period. Rather than distributing the impact of ADL back evenly over the 20 year period, from 2011 back to 1991, the recasting process differentially adjusted each of the four intercensal periods, based on information available for that period. This resulted in a greater impact on the data around the 2006 Census point, gradually decreasing to a minimal impact on the 1996–1991 data.

The recasting process involved calculating revised undercount adjustments at each census rebasing point (1996, 2001 and 2006) in order to create more plausible estimates of population change between censuses. These points were then used as the base population for new rebasing of the intercensal estimates between the recast

Table 19.1 Release timing of various versions of 1991–2011 ERP

ERP version: 2011/1991–2011	Australia, states/territories	Sub-state
Unrebased/before recasting	Dec 2011 ABS (2011)	Mar 2012 ABS (2012a)
Preliminary rebased/before recasting	Jun 2012 ABS (2012c)	Jul 2012 ABS (2012d)
Final rebased/after recasting	Jun 2013 ABS (2013b)	Aug 2013 ABS (2013c)

base points – quarterly for national and state/territory estimates, annually for the sub-state estimates.

Adjustment calculations followed a top-down approach. The first step was to calculate the total adjustment to undercount at the national level for each census point, which was then apportioned to the states and territories, followed by the part of state (ie. capital city/rest of state) level, and finally to sub-part of state levels of geography. Age and sex profiles were applied following the same top-down sequence, based on the age/sex breakdown of the previous version of the population estimates.

The final rebased version of the 2011 ERPs were released at the same time as the recast 1991– 2011 series. The recast 2006–2011 series therefore includes not only adjustments made to the 2006 Census start point due to recasting, but also (much smaller) differences in the 2011 end point due to final rebasing of the 2011 estimates. The differences between preliminary and final rebased 2011 ERPs were marginal: at the national level, final rebased 30 June ERP was 16,091 people or 0.07 % higher than preliminary rebased ERP (Table 19.1).

19.4 Method of Census Point Adjustments

National The national recasting adjustment in 2006 was taken directly from the ADL statistical impact study results, which was the best estimate of the total impact of implementing ADL compared to previous Post Enumeration Surveys. For 2001, the reduction was based on the 2006 reduction and an offsetting impact of a change in PES methodology made between the 2001 and 2006 surveys (ABS 2007a). The 1996 national adjustment was derived to minimise the change between the previous version of the estimates and the recast estimates between 1991 and 2001, and which would maintain growth rates over the 1991–1996 and 1996–2001 periods as closely as possible (ABS 2013b).

State/Territory The apportionment of the 2006 Census point national adjustment to the state and territory level was calculated based on a combination of intercensal difference and the 2006 PES adjustment. This reflected the strength of each source – the PES estimate directly links the adjustment to the original state/territory allocation of net undercount, while component growth and related error is a reliable indicator of changing state/territory distribution over time. The use of intercensal difference also enabled the process to directly address an observable degree of implausible variability in state and territory level intercensal difference over time. To derive the state and territory level adjustments for 2001, the 2006 state/territory split was offset by the state/territory-specific impact of a change in PES methodology made between the 2001 and 2006 surveys, where an improved estimator (Prediction Regression) was implemented (ABS 2007b). The 2001 state and territory split was then multiplied by the ratio between the 1996 and 2001 magnitude adjustments to derive the 1996 state and territory split.

Part of State Component growth – the combined effect of births, deaths and migration – was used to distribute the state/territory recasting adjustments to capital city/rest of state areas. This involved calculating new annual estimates of internal and overseas migration at the part of state level (births and deaths data at the part of state level already existed) and applying these back from 2011. Recast ERP for 30 June 2006, 2001 and 1996 was prepared for each part of state based on a methodology that drew upon a combination of this component growth for each intercensal period (2006–2011, 2001–2006, 1996–2001, 1991–1996), and changes in census counts for these periods supplemented by estimates of residents temporarily overseas on each census night (ABS 2013c). This approach was taken due to the data used to adjust at the national and state/territory level not being available at the part of state level.

Sub-part of State The part of state adjustment factors were applied consistently to all sub-part of state areas, on a pro-rata basis, based on the original rebased 2006, 2001 and 1996 sub-part of state estimates. For example, where Walkerville LGA had 0.603 % of its capital city (Greater Adelaide) population in 2006 before recasting, after recasting it still had 0.603 % of the Greater Adelaide population.

The methods used to adjust census point estimates at the national, state/territory and part of state levels were chosen as they provided the best estimate of the impact of the improved 2011 undercount adjustment for previous censuses, taking into account that estimates for the starting point of the recasting period (1991) were not being adjusted. The method of using the original sub-part of state estimates to guide the recast sub-part of state series was chosen as no further information was available to improve the distribution of population below the part of state level. For all recast census point estimates, the 1996, 2001 and 2006 Census counts continued to be the foundation on which the recast estimates were calculated.

19.5 Method of Non-census Point Adjustments

With the adjusted population levels set for each census point, the quarterly national and state/territory intercensal estimates were then recast progressively at these levels from the 1991–1996 to 2006–2011 periods according to the standard rebasing method. This method continued to assume that the intercensal difference should be

Table 19.2 Adjustments to ERP, by age group, 1996–2006

	Adjustment to ERP 1996		Adjustment to ERP 2001		Adjustment to ERP 2006	
Age group	Number	Percent	Number	Percent	Number	Percent
0–14	−16,729	−0.43	−26,965	−0.68	−48,058	−1.19
15–29	−23,770	−0.58	−38,315	−0.94	−68,288	−1.59
30–49	−22,102	−0.40	−35,627	−0.61	−63,496	−1.05
50 & over	−23,346	−0.49	−37,632	−0.68	−67,072	−1.06
All ages	−85,947	−0.47	−138,539	−0.71	−246,914	−1.19

apportioned equally over the 5-year intercensal period in the absence of information about how the error accrues within this period.

At the part of state level, new component-based estimates of annual population change between census years were applied to split the recast state/territory 30 June population estimates to obtain the recast part of state ERP series for the intercensal years. The new part of state estimates were then disaggregated on a pro-rata basis into sub-part of state areas based on the original rebased sub-part of state ERPs to obtain the new sub-part of state ERP series for the intercensal years.

19.6 Method of Adjusting Age/Sex Components

At the national and state/territory level, the age structure for each adjustment was adopted directly from the ADL statistical impact study results, grouped into ages 0–14, 15–29, 30–49, and over 50 years. These age groups were chosen to mitigate the high standard errors of finer age groupings, but also reflected the relative consistency of the ADL impact within these age groups. The greatest adjustment was required for the 15–29 year age group, reflecting both the relatively high undercount in each PES, and the higher impact of ADL on this group. The existing sex ratios were maintained at the national and state/territory level for all 5 year age groups (Table 19.2).

At the sub-state level, the recast total population estimates for each area were in turn broken down into age and sex based on the previous version of the age/sex structure of that area.

19.7 Effect on Population Levels

At the national level, recasting resulted in a reduction of Australia's official population estimates between 1991 and 2011, with the largest adjustment for 2006 (−246,900 people, or −1.2 %). The downwards revisions were progressively smaller going back to 1991.

Table 19.3 Adjustments to ERP, by part of state, 1996–2006

	Adjustment to ERP 1996		Adjustment to ERP 2001		Adjustment to ERP 2006	
Area (part of state)	Number	Percent	Number	Percent	Number	Percent
Capital cities	−71,419	−0.60	−68,888	−0.55	−95,886	−0.71
Remainder	−14,528	−0.23	−69,651	−1.02	−151,028	−2.08
Australia	−85,947	−0.47	−138,539	−0.71	−246,914	−1.19

At the state and territory level, the effect of recasting was also to adjust the estimates downwards for most states and territories, in particular the largest jurisdictions. However, ERP for some states and territories were adjusted slightly upwards for some years. The state/territory that received the largest adjustment was Queensland's 2006 estimate, which was revised down by 82,900 people, or −2.03 %.

For most regions within Australia, adjustments were largest for 2006. Despite having only 35 % of the total population, the combined non-capital city (ie. rest of state) areas received 61 % of the 2006 downwards revision, or −151,000 people. The combined capital cities were also revised downwards, by 95,900 people for the same point in time (Table 19.3).

The individual part of state with the largest adjustment for 2006 was rest of Queensland, which was revised down by 76,300 people, or −3.51 %.[4]

[4] Adjustments to estimated resident population, by part of state

	Adjustment to ERP 1996		Adjustment to ERP 2001		Adjustment to ERP 2006	
Part of state	Number	Percent	Number	Percent	Number	Percent
NSW capital city	−24,484	−0.63	−25,767	−0.62	−25,900	−0.60
Rest of NSW	−3783	−0.16	−19,101	−0.78	−47,497	−1.87
VIC capital city	−23,800	−0.71	−21,690	−0.62	−38,228	−1.01
Rest of VIC	−1371	−0.11	−19,421	−1.51	−27,046	−2.04
QLD capital city	−20,786	−1.31	−20,764	−1.21	−6611	−0.35
Rest of QLD	−14,712	−0.84	−36,713	−1.92	−76,305	−3.51
SA capital city	−4527	−0.40	−6736	−0.58	−10,362	−0.86
Rest of SA	−647	−0.18	−1531	−0.43	−4997	−1.36
WA capital city	−1023	−0.08	3303	0.23	−13,095	−0.82
Rest of WA	3973	0.94	1812	0.40	4295	0.92
TAS capital city	405	0.21	−893	−0.45	−1840	−0.89
Rest of TAS	757	0.27	2766	1.01	1191	0.42
NT capital city	1418	1.48	1438	1.35	−901	−0.79
Rest of NT	1255	1.46	2537	2.79	−669	−0.69
ACT	*1378*	*0.45*	*2221*	*0.70*	*1051*	*0.31*

As the part of state region proportional adjustment was applied consistently to all regions within each part of state, it was all sub-part of state areas in rest of Queensland that had the equal largest percentage adjustments downwards of all sub-state regions in Australia in 2006, at −3.51 %. The most populated areas within each part of state at the time therefore received the largest numeric adjustments. Southport SA2 (−909 people), Townsville SA3 (5797), Gold Coast SA4 (−16,812) and Gold Coast LGA (−16,350), all located in rest of Queensland, received the largest adjustments of all SA2s, SA3s, SA4s and LGAs in Australia respectively in 2006.

Recasting adjustments for all areas approached zero from 2006 back to 1991.

19.8 Effect on Population Growth

To distinguish between population change over time (for the same version of population estimates), and the changes in population levels for the same point of time as a result of recasting, this section refers to population change over time as population growth, whether positive or negative.

After making the recasting adjustments, all regions of Australia had new historical estimates, and therefore new growth rates. As the 2006 adjustments for most regions are downwards, 2006–2011 population growth rates are now higher for those regions after recasting.

Reflecting the largest adjustments generally taking place to the 2006 estimates, growth rates for the 2006–2011 period changed more than for any other 5 year span of the 20 year recasting period.

For Australia, total population growth between 2006 and 2011 was estimated to be 263,000 higher than before recasting, or 16.2 % higher than previously estimated. The non-capital city areas grew by a combined 160,200 more than previously estimated over the 5 year period, or by 47.5 % more. The eight capital cities grew by a combined 102,800 people, or by 8.0 % more than before recasting (Table 19.4).

Table 19.4 Changes to population growth, by part of state, 2006–2011

	Population growth 2006–2011, number				Population growth 2006–2011, percent		
Area (part of state)	Before recasting	After recasting	Change, number	Change, percent	Before recasting	After recasting	Change
Capital cities	1,289,205	1,392,048	102,843	8.0	9.6	10.4	0.8
Remainder	336,848	497,010	160,162	47.5	4.6	7.0	2.4
Australia	1,626,053	1,889,058	263,005	16.2	7.9	9.2	1.4

At the individual part of state level,[5] the largest change in 5 year growth as a result of recasting was for rest of Victoria 2006 to 2011, where the population growth estimate of 37,900 before recasting was adjusted upwards to 69,900, or by 79 %. Although rest of Queensland experienced the largest change of all parts of state populations in 2006 (78,100 people), its relatively high underlying 2006–2011 growth rate compared to rest of Victoria reduced the effect of recasting on rest of Queensland's growth.

The changes in population growth due to recasting varied considerably across Australia, and even within each part of state where growth rates were affected

[5] Changes to population growth, by part of state

	Population growth 2006–11, number				Population growth 2006–11, percent		
Part of state	Before recasting	After recasting	Change, number	Change, percent	Before recasting	After recasting	Change
NSW capital city	323,931	352,788	28,857	8.9	7.6	8.3	0.7
Rest of NSW	71,450	123,051	51,601	72.2	2.8	4.9	2.1
VIC capital city	370,115	408,606	38,491	10.4	9.7	10.9	1.1
Rest of VIC	37,871	67,945	30,074	79.4	2.9	5.2	2.4
QLD capital city	231,701	239,171	7470	3.2	12.1	12.5	0.4
Rest of QLD	151,489	229,615	78,126	51.6	7.0	10.9	4.0
SA capital city	63,335	74,848	11,513	18.2	5.3	6.3	1.0
Rest of SA	7009	12,237	5228	74.6	1.9	3.4	1.5
WA capital city	242,107	256,655	14,548	6.0	15.2	16.3	1.0
Rest of WA	50,727	46,173	−4554	−9.0	10.8	9.7	−1.1
TAS capital city	9683	11,520	1837	19.0	4.7	5.6	0.9
Rest of TAS	11,561	10,661	−900	−7.8	4.1	3.7	−0.3
NT capital city	14,700	15,645	945	6.4	12.9	13.8	0.9
Rest of NT	6004	6590	586	9.8	6.2	6.9	0.7
ACT	33,633	32,815	−818	−2.4	10.1	9.8	−0.3

differently despite the same proportional (pro-rata) adjustments made to population levels within the part of state. This is due to the different underlying growth rates of these regions – while the percentage adjustment to each area within each part of state was the same, recasting gave the impression of a larger growth impact in low growth areas. Areas with large population growth before recasting had relatively small changes to population growth rates, due to the scale of the recasting adjustment being smaller than the underlying population growth occurring in these areas. Examples showing a range of differences in 2006–2011 growth for three areas within one part of state, where the same percentage recasting adjustment was made to the 2006 ERPs, are presented in Table 19.5. The new growth estimate for Hume (5230 people) was 170 % higher than the previous figure (1938), while for Geelong the new growth estimate (20,525) was only around 40 % higher than the previous version (14,720).

Around 37 % of SA2s in Australia with declining 2006–2011 populations before recasting reverted to having increasing populations after recasting. Table 19.6 presents ranges of 2006 to 2011 population growth for SA2s before and after recasting.

Table 19.5 Population estimates before and after recasting, selected SA4s in rest of Victoria, 2006–2011

Area (SA4)	Population as at 2006	Population as at 2011	Growth 2006–2011, number	Growth 2006–2011, per cent
Before recasting and final rebasing				
Geelong	240,966	255,686	14,720	6.1
Hume	159,394	161,287	1938	1.2
Warrnambool & South West	123,576	122,223	−1353	−1.1
After recasting and final rebasing				
Geelong	236,055	256,580	20,525	8.7
Hume	156,105	161,335	5230	3.4
Warrnambool & South West	121,058	122,599	1541	1.3

Table 19.6 Population growth rates before and after recasting, SA2s[a], 2006–2011

No. of SA2s	After recasting					
Before recasting	Less than 0 %	0–< 2 %	2–< 5 %	5–< 10 %	10 % & over	Total
Less than 0 %	280	122	35	4	0	441
0–< 2 %	6	80	142	20	0	248
2–< 5 %	3	7	225	180	1	416
5–< 10 %	0	1	6	378	106	491
10 % & over	0	0	2	2	494	498
Total	289	210	410	584	601	2094

[a]With base 2006 population above 50 people

Table 19.7 Largest changes to population growth, local government areas, 2006–2011

	Population growth 2006–2011, number			Population growth 2006–2011, per cent		
Area (LGA)	Before recasting	After recasting	Change	Before recasting	After recasting	Change
Gold Coast	47,521	65,127	17,606	10.2	14.5	4.3
Sunshine Coast	21,774	33,543	11,769	7.4	11.8	4.4
Toowoomba	3634	9481	5847	2.4	6.5	4.1
Townsville	15,111	20,632	5521	9.1	12.9	3.8
Greater Geelong	9465	14,342	4877	4.6	7.1	2.5

At the SA2 level, the largest numeric change in 5 year growth was for Noosa Hinterland (Queensland) 2006 to 2011, where the population growth estimate of 795 before recasting was adjusted upwards to 1801, or by 1006 people after recasting.

At the LGA level, the largest numeric change in 5 year growth was for Gold Coast (Queensland) 2006 to 2011, where the population growth estimate of 47,521 before recasting was adjusted upwards to 65,119, or by 17,598 people. Table 19.7 shows the five LGAs in Australia with the largest changes in 2006–2011 numeric growth as a result of recasting.

19.9 The Recast Population Series – More Realistic Population Change

The recast 1991–2011 series of population estimates, which effectively incorporated new undercount adjustments for the rebased 1996, 2001 and 2006 estimates, provides a more realistic indication of population change, especially below the state/territory level.

Figure 19.1 illustrates that at the combined capital level the 20 year recast series is noticeably less variable than the previously published series before recasting (ie. the series determined from the original rebased estimates from the 1996, 2001 and 2006 Censuses), in terms of share of the state's population. This has resulted in a more stable change in share of the state population for capital cities and rest of state areas. The improved part of state series is especially apparent in Queensland, where a suspiciously high undercount adjustment for rest of Queensland in 2006 (5.1 %) was reduced to a rate more in line with undercount adjustments for rest of Queensland in 2001 (2.7 %) and 2011 (1.5 %) for recasting.

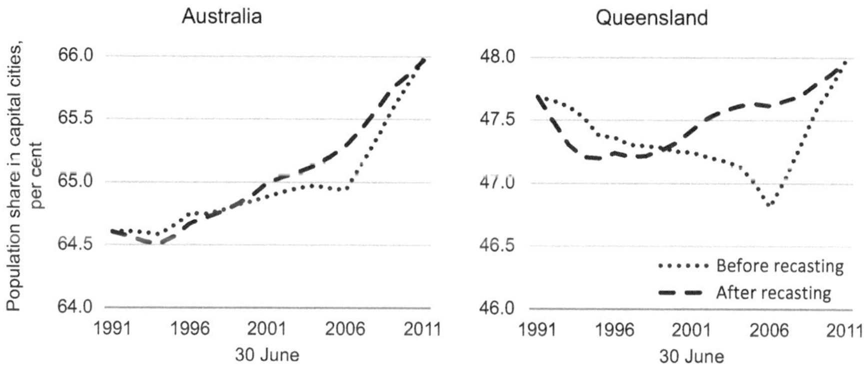

Fig. 19.1 Population change in capital cities, 1991–2011

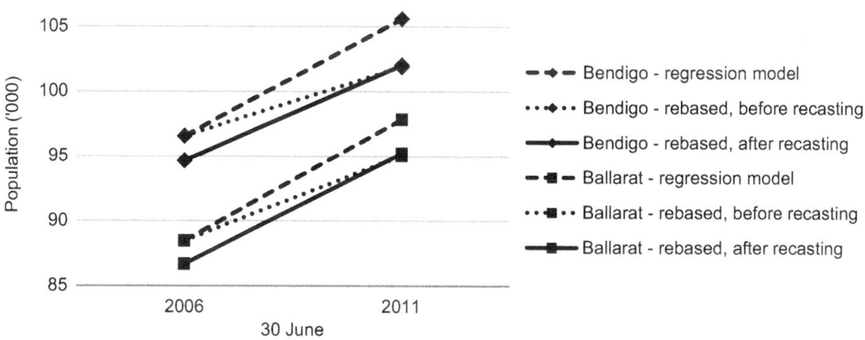

Fig. 19.2 Population growth, selected local government areas, 2006–2011

The recast sub-state series has also resulted in estimates of population growth more in line with the regression-modelled estimates of growth for the intercensal periods, especially for the 2006–2011 period, compared with the rebased series before recasting. Examples for two major population centres in rest of Victoria are provided in Fig. 19.2.

Recasting provides a new perspective on intercensal difference analysis conducted on the 2001–2006 (ABS 2008), 1996–2001 (ABS 2004) and 1991–1996 (ABS 1998) regression-modelled estimates. These analyses effectively concluded that differences between regression-modelled and rebased ERP for 2006, 2001 and 1996 were due to errors in the regression-modelled estimates. It now appears that the previous census-based estimates, which had been derived using

Table 19.8 Changes in net census undercount, and intercensal differences, Victoria capital city

2001–2006 intercensal period			
Net undercount adjustment, 2001	34,074	Regression-modelled ERP, 2006	3,685,029
Net undercount adjustment, 2006	86,635	Rebased ERP, 2006	3,743,635
Change in net census undercount	+52,561	Intercensal difference 2006	−58,606
2006–2011 intercensal period			
Net undercount adjustment, 2006	86,635	Regression-modelled ERP, 2011	4,138,050
Net undercount adjustment, 2011	55,079	Rebasesd ERP, 2011	4,108,541
Change in net census undercount	−31,556	Intercensal difference 2011	+29,509

volatile net undercount adjustment rates for the same part of state regions over time,[6,7] explains a significant component of the intercensal differences.[8]

Table 19.8 presents an example of the calculation of change in net undercount between successive censuses, and intercensal difference.

Wider analysis of intercensal differences at the part of state level shows that intercensal differences from 1996 to 2011 were inversely related to the change in net undercount between censuses. This strongly implies that the intercensal differences for sub-state ERPs for these intercensal periods were not so much the fault of the regression-modelled estimates, but that they were largely due to the variable net undercount rates calculated at part of state level at each census rebasing time. Had more consistent part of state undercount rates over time been applied to the 1996,

[6] Net undercount rates, by part of state, 1991–2011 censuses

Area	1991 percent	1996 percent	2001 percent	2006 percent	2011 percent
NSW capital city	2.0 (0.2)	1.8 (0.2)	1.5 (0.2)	3.0 (0.5)	1.8 (0.5)
Rest of NSW	1.7 (0.2)	1.1 (0.4)	2.8 (0.4)	1.3 (0.6)	2.2 (0.7)
VIC capital city	1.7 (0.2)	1.7 (0.3)	1.0 (0.2)	2.4 (0.5)	1.4 (0.4)
Rest of VIC	2.0 (0.3)	1.5 (0.5)	2.5 (0.4)	2.0 (0.7)	0.1 (0.7)
QLD capital city	1.4 (0.1)	1.4 (0.3)	1.0 (0.2)	1.9 (0.8)	2.0 (0.5)
Rest of QLD	2.6 (0.2)	2.1 (0.5)	2.7 (0.4)	5.1 (0.8)	1.5 (0.6)
SA capital city	1.2 (0.1)	0.7 (0.3)	1.5 (0.2)	2.2 (0.5)	1.1 (0.4)
Rest of SA	2.2 (0.4)	2.3 (0.9)	1.9 (0.5)	2.8 (0.8)	0.9 (0.8)
WA capital city	1.4 (0.2)	1.2 (0.2)	1.6 (0.3)	2.7 (0.7)	2.9 (0.6)
Rest of WA	4.1 (0.5)	2.5 (1.3)	3.1 (0.7)	4.4 (1.3)	1.2 (1.3)
TAS capital city	1.6 (0.2)	1.5 (0.9)	1.5 (0.4)	1.3 (1.1)	0.6 (1.0)
Rest of TAS	1.9 (0.2)	1.4 (0.5)	1.6 (0.7)	2.5 (0.0)	3.1 (1.0)
NT capital city	n.a.	n.a.	n.a.	5.5 (1.7)	3.7 (1.7)
Rest of NT	n.a.	n.a.	n.a.	10.1 (2.4)	10.9 (1.9)
ACT	*1.8 (0.3)*	*1.6 (0.3)*	*1.0 (0.4)*	*1.2 (1.0)*	*0.7 (0.8)*
Total capital city	1.7 (0.1)	1.5 (0.1)	1.3 (0.1)	2.5 (0.3)	1.8 (0.2)
Total rest of state	2.2 (0.1)	1.7 (0.1)	2.6 (0.2)	3.0 (0.4)	1.7 (0.3)
Australia	1.8 (0.1)	1.6 (0.1)	1.8 (0.1)	2.7 (0.2)	1.7 (0.3)

n.a. not available. Standard error shown in *brackets*

2001, 2006 and 2011 Census counts for rebasing at the time, the regression-modelled part of state ERP for the four intercensal periods would have aligned much closer with rebased ERP. This is evident in Fig. 19.3, which plots the changes in net undercount with the intercensal differences for that period for selected parts of states, as well as for the combined non-capital city regions.

Below part of state, evidence also suggests that the inconsistent census undercount adjustment rates applied to the same areas over time led to less plausible population change in the original rebased intercensal estimates, and overwrote reasonable regression-modelled estimates based on population indicator data for each area.

At the LGA level, intercensal differences calculated after recasting were generally lower than before recasting, especially for the 2006–2011 period. For example Gold Coast (which had the largest change in 2006–2011 growth before and after recasting of all LGAs in Australia, as shown in Table 19.7) had its 2006–2011 intercensal difference decline from +22,526 to +4920 people after recasting, or from +4.8 to +0.5 %. For the 2001–2006 period, intercensal difference also improved substantially for Gold Coast, from −9013 (or −2.3 %) to −78 (−0.4 %) after recasting.

The overall performance of the regression-modelled growth estimates can be summarised by calculating averages of the absolute values of the intercensal differences, in numeric and percentage terms. Table 19.9 shows the average

---

[7] Change in net undercount, by part of state, 1991–2011 censuses

Area	1991–1996 number ('000)	1996–2001 number ('000)	2001–2006 number ('000)	2006–2011 number ('000)
NSW capital city	−2.6	−6.4	63.2	−45.1
Rest of NSW	−12.1	42.9	−35.7	24.2
VIC capital city	3.3	−18.8	52.6	−31.6
Rest of VIC	−5.9	14.3	−6.2	−25.1
QLD capital city	2.9	−5.5	18.7	7.6
Rest of QLD	−1.8	12.6	61.2	−78.8
SA capital city	−4.5	8.9	8.0	−10.4
Rest of SA	0.8	−1.3	4.0	−8.6
WA capital city	−0.9	6.3	18.8	11.0
Rest of WA	−5.9	3.4	7.9	−17.2
TAS capital city	−0.1	0.2	−0.4	−1.2
Rest of TAS	−1.3	0.6	2.6	2.0
NT capital city	n.a.	n.a.	n.a.	−1.6
Rest of NT	n.a.	n.a.	n.a.	1.4
ACT	*0.8*	*−0.6*	*0.7*	*−1.4*
Total capital city	−1.9	−12.1	167.9	−72.7
Total rest of state	−26.3	72.5	43.3	−102.2
Australia	−28.3	57.7	203.4	−174.9

n.a. not available

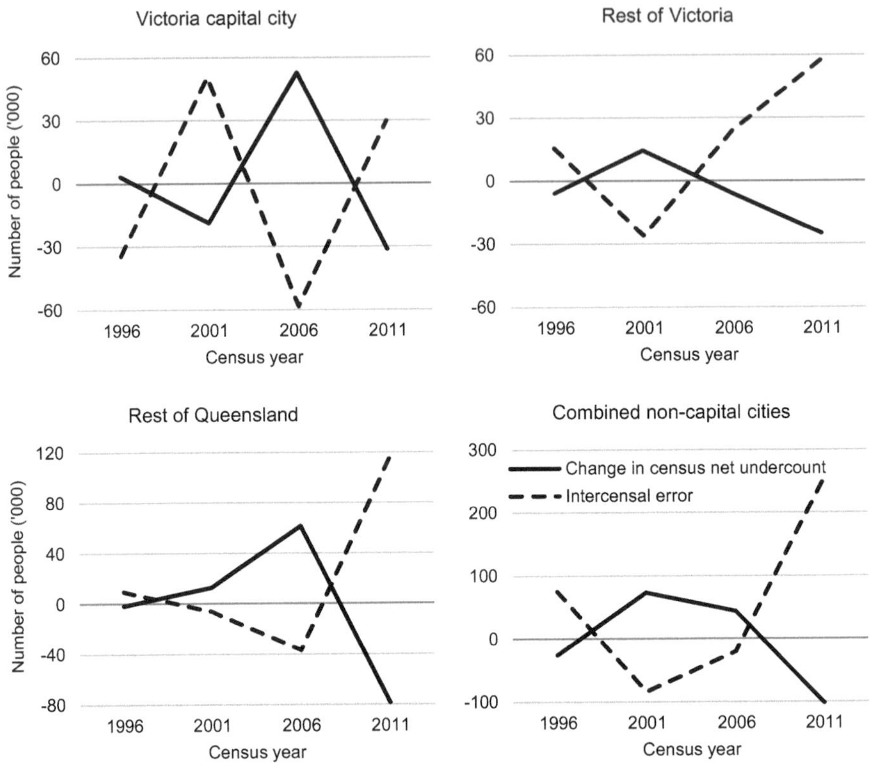

Fig. 19.3 Change in census undercount and intercensal difference, 1996–2011

[8] Intercensal difference, by part of state, 1996–2011

Area	1996 number ('000)	2001 number ('000)	2006 number ('000)	2011 number ('000)
NSW capital city	25.7	−59.7	12.5	11.1
Rest of NSW	−23.0	45.1	−55.2	0.5
VIC capital city	−3.4	−34.6	50.5	−58.6
Rest of VIC	10.2	15.5	−26.2	23.7
QLD capital city	11.1	6.4	5.2	−0.2
Rest of QLD	−0.9	9.3	−6.3	−37.2
SA capital city	8.0	7.4	−8.1	−7.4
Rest of SA	1.6	−2.8	−0.8	−5.9
WA capital city	21.7	−12.3	7.5	−10.8
Rest of WA	8.2	9.8	1.1	2.3
TAS capital city	−2.4	−0.8	−2.7	0.0
Rest of TAS	−3.8	−0.2	1.2	−1.0
NT capital city	−2.1	−1.5	0.8	−0.3
Rest of NT	−1.8	−1.6	1.6	−3.6
ACT	*4.2*	*−1.2*	*−4.3*	*−5.4*
Total capital city	58.6	−95.2	65.8	−66.2
Total rest of state	−9.6	75.0	−84.7	−21.2
Australia	49.1	−20.2	−18.9	−87.4

Table 19.9 Average absolute differences[a] between regression- and census-based estimates of population growth, before and after recasting, local government areas

	Average absolute difference 2001–2006		Average absolute difference 2006–2011	
Version of ERP	Number	Percent[b]	Number	Percent[b]
Before recasting	673	3.4	863	4.4
After recasting	654	3.5	559	3.3

[a]Average of the absolute values of the differences
[b]Excludes areas with population under 1000

difference in growth between the regression-modelled estimates and the rebased estimates before and after recasting, for 2001–2006 and 2006–2011, in terms of the absolute values of these differences in growth.[9] Overall, intercensal differences for after recasting were well below those before recasting for the 2006–2011 series of estimates, and around the same for the 2001–2006 estimates.

19.10 Conclusion

Rebasing is the process of turning population counts from the census into ERP, and incorporates net undercount adjustments derived from the census Post Enumeration Survey (PES).

The PES is an essential operation that delivers critical information about the coverage of the Australian census and whose results provide reliable net undercount rates for the country's population and its significantly large population subgroups. However some relatively high standard errors associated with these rates for smaller populations[6] indicate that the undercount estimates should not be the sole determinant of the undercount adjustment for the rebased population estimates.

The rebasing process replaces regression-modelled estimates of intercensal population change with new census-based estimates. The area-specific information used in the regression estimation model is all but ignored when the rebased census year estimates are calculated. Applying the published part of state net undercount rates to census counts to calculate the original rebased ERP in 1996, 2001 and 2006 led to implausible population change between censuses, and a detrimental perception of the quality of the regression-modelled intercensal estimates that they replaced.

The new automated data linking method used in the 2011 PES explained irregularities in the rebased estimates of population change out from the 2006

[9] Analysis of intercensal difference at the LGA level prior to 2001–2006 based on the recast series was not able to be made at the LGA level as regression-based LGA-level estimates prior to 2001–2006 were not prepared on the same boundaries as the recast series. Regression-based population estimates prior to 2011 were not prepared due to the SA2 geography not being established until 2011.

Census. However, inconsistencies in undercount rates between previous censuses, which were also subject to smaller methodological changes, were also shown to have a detrimental effect on the rebased ERP series originally derived from these censuses.

The extraordinary recasting revision to 20 years of ERP which took place in 2013 provided the opportunity to create more plausible official estimates of population change for Australia's regions. Recasting effectively allowed more consistent undercount adjustments over time to be retrospectively applied to census counts for the calculation of ERP back three previous censuses.

It is expected that in the future, Australia's population estimates will continue to be rebased at each census back to the previous census only. It is not anticipated that the recasting process undertaken after the 2011 Census will be repeated in relation to revising long-term historical ERP. Recasting should be seen as an exceptional event made necessary by significant methodological changes in 2011.

However the experience gained from the recasting exercise and its impact on ERP could be used to inform plans for 2016 rebasing and beyond, in particular the application of net undercount rates for sub-state areas. Should net undercount rates in future census continue to show large variations for the same regions, whether or not due to methodological changes, demographic analysis should be undertaken to determine the consequential undercount adjustments. Analysis could be in the form of taking into account the regression-modelled ERPs out from the previous census, the indicator data used in these models, and the component data (births, deaths and migration) including a new series of sub-state internal migration estimates (ABS 2015). The use of this data, and incorporating advice on regional population change from independent experts, could assist in calculating rebased population change from the previous census at rebasing time, within bounds determined by the standard errors of the undercount rates. Making better use of the rich demographic time-series data that exists for Australia's sub-state areas would create a higher quality and less volatile time series of regional population estimates for Australia in future.

Acknowledgements With thanks to Phil Browning, Tricia Chester, Beidar Cho, AJ Lanyon and Michael Roden for their useful comments in the production of this chapter.

References

ABS. (1998). Issues in estimating small area populations, Australian Bureau of Statistics Working Paper 98/1.

ABS. (2004). *Assessing the accuracy of Australia's small area population estimates, 2001*. Paper presented at the Australian Population Association conference, Canberra, 2004.

ABS. (2007a). Census of population and housing details of undercount, August 2006, cat. no. 2940.0.

ABS. (2007b). An estimating equation approach to census coverage adjustment, cat. no. 1351.0.55.019.

ABS. (2008). *Assessing the accuracy of Australia's regional population estimates, 2006*. Paper presented at the Australian Population Association conference, Alice Springs, 2008.
ABS. (2009). Population estimates: Concepts, sources and methods, 2009, cat. no. 3228.0.55.001.
ABS. (2011). Australian Bureau of Statistics 2011, Australian demographic statistics, June 2011, cat. no. 3101.0.
ABS. (2012a). Australian Bureau of Statistics 2012(a), Regional population growth, Australia, 2010–11, cat. no. 3218.0.
ABS. (2012b). Australian Bureau of Statistics 2012(b), census of population and housing, details of undercount, 2011, cat. no. 2940.0.
ABS. (2012c). Australian Bureau of Statistics 2012(c), Australian demographic statistics, December 2011, cat. no. 3101.0.
ABS. (2012d). Australian Bureau of Statistics 2012(d), Regional population growth, Australia, 2011, cat. no. 3218.0.
ABS. (2012e). Australian Bureau of Statistics 2012(e), Australian demographic statistics, March 2012, cat. no. 3101.0.
ABS. (2012f). Australian Bureau of Statistics 2012(f), Australian demographic statistics, June 2012, cat. no. 3101.0.
ABS. (2013a). Australian Bureau of Statistics 2013(a), Regional population growth, Australia, 2011–12, cat. no. 3218.0.
ABS. (2013b). Australian Bureau of Statistics 2013(b), Australian demographic statistics, December 2012, cat. no. 3101.0.
ABS. (2013c). Australian Bureau of Statistics 2013(c), Regional population growth, Australia, 2012, cat. no. 3218.0.
ABS. (2015). Australian Bureau of Statistics 2015, Migration, Australia, 2013–14, cat. no. 3412.0.

Chapter 20
Creating Population Surfaces for the Analysis of Small Area Change

Christopher D. Lloyd

Abstract Assessment of population change over small geographical areas relies on access to consistent geographical zones. Where these are not available, some form of areal interpolation or surface modelling procedure provides a solution. This chapter takes as its focus the analysis of population change over small areas and, building on previous research, it makes use of a combined polygon overlay and cell smoothing procedure to derive estimates of population counts for a 1 km grid to facilitate assessment of change through time. A case study is provided which focuses on England and Wales. This study is used to illustrate the nature of the population surfaces derived and their capacity for exploring change in population sub-groups locally, using the example of differences between populations in 2001 and 2011. Two groups with very different spatial distributions are included – the population by ethnicity (using a simple White/non-White categorisation) and the population by limiting long term illness (LLTI). The results suggest that the optimal amount of smoothing is variable dependent but use of ancillary data (here, in the form of landuse), rather than smoothing is most important. It is argued that population surfaces derived in this way make the most of the available information and provide a sound basis from which to explore longer term change over very small areas. In addition, the development of population surfaces open up the possibility to make use of a wide of tools for the analysis of raster data.

Keywords Population change • Areal interpolation • England and Wales

20.1 Introduction

Most geographical analyses of populations are based on counts for areas or spatial aggregations; for example, counts of unemployed persons derived from censuses may be provided for zones such as census tracts (USA) or wards (UK). The analysis of change through time using such data is often hampered by a set of problems including change in the size and shape of the zones used to report counts at each

C.D. Lloyd (✉)
University of Liverpool, Liverpool, UK
e-mail: c.d.lloyd@liverpool.ac.uk

D.A. Swanson (ed.), *The Frontiers of Applied Demography*, Applied Demography Series 9, DOI 10.1007/978-3-319-43329-5_20

time point, change in the questions asked and change in the output variables (e.g., ranges of ages). Solutions to the first problem include (i) converting counts from irregular zones to a surface, (ii) transferring counts from one set of zones to another using areal interpolation, (iii) transferring counts from one set of zones to another on a best-fit basis (Martin et al. 2002; Lloyd 2014). The approach used in the current study combines the first two of these. Here, the first approach is considered as a subset of the second, as discussed below. Specifically, counts for irregular zones are transferred to grids using a procedure which could be applied to any dataset and which allows direct comparisons of population counts over small areas over potentially very long periods.

Areal interpolation allows for the transfer of population counts to common geographies and thus enables assessment of small area change (e.g., by subtracting the population total in 1 year from the equivalent in another). However, any analysis based on pre-defined units suffers from the modifiable areal unit problem (MAUP; Openshaw and Taylor 1979; Openshaw 1984; Wong 2009). The MAUP comprises scale effects (the size of the zones determines analysis results) and zonation effects (the shape of the zones impacts on results). Analyses based on gridded counts are still subject to the MAUP but, in this case, the zones are of constant size. While areal interpolation does not solve the MAUP, it does allow the exploration of changes in results as different output zonal systems (irregular zones or grids) are produced.

The transfer of counts between zonal systems is, as noted above, referred to as areal interpolation. Most applications of areal interpolation entail the transfer of counts from one set of irregular zones (e.g., 1991 wards [source zones]) to another set of irregular zones (e.g., 2011 wards [target zones]) so that counts for the two sets of years can be compared (see Flowerdew and Green 1994 for a review of methods). The simplest means of reallocating counts is to use areal weighting – the zones are overlaid and the proportion of a source zone which overlies a target zone is computed. The equivalent proportion of the population of the source zone is then assigned to the target zone and the same process is followed for all source zones which overlie each target zone. More sophisticated approaches use ancillary data, such as land use data, to inform the reallocation of counts. For example, all areas with a land use of water or forest might be assigned a count of zero. Approaches which use information within source zones to help reallocation counts in this way are sometimes referred to as dasymetric, while control zone approaches use ancillary data for zones other than the source zones (e.g., target zones or another set of zones) (Gregory and Ell 2005). In the present study, a control zone approach is used since the land use zones differ from source zones.

The term areal interpolation is taken here to also refer to surface modelling approaches whereby counts are transferred from irregular zones to a regular grid. Several such approaches exist. These include the pycnophylactic interpolation approach of Tobler (1979) which entails overlay of irregular zones with a regular grid and proportional allocation of zone counts to cells such that cells within zones have the same total population as the zone in which they lie. Following this, the cells are smoothed using a filter so as to make neighbouring cells from different zones more similar so as to reduce 'steps' in values between zones following the assumption that population counts are generally smoothly-varying across space. The smoothed counts are then rescaled to ensure that the cells within zones still sum to the zone population. This is termed the pycnophylactic (or mass preserving) condition. Martin (1989) developed a smoothing approach which uses the centroids of zones as the locations of zones. The original version of this method was not locally mass preserving (there was no guarantee that counts of zones within zones had the same total value as the zone population) and a locally mass preserving version of the method was developed later (Martin 1996). These approaches are based on smoothing of counts, but they do not incorporate ancillary data. These approaches make use only of population data; other research has suggested that the use of ancillary data (e.g., land use data) is likely to be beneficial in reallocating counts from one set of zones to another (e.g., Kim and Yao 2010; and see Mennis 2003 for a population surface modelling application). Lloyd and Firoozi Nejad (2014), in a study which used 100 m grid cell population counts provided as an output from the Census of Population for Northern Ireland to assess the accuracy of surface modelling outputs, generated surfaces with a combination of ancillary data and smoothing of gridded estimates. The process was three step and entailed (i) overlay of grid cell centroids onto the source zones and proportional allocation of the source zone population to the cell centroids (e.g., if ten cell centroids fell within a source zone then they were each assigned one tenth of the source zone population) followed by (ii) smoothing of the population values attached to the cell centroids using a square filter and (iii) all smoothed cell counts are rescaled such that total counts for cells which fall within source zones are the same as the counts for the corresponding source zones (the method is locally mass preserving). Different window sizes were applied (three-by-three to 15 by 15 cells) to generate population surfaces for numbers of Catholics and for numbers of persons with a limiting long term illness (LLTI). The research suggested that smoothing provided greater benefits in mapping Catholics than LLTI and that this reflected the fact that counts of Catholics are less spatially variable (more spatially dependent) than are counts by LLTI. The results indicated that the optimal window size was a function of the population sub-group characteristics *and* the size and shape of the input zones. The capacity to easily smooth reallocated values is a particular benefit of surface modelling approaches – with areal interpolation approaches which produce reallocations of counts to irregular zones, it would be much more difficult to adjust counts to reflect neighbouring values in a systematic way. More generally, the production of gridded population counts makes available the rich array of methods for the analysis and processing of raster data (e.g., see Sonka et al. 2015). Gallego (2010) generates a population surface for the EU and considers the value of such a resource for exploring the relationship between the population and environmental characteristics of areas.

The approaches to population surface modeling presented by Martin (1989) and Tobler (1979) are based on overlay of source zones and grid cell centroids with a proportional share of the populations assigned to cell centroids from the source zone in which they fall. Such an approach is efficient but it is also problematic. As an example, if the cell spacing is large relative to the sizes of (at least some of) the source zones, then some source zones may not contain any grid cell centroids and thus their populations are 'lost' and not transferred to the population surface. One solution to this problem is to overlay a grid which is finer than required and then aggregate the cells to the level required at the final stage (e.g., 25 m cells aggregated to a final output with 100 m cells). A better solution is to not use centroids at all and instead to overlay polygon grid cells onto the source zones. This approach is used in the current paper and it is argued to make best use of the data. Following Lloyd and Firoozi Nejad (2014), two variables are selected – in this case ethnicity (White and non-White persons) and health status (persons with or without a limiting long term illness; LLTI), extracted from the 2001 and 2011 Censuses for England and Wales. Values of the Moran's I autocorrelation coefficient computed from the counts justify the selection of these counts by highlighting their very different degrees of spatial continuity. Population surfaces are generated for each of these four sets of counts. The selection of the variables is on the grounds that the population is relatively spatially continuous by ethnicity whereas it is quite spatially discontinuous by LLTI (Lloyd 2015) and so the optimal amount of smoothing in each case is likely to differ. The paper presents an approach to surface model generation which is adapted to a particular set of counts and which combines information on land use types and smoothing to produce population sub-group surfaces which can be used to explore change through time. The main focus is on surface model generation, but the examples are used to demonstrate some of the benefits of population surfaces and some of the ways in which local changes in population between (in this case) Census years can be explored.

20.2 Data and Methods

The case study focuses on England and Wales and makes use of the smallest available Census zone, output areas (OAs), as source zones to generate population surfaces. In 2001 there were 175,434 OAs in England and Wales (mean population = 297); while in 2011 there were 181,408 OAs (mean population = 309). Direct comparisons of many 2001 and 2011 OAs is possible since only some 2.6% of 2001 OAs have been changed as a result of the 2011 Census. However, the concern here is with developing an approach which allows for assessment of change for all areas for any time periods. Note than the 1 km cells generated from the OAs will, in some cases, be larger than the OAs and, in such cases, cells values will be aggregations of (parts of) OAs. The variables, and the tables from which they are taken, are listed in Table 20.1.

Table 20.1 Key statistics census tables and derived variables

Table 2001	Table 2011	Table description	Description
KS006	KS201	Ethnic group	All whites; Non-whites
KS008	KS301	Health and provision of unpaid care	LLTI; Non LLTI

The transfer of population counts between incompatible zones can be conducted using areal weighting. In essence, this entails the overlay of the source zones, s, and target zones, t, and the proportional allocation of the source zone population to the target zones which it overlays:

$$\hat{z}_t = \sum_{s=1}^{N} \frac{A_{st}}{A_s} z_s \tag{20.1}$$

Where $\hat{\mathbf{z}}_t$ is the estimated population for the target zone t, A_{st} is the area of the zone of intersection between s and t and A_s is the area of source zone s. Overlay provides the basis of the first stage of the procedure used in this paper.

The next stage to the method applied here is a union overlay of the selected OAs with Strategi® landuse polygons (version 01/2014; Ordnance Survey 2015) ('Lake/other inland water polygon'; 'Large Urban Area Polygon'; 'Small Urban Area Polygon'; 'Wood/Forest polygon') to give a new layer (here called 'OALU'; where LU indicates landuse). It should be noted that the same landuse layer was used to redistribute counts from 2001 and 2011 Census data.

The weights assigned to urban and non-urban areas follow Walford and Hayles (2012). That is, if polygons from the overlay (OALU) include either urban classification (some land use types have overlaps and thus an area may be associated with one than one land use), a weight of 0.9 is assigned; if no land use type overlaps then a weight of 0.1 is assigned (considered 'rural'). Where a (part of a) OA overlaps only a 'Lake/other inland water' or a 'Wood/Forest polygon' then a weight of zero is assigned. All OALU polygons which do not represent OAs are removed (use of union overlay means that LU areas which don't overlap OAs (e.g., parts of the coast) are retained). Alternative weighting schemes are being assessed as a part of future work.

The first stage of the estimation procedure applies these weights, λ, as follows (Gregory and Ell 2005):

$$\hat{z}_{st} = \frac{\lambda_{j(c)} A_{st}}{\sum_{k=1}^{N} \lambda_{j(k)} A_{sk}} z_s \tag{20.2}$$

where $\hat{\mathbf{z}}_{st}$ is the estimated population for the zone of intersection between s and t and A_{st} is its area; $\lambda_{j(c)}$ is the weight for the specified control zone type; $\lambda_{j(k)}$ is the weight for zone of intersection k and A_{sk} is the area of the zone of intersection k within source zone s and there are N zones of intersection. In practice, the denominator was constructed by aggregating the product of $\lambda_{j(k)} A_{sk}$ for each source zone and this aggregated file was joined to OALU. Given this, values of $\hat{\mathbf{z}}_{st}$ can then be derived.

The areal weighting procedure applied in this paper is as follows:

1. Using a union overlay of OA and LU, Produce OALU

2. Compute populations (OALU_Estimate) for each OALU segment using Eq. (20.2) (WtArea = Wt × OALUArea; WtAreaSum = WtArea summed by OA; Join WtAreaSum back to OALU using the OA code; OALU_Estimate = WtArea/WtAreaSum × OAPop)
3. Union overlay grid (e.g., 1 km for E & W) with OALU = OALUG
4. Compute populations for each OALUG area using Eq. (20.1) (AreaOALUG/OALUArea × OALU_Estimate)
5. Aggregate these populations by grid cell (giving gridded values GRIDEST)

An additional step is to smooth cells (with the pycnophylactic condition); this is problematic since some cells comprise parts of several OAs and smoothing the values in these cells would result in the need for an unnecessarily complex procedure to ensure global mass preservation. A solution to this problem which allows for mass preservation is to (i) group adjacent cells which fall entirely within a OALU zone and (ii) compute their summed population. Then (iii) only the cells identified in (i) are smoothed and, at each stage, individual cell values within groups identified in (i) are rescaled so that their collective population remains the same.

The smoothing procedure was implemented as follows: The areas of OALUG were used to determine all 1 km cells which fell completely within a OALU polygon – these cells as labelled '1'; all other 1 km cells are labelled '0' (Fig. 20.1 shows an example of cells within an OALU boundary – three are shaded dark grey and are labelled '1'). The internal boundaries of the grid cell polygons are dissolved, thus creating contiguous areas of cells labelled '0' or '1', each new area has a unique ID; and the total population of each new zone (NewZonePop) is determined. These values are attached to GRIDEST using a point in polygon operation (grid cell centroids are linked to the new 0/1 areas) and the grid is smoothed using a three-by-three cell window. Only for cells labelled '1' is the smoothed value (SmoothGRIDEST) retained; for cells labelled '0' the original cell value is retained. After smoothing, the smoothed values for all neighbouring cells labelled '1' are rescaled so that the sum of smoothed cell values within each zone remains the same (and thus mass is preserved). This rescaling is done by computing the sum of all SmoothGRIDEST values within each new zone = SumSmoothGRIDEST and the rescaled values are then given by: SmoothGRIDEST/SumSmoothGRIDEST × NewZonePop. The grid was updated to contain the rescaled smoothed values (for cells labelled '1') and the original cell values (for cells labelled '0'). Building on the approach of Tobler (1979), this grid was smoothed again iteratively and the root mean square (RMS) difference between output grid values computed at each stage. The iterative smoothing process stopped once the RMS difference between '1' cells in the most recently smoothed grid and in the previous smoothed grid became smaller than 0.001.

Figure 20.1 indicates the selection of cells for smoothing. Dark cells are contiguous and un-split by OAs or LU boundaries. These three cells are grouped and their total population recorded. Each of these cells is smoothed using a three-by-three pixel moving window and then their populations are rescaled such that their total combined population remains constant, as described above. For clarity, the

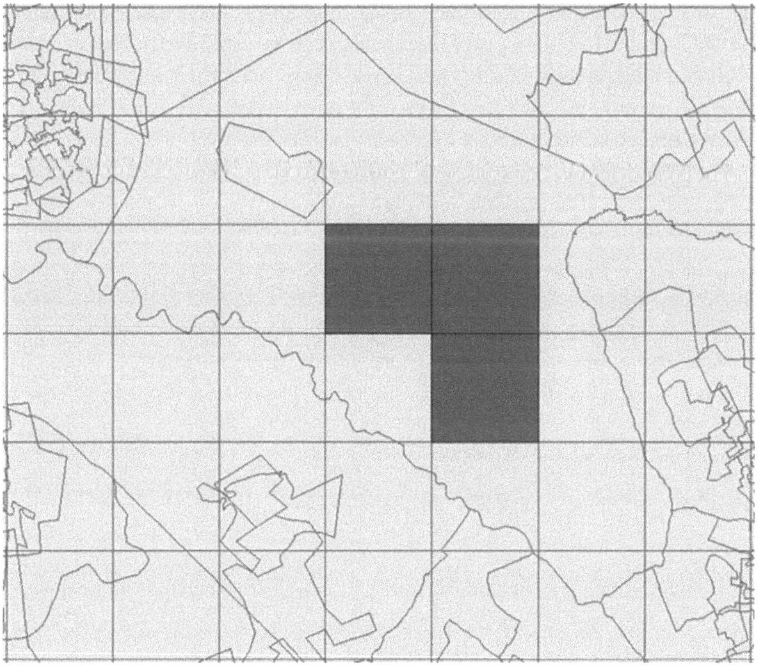

Fig. 20.1 Example of selection of cells for smoothing

populations attached to the three individual cells are the same before smoothing but are likely to differ after smoothing while their total combined population remains constant. As an example, if there are three contiguous '1' cells with a total population of 120 (by definition, these are equal at first: 40, 40, 40) and each is smoothed and become, say, 35, 45, 46 then the total population has increased by 6. The *total* population is required to remain the same yet the *proportion* of the total population should be the same as the new smoothed values. Therefore: $35 + 45 + 46 = 126$. Then, $35/126 \times 100 = 27.78$; $45/126 \times 100 = 35.71$; $46/126 \times 100 = 36.51$. Next, we assign each of the three cells the relevant percentage of the total population (120): $120/100 \times 27.78 = 33.33$; $120/100 \times 35.71 = 42.86$; $120/100 \times 36.51 = 43.81$. Now, the total populations of the three cells are $33.33 + 42.86 + 43.81 = 120$. The cells are iteratively smoothed again and rescaled until the values change by a sufficiently small amount (i.e., below the RMS threshold). Smoothing cells which overlap two or more source zones was avoided as this would entail transfer of people outside of their original source area.

The approach is globally mass preserving since the smoothed cells are appropriately rescaled. But, given the splitting of source zones by target zone boundaries, it is not locally mass preserving. Local mass preservation is possible if grid centroids are used to link cell locations to source zones. However, such an approach is sub-optimal; Martin et al. (2011) generated population surfaces with 100 m cells but they did so by making estimates to 25 m or 50 m cells and then aggregating these to 100 m cells. This is superior to estimating directly to 100 m cells since some 100 m cell boundaries are likely to overlap more than one source zone and thus assigning counts to each cell from only one source zone is unrealistic and can result in locally large errors. The approach is, like that presented here, not locally mass preserving since the final output includes 100 m cells which may take values from one than one source zone and thus summing populations for cells within source zones will not result in the same total populations. It is argued that an approach based on polygon overlay rather than centroid linkage is superior since it enables proper proportional reallocation of counts. It is also argued that *local* mass preservation is not a sensible objective in population surface modelling as grid cells should have the capacity to take a share of the population from more than one source zone.

20.3 Assessing Spatial Variation in Counts

The iterative smoothing approach is used on the assumption that the optimal degree of smoothing will vary by population sub-group (see Lloyd and Firoozi Nejad 2014 for a related study). The optimal degree of smoothing will be a function of the degree of spatial variability (spatial dependence) in counts. Here, the Moran's I autocorrelation coefficient is used to measure the degree of spatial dependence in the counts used in the case study. Moran's I (Moran 1950; Cliff and Ord 1973) with weights, w_{ij}, between locations $\mathbf{s}_i$ and $\mathbf{s}_j$ row-standardised (i.e., the weights for each i sum to one) can be given by:

$$I = \frac{\sum_{i=1}^{n}\sum_{j=1}^{n} w_{ij}\left(z(\mathbf{s}_i) - \bar{z}\right)\left(z(\mathbf{s}_j) - \bar{z}\right)}{\sum_{i=1}^{n}\left(z(\mathbf{s}_i) - \bar{z}\right)^2} \quad (20.3)$$

where the values $z(\mathbf{s}_i)$ have the mean $\bar{z}$. Positive values of I indicate positive spatial autocorrelation (spatial dependence), while negative values indicate negative spatial autocorrelation. The weights are here based on queen contiguity, whereby all adjacent zones (connected by edges or vertices) receive an equal weight and then these are row standardised. Thus, if there are five neighbouring zones then each receives a weight of 0.2. Moran's I was computed using the GeoDa software (Anselin et al. 2006). GeoDa incorporates a permutation approach to determine the pseudo significance levels for global and local spatial autocorrelation statistics. In GeoDa, the data values are spatially reconfigured randomly m times and I is computed given each random configuration. The reference distribution of I derived in this way is then compared to I for the observed data and the probability that the observed value could be derived from a random distribution is ascertained. In this analysis 999 permutations (corresponding to a pseudo-significance level of $p = 0.001$; as $1/999 = 0.001$) were used.

Table 20.2 Moran's *I* for each of the sets of counts by OA. All values have a pseudo *p* value of 0.001

Counts	2001	2011
White	0.369	0.413
Non-white	0.819	0.838
LLTI	0.387	0.356
No LLTI	0.208	0.215

Lloyd (2015) analysed clustering in a set of variables (log-ratios) derived from percentages of the population by sub-groups including by ethnicity (White/non-White) and health status (No LLTI/LLTI). That analysis was based on Moran's *I* for a geographically weighted neighbourhood (20 nearest neighbours) using OAs. The values of *I* for White/non-White were 0.754 for 2001 and 0.841 for 2011; for No LLTI/LLTI the value for 2001 was 0.397, while for 2011 it was 0.411. Values of *I* by OA for the counts of persons in the four groups used in this study are given in Table 20.2; all values have a pseudo *p* value of 0.001. The large differences in values for the counts, with large values for Non-White and small or moderate values for the other counts, provides justification for an iterative smoothing approach.

20.4 Generation of Population Surfaces

The key focus of the paper is on the method used to generate population surfaces and examples are used primarily to illustrate the method rather than because of what they might tell us about population change. The overlay of OAs, land use and grid cells is the same for all population sub-groups and it is only in the number of smoothing iterations that the approaches differ. Table 20.3 indicates the number of smoothing iterations that was required to reach convergence (i.e., where the RMS difference between surfaces became less than 0.001) in the case of the four sets of counts for the 2001 and 2011 Census.

The figures suggest that a smaller number of iterations is required to reach convergence for 'smoother' counts (i.e., those which are more spatially continuous) than is the case for less-smooth counts and the Moran's *I* values (Table 20.2) are negatively correlated with the numbers of smoothing iterations. The Moran's *I* values indicate that counts of non-White persons are relatively spatially continuous and this reflects larger numbers in urban areas and consistently smaller numbers in more rural areas. In this case, the counts are not changed markedly by smoothing as they are already highly spatially continuous. In contrast, the other three sets of counts are subject to a larger degree of spatial variation. As an example, *proportions* of White persons are generally high in most areas with smaller proportions in urban areas but *counts* are spatially variable and much more so than counts of non-White persons, which may be close to zero in many areas. As such, the value of results obtained using a iterative smoothing approach, like that employed here, may be seen as counter-logical in that less spatially-continuous counts are smoothed more (in the sense of numbers of iterations) than

Table 20.3 (i) The number (n) of smoothing iterations for each count until the RMS difference decreased to less than 0.001 and (ii) RMS values for the 1st versus the 2nd smoothed grids. Note that, for example, 3 iterations refers to 4 smoothing steps with the 1st RMS calculation comparing the 1st and 2nd stage smoothed grids

	<0.001			
	2001	2011	2001	2011
Counts	n iterations		RMS (1st vs 2nd smoothed grids)	
White	16	11	0.526	0.564
Non-white	3	3	0.009	0.012
LLTI	10	7	0.105	0.117
No LLTI	15	10	0.431	0.463

are more spatially continuous counts. This suggests that a simple iterative approach may be problematic if applied in isolation and that the approach could usefully be modified. It is assumed that some smoothing is desirable so as to avoid clumps of identical values (particularly within large rural OAs), and the question is not whether to smooth, but how much. One possible adaptation of the approach is to constrain the RMS change allowed to some prior maximum such that, for example, the smoothing process stops if the RMS becomes less than 0.001 *or* if it is greater than 0.1. Table 20.3 also includes the RMS values computed for the first smoothing iteration with respect to the second. In this case, with an upper limit of 0.1 for the RMS, grids produced for White, LLTI and No LLTI would be smoothed only once while the grid for Non-White would reach three iterations before the RMS values meet the minimum value criterion. Of course, the RMS values are a function of the counts in each population subgroup and so these could be rescaled so that they are proportions, for example, rather than raw counts. In the following examples, the full set of iterations is used to generate grids (e.g., 16 for the White group in 2001) as the examples are illustrative and the amount of smoothing will have little impact on national-level trends.

20.5 Analysis of Population Change

Population surfaces generated using the combined overlay and iterative smoothing approach are here used to illustrate some approaches to analysis of population change over small areas. Figure 20.2 shows non-White persons as a percentage of all persons in 2011, while Fig. 20.3 shows these percentages minus the equivalent figures for 2001. Comparison of Figs. 20.2 and 20.3 suggests that the largest percentage point increases in non-White persons were in areas with larger shares of non-White persons. This is consistent with an observation of higher birth rates among non-White groups (or at least non-White British) and growth of these groups *in situ*. But there is proportional growth in other areas as well and this supports the idea of a growth in diversity in areas which were formally fairly ethnically

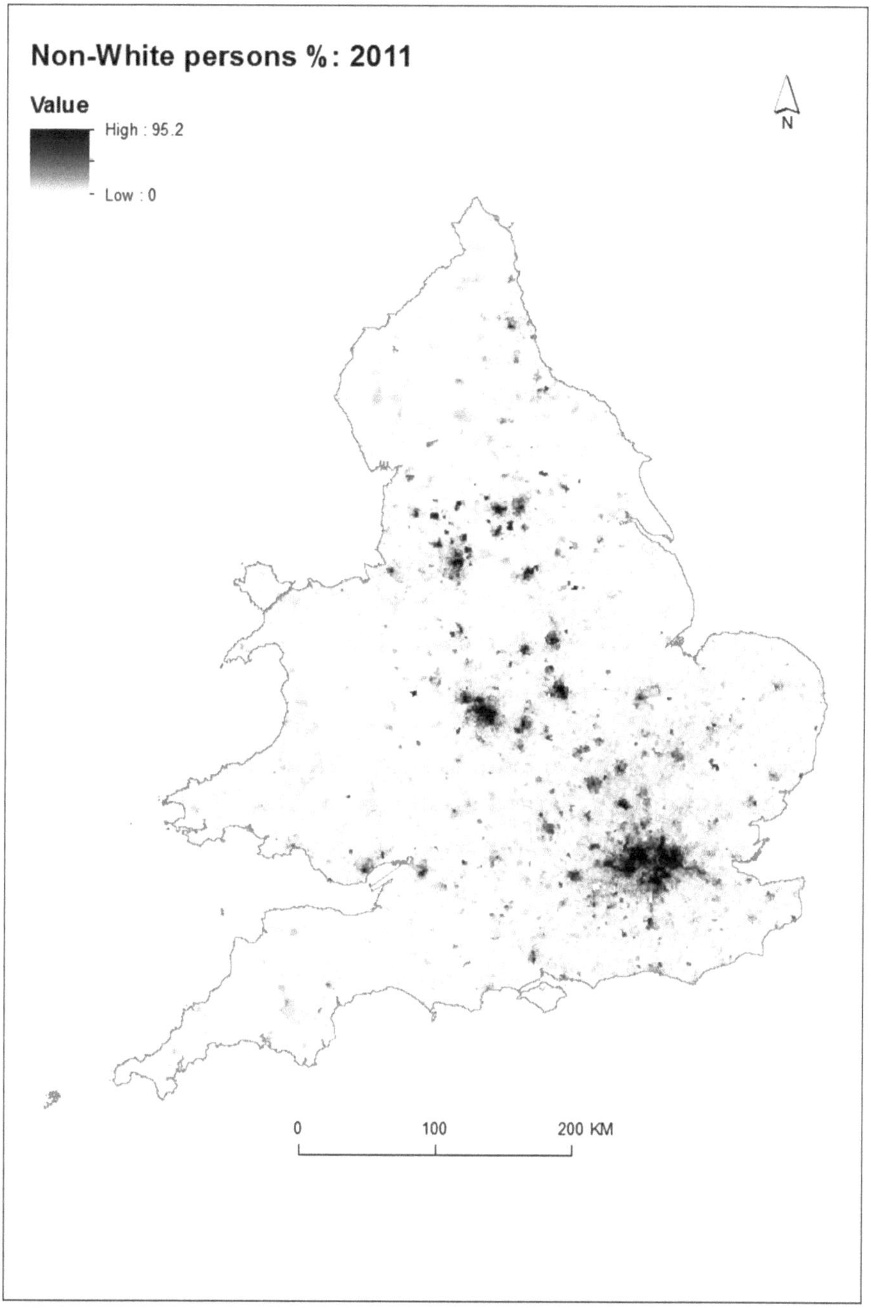

Fig. 20.2 Non-white persons (%) in 2011

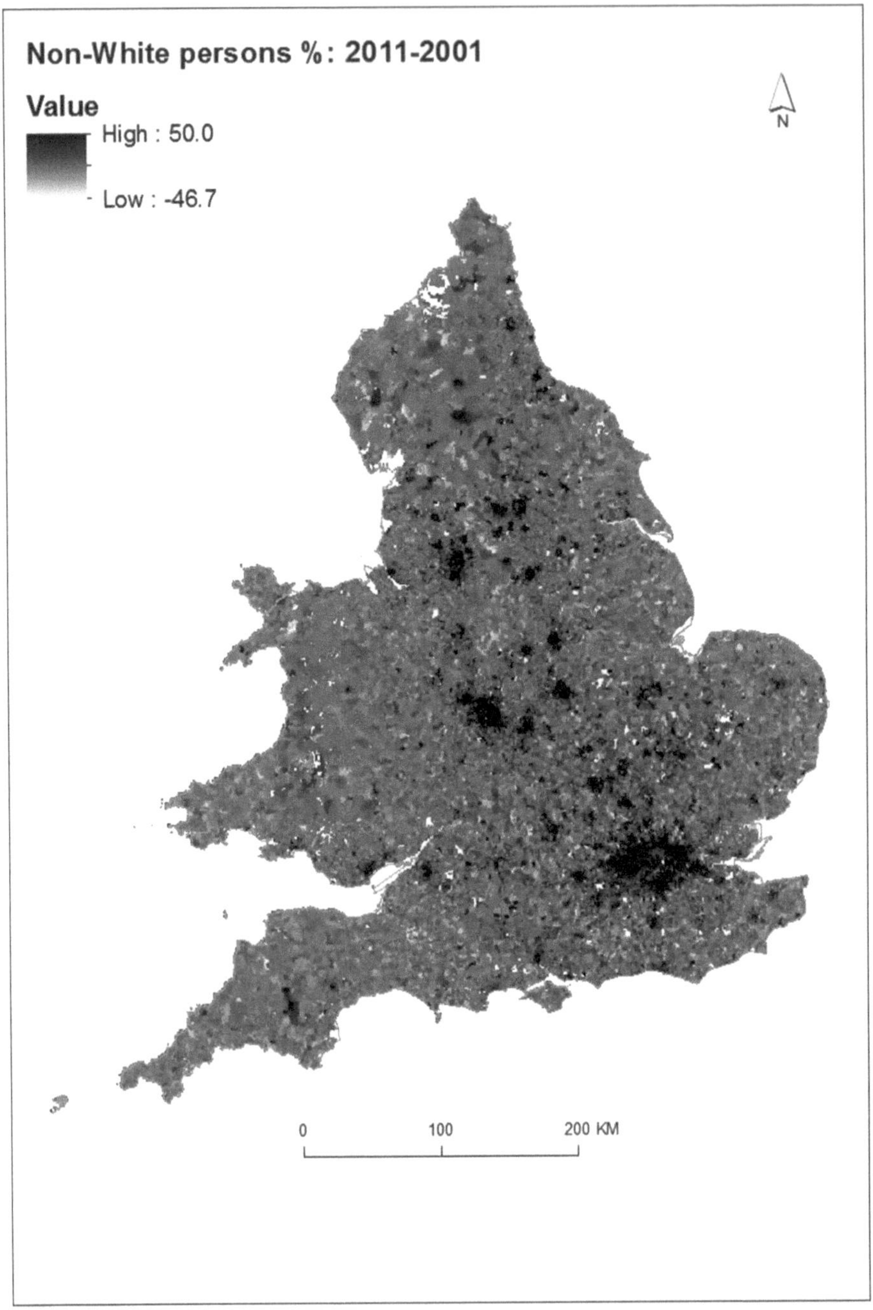

Fig. 20.3 Non-white persons (%) in 2011 – Non-white persons (%) in 2001

homogenous (Catney 2015a). The maximum increase is 50 %, but most increases are much smaller. The large majority of increases are by less than 5 % and the equivalent is true of decreases (i.e., most decreases are by less than 5 %). Catney (2015a) maps change in the non-White population 1991–2011 by wards and observes similar patterns of growth in urban areas, but also in other areas than had been predominantly White in 1991. It is worth nothing that counts for some cells are very small (and may be smaller than one) and smaller counts could be suppressed. For the Northern Ireland Census grid square product (see Shuttleworth and Lloyd 2009 for a summary), for grid squares containing at least 25 persons and eight or more households (both criteria must be satisfied) the full range of counts by sub-group are reported, while for cells which do not meet these criteria the outputs are restricted to total males, total females, total persons and total households.

In the case of percentages of persons with a LLTI (not shown for reasons of space), there no obvious spatial trends in the differences. However, there are decreases visible in some urban areas, including parts of London and this may be associated with a younger age profile in such areas in 2011 than in 2001. The large majority of increases and decreases are by less than 5 %. Very large increases may be of most interest and these could be extracted and highlighted. Of course, the differences considered here cover only a 10 year period and maps representing longer time periods would be expected to show larger changes and perhaps stronger geographical trends.

Maps of percentage point change, as presented above, provide one perspective. An alternative is to instead explore growth or loss in a group as a percentage of its population in the first of two Census years. Figure 20.4 shows the Non-White population in 2011 minus the equivalent in 2001 divided by the 2001 count, and finally multiplied by 100 to give a percentage. The map shows only cells with a Non-White population $>=$ 100 in 2001. The ring of relatively larger increases around London as opposed to more centrally is suggestive of a process of movement by members of some non-White groups from urban core areas to suburban areas (Catney 2015b) but only migration data can confirm this.

The use of grids rather than irregular zones allows for a rich exploration of population change over small areas unhampered by the visual and analytical limitations of zones which may vary according to, for example, population size or administrative whimsy. In addition, gridded population estimates offer advantage in that links to environmental data sets, which are often gridded, are possible thus allowing the exploration of the relationships between the populations of areas and their environmental characteristics.

20.6 Discussion and Conclusions

The approach presented here differs to other surface modelling approaches in that populations of cells are determined using intersections with zones rather than simple allocation based on grid centroids. Martin et al. (2011) used the

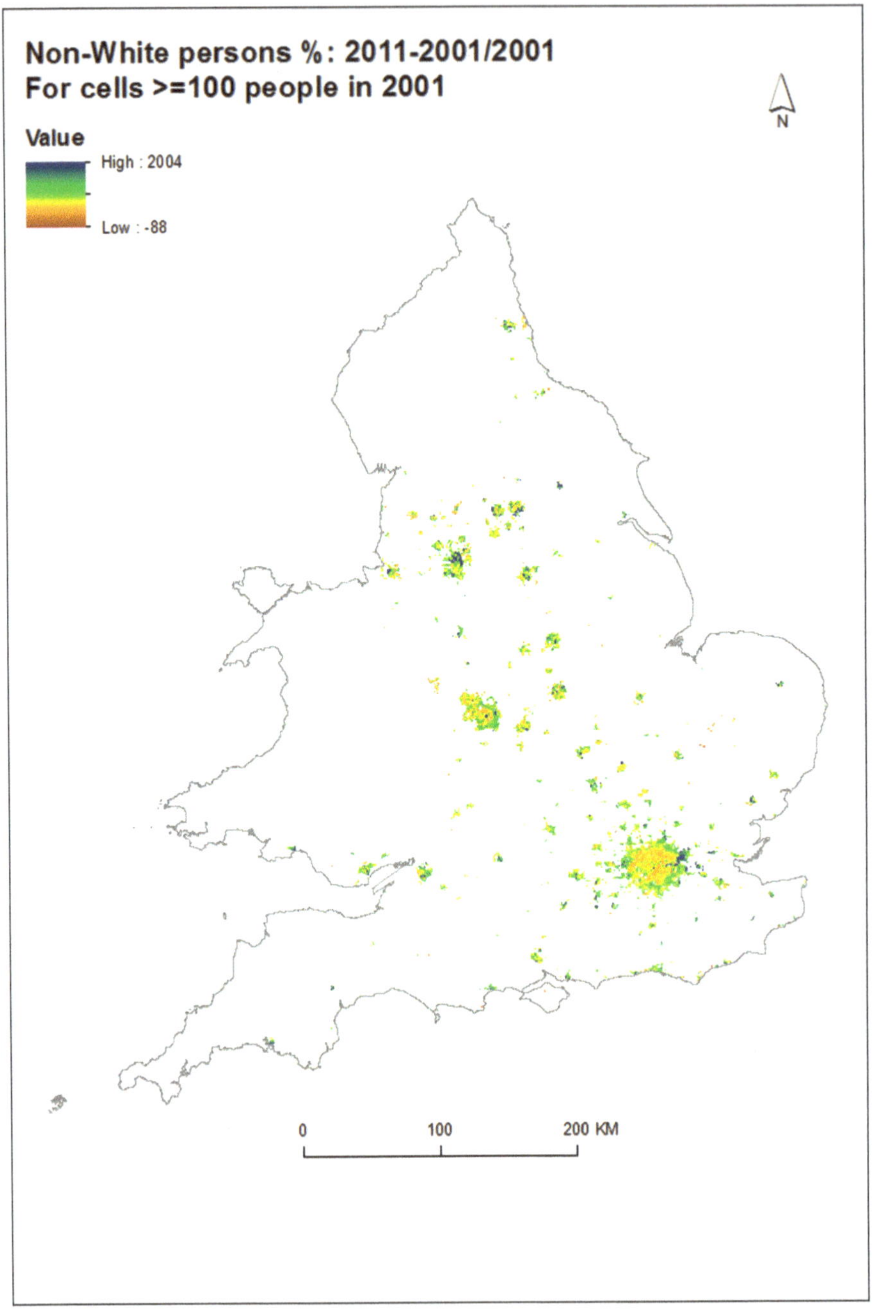

Fig. 20.4 Non-white persons (%) in (2011–2001)/2001. For cells with > = 100 non-white people in 2001

SurfaceBuilder software[1] to construct a series of gridded population models for Northern Ireland, using grid cell sizes of 25 m and 50 m; these were then aggregated to 100 m cells matching the cells for which true counts are available from the Northern Ireland Census. While such an approach can provide accurate estimates, in the present analysis an intersection-based approach was preferred since it does not approximate linkage between grid cells and source zones.

The present method differs conceptually from the method of Tobler (1979) in that Tobler's method seeks to reduce differences between neighbouring cells which fall within different source zones while, in the current study, cells at the edge of source zones (and which are split by more than two source zones) are not smoothed and the focus is instead on smoothing cells completely contained by source zones but which neighbour cells which are split by source zones thus reducing differences between cells which are 'central' to a source zone and those at the edge of each source zone.

Building on the work of Lloyd and Firoozi Nejad (2014) the analysis shows that the optimal amount of smoothing differs for different variables. This is intuitively sensible since smoothing spatially discontinuous counts results in an appropriate reduction in differences between neighbouring areas while, with spatially continuous counts, smoothing is beneficial as it makes cells located completely within source zones more similar to neighbouring cells at the edge the source zone reducing the likelihood of a 'step' between one cell and another.

The present approach is based on iterative smoothing of gridded counts obtained using overlays of OA-level data and landuse data. It is shown that such an approach has a greater impact on counts which are less spatially-continuous and this runs counter to the aim of adapting the process such that spatially continuous counts are smoothed more than are counts which vary more between adjacent areas. Some smoothing of gridded counts is likely to be conceptually beneficial in most cases as this means that identical counts in adjacent cells which are completely-contained by source zones (overlay of OAs and landuse areas) are avoided and this is supported by the work of Lloyd and Firoozi Nejad (2014) using gridded population counts for Northern Ireland. An approach is suggested which iteratively smoothes the output grids until the RMS difference between grids reduces to less than 0.001. An adaptation, whereby an RMS value of greater than 0.1 also stops the smoothing process, was also considered as this prevents over-smoothing of spatially discontinuous counts. The complexities which arise for assessing different population sub-groups suggest that an iterative smoothing approach is not intuitive and may not produce logical results (e.g., by over-smoothing in the case of groups where marked differences between neighbourhoods are found) and, as such, alternatives should be explored.

One additional adaptation to the smoothing approach outlined here is to use, for example, the local variance or Moran's I to measure how much spatial variation changes at each smoothing iteration and to increase the number of iterations only if

[1] http://www.public.geog.soton.ac.uk/users/martindj/davehome/software.htm

the measured spatial variation changes by a sufficiently small amount while the RMS difference, as outlined above, must be sufficiently small for another smoothing iteration to take place. The most conservative approach would be to use a single smoothing iteration for all counts and the assumption is made that some smoothing is beneficial for all sub group counts as this prevents clumps of identical values appearing in outputs and clearly such clumps are unlikely to exist in reality. In practice (and as suggested by Lloyd and Firoozi Nejad 2014), if there are sufficiently spatially detailed ancillary information (e.g., post codes or land use), then the impact of smoothing is likely to be negligible. In the case study presented here, OAs are often small in relation to 1 km grid cells and the landuse data provide a high level of spatial detail. An alternative approach to iterative smoothing is based on variogram deconvolution and area-to-point kriging (e.g., Kyriakidis 2004; Goovaerts 2008). Yoo et al. (2010) present a comparison of Tobler's pycnophylactic method and area-to-point kriging. Future work will assess what might be added by replacing the smoothing process utilised here with a variogram deconvolution approach; although, as suggested here, the impact is likely to be small where ancillary data are spatially detailed and source zones are often small relative to target zones.

The optimal approach to generation of population surfaces in the UK may be one based on postcode data combined with land use information. The purpose of the present study was to develop an approach which was as transferable (geographically and temporally) as possible and thus land use data are utilised in reallocating counts. The accuracy of estimates made in this study cannot be tested directly but the approach builds directly on the work of Lloyd and Firoozi Nejad (2014) which assessed accuracy using grid square data released as an output from the 2001 Northern Ireland Census. The latter research justifies the use of ancillary data (landuse) and a smoothing approach in combination. In the present study, 1 km grids were generated and in many cases these are larger than the source zones. Where the aim is to generate finer grids (e.g., 100 m) the challenge is clearly greater. But, in more densely populated areas in particular, an approach like the one detailed here should prove adequate; future work will test this assertion.

Population surfaces allow for detailed assessment of geographical changes in individual variables and in relationships between variables. They open up the capacity to explore long-term deprivation patterns, to produce geodemographic classifications and assess how these change over small areas, and to profile area change in many different ways. In addition, gridded population data can be interrogated with a rich array of methods which have been developed for the processing and analysis of raster grids (e.g., see Sonka et al. 2015). A key benefit for statistical analyses is that the zonal units are of equal size and thus results are comparable between regions in a way which may not be strictly true with irregular regions.

The paper presents an approach for population surface modelling which is based on intersection of grid cells with source zones rather than use of a point in polygon operation which links grid cell centroids to source zones. The smoothing of counts reallocated to grid cells is then implemented using an iterative approach. The approach is flexible and can be applied using any suitable ancillary data. The next phase of the work is being supported by the award of an Economic and Social Research Council (ESRC) grant under the Secondary Data Analysis Initiative Phrase 2 (grant ES/L014769/1) and the approach is being extended to the whole of the UK for each Census year from 1971–2011. Surface models (1 km grids and 100 m grid for at least some areas) are being generated for all available comparable variables and other variables available only for some Census years. This resource will open up the potential for major studies of small area population change in the UK over 40 years as well as developing approaches which can be applied internationally to facilitate studies of population change generally and geographic inequalities specifically.

Acknowledgments The Office for National Statistics is thanked for the provision of data. Office for National Statistics, 2001 and 2011 Census: Aggregate data (England and Wales) [computer file]. UK Data Service Census Support. Downloaded from: http://casweb.mimas.ac.uk/http://infuse.mimas.ac.uk. This information is licensed under the terms of the Open Government Licence [http://www.nationalarchives.gov.uk/doc/open-government-licence/version/2]. Office for National Statistics, 2001 and 2011 Census: Digitised Boundary Data (England and Wales) [computer file]. UK Data Service Census Support. Downloaded from: http://casweb.mimas.ac.uk/http://edina.ac.uk/census. Census output is Crown copyright and is reproduced with the permission of the Controller of HMSO and the Queen's Printer for Scotland. This research was supported by the British Academy/Leverhulme Trust (Small Research Grant award SG121849) and the Economic and Social Research Council (Grant ES/L014769/1), and this funding is acknowledged gratefully.

References

Anselin, L., Syabri, I., & Kho, Y. (2006). GeoDa: An introduction to spatial data analysis. *Geographical Analysis, 38*, 5–22.

Catney, G. (2015a). The changing geographies of ethnic diversity in England and Wales, 1991–2011. *Population, Space and Place* (in press).

Catney, G. (2015b). Exploring a decade of small area ethnic (de-)segregation in England and Wales. *Urban Studies* (in press).

Cliff, A. D., & Ord, J. K. (1973). *Spatial autocorrelation*. London: Pion.

Flowerdew, R., & Green, M. (1994). Areal interpolation and types of data. In S. Fotheringham & P. Rogerson (Eds.), *Spatial analysis and GIS* (pp. 121–145). London: Taylor and Francis.

Gallego, J. (2010). A population density grid of the European Union. *Population and Environment, 31*, 460–473.

Goovaerts, P. (2008). Kriging and semivariogram deconvolution in the presence of irregular geographical units. *Mathematical Geosciences, 40*, 101–128.

Gregory, I., & Ell, P. (2005). Breaking the boundaries: Geographical approaches to integrating 200 years of the census. *Journal of the Royal Statistical Society, Series A, 168*, 419–437.

Kim, H., & Yao, X. (2010). Pycnophylactic interpolation revisited: Integration with the dasymetric-mapping method. *International Journal of Remote Sensing, 31*, 5657–5671.

Kyriakidis, P. C. (2004). A geostatistical framework for area-to-point spatial interpolation. *Geographical Analysis, 36*, 259–289.

Lloyd, C. D. (2014). *Exploring spatial scale in geography*. Chichester: Wiley.

Lloyd, C. D. (2015). Assessing the spatial structure of population variables in England and Wales. *Transactions of the Institute of British Geographers, 40*, 28–43.

Lloyd, C. D., & Firoozi Nejad, B. (2014). Surface models and the spatial structure of population variables: Exploring smoothing effects using Northern Ireland grid square data. *Computers, Environment and Urban Systems, 48*, 64–72.

Martin, D. (1989). Mapping population data from zone centroid locations. *Transactions of the Institute of British Geographers, New Series, 14*, 90–97.

Martin, D. (1996). An assessment of surface and zonal models of population. *International Journal of Geographical Information Systems, 10*, 973–989.

Martin, D., Dorling, D., & Mitchell, R. (2002). Linking censuses through time: Problems and solutions. *Area, 34*, 82–91.

Martin, D., Lloyd, C., & Shuttleworth, I. (2011). Evaluation of gridded population models using 2001 Northern Ireland census data. *Environment and Planning A, 43*(8), 1965–1980.

Mennis, J. (2003). Generating surface models of population using dasymetric mapping. *The Professional Geographer, 55*, 31–42.

Moran, P. A. P. (1950). Notes on continuous stochastic phenomena. *Biometrika, 37*, 17–23.

Openshaw, S. (1984). *The modifiable areal unit problem* (Concepts and techniques in modern geography, Vol. 38). Norwich: GeoBooks.

Openshaw, S., & Taylor, P. J. (1979). A million or so correlation coefficients: Three experiments on the modifiable areal unit problem. In N. Wrigley (Ed.), *Statistical applications in the spatial sciences* (pp. 127–144). London: Pion.

Ordnance Survey. (2015). Strategic user guide and technical specification, version 9.2. Southampton: Ordnance Survey. https://www.ordnancesurvey.co.uk/docs/user-guides/strategi-user-guide.pdf. Accessed 20 Oct 2015.

Shuttleworth, I. G., & Lloyd, C. D. (2009). Are Northern Ireland's communities dividing? Evidence from geographically consistent census of population data, 1971–2001. *Environment and Planning A, 41*, 213–229.

Sonka, M., Hlavac, V., & Boyle, P. (2015). *Image processing, analysis and machine vision* (4th ed.). Stanford: Cengage Learning.

Tobler, W. R. (1979). Smooth pycnophylactic interpolation for geographical regions. *Journal of the American Statistical Association, 74*, 519–530.

Walford, N. S., & Hayles, K. N. (2012). Thirty years of geographical (in)consitency in the British population census: Steps towards harmonisation of small-area census geography. *Population, Space and Place, 18*, 295–313.

Wong, D. (2009). The modifiable areal unit problem (MAUP). In A. S. Fotheringham & P. A. Rogerson (Eds.), *The SAGE handbook of spatial analysis* (pp. 105–123). London: SAGE.

Yoo, E. H., Kyriakidis, P. C., & Tobler, W. (2010). Reconstructing population density surfaces from areal data: A comparison of Tobler's pycnophylactic interpolation method and area-to-point kriging. *Geographical Analysis, 42*, 78–98.

Chapter 21
Small-Area Population Forecasting: A Geographically Weighted Regression Approach

Guangqing Chi and Donghui Wang

Abstract The regression approach for small-area population forecasting is increasingly used in urban planning and emerging research in climate change and infrastructure systems in response to disasters because the regression approach can not only provide population projections but also estimate the relationships between population change and possible driving factors. In this research, we use the geographically weighted regression (GWR) method to estimate relationships between population change and a variety of driving factors and consider possible spatial variations of the relationships for small-area population forecasting using 1990–2010 data at the minor civil division level in Wisconsin, USA. The results indicate that the GWR method provides an elegant estimation of the relationships between population change and its driving factors, but it underperforms traditional extrapolation projections. The findings have important implications about the need for more accurate population projections versus more accurate estimation of the relationships between population and its driving factors.

Keywords Population forecasting • Small area • GWR • Regression

21.1 Introduction

Population forecasts at subcounty levels have long been used for urban planning pertaining to such as smart growth, comprehensive planning, growth management, federal transportation legislation, and other areas (Chi 2009). The use of population forecasts at subcounty levels are also increasingly in demand by the research community, especially in research related to climate change and infrastructure systems in response to disasters. For example, climate change research often treats

G. Chi (✉) • D. Wang
Department of Agricultural Economics, Sociology, and Education, The Pennsylavania State University, State College, PA, USA

Computational and Spatial Analysis Core, Population Research Institute and Social Science Research Institute, The Pennsylvania State University, State College, PA, USA
e-mail: gchi@psu.edu; duw168@psu.edu

D.A. Swanson (ed.), *The Frontiers of Applied Demography*, Applied Demography Series 9, DOI 10.1007/978-3-319-43329-5_21

population growth as a causal factor of climate change and incorporates population forecasts as an element in climate projection (Lutz and Striessnig 2015; USGCRP 2015) Also, emerging research in interdependent infrastructure systems in response to disasters requires population forecasts at subcounty levels to prepare for possible disasters in various scenarios (Cutter and Smith 2009; Wood et al. 2009).

There are many methods for small-area population forecasting purposes, including extrapolation projections (Armstrong 2001), time-series models (Armstrong et al. 2001), postcensal population estimation models (Espenshade and Tayman 1982; Swanson and Beck 1994; Swanson and Tayman 2014; Swanson et al. 2010) conditional probabilistic models (Alho 1997; O'Neill 2004; Sanderson et al. 2004), integrated land use models (Agarwal et al. 2002; Deal et al. 2005; Hunt and Simmons 1993), population forecasting by grid cells (Riahi and Nakicenovic 2007), spatial Bayesian models (Anselin 2001), and knowledge-based regression models (Ahlburg 1987; Lutz and Goldstein 2004; Meadows et al. 1972, 1992; Sanderson 1998). Chi (2009) and Wilson and Rees (2005) provide a review of the literature.

Among the small-area population forecasting methods, the knowledge-based regression approach has received increasing attention because it not only provides population forecasts, but more important, it can estimate the relationships among population growth, traffic flow, land use, infrastructure, climate change, and many other demographic, socioeconomic, and built and natural environmental factors. The latter are important for addressing traditional demand in urban planning and in meeting the increasing demand in research on climate change and disasters.

The existing knowledge-based regression approach for small-area population forecasts typically estimates the relationship between population change and causal factors in the estimation period and then applies the estimated parameters to the projection period. The simplest regression forecasting model is the standard ordinary least squares (OLS) regression forecasting model, which considers a variety of driving factors of population change (e.g., Chi 2009). A spatial regression forecasting model has also been developed (e.g., Chi and Voss 2011) to include the impact of population growth in neighboring geographic units. For example, the opening of a factory in Town A could add several hundred, or even a thousand, employment opportunities. The inflow of new employees and their families to Town A would increase housing prices of Town A. Some new employees and their families may choose to live in Town A's neighbor Town B, where housing prices are relatively lower. In that case, population growth in Town B is due to population growth in Town A, but it has nothing to do with Town B's characteristics. The only reason that Town B receives this benefit is because it is geographically close to Town A. This is a phenomenon that can be well explained by Tobler's First Law of Geography (1970), which states that everything relates to everything else, but the nearer one does more. In addition, the spatial regression forecasting approach could further include the impacts of driving factors in neighboring geographic units. In the example given, the size of the factory in Town A could have an impact on population growth in Town B (i.e., the larger the factory, the more workers it would hire—and the more migrants Town B might receive).

When we use regression methods for population forecasting, whether standard regression or spatial regression, there is only one coefficient to be estimated for each explanatory variable. This coefficient, often called the global coefficient, applies to all geographic units of the study area. If opening a factory in Town A adds 1000 people to Town A, opening a factory in Town C would add 1000 people in Town C, regardless where the two towns are. But what if Town A is a Dallas-type suburb and Town C is a small town in the South? Town A could consume the job opportunities created by the new factory because it is large and has an adequate labor force. Town C, however, may have to encourage workers to move there if it does not have a skilled labor force. In that case, the effect of opening a factory on population growth would be stronger in Town C than in Town A. Therefore, the effect of opening a factory would differ from one location to another. The possible variation of the effects that explanatory variables have on population change can occur for many other driving factors. However, the spatial regression forecasting models are not capable of capturing the possible spatial variations of the effects that the driving factors have on population change.

In this study, we adopted a geographically weighted regression (GWR) approach for small-area population forecasting. The GWR method allows us to capture the possible variation of effects that the explanatory variables have on the response variable across space. We hypothesize that the delicate configuration of the effects can provide a more accurate estimate of the relationship between population change and its related factors, which would in turn improve the accuracy of population projections. More importantly, the spatial variation of the effects captured by the GWR method could provide further insights for urban and regional planning purposes as well as research in climate change and disasters.

21.2 The GWR Approach for Population Forecasting

21.2.1 The GWR Model

The principles of GWR models are described in the book *Geographically Weighted Regression: The Analysis of Spatially Varying Relationships* (Fotheringham et al. 2002). GWR models have received growing attention from social scientists as a means to address the spatial heterogeneity of social science phenomena, such as rural development (Ali et al. 2007), environmental justice (Mennis and Jordan 2005), urban studies (Yu et al. 2007), poverty research (Longley and Tobon 2004), public health and epidemiology (Fraser et al. 2012; Matthews and Yang 2012; Yang and Matthews 2012), and others.

Here we provide a brief description of basic GWR models.

A basic GWR model is specified as:

$$Y = X\beta(u, v) + \varepsilon \tag{21.1}$$

where

Y is an n by 1 vector of response variables,
X is an n by p design matrix of explanatory variables,
(u,v) is the coordinate of each location in space,
$\beta(u,v)$ is a p by n design matrix of explanatory variables by a continuous function across each location in space, and
ε is an n by 1 vector of error terms that are independently and identically distributed.

The terms Y and X are denoted as in a standard linear regression model (Eq. 21.1), but the remaining terms of the model, $\beta(u,v)$ and ε, are not.

The key difference between GWR models and other regression models, spatial or aspatial, is in $\beta(u,v)$. $\beta(u,v)$ is a continuous function across each location in space, meaning that the coefficients (1) vary across space and (2) do so continuously (or smoothly). To understand this more clearly, we rewrite the model in a more common way, which happens to be the original way:

$$y_i = \sum_{k=0}^{p} \beta_k(u_i, v_i)x_{ik} + \varepsilon_i \tag{21.2}$$

where

i denotes the ith location (or observation) in space,
y_i denotes the response variable for the ith observation,
(u_i,v_i) denotes the coordinate of the ith location in space,
k denotes the number of explanatory variables including the constant ($k = 0$),
x_{ik} denotes the explanatory variable k for the ith observation,
$\beta_k(u_i,v_i)$ is the realization of the continuous function $\beta_k(u,v)$ at location i, and
ε is an n by 1 vector of error terms that are independently and identically distributed.

This equation suggests that $\beta_k(u_i,v_i)$ could vary for each location i. If we assume that the coefficients vary randomly, that leaves us with $(p+1)\cdot n+1$ parameters to estimate, which is more than what the n observations allow us to do. To solve this dilemma, the GWR model does not assume the coefficients to be random; rather, it assumes them to be deterministic functions of the corresponding X and Y in location i and its neighbors. This is similar to the spatial lag and spatial error models in which we have a neighborhood structure to define the neighbors and a spatial weight matrix to quantify the effects of each neighbor.

The spatial weight matrices used for GWR models are different from those used for spatial lag or spatial error models, which are the two models used for spatial regression forecasting as discussed in the Introduction section: the latter are global matrices, but the former are local matrices. In a GWR model, the local weight matrix is a kernel function that (1) has a bandwidth (or threshold value) for selecting a subset of the observations and (2) gives the nearer neighbors more weight than the farther ones in an inversed distance function.

Just as there are many spatial weight matrices for spatial lag or spatial error models, the GWR local weight matrix can vary. Most GWR applications use the continuous functions that produce the smoothly decreasing weights as distance increases, such as the adaptive bi-square kernel function:

$$w_{ij} = \left(1 - \frac{d_{ij}^2}{\theta_{i(k)}}\right)^2 \text{ if } d_{ij} < \theta_{i(k)} \tag{21.3}$$

where

w_{ij} is the weight value of observation at location j for estimating the coefficient at location i,
d_{ij} is the flight distance between i and j, and
$\theta_{i(k)}$ is an adaptive bandwidth size defined as the kth nearest neighbor distance.

The biggest difference between a GWR model and a spatial lag or spatial error model is that in the latter all observations are used to estimate the coefficients, while in the former only a subset of observations is used to estimate each coefficient. For example, when we estimate $\beta(u_I,v_I)$, we are using the explanatory variable X_2 and response variable Y in location i as well as in its neighbors, which are selected based on a weight function w_{ij}.

GWR models are fitted based on a local log-likelihood function that minimizes the total squares of the difference between the actual Y value and the estimated Y value (Fotheringham et al. 2002). GWR model-fitting to data balanced with model parsimony can be measured by Akaike's Information Criterion (AICc, which is AIC with adjustment for finite sample sizes) and Schwartz's Bayesian Information Criterion (BIC). A smaller AICc or BIC value indicates a better-fitted GWR model. The major diagnostic for GWR models is a test for spatial non-stationarity (i.e., heterogeneity). An essential assumption of GWR models is that the effects that explanatory variables have on the response variable vary spatially. After we fit a GWR model, we need to test whether the spatial variations of the effects are statistically significant. In the standard GWR software package, there is a statistic called diff-criterion that is based on a set of model-fitting measures including AICc, BIC, and cross-validation (CV). A negative diff-criterion value suggests spatial heterogeneity in terms of the effects. In contrast, a positive diff-criterion value suggests spatial homogeneity in terms of the effects. In the latter case, the corresponding explanatory variable should be treated as having a spatially homogenous effect on the response variable, and the GWR model should be re-fit.

21.2.2 The GWR Forecasting Model

Just as spatial lag and spatial error models can be used for forecasting purposes (Chi and Voss 2011), geographically weighted regression models can be used for forecasting as well. The basic idea is to apply the parameters estimated from the estimate period to the model in the forecasting period. This is typically done in two steps.

In step one, we establish a relationship between the response variable and its explanatory variables measured at an earlier time point, because forecasting is performed to project the future with current and past information. The first part of a GWR forecasting model is specified as:

$$Y_{it} = \sum_{k=0}^{p} \beta_{k(t-1)}(u_i, v_i)\left(X_{ik(t-1)} + Y_{i(t-1)}\right) + \varepsilon_i \tag{21.4}$$

where

i denotes the ith location (or observation) in space,
t denotes time t,
t–1 denotes time t–1,
Y_{it} denotes the response variable for the ith observation at time t,
k denotes the number of explanatory variables, including the constant ($k = 0$),
(u_i,v_i) denotes the coordinate of the ith location in space,
$X_{ik\cdot(t-1)}$ denotes the explanatory variable k for the ith observation at time t–1,
$Y_{i\ (t-1)}$ denotes the response variable for the ith observation at time t–1,
$\beta_k \cdot (t–1)(u_i,v_i)$ is the realization of the continuous function $\beta_k(u,v)$ at location i at time t–1, and
ε is an n by 1 vector of error terms that are independently and identically distributed.

We include Y_{t-1} because the target to be forecasted, the response variable, is often affected by its previous existence and thus is often used in regression forecasting models.

If any variable on the right side of the model does not exhibit statistically significant spatial heterogeneity, we re-fit the model by treating that variable's effects as spatially homogeneous, which would result in a global coefficient (i.e., a single coefficient value) for that particular variable.

In step two, we use the estimated coefficients, the explanatory variables at time t, and the response variable at time t to forecast the response variable at time $t + 1$:

$$\hat{Y}_{i(t+1)} = \sum_{k=0}^{p} \hat{\beta}_{k\cdot t}(u_i, v_i)(X_{i\cdot k\cdot t} + Y_{i\cdot t}) \tag{21.5}$$

where

$\hat{Y}_{i(t+1)}$ is the response variable for the ith observation at time $t + 1$,
$Y_{i\cdot t}$ is the response variable for the ith observation at time t,
$X_{i\cdot k\cdot t}$ is the explanatory variable k for the ith observation at time t, and
$\hat{\beta}_{k\cdot t}(u_i,v_i)$ are coefficients estimated from step one.

21.3 Data

In this study, we focus on the state of Wisconsin in the United States to demonstrate the use of GWR for population forecasts and test its performance. We conducted a population projection[1] for all minor civil divisions (MCDs) in Wisconsin (Fig. 21.1). Wisconsin is a strong MCD state: MCDs are the smallest governmentally functioning units that collect taxes and provide public services. MCDs have social and political meanings; conducting population projections at the MCD level has an advantage of linking to planning and policy making. MCDs in Wisconsin are composed of cities, villages, and towns; they have an average geographic area of 76.57 square miles and an average population of 3097 as of 2010.

We also take a knowledge-based approach for the GWR forecasting method by incorporating a variety of factors that could be related to population change. We used thirty-three explanatory variables that include demographic characteristics (previous growth rate; population density; proportions of young population, old population, black population, Hispanic population, college population, population who finished high school, population with bachelor's degree, non-movers, female household heads with children under 18 years, seasonal housing units, workers in retail industry, and workers in agricultural industry) socioeconomic status (median household income, crime rate, housing units using public water, new housing units, median house value, workers using public transportation to travel to work, access to urban buses, county seat status), transportation accessibility (the inverse distance to the centroid of central cities, the inverse distance from the centroid of a MCD to its nearest major airport, the inverse distance to interchange of interstate highways, highway lengths, journey to work), natural amenities (proportion of forested areas, proportion of water areas, the lengths of lakeshore/riverbank/coastline adjusted by the MCD's area, golf courses, proportion of areas with slope between 12.5 % and 20 %), and land use and development. The data for these variables come from a variety of federal and state governmental agencies including the U.S. Census Bureau; the U.S. Geological Survey; the Federal Bureau of Investigation; the National Atlas of the United States; the Wisconsin Departments of Transportation, Natural Resources, and Public Instruction; and several research units of the University of Wisconsin–Madison. A description of these variables and data sources can be found in Chi (2009).

---

[1] A projection includes one or more assumptions, while a forecast is a projection that is most likely to occur based on judgments. However, we use "projection" and "forecast" interchangeably in this chapter.

Fig. 21.1 Counties and minor civil divisions in Wisconsin (Source: Chi and Marcouiller (2011))

21.4 Analytical Approaches

21.4.1 Regression Projections

In general, a regression forecasting approach is composed of three steps: (1) estimating the parameters in the estimation period—typically by moving the time point backward, (2) applying the estimated parameters to the projection period for population projection, and (3) evaluating the performance of the projection by comparing the projected population with the actual population.

We started with the standard regression (Model 1) approach by including all thirty-three explanatory variables. We used the backward elimination approach (Agresti and Finlay 2009) to remove variables that are not significant at the $p \leq 0.05$ level. Our refined Model 1 includes twelve explanatory variables: *previous growth* (population growth rate in the previous decade), *population density*, *old* (proportion of the old population aged 65+), *black* (proportion of the black population), *stayers* (proportion of non-movers), *unemployment rate*, *seasonal housing* (proportion of seasonal housing units), *agricultural employment* (proportion of workers in the agricultural industry), *water* (proportion of water areas), *forest* (proportion of forestry areas), *accessibility to work* and *new housing units* (proportion of housing units that are 40 years old and less).

In Model 2 (regression with neighbor growth), we added the weighted (i.e., average) population growth rate of neighbors as an explanatory variable to the model. In Model 3 (regression with neighbor growth and characteristics), we further added the weighted (i.e., average) explanatory variables of neighbors as explanatory variables to the Model 2. Model 3 goes through the backward elimination process, and only the statistically significant explanatory variables or weighted explanatory variables of neighbors were retained; the latter includes *previous growth, population density, stayers, unemployment rate*, and *forest*. The coefficients estimated in the 1990–2000 period are used to project population growth in 2000–2010.

The procedures for using the GWR method for population forecasting (Model 4) is similar to those for using standard regression and spatial regression methods. In step one, we fit a GWR model to the data in the estimation period. The response variable is population growth from 1990–2000, and the explanatory variables are measured in 1990 as retained from Model 1, in the function of:

$$Y_{i(1990,2000)} = \sum_{k=0}^{p} \beta_k(u_i, v_i)\left(X_{ik\cdot 1990} + Y_{i\cdot(1980,1990)}\right) + \varepsilon_i \tag{21.6}$$

Based on the diff-criteria, the variables that do not exhibit statistically significant spatially heterogeneous effects on population growth will be treated as having spatially homogeneous effects. The GWR model will then be re-fit to the data.

In step two, we use the estimated parameters (including the local coefficients and global coefficients) to project population growth in 2000–2010. The explanatory variables are measured in 2000. The function for the projection is:

$$\hat{Y}_{i\cdot(2000,2010)} = \sum_{k=0}^{p} \hat{\beta}_k(u_i, v_i)\left(X_{i\cdot k\cdot 2000} + Y_{i\cdot(1990,2000)}\right) \tag{21.7}$$

Based on the predicted population growth rate from 2000–2010 and the population size in 2000, we can calculate the predicted 2010 population size.

For comparison purposes, we also conducted projections employing four simple but widely used baseline projections: a 10-year-based linear extrapolation (Model A), a 20-year-based linear extrapolation (Model B), a 10-year-based exponential extrapolation (Model C), and a 20-year-based exponential extrapolation (Model D). They are established and fundamental population projection methods that have been used for small geographic areas in many states for many years (Chi 2009).

21.4.2 Projection Evaluations

To evaluate the performance of the GWR forecasting approach (Model 4), we compare it to Models 1, 2, and 3 as well as to Models A, B, C, and D by four quantitative measures: mean algebraic percentage error (MALPE), mean absolute percentage error (MAPE), median algebraic percentage error (MedALPE), and median absolute percentage error (MedAPE). MALPE is a measure in which the positive and negative values can cancel each other, and therefore it is used as a measure of projection bias. MAPE is a measure of the average percentage difference between the forecasted population and the actual population, regardless of over- or under-projection; therefore, it is used as a measure of projection precision. MedALPE and MedAPE measure the "typical" projection errors and ignore the outliers. Although there additional ones, these four measures collectively provide an evaluation of projection performance (Smith et al. 2013: 324–327; Swanson and Tayman 2012: 268–274).

The eight projections were further evaluated by population size in 2010 and population growth rate in 2000–2010. Existing studies (e.g., Smith 1987) suggest that population projection accuracy is positively associated with population size and negatively associated with population growth rate (in absolute terms).

21.5 Results

21.5.1 Estimates of Regression Models

Using the backward elimination approach, twelve variables are retained in Model 1 (Table 21.1). Model 2 adds an additional variable, the average population growth rate of neighboring MCDs. Model 3 includes both the average population growth rate of neighboring MCDs and characteristics of neighboring MCDs. We initially included all twelve predictors and their neighboring average, which results in 24 explanatory variables in total. Using the backward elimination approach, we retained twelve explanatory variables and four weighted neighbor characteristics including *population density*, *stayers*, *unemployment rate*, and *forest*.

Table 21.1 Estimations of refined regression models (1990–2000)

	Model 1			Model 2			Model 3		
	(standard regression)			(regression with neighbor growth)			(regression with neighbor growth and characteristics)		
	Coef.	S.D.		Coef.	S.D.		Coef.	S.D.	
Intercept	0.276	0.042	***	0.241	0.042	***	0.352	0.061	***
Previous growth	0.116	0.030	***	0.089	0.030	***	0.086	0.03	***
Population density in 1990	−5.15E-5	0.000	***	−4.66E-5	2.00E-5	***	/	–	–
Old in1990	−0.207	0.063	***	−0.135	0.063	***	/	–	–
Black in 1990	−0.892	0.217	***	−0.845	0.215	***	−0.870	0.210	***
Stayers in 1990	−0.215	0.041	***	−0.197	0.041	***	−0.170	0.038	***
Seasonal housing in 1990	0.176	0.025	***	0.165	0.025	***	0.154	0.023	***
Agricultural employment in 1990	−0.151	0.036	***	−0.102	0.037	***	–	–	–
Unemployment rate in 1990	−0.310	0.094	***	−0.263	0.094	***	−0.193	0.097	*
New housing in 1990	0.218	0.028	***	0.199	0.028	***	0.246	0.026	***
Accessibility to work	−0.114	0.028	***	−0.097	0.028	***	−0.093	0.027	***
Forest	−0.075	0.019	***	−0.070	0.019	***	/	–	–
Water	−0.180	0.051	***	−0.201	0.051	***	−0.184	0.050	***
Previous growth 1980–1990 (neighbor average)			–	0.330	0.052	***	0.294	0.056	***
Population density in 1990 (neighbor average)			–	–	–	–	0.000	0.000	***
Stayers in 1990 (neighbor average)			–	–	–	–	−0.242	0.074	***
Unemployment rate in 1990 (neighbor average)			–	–	–	–	−0.370	0.182	*
Forest (neighbor average)			–	–	–	–	−0.097	0.024	***
Measures of fit									
Adjusted R^2				0.216	–	–	0.232	0.239	
AIC				−2230.190	–	–	−2168.833	−2184.497	
AICc				−2129.990	–	–	−2168.602	−2184.266	

Notes: *** $p \leq 0.001$, *$p \leq 0.05$
Coef. coefficient, *S.D.* standard error
The neighbor averages are calculated based on the first-order queen continuity matrix

Table 21.1 shows the estimated coefficients of the three models. All the retained explanatory variables are statistically significant in explaining population growth. The AICc and adjusted R^2 statistics suggest that Model 2 is better fitted to data than Model 1, and Model 3 is better fitted to data than Model 2.

Using the explanatory variables retained in Model 1, we fit a GWR model. We applied the adaptive bi-square kernel function and the smallest AICc selection criterion to find the optimal bandwidth. Table 21.2 presents the result of the initial GWR model. As discussed earlier, a GWR model allows the coefficients to vary spatially; thus, the coefficients vary from one MCD to another. In Table 21.2, for each explanatory variable we presented their minimum, lower quantile, median, upper quantile, and maximum values. The last column of Table 21.2 presents the diff-criterion that measures spatial non-stationarity; a positive diff-criterion value suggests no spatial heterogeneity, but a negative diff-criterion value suggests the existence of spatial heterogeneity.

Compared to the three models presented in Table 21.1, the initial GWR model has higher adjusted R^2 and smaller AICc, which suggest that the GWR model fits to data better than Models 1, 2, and 3. The diff-criterion indicates that the coefficients for two variables—old and seasonal housing—are spatially homogenous, while the coefficients for the remaining explanatory variables are spatially heterogeneous. Therefore, we must treat the two variables as having spatially homogenous effects

Table 21.2 Estimations of the initial geographically weighted regression model (1990–2000) and the test for spatial heterogeneity

	Min	Lower quantile	Median	Upper quantile	Max	Diff-criterion
Intercept	−0.344	0.137	0.254	0.400	0.853	−176.79
Previous growth	−0.122	0.011	0.090	0.188	0.370	−0.17
Population density in 1990	−2.69E-4	−1.38E-4	−9.10E-5	3.00E-6	1.55E-4	−0.90
Old in 1990	−0.573	−0.293	−0.198	−0.086	0.157	1.37
Black in 1990	−6.039	−1.174	−0.432	0.630	5.696	−18.84
Stayers in 1990	−0.525	−0.326	−0.203	−0.120	0.043	−72.22
Seasonal housing in 1990	−0.739	−0.303	−0.158	−0.076	0.167	3.30
Agricultural employment in 1990	−0.064	0.093	0.155	0.233	0.366	−5.90
Unemployment rate in 1990	−0.457	−0.243	−0.085	−0.012	0.219	−4.62
New housing in 1990	−0.097	0.126	0.220	0.298	0.438	−70.76
Accessibility to work	−2.283	−0.527	−0.242	−0.098	0.448	−52.40
Forest	−0.722	−0.333	−0.172	−0.048	0.520	−7.50
Water	−0.401	−0.097	−0.051	0.005	0.168	−1.43
Measures of fit						
Adjusted R^2	0.290					
AICc	−2216.6					
Best bandwidth size	403					

on population growth. We re-fit the GWR model by estimating global coefficients for the two variables but local coefficients for the other variables. Table 21.3 shows the estimated coefficients for the refined GWR model. We also mapped the spatial variation of the coefficients for the explanatory variables that have spatially heterogeneous effects on population growth, in Fig. 21.2. To make the maps easier to read, we laid county boundaries instead of MCD boundaries on top of the maps. This also applies to the other maps in this chapter.

The results clearly show that the relationships between population growth and the various predictors differ across space. For example, positive association between population growth and *previous growth* tends to concentrate in the eastern area of Wisconsin, while the negative association between the two tends to be clustered in a belt shape across the northern part of Wisconsin. The results also show that negative relationships between population growth and several socioeconomic predictors, such as *stayers*, *agricultural employment* and *new housing* tend to concentrate in Milwaukee and its surrounding areas. The only exception is the local coefficients of *unemployment rate*, which are positively clustered in Milwaukee and its surrounding areas. In addition, the two predictors that capture the natural resources—*water* and *forest*—have positive coefficients in the upper northern areas and negative coefficients in the southern areas of Wisconsin.

Table 21.3 Estimations of the refined geographically weighted regression model (1990–2000)

	Min	Lower quantile	Median	Upper quantile	Max
Local					
Intercept	−0.321	0.161	0.294	0.446	0.876
Previous growth	−0.157	0.000	0.080	0.181	0.350
Population density in 1990	−2.48E-4	−1.17E-4	−6.40E-5	1.70E-5	1.56E-4
Black in 1990	−7.428	−1.237	−0.477	0.414	6.148
Stayers in 1990	−0.552	−0.342	−0.202	−0.110	0.082
Agricultural employment in 1990	−0.799	−0.310	−0.191	−0.084	0.166
Unemployment rate	−2.551	−0.478	−0.216	−0.044	0.623
New housing in 1990	−0.115	0.115	0.204	0.286	0.430
Accessibility to work	−0.489	−0.247	−0.095	−0.011	0.216
Forest	−0.485	−0.094	−0.038	0.015	0.254
Water	−0.788	−0.308	−0.173	0.000	0.442
Global	Estimate	SD			
Old in 1990	−0.339	0.076			
Seasonal housing in 1990	0.158	0.030			
Measures of fit					
Adjusted R^2	0.215	–			
AICc	−2128.00				
Best bandwidth size	365	–			

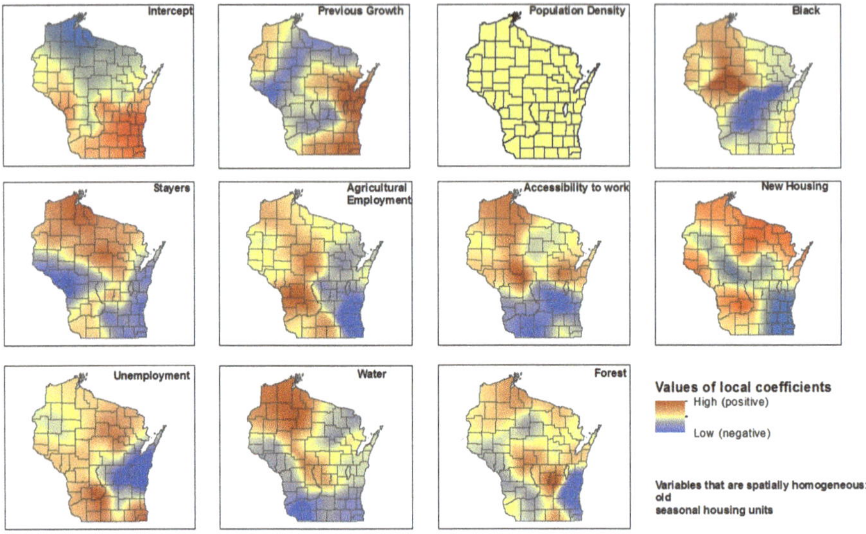

Fig. 21.2 Coefficients of the refined geographically weighted regression model (1990–2000). Note: The units of analysis are minor civil divisions (MCDs). To make the maps easier to read, we laid county boundaries instead of MCD boundaries on top of the maps. This applies to Figs. 21.4, 21.5, 21.6, and 21.7 as well

The GWR model provides elegant estimates of the coefficients by allowing them to vary spatially. The GWR model is also better fitted to data than the standard regression and spatial regression methods. Below we use the estimated coefficients to project population in 2010 and compare it to the actual 2010 population. We do this for all four regression methods and all four baseline projections.

21.5.2 Evaluation of Population Projections

Table 21.4 presents the measures of projection evaluations for all methods. The GWR model (Model 4) achieves slightly better projection than Models 2 and 3 (the two spatial regression models) but underperforms Model 1, which is a simpler standard regression model. Further, the GWR method and all three regression models underperform the extrapolation projections except that Model 1 slightly outperforms Model C (10-year-based exponential extrapolation).

We further evaluate population projections by population size in 2010 and population growth rate from 2000–2010. The results are presented in Fig. 21.3. The solid lines represent regression methods, and dashed lines are the results from extrapolation methods. When breaking down by population size, the extrapolation methods yield less-biased results—all four extrapolation methods have smaller MALPEs compared to the regression models (Fig. 21.3a). The only exception is that the standard regression (Model 1) yields the lowest MALPE value among large

Table 21.4 Evaluating population projection by quantitative measures

Model	MALPE	MAPE	MedALPE	MedAPE
Regression methods				
Model 1: Standard regression	7.58	12.18	7.50	9.82
Model 2: Regression with neighbor growth	15.44	17.82	13.94	14.61
Model 3: Regression with neighbor growth and characteristics	11.79	14.89	10.66	11.84
Model 4: GWR	11.47	14.82	10.59	11.88
Extrapolation methods				
Model A (10-year-based linear)	5.63	13.15	4.99	9.36
Model B (20-year-based linear)	1.42	10.40	1.35	7.39
Model C (10-year-based exponential)	8.77	15.36	6.49	10.50
Model D (20-year-based exponential)	3.39	11.32	2.52	7.85

MCDs (with a population size of 8663 and above). Also, the GWR model does not achieve more precise projections by population size. The only exception is that when the population size is 552 or less, GWR projections are more precise than the three regression models and the two 10-year-based extrapolation projections, but they still do not outperform the two 20-year-based extrapolation projections (Fig. 21.3b).

The GWR model does not outperform extrapolation projections by population growth rate. The extrapolation methods in general have better performance based on the smaller MALPE criterion (Fig. 21.3c). The only exception is that regression models provide more precise projections for the MCDs where the population growth rate is 10% and above (Fig. 21.3d). Overall, the GWR method as well as the three regression models underperform the simple extrapolation projections when we look into different segments of population size and population growth rate.

21.6 Why the GWR Approach Underperforms

This finding is disappointing because we expected that our better-fitted GWR model would achieve higher levels of accuracy. Unfortunately, our delicate GWR model does not work well, nor do the standard regression and spatial regression models. For those models, we took a knowledge-based approach, meaning that we considered a variety of factors that have been argued theoretically and/or found empirically to affect population change.

But why does more knowledge not produce more accurate population projections for small areas? Why do sophisticated models not outperform simple extrapolation projections? It is a traditional belief that the more we know about our society and environment, and the more advanced and sophisticated our models become, the better we can predict the future. A third of a century ago, Keyfitz

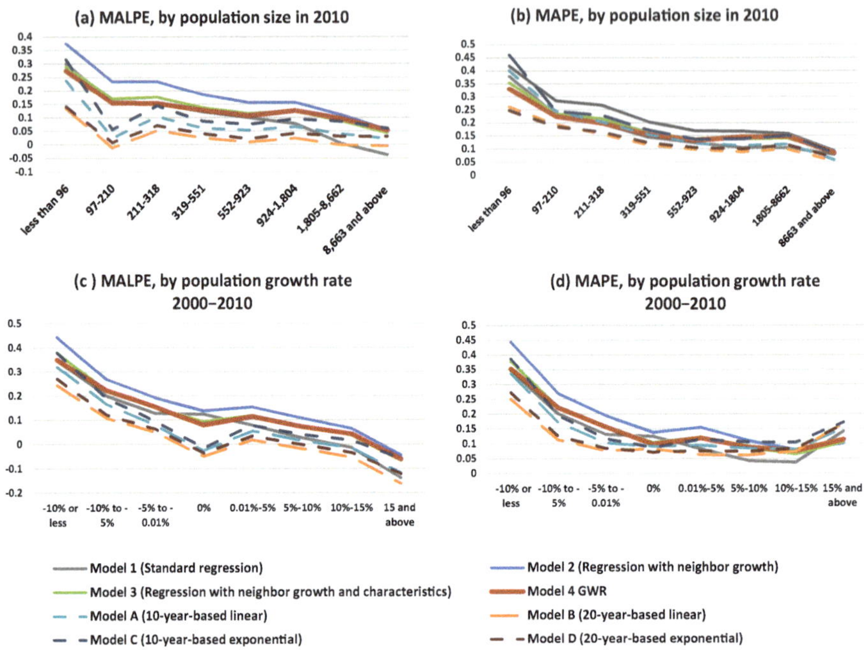

Fig. 21.3 Evaluating population projection by population size in 2010 and population growth rate 2000–2010

(1982) argued that temporal instability causes the knowledge-based regression approach to underperform extrapolation projections. When we use the regression approach for population projection, we have to assume that the effects of the explanatory variables on population change are consistent between the estimation period and the projection period. However, this assumption is not necessarily true. The pattern of population distribution can easily change from one decade to another. The United States in the 1990s experienced strong growth but suffered great economic recession in the late 2000s. Some explanatory variables could have very different effects on population change in the two periods.

To illustrate the impact that temporal instability has on the performance of the regression approach, we fit the models using the data in the 2000–2010 with all the same explanatory variables. We conducted the Chow (1960) test to examine whether the coefficients are different between the estimation period and the projection period (Table 21.5). The F-statistics of all three regression models are substantially higher than the critical F value at the $p \leq 0.001$ significance level, suggesting that the coefficients in the three regression models are statistically different between the two decades.

We also tested the possible temporal instability of the GWR model. We first fit the GWR model to the 2000–2010 period using the same twelve predictors as used in the estimation period. We repeated the analytical process as we did for 1990–2000 period (i.e., we first treat all the predictors as local to fit the model,

Table 21.5 Chow test of regression stability

Model	Degrees of freedom	Critical *F* value (at the 0.001 significance level)	*F* statistic
Model 1: Standard regression	(11, 3652)	2.851	42.51
Model2: Regression with neighbor growth	(14, 3646)	2.589	33.95
Model3: Regression with neighbor growth and characteristics	(18, 3638)	2.359	24.12

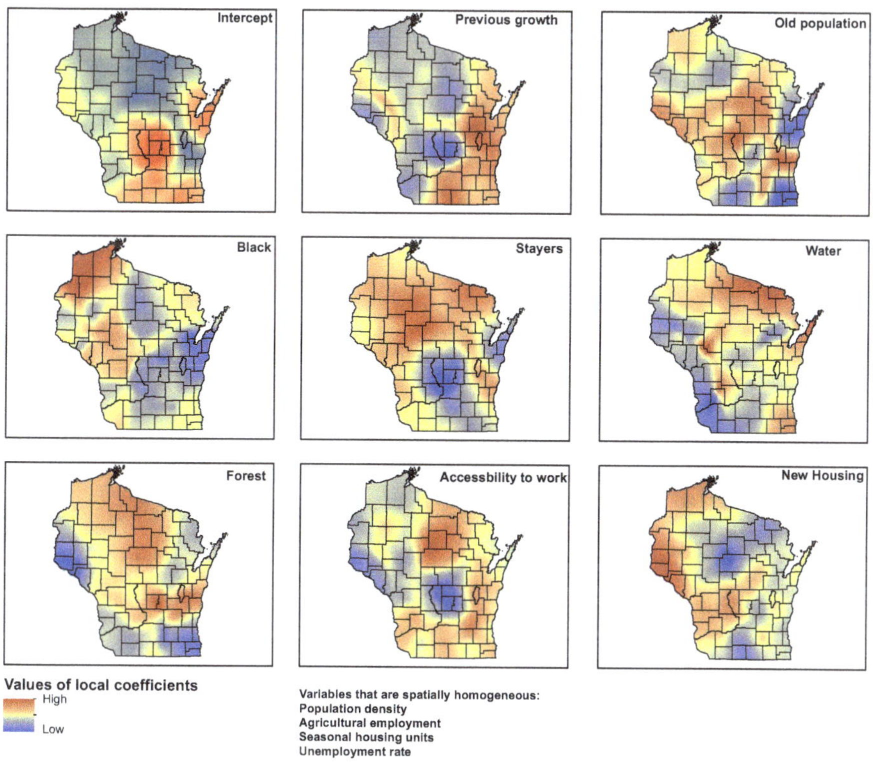

Fig. 21.4 Coefficients of the refined geographically weighted regression (2000–2010)

and then re-fit the model based on the diff-criterion). In this analysis, four predictors—*population density, agricultural employment, seasonal housing* and *unemployment rate*—are spatially homogenous. The results of the local coefficients are mapped in Fig. 21.4.

Figure 21.5 shows the estimated local R^2 in the estimation and projection periods. The comparison between the two maps suggests that the goodness-of-fit of the GWR models varies across space and the two periods. The 1990–2000 GWR

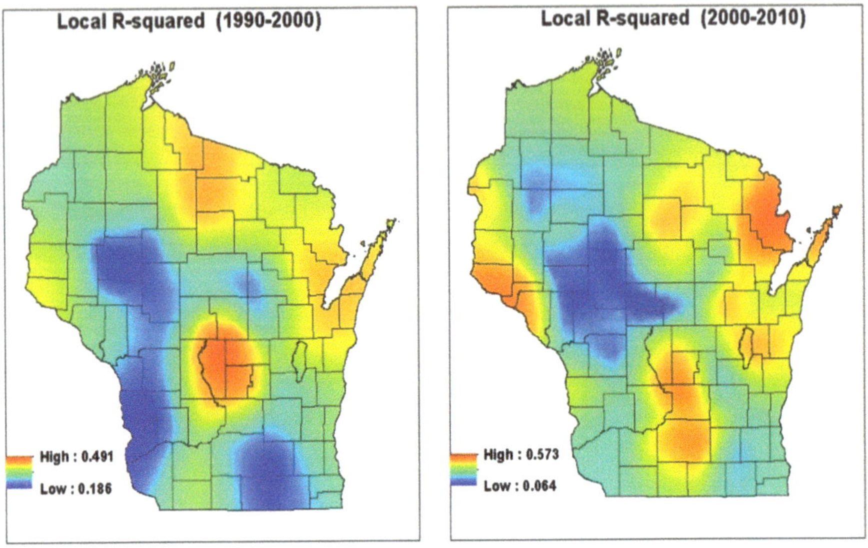

Fig. 21.5 Overall goodness-of-fits of the geographically weighted regression models

model yields better model fit in the middle and upper north areas of Wisconsin, while it fits poorly in the southwest and in the bottom south. The 2000–2010 GWR model still fits well in the upper north, especially in the northeast corner, with the local R^2 as high as 0.573, while the model yields a lower local R^2 value in the middle-west area of the state.

For the explanatory variables that have spatially heterogeneous effects on population change in both periods, we further calculated coefficient differences between the same GWR local predictors across the two periods. The result is presented in Fig. 21.6. The differences vary greatly across space. For example, water has a stronger effect on population change in the projection period than in the estimation period in the north, northeast peninsular, and southeast corner of Wisconsin, but vice versa in the central and west.

We present the areas for which the coefficient differences are statistically significant at the $p \leq 0.05$ level in Fig. 21.7. For each explanatory variable, the coefficient difference is significant in some areas. For example, *previous growth* has a stronger effect on population growth in the estimation period than in the projection period in all the areas. *Accessibility to work* has a stronger effect on population growth in the estimation period than in the projection period only in the north corner but has a weaker effect in the estimation period than in the projection period in the remaining areas.

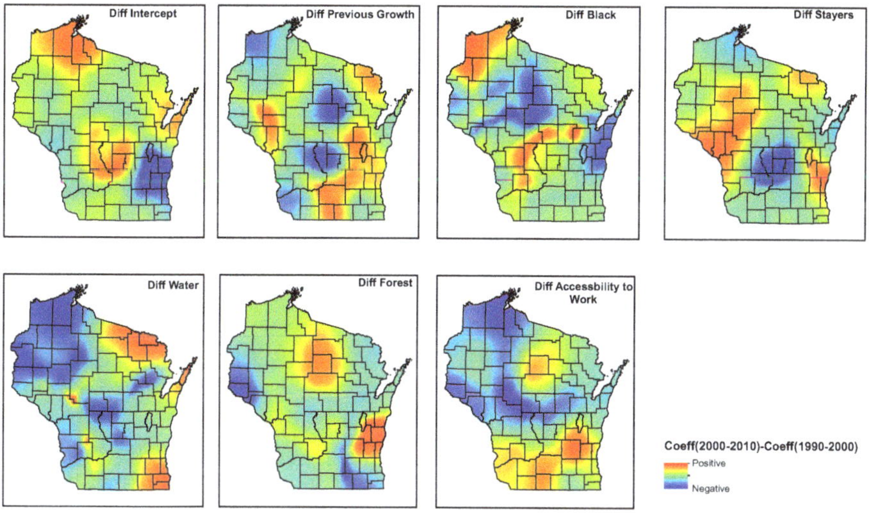

Fig. 21.6 Comparing local coefficients between the two time periods (1990–2000 and 2000–2010)

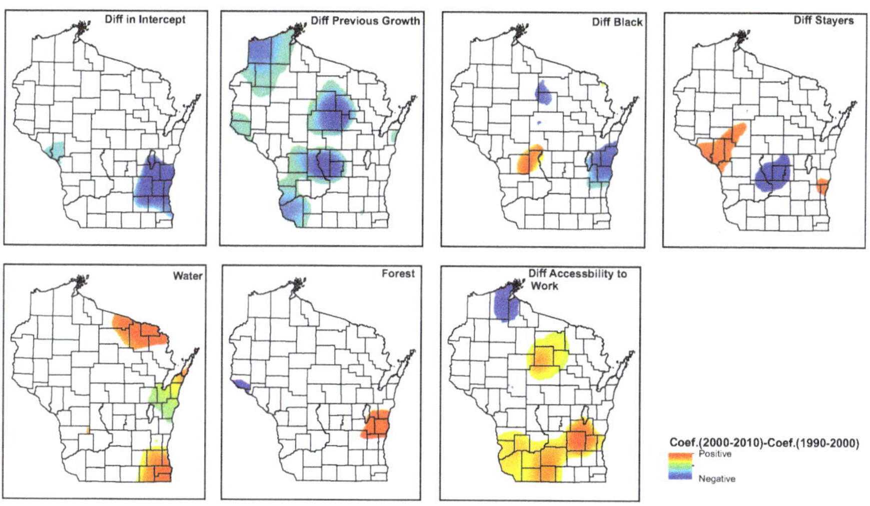

Fig. 21.7 Comparing local coefficients between the two time periods (1990–2000 and 2000–2010), highlighting areas where the differences are statistically significant at $P \leq 0.05$ level)

21.7 Rethinking the Regression Approach for Population Forecasting

Overall, our results did not turn out as well as we hypothesized (and hoped) they would. Neither our knowledge-based GWR regression model nor the standard regression and spatial regression models outperform the simple extrapolation projections that are based purely on population growth trends from the past.

As discussed, temporal instability accounts for the underperformance of the regression approach. When we use the regression approach for population projection, we assume that the coefficients are consistent across the estimation period and the projection period, but that assumption is not often true for small areas. An explanatory variable could have an effect on population growth in one time period but not another, could have a stronger effect on population growth in one time period than another, or could have opposite effects in the two time periods. Further, these differences between the two time periods could vary from one location to another. What's more, population change is more dramatic in small areas than in bigger areas, as changes in the former can be canceled out when aggregated to bigger areas. These violations of the temporal stability assumption for using regression for projection reduce the robustness of the regression approach for small-area population projections. This suggests that as is the case with time-series models, it may be important for temporally-based regression models to be stationary (Swanson 2004). However, in contrast, simple extrapolation projection, which relies only on past population growth trends, has consistently been the most successful method, for at least a half century (see, e.g., Smith et al. 2013: 331–336).

The discovery that knowledge-based sophisticated regression methods do not outperform simple extrapolation projections makes us wonder whether we use the regression methods appropriately for small-area population projections. The recent exploding developments in Big Data might provide a hint. Among its many uses, Big Data has been employed for prediction purposes. For example, a Google search was successfully used to predict the outbreak of H1N1 in 2009, much faster and more efficiently than the traditional reporting mechanisms used by the Centers for Disease Control and Prevention (Ginsberg et al. 2009). With Big Data, machine learning algorithms are developed to find patterns in data. All possible data are used, regardless of how theoretically or conceptually unrelated they are. Such research focuses more on the statistical correlation than the causality of the data.

Extrapolation projections consider only the temporal correlation of population growth in the past. The regression projection approaches consider the causality (i.e., whether the factors included have a meaningful effect on population growth). The regression approach has two goals: one is to find reasonable causality; the other is to perform the projection. Unfortunately, the two goals can be in conflict, as addressed earlier in this chapter.

If, however, the ultimate goal is not to identify causality but to provide more accurate projections, then why the concern about causality? What if we demographers set causality aside and employ all the data that we collect, fit all types of regression models with all possible combinations of the factors, and find the model that produces the most accurate projections? Further research could be undertaken to determine whether that approach would work.

Acknowledgements We thank Annelise Hagedorn and Xuan Zhou for assistance in data collection and cleanup. Appreciation is extended to David Swanson and Jacques Poot for providing comments on earlier drafts of this chapter. This research was supported by the National Science Foundation (Award # 1541136).

References

Agarwal, C., Green, G. M., Grove, J. M., Evans, T. P., & Schweik, C. M. (2002). *A review and assessment of land-use change models: Dynamics of space, time, and human choice* (General technical report NE-297). Newtown Square: USDA Forest Service Northeastern Research Station and Indiana University Center for the Study of Institutions, Populations, and Environmental Change.

Agresti, A., & Finlay, B. (2009). *Statistical methods for the social sciences*. Upper Saddle River: Pearson Prentice Hall.

Ahlburg, D. A. (1987). The impact of population growth on economic growth in developing nations: The evidence from macroeconomic-demographic models. In D. G. Johnson & R. D. Lee (Eds.), *Population growth and economic development: Issues and evidence* (pp. 492–522). Madison: University of Wisconsin Press.

Alho, J. M. (1997). Scenarios, uncertainty and conditional forecasts of the world population. *Journal of the Royal Statistical Society: Series A (Statistics in Society), 160*(1), 71–85.

Ali, K., Partridge, M. D., & Olfert, M. R. (2007). Can geographically weighted regressions improve regional analysis and policy making? *International Regional Science Review, 30*(3), 300–329.

Anselin, L. (2001). Spatial econometrics. In B. Baltagi (Ed.), *Companion to econometrics* (pp. 310–330). Oxford: Blackwell.

Armstrong, J. S. (2001). Extrapolation of time-series and cross-sectional data. In J. S. Armstrong (Ed.), *Principles of forecasting: A handbook for researchers and practitioners* (pp. 217–244). Boston: Kluwer.

Armstrong, J. S., Adya, M., & Collopy, F. (2001). Rule-based forecasting: Using judgment in time-series extrapolation. In J. S. Armstrong (Ed.), *Principles of forecasting: A handbook for researchers and practitioners* (pp. 259–282). Boston: Kluwer.

Chi, G. (2009). Can knowledge improve population forecasts at subcounty levels? *Demography, 46*(2), 405–427. doi:10.1353/dem.0.0059.

Chi, G., & Marcouiller, D. W. (2011). Isolating the effect of natural amenities on population change at the local level. *Regional Studies, 45*(4), 491–505.

Chi, G., & Voss, P. R. (2011). Small-area population forecasting: Borrowing strength across space and time. *Population, Space and Place, 17*(5), 505–520.

Chow, G. C. (1960). Tests of equality between sets of coefficients in two linear regressions. *Econometrica, 28*(3), 591–605.

Cutter, S. L., & Smith, M. M. (2009). Fleeing from the hurricane's wrath: Evacuation and the two Americas. *Environment: Science and Policy for Sustainable Development, 51*(2), 26–36. doi:10.3200/envt.51.2.26-36.

Deal, B., Pallathucheril, V. G., Sun, Z., Terstriep, J., & Hartel, W. (2005). *LEAM technical document: Overview of the LEAM approach*. Urbana: Department of Urban and Regional Planning, University of Illinois at Urbana-Champaign.

Espenshade, T. J., & Tayman, J. M. (1982). Confidence intervals for postcensal state population estimates. *Demography, 19*(2), 191–210.
Fotheringham, A. S., Brunsdon, C., & Charlton, M. (2002). *Geographically weighted regression*. West Sussex: Wiley.
Fraser, L. K., Clarke, G. P., Cade, J. E., & Edwards, K. L. (2012). Fast food and obesity: A spatial analysis in a large United Kingdom population of children aged 13–15. *American Journal of Preventive Medicine, 42*(5), e77–e85. doi:10.1016/j.amepre.2012.02.007.
Ginsberg, J., Mohebbi, M. H., Patel, R. S., Smolinski, M. S., Brillian, L., & Brammer, L. (2009). Detecting influenza epidemics using search engine query data. *Nature, 457*(7232), 1012–1014. doi:10.1038/nature07634.
Hunt, D., & Simmons, D. (1993). Theory and application of an integrated land-use and transport modeling framework. *Environment and Planning B, 20*, 221–244.
Keyfitz, N. (1982). Can knowledge improve forecasts? *Population and Development Review, 8*(4), 729–751.
Longley, P. A., & Tobon, C. (2004). Spatial dependence and heterogeneity in patterns of hardship: An intra-urban analysis. *Annals of the Association of American Geographers, 94*(3), 503–519.
Lutz, W., & Goldstein, J. R. (Guest Eds.). (2004). How to deal with uncertainty in population forecasting. *International Statistical Review, 72*(1 and 2), 1–106.
Lutz, W., & Striessnig, E. (2015). Demographic aspects of climate change mitigation and adaptation. *Population Studies: A Journal of Demography, 69*(S1), S69–S76. doi:10.1080/00324728.2014.969929.
Matthews, S. A., & Yang, T. C. (2012). Mapping the results of local statistics: Using geographically weighted regression. *Demographic Research, 26*, 151–166.
Meadows, D. H., Meadows, D. L., Randers, J., & Behrens, W. I. (1972). *The limits to growth*. New York: Potomac Associations.
Meadows, D. H., Meadows, D. L., & Randers, J. (1992). *Beyond the limits: Confronting global collapse, envisioning a sustainable future*. London: Earthscan.
Mennis, J. L., & Jordan, L. (2005). The distribution of environmental equity: Exploring spatial nonstationarity in multivariate models of air toxic releases. *Annals of the Association of American Geographers, 95*(2), 249–268.
O'Neill, B. C. (2004). Conditional probabilistic population projections: An application to climate change. *International Statistical Review, 72*(2), 167–184.
Riahi, K., & Nakicenovic, N. (Eds.). (2007). *Greenhouse gases—Integrated assessment, technological forecasting and social change, Special Issue, 74*(7), 873–886.
Sanderson, W. C. (1998). Knowledge can improve forecasts: a review of selected socioeconomic population projection models. *Population and Development Review 24*(Suppl), 88–117.
Sanderson, W. C., Scherbov, S., O'Neill, B. C., & Lutz, W. (2004). Conditional probabilistic population forecasting. *International Statistical Review, 72*(2), 157–166.
Smith, S. K. (1987). Tests of forecast accuracy and bias for county population projections. *Journal of the American Statistical Association, 82*, 991–1003.
Smith, S. K., Tayman, J. M., & Swanson, D. A. (2013). *A practitioner's guide to state and local projections*. Dordrecht: Springer.
Swanson, D. A. (2004). Advancing methodological knowledge within state and local demography: A case study. *Population Research and Policy Review, 23*(August), 379–398.
Swanson, D. A., & Beck, D. (1994). A new short-term county population projection method. *Journal of Economic and Social Measurement, 20*(1), 25–50.
Swanson, D. A., & Tayman, J. M. (2012). *Subnational population estimates*. Dordrecht: Springer.
Swanson, D. A., & Tayman, J. M. (2014). Measuring uncertainty in population forecasts: A new approach. In Marsili M., & Capacci G (Eds.), *Proceedings of the 6th EUROSTAT/UNECE work session on demographic projections*. Rome: Italian National Institute of Statistics.
Swanson, D. A., Schlottman, A., & Schmidt, R. (2010). Forecasting the population of census tracts by age and sex: An example of the Hamilton-Perry method in action. *Population Research and Policy Review, 29*(1), 47–63.

Tobler, W. (1970). A Computer movie simulating urban growth in the Detroit region. *Economic Geography, 46*, 234–240.

U.S. Global Change Research Program (USGCRP). (2015). *Towards scenarios of U.S. demographic change: Workshop report*. Washington, DC: USGCRP. Available at http://www.globalchange.gov/sites/globalchange/files/USGCRP_Workshop_Report-Towards_Scenarios_of_US_Demographic_Change.pdf as of September 2015

Wilson, T., & Rees, P. (2005). Recent developments in population projection methodology: A review. *Population, Space and Place, 11*, 337–360.

Wood, N. J., Burton, C. G., & Cutter, S. L. (2009). Community variations in social vulnerability to Cascadia-related tsunamis in the U.S. Pacific Northwest. *Natural Hazards, 52*(2), 369–389. doi:10.1007/s11069-009-9376-1.

Yang, T. C., & Matthews, S. A. (2012). Understanding the non-stationary associations between distrust of the health care system, health conditions, and self-rated health in the elderly: A geographically weighted regression approach. *Health and Place, 18*(3), 576–585. doi:10.1016/j.healthplace.2012.01.007.

Yu, D. L., Wei, Y. D., & Wu, C. (2007). Modeling spatial dimensions of housing prices in Milwaukee, WI. *Environment and Planning B, 34*, 1085–1102. doi:10.1068/b32119.

Chapter 22
A New Method for Estimating Small Area Demographics and Its Application to Long-Term Population Projection

Takashi Inoue

Abstract This chapter discusses a newly proposed method by the author for estimating small area demographics, and explains how to apply this method to long-term small area population projection. Since small area demographics are generally very unstable, various methods of estimating the true values of such data have been developed or considered, chiefly by statisticians and demographers. Most previous methods essentially perform data smoothing by using demographics from adjacent small areas. Empirical Bayes estimation is a representative smoothing method for small area demographics and has been widely applied to mortality. Although the theoretical precision of this method has already been proven, we should pay attention to the fact that there are a few disadvantages in applying it to population projection. In order to overcome those disadvantages, the author proposed a new estimation method of small area demographics in 2014, using the concept of population potential developed by Stewart in 1947. The new method consists of six formulas. This chapter examines the efficacy of the new method by applying two of those six formulas, and the empirical Bayes estimator, to actual mortality data, and then explains how to apply the new method to long-term population projection. A brief introduction of the original population projection system constructed on the internet using the new method is also included in the chapter. These discussions enable us to understand that the new method has the following advantages: first, it satisfactorily utilizes position or coordinate data of adjacent areas; second, the strength of its smoothing is adequate to the avoidance of yielding extraordinary values in projected population, especially in the case of long-term projection; and third, it does not require that in order for demographics to smooth they must follow the specific distribution.

Keywords Small area population projection • Population potential • Data smoothing • Empirical Bayes estimation

T. Inoue (✉)
Aoyama Gakuin University, Shibuya, Tokyo 150-8366, Japan
e-mail: t-inoue@cc.aoyama.ac.jp

D.A. Swanson (ed.), *The Frontiers of Applied Demography*, Applied Demography Series 9, DOI 10.1007/978-3-319-43329-5_22

22.1 Introduction

This chapter defines two important terms, "small area" and "demographics", before referring to or handling them. The term "small area" indicates a district much smaller in size than municipalities; although in previous studies, it indicates not only such a district, but also an area similar to or larger in size than municipalities. The demographics are assumed to be population statistics giving ratios (or rates and proportions) of two datasets: e.g., mortality rates that denote the number of deaths to the total population, cohort change ratios (CCRs) that denote the survival rate for a cohort from two censuses, and child-woman ratios (CWRs) that denote the population under 5 years of age to the female population aged 15–49.

Since small area demographics are generally very unstable,[1] various methods of estimating the true value of such population statistics have been developed or considered, chiefly by statisticians and demographers (e.g., Gonzalez and Hoza 1978; Fay and Herriot 1979; Holt et al. 1979; Laake 1979; Tsutakawa et al. 1985; Clayton and Kaldor 1987; Tango 1988; Farrell et al. 1997; Anselin et al. 2006; Datta and Ghosh 2012). These previous methods essentially perform data smoothing by using population statistics from neighboring or adjoining small areas. Anselin et al. (2006) classified these methods largely into four types: (1) mean and median based smoothing; (2) nonparametric smoothing; (3) empirical Bayes smoothers; and (4) fully Bayes smoothers. This chapter focuses on types 1 and 3, because type 1 includes a new method proposed by the author, and type 3 has been employed in the most studies and is a representative smoothing method for small area demographics. Type 1, mean and median based smoothing, which is mentioned in the next section in detail, is used to spatially take a weighted moving average, and it corresponds to one of spatial filtering techniques used in the analysis of socio-economic phenomena as well as in image processing.[2] Most methods of type 1 require a large dataset for distances between all small areas, because weights used in moving averages are calculated chiefly from such distances. Meanwhile, type 3, empirical Bayes estimation, has already been proven to be theoretically precise by many statisticians (e.g., Maritz 1969; Efron and Morris 1973, 1975; Morris 1983; Louis 1984; MacGibbon and Tomberlin 1989; Ghosh et al. 2008; Kubokawa 2013a), and has been widely applied to directly age-adjusted mortality rates (DARs)[3] or standardized mortality ratios (SMRs) (e.g., Tsutakawa et al. 1985; Tango 1988; Marshall 1991; Waller and Gotway 2004; Kubokawa 2013b). Nevertheless, we should pay attention to the fact that there are the following disadvantages of applying type 3 to population projection. First, it does not efficiently utilize position or coordinate data of adjacent areas; second, the strength of its smoothing

[1] This is called "small number problem" in spatial epidemiology (Haining et al. 2010).

[2] There are some studies that analyze geographical phenomena by using spatial filtering techniques different from taking simply moving averages. Please see, for example, Griffith (2003) for details.

[3] The DAR is calculated by directly using mortality observed in each area, and often shows an extraordinary value. This undesirable phenomenon was called ill-determined rate by Mosteller and Tukey (1977); (Tango 1988).

is not necessarily adequate to the avoidance of an extraordinary increase in projected population, especially in the case of long-term projection; third, it can only be used under the condition that demographics to smooth follow the specific distribution.[4] On the other hand, these disadvantages do not occur in applying type 1 to population projection, for reasons mentioned later.

Consequently, the author proposed a new estimation method of type 1 for small area demographics in 2014, using the concept of population potential developed by Stewart in 1947 (Inoue 2014). The population potential, one of the most important measures in population geography, has been discussed in geographical studies (e.g., Warntz 1964; Tocalis 1978; Rich 1980; Mfungahema and Kitamura 1997; Inoue 2007). The new method consists of one basic formula and five applied formulas. It is unique among methods of type 1 in obtaining weights from the population potential, and has an advantage over other methods of type 1 in simplicity of calculation because the applied formulas do not require any distance datasets.

With regards to small area population projection, we have usually employed the cohort change ratio method formulated by Hamilton and Perry (1962), which many demographers have discussed (e.g., Smith et al. 2002; Swanson et al. 2010; Swanson and Tayman 2012; Smith et al. 2013). The reason for this is that almost all governments in the world have failed to provide small area vital and migration statistics, which are indispensable to the standard population projection method, namely, the cohort component method. Accordingly, this chapter deals only with CCRs and CWRs as demographics to smooth, since these two kinds of demographics are both necessary for the cohort change ratio method.

This chapter refers to existing methods of type 1 in Sect. 22.2, considers the new method in Sect. 22.3, examines the efficacy of this method in Sect. 22.4 by comparing it with the empirical Bayes method using actual mortality data, explains how to apply this method to long-term small area population projection in Sect. 22.5, and finally provides a comprehensive discussion in Sect. 22.6.

22.2 Existing Methods of Mean and Median Based Smoothing

The method of mean and median based smoothing (method of type 1) is to calculate an average or median value of the demographics of a neighborhood that includes a small area designated as the object of estimation.[5] The method is to spatially take a

---

[4] When smoothing demographics by the method of type 3, that is, empirical Bayes estimation, we usually assume that the demographics follow Poisson distribution that shows the probability of very rare phenomenon. DARs and SMRs obviously follow Poisson distribution. As mentioned later, however, since CCRs indispensable for small area population projection cannot be considered as such probability, the smoothing of CCRs through type 3 requires close attention.

[5] The method of "median" based smoothing does not necessarily reflect demographics of an object area and cannot be considered as a desirable one, therefore this chapter excludes this method from discussion.

weighted moving average corresponding to one of the spatial filtering techniques used in the analysis of socio-economic phenomena as well as in image processing.

This method has the advantage that it is possible to satisfactorily utilize the position or coordinate data of adjacent areas, if given appropriate weights in the calculation of moving averages. However, when conducting the simplest weighting, that is, in the case where all weights are equal to 1, this method is equivalent to calculating simple moving averages and loses the advantage mentioned above. In this case, the method of mean based smoothing is formulated as follows:

let x_i be the demographics of object area i; estimator $\dot{x}_i$ of x_i is expressed by

$$\dot{x}_i = \frac{1}{n_i} \sum_{j \in V_i} x_j \ . \tag{22.1}$$

Here, V_i is a set of area numbers selected according to a criterion, such as; to be adjoining to area i; to be located within a given distance from area i; or to belong to a municipality that includes area i, while n_i is the number of elements of V_i. In a word, V_i denotes the vicinity of area i. Meanwhile, if the weight is given by the function of the distance between two small areas, for example, by the inverse of the distance, estimator $\dot{x}_i$ is written as

$$\dot{x}_i = \frac{\sum_{j \in V_i} w_{ij} x_j}{\sum_{j \in V_i} w_{ij}} \ . \tag{22.2}$$

Here, w_{ij} is the weight given to area j regarding area i. Obviously, this case holds the above-mentioned advantage, although we need a large distance dataset. If $w_{ij} = 1$ is substituted into this equation, it conforms to Eq. (22.1).

Figure 22.1 shows an example of the relationship between an object area and its vicinity in the case where V_i is defined according to a criterion such as to be adjoining to area i at some point or line.

In Fig. 22.1, each cell represents a small area with the integers denoting the area number. The dark shaded cell indicates the object area while the dark and light shaded cells together indicate its vicinity. In this case, that is, the case of $i = 57$, V_{57} and n_{57} are written as $V_{57} = \{46, 47, 48, 56, 57, 58, 66, 67, 68\}$ and $n_{57} = 9$, respectively. Furthermore, if all weights are 1.0, then $\dot{x}_{57}$ is given by $\dot{x}_{57} = \frac{1}{9}(x_{46} + x_{47} + x_{48} + x_{56} + x_{57} + x_{58} + x_{66} + x_{67} + x_{68})$; if weights follow the rule that (1) the object area is quadruply weighted, (2) areas adjoining to the object area at a line are double weighted, (3) areas adjoining to the object area at a point are single weighted, then $\dot{x}_{57}$ is given by $\dot{x}_{57} = \frac{1}{4}x_{57} + \frac{1}{8}(x_{47} + x_{56} + x_{58} + x_{67}) + \frac{1}{16}(x_{46} + x_{48} + x_{66} + x_{68})$.

Obviously, we can control the strength of smoothing using this method by changing the extent of V_i, so that this method can avoid yielding extraordinary

0	1	2	3	4	5	6	7	8	9
10	11	12	13	14	15	16	17	18	19
20	21	22	23	24	25	26	27	28	29
30	31	32	33	34	35	36	37	38	39
40	41	42	43	44	45	46	47	48	49
50	51	52	53	54	55	56	57	58	59
60	61	62	63	64	65	66	67	68	69
70	71	72	73	74	75	76	77	78	79
80	81	82	83	84	85	86	87	88	89
90	91	92	93	94	95	96	97	98	99

Fig. 22.1 An object area and its vicinity (Source: Inoue (2014))

values in the projected population. In addition, this method does not require that the demographics to smooth follow the specific distribution, making a difference from the empirical Bayes method.

22.3 The Proposed New Method

After describing the fundamental principle of our proposed new method, this section introduces the concept of population potential in Sect. 22.3.1, which is crucially important in developing the method. Moreover, this section derives the basic formula (in Sect. 22.3.2) as well as the applied formulas (in Sects. 22.3.3, 22.3.4, and 22.3.5) from the above fundamental idea.

22.3.1 Fundamental Idea of the New Method

When estimating the demographics of an object area, the reason we use the statistics of its vicinity is that there is a universal principle, namely, "the demographics of an object area approximate those of an area located more closely to it." In addition, "if different areas are the same distance from the object area, their demographics approximate those of the more populated area." The new method discussed in this chapter is to estimate the true value from a weighted moving average based on the above two principles. Consequently, since the weight is defined according to the two principles, it becomes larger when neighboring areas are closer and more populated. This essentially matches the concept of population potential, which is a crucially important dynamic in our new method. Therefore, we can obtain the formulas of our new method if utilizing the population potential as the weight in calculating moving averages.

The population potential, which was developed by Stewart (1947), is defined as "the amount of population energy acting on an area from its neighborhood". Let e_{ij} be the population energy acting on area i from area j and E_i be the population potential of area i, then

$$E_i = \sum_j e_{ij} \ . \tag{22.3}$$

The variable e_{ij} is given by the two different forms as shown in Eq. (22.4)[6] as follows:

$$e_{ij} = 2kp_i\sqrt{\frac{\pi}{a_i}} \quad (\text{if } j = i); \qquad e_{ij} = k\frac{p_j}{d_{ij}} \quad (\text{otherwise}). \tag{22.4}$$

Here k is a constant; p_i and a_i indicate the population and size of area i respectively; and d_{ij} denotes the distance between areas i and j.

22.3.2 Basic Formula

This section derives the basic formula from Eqs. (22.2) and (22.4). Let x_i be the demographics of object area i, and $\hat{x}_i$ be an estimator of x_i. We obtain $\hat{x}_i$ from Eq. (22.2) by replacing w_{ij} with e_{ij} (Eq. 22.5) as follows:

$$\hat{x}_i = \frac{\sum\limits_{j \in V_i} e_{ij}x_j}{\sum\limits_{j \in V_i} e_{ij}} = \frac{e_{ii}x_i + \sum\limits_{j \in V_i, j \neq i} e_{ij}x_j}{e_{ii} + \sum\limits_{j \in V_i, j \neq i} e_{ij}} \ . \tag{22.5}$$

Substituting Eq. (22.4) into Eq. (22.5), we get

$$\hat{x}_i = \frac{2p_i x_i\sqrt{\frac{\pi}{a_i}} + \sum\limits_{j \in V_i, j \neq i} \frac{p_j x_j}{d_{ij}}}{2p_i\sqrt{\frac{\pi}{a_i}} + \sum\limits_{j \in V_i, j \neq i} \frac{p_j}{d_{ij}}} \ . \tag{22.6}$$

[6] The reason this variable takes the two different forms is that we must avoid division by zero in calculating the population potential of area i, whose distance from area i is naturally zero.

Furthermore, assuming that $x_i = q_i/p_i$ (where q_i is the population of area i, differing from p_i, e.g., a sub-population such as the elderly population), Eq. (22.6) leads to Eq. (22.7) as follows:

$$\hat{x}_i = \frac{2q_i\sqrt{\frac{\pi}{u_i}} + \sum_{j \in V_i, j \neq i} \frac{q_j}{d_{ij}}}{2p_i\sqrt{\frac{\pi}{a_i}} + \sum_{j \in V_i, j \neq i} \frac{p_j}{d_{ij}}} \,. \tag{22.7}$$

Equation (22.7) represents the basic formula of our new method.

22.3.3 Applied Formula 1

The basic formula requires an extensive amount of distance data between all combinations of small areas. It also entails complicated numerical calculations even when such distance data are obtained. To address these issues, this section derives three kinds of more practical applied formulas by modifying the basic formula given by Eq. (22.7).

First, we assume that: (1) both areas i and V_i take the form of a precise circle; and (2) population density shows a uniform distribution both in area i and in the donut-shaped area that is generated by removing area i from area V_i (Fig. 22.2).

Second, we define several aggregate variables on area V_i through the following equations: $P_i = \sum_{j \in V_i} p_j$, $Q_i = \sum_{j \in V_i} q_j$, and $A_i = \sum_{j \in V_i} a_j$. If using these variables, the population energies acting on the center of area i from area i and from the donut-shaped area are given in Eqs. (22.8) and (22.9) respectively.

$$k\int_0^{\sqrt{\frac{a_i}{\pi}}} \frac{1}{r} \times \frac{p_i}{a_i} \times 2\pi r dr = 2kp_i\sqrt{\frac{\pi}{a_i}} \,. \tag{22.8}$$

$$k\int_{\sqrt{\frac{a_i}{\pi}}}^{\sqrt{\frac{A_i}{\pi}}} \frac{1}{r} \times \frac{P_i - p_i}{A_i - a_i} \times 2\pi r dr = 2k(P_i - p_i)\frac{\sqrt{\pi}}{\sqrt{A_i} + \sqrt{a_i}} \,. \tag{22.9}$$

Meanwhile, as the demographics x_j of area j, constituting the donut-shaped area, become uniform regardless of j—although still being dependent on i, being expressed as $(Q_i - q_i)/(P_i - p_i)$—Eq. (22.5) is transformed into

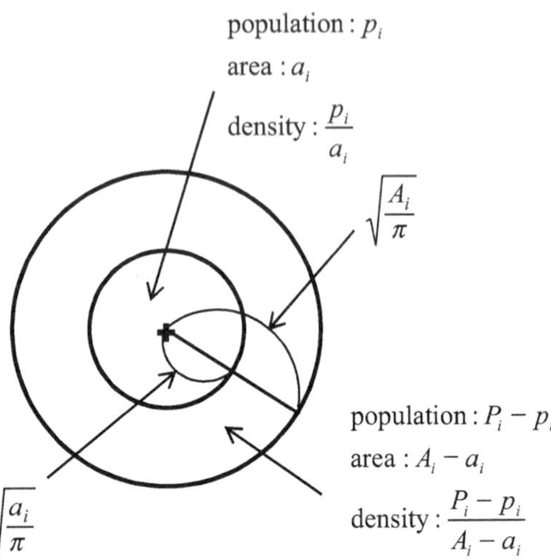

Fig. 22.2 Positional relationship between area i and the donut-shaped area (Source: Inoue (2014))

$$\hat{x}_i = \frac{\sum_{j\in V_i} e_{ij}x_j}{\sum_{j\in V_i} e_{ij}} = \frac{e_{ii} \times \frac{q_i}{p_i} + \frac{Q_i - q_i}{P_i - p_i} \times \sum_{j\in V_i, j\neq i} e_{ij}}{e_{ii} + \sum_{j\in V_i, j\neq i} e_{ij}} . \tag{22.10}$$

Equations (22.8) and (22.9) provide the weights of the moving averages corresponding to e_{ii} and $\sum_{j\in V_i, j\neq i} e_{ij}$ in Eq. (22.10) respectively. Therefore, we can obtain the following equation.

$$\hat{x}_i = \frac{2kp_i\sqrt{\frac{\pi}{a_i}} \times \frac{q_i}{p_i} + 2k(P_i - p_i)\frac{\sqrt{\pi}}{\sqrt{A_i}+\sqrt{a_i}} \times \frac{Q_i - q_i}{P_i - p_i}}{2kp_i\sqrt{\frac{\pi}{a_i}} + 2k(P_i - p_i)\frac{\sqrt{\pi}}{\sqrt{A_i}+\sqrt{a_i}}} . \tag{22.11}$$

Equation (22.11) reduces to the applied formula 1 as follows:

$$\hat{x}_i = \frac{q_i\sqrt{A_i} + Q_i\sqrt{a_i}}{p_i\sqrt{A_i} + P_i\sqrt{a_i}} . \tag{22.12}$$

22.3.4 Applied Formula 2

The applied formula 1 requires the size data of areas i and V_i, however, this data is not always attached to population statistics. Therefore, we make the assumption that the population density of the vicinity (area V_i including area i) shows a uniform distribution, that is, $p_i/a_i = P_i/A_i$, and that then transforms Eq. (22.12) into Eq. (22.13) as follows:

$$\hat{x}_i = \frac{q_i\sqrt{P_i} + Q_i\sqrt{p_i}}{p_i\sqrt{P_i} + P_i\sqrt{p_i}} \ . \tag{22.13}$$

Rearranging this, Eq. (22.13) leads to the applied formula 2a (Eq. 22.14).

$$\hat{x}_i = \frac{\sqrt{p_i}}{\sqrt{p_i} + \sqrt{P_i}} \cdot \frac{q_i}{p_i} + \frac{\sqrt{P_i}}{\sqrt{p_i} + \sqrt{P_i}} \cdot \frac{Q_i}{P_i} \ . \tag{22.14}$$

Because $\frac{\sqrt{p_i}}{\sqrt{p_i}+\sqrt{P_i}} + \frac{\sqrt{P_i}}{\sqrt{p_i}+\sqrt{P_i}} = 1$, Eq. (22.14) means that estimator $\hat{x}_i$ becomes the weighted average of demographics q_i/p_i of area i and demographics Q_i/P_i of area V_i.

Furthermore, assuming that V_i is fixed, regardless of i, and expressed as V (e.g., V represents a municipality including all small areas), P_i and Q_i become constants, given by $P = \sum_{i \in V} p_i$ and $Q = \sum_{i \in V} q_i$ respectively, and Q/P becomes the weighted average of q_i/p_i. Accordingly, Eq. (22.14) is rewritten as the applied formula 2b (Eq. 22.15).

$$\hat{x}_i = \frac{\sqrt{p_i}}{\sqrt{p_i} + \sqrt{P}} \cdot \frac{q_i}{p_i} + \frac{\sqrt{P}}{\sqrt{p_i} + \sqrt{P}} \cdot \frac{Q}{P} \ . \tag{22.15}$$

In Eq. (22.15), if q_i indicates the elderly population of small area i, q_i/p_i and Q/P represent the proportions of elderly population of area i and of the municipality including area i, respectively. Estimator $\hat{x}_i$ becomes the weighted average of the two rates. The weight is given by the ratio of the root of the population to the sum of the roots of the populations. As a result, the applied formula 2b shows a numerical form similar to the empirical Bayes estimator (EBE) or the Stein-type shrinkage estimator (SE) (Efron and Morris 1973, 1975; Shinozaki 1991; Datta and Ghosh 2012), although the weight is changeable according to i, in contrast with the weights obtained by the EBE and SE.

22.3.5 Applied Formula 3

Another estimator can be obtained by reciprocal transformation of the estimator of p_i/q_i (equal to the inverse of q_i/p_i). The estimator of p_i/q_i is calculated similarly to that of q_i/p_i, if the population energy is calculated from not p_i but q_i. We can easily derive another estimator from the applied formula 2 (Eq. 22.13) by interchanging p_i and q_i as well as P_i and Q_i and by taking the inverse of the formula. As a result, the estimator is calculated by the following equation, which leads to the applied formulas 3a and 3b.

$$\hat{x}_i = \frac{q_i\sqrt{Q_i} + Q_i\sqrt{q_i}}{p_i\sqrt{Q_i} + P_i\sqrt{q_i}} \quad . \tag{22.16}$$

Similarly to the applied formulas 2a and 2b, Eq. (22.16) can be transposed into the applied formulas 3a (Eq. 22.17) and 3b (Eq. 22.18) as follows:

$$\hat{x}_i = \frac{p_i\sqrt{Q_i}}{p_i\sqrt{Q_i} + P_i\sqrt{q_i}} \cdot \frac{q_i}{p_i} + \frac{P_i\sqrt{q_i}}{p_i\sqrt{Q_i} + P_i\sqrt{q_i}} \cdot \frac{Q_i}{P_i} \quad . \tag{22.17}$$

$$\hat{x}_i = \frac{p_i\sqrt{Q}}{p_i\sqrt{Q} + P\sqrt{q_i}} \cdot \frac{q_i}{p_i} + \frac{P\sqrt{q_i}}{p_i\sqrt{Q} + P\sqrt{q_i}} \cdot \frac{Q}{P} \quad . \tag{22.18}$$

Here, as mentioned above, P and Q are given by $\sum_{i\in V} p_i$ and $\sum_{i\in V} q_i$ respectively.

22.4 Examination of the New Method

This section examines the efficacy of our new method by applying both new and existing approaches to the same mortality statistics and comparing the results obtained. Three types of estimators are examined. One is the empirical Bayes estimator (EBE), based on the Poisson-gamma model, which has been widely adopted in estimations of mortality (Tango 1988). The other two are new estimators calculated by the applied formulas 2b and 3b. As is well known, the mortality statistics follow the Poisson distribution that shows the probability of very rare phenomenon, and the EBE considered in this section is derived based on the hypothesis that the demographics follow the Poisson distribution. Therefore, in the case of handling the mortality statistics, that hypothesis does not give any disadvantage, differing from the below-mentioned case of handling the CCR and the CWR.

Data used for examination are the SMR (standardized mortality ratio) for female deaths from gastric cancer in Saitama Prefecture, Japan by municipality from 1995 to 1999, calculated by Kubokawa (2013b). Table 22.1 shows the computation results. In this table, four columns on the left-hand side are Kubokawa's original data and the other two columns are newly added data. Expected deaths correspond to ones estimated on the assumption that the mortality of each municipality equals that of the whole of Saitama. The SMR is the ratio of actual deaths to expected deaths. The weighted average of the SMRs is exactly 100.0 in theory. Data in all three columns (EBE, applied formula 2b, applied formula 3b) indicate estimators of the SMR.

According to Table 22.1, every estimator approaches 100.0 more closely than the SMR and therefore, every type of estimation shows a remarkable tendency to smooth the variations in the SMR. This suggests that the new method is effective in similar ways to an existing method (i.e., the empirical Bayes method), at least in terms of smoothing. The intensity of the smoothing, however, differs greatly among the methods. Estimation by the applied formula 2b smooths the variation of SMR

Table 22.1 Female mortality from gastric cancer in Saitama Prefecture, Japan, by municipality, 1995–1999

Municipality	Actual deaths	Expected deaths	SMR	EBE	Applied formula 2b	Applied formula 3b
Kawagoe	206	192.1	107.2	105	101.3	100.1
Kumagaya	136	102.7	132.4	112	104.5	100.4
Kawaguchi	253	242.8	104.2	104	100.8	100.1
Urawa	244	256.7	95.1	98	99.0	99.9
Oomiya	244	264.8	92.1	97	98.4	99.9
Gyouda	69	61.2	112.7	103	101.4	100.2
Tokorozawa	174	179.6	96.9	100	99.5	100.0
Honjyo	39	45.9	85.0	97	98.5	99.7
Higashimatsuyama	41	52.7	77.8	95	97.7	99.6
Kasukabe	121	105.5	114.7	106	102.1	100.2
Yono	58	48.3	120.1	104	102.0	100.3
Koshigawa	130	153.1	84.9	94	97.5	99.7
Hatogaya	46	35.2	130.7	105	102.6	100.4
Sakado	67	51.6	129.8	107	103.1	100.4
Satte	44	34.9	126.1	104	102.2	100.4
Yoshikawa	18	27.9	64.5	95	97.2	99.3
Ogose	15	10.8	138.9	102	101.9	100.5
Naguri	5	3.5	142.9	100	101.2	100.6
Namekawa	6	8.9	67.4	98	98.5	99.4
Ogawa	19	27	70.4	96	97.7	99.4
Kawashima	21	16.5	127.3	102	101.7	100.4
Yoshimi	28	15.3	183.0	106	104.9	101.0
Nagatoro	10	8	125.0	101	101.1	100.4
Okano	4	11.7	34.2	95	96.6	98.3
Ryoujin	1	3.2	31.3	98	98.1	98.1
Higashichichibu	6	4.4	136.4	100	101.2	100.5
Kamiizumi	5	1.5	333.3	101	104.5	102.0
Kamisato	25	18.3	136.6	103	102.3	100.5
Oosoto	11	6.3	174.6	102	102.9	100.9
Okabe	17	13.9	122.3	101	101.2	100.3
Kawamoto	15	9.1	164.8	103	103.0	100.8
Shiraoka	20	26.1	76.6	97	98.2	99.6
Shoubu	18	15.8	113.9	101	100.8	100.2

Source: Kubokawa (2013b) (partly revised)

more intensely than does the empirical Bayes method. Estimation by the applied formula 3b smooths it the most intensely and reduces the range of the SMR to less than 4.0. Such strong smoothing has both an advantage and a disadvantage, although estimation particularly by the applied formula 3b gives us an impression of excessive shrinkage. The advantage is that it can avoid yielding extraordinary values, especially in the case where an estimator needs to be used iteratively, like the CCR in long-term population projection. The disadvantage is the possible loss of the true variation in the original data. Further studies are needed to evaluate these advantages and disadvantages mathematically.

22.5 Application of the New Method to Population Projection

After explaining how to apply our new method to small area population projection, this section undertakes a simple simulation by applying it to actual census data. Additionally, this section briefly introduces the original web system "The Web System of Small Area Population Projection for the Whole Japan," which the author developed and published by applying our new method to small area population statistics from Japan census.

As mentioned before, almost all governments in the world have failed to provide vital and migration statistics by small area. For this reason, this section adopts the cohort change ratio method formulated by Hamilton and Perry (1962) that does not employ such statistics. Since the cohort change ratio method requires only two demographics, CCRs and CWRs, or any birth statistics that can replace CWRs, this section explains how to calculate these two demographics by using our new method. In that explanation, we take the applied formula 2b (Eq. 22.15) that shows a simpler form than other formulas, as an example.

First, assuming that the census interval is 5 years, we prepare census population data by small area, sex, and age class at two points of time. Second, let $p_{i,t}(u,\ v)$, $f_{i,t}(u,\ v)$, $P_t(u,\ v)$, and $F_t(u,\ v)$ be as follows:

$p_{i,t}(u,\ v)$: the population aged from u to v, of area i, and at census year t,
$f_{i,t}(u,\ v)$: the female population aged from u to v, of area i, and at census year t,
$P_t(u,\ v)$: the population aged from u to v, of the municipality including area i, and at census year t,
$F_t(u,\ v)$: the female population aged from u to v, of the municipality including area i, and at census year t,
estimator $\hat{y}_{i,b}$ of the CCR and estimator $\hat{z}_i$ of the CWR are derived from Eq. (22.15) as follows[7]:

[7] Regarding the oldest age class, we must utilize the formula different slightly from equation [19]. Please see, for example, Smith et al. 2002 for details. If the fertility of women aged 15–19 and 40–49 shows a very low value as observed in Japan, it will be better to replace $f_{i,t}(15,\ 49)$ with $f_{i,t}(20,\ 39)$ in the calculation of CWRs.

$$
\begin{aligned}
\hat{y}_{i,b} = & \frac{\sqrt{p_{i,s-5}(b-5,\ b-1)}}{\sqrt{p_{i,s-5}(b-5,\ b-1)} + \sqrt{P_{s-5}(b-5,\ b-1)}} \\
& \cdot \frac{p_{i,s}(b,\ b+4)}{p_{i,s\ 5}(b-5,\ b-1)} \\
& + \frac{\sqrt{P_{s-5}(b-5,\ b-1)}}{\sqrt{p_{i,s-5}(b-5,\ b-1)} + \sqrt{P_{s-5}(b-5,\ b-1)}} \\
& \cdot \frac{P_s(b,\ b+4)}{P_{s-5}(b-5,\ b-1)},
\end{aligned}
\tag{22.19}
$$

$$
\begin{aligned}
\hat{z}_i = & \frac{\sqrt{f_{i,s}(15,\ 49)}}{\sqrt{f_{i,s}(15,\ 49)} + \sqrt{F_s(15,\ 49)}} \cdot \frac{p_{i,s}(0,\ 4)}{f_{i,s}(15,\ 49)} \\
& + \frac{\sqrt{F_{s-m}(15,\ 49)}}{\sqrt{f_{i,s}(15,\ 49)} + \sqrt{F_s(15,\ 19)}} \cdot \frac{P_s(0,\ 4)}{F_s(15,\ 49)}.
\end{aligned}
\tag{22.20}
$$

Here, s denotes the latest census year and therefore is the fixed value. b denotes the youngest age of an age class as of year s.

The procedures for conducting long-term population projection using Eqs. (22.19) and (22.20) are as follows: (1) we prepare the census population statistics of year $s-5$ and year s; (2) we project the population of year $s+5$ from the population of year $s-5$ and year s using those two equations; (3) we project the population of year $s+10$ from the population of year s and the newly projected population of year $s+5$ using those two equations; and (4) we iterate the procedure (3). These procedures enable us to understand that slightly high CCRs and CWRs have a risk of generating an extremely huge population by amplifying it. For example, if the consecutive four CCRs (ratio of 20–24 age class to 15–19 age class, . . ., ratio of 35–39 age class to 30–34 age class) are all equal to 2.0, the size of the cohort aged 15–19 at the first time increases by 16 times in only 20 years.

Third, we undertake a simple simulation to examine to what extent our new method is effective, especially when applying it to small area population statistics. Formulas used for the simulation are the above two Eqs. (22.19) and (22.20), and data used for the simulation are the 2000, 2005 and 2010 censual small area population statistics of Shibuya Ward, Tokyo, Japan[8]. The concrete simulation procedure is as follows: (1) by applying four methods (no smoothing method, empirical Bayes method, uniform distribution based method,[9] and our new method) to the 2000 and 2005 censual population, we simulate the 2010 population; (2) we calculate the degree of dissociation between the simulated 2010 population and the

[8] In this simulation, $f_{i,t}(20,\ 39)$ is used for the calculation of CWRs for the reason mentioned in footnote 7.

[9] This is the method of projecting population on the hypothesis that CCRs and CWRs of all small areas conform entirely to those of the municipality constituted by them.

Table 22.2 Degree of dissociation

Method of smoothing	Degree
No smoothing method	38,718
Empirical Bayes method	34,222
Uniform distribution based method	29,748
New method	29,488

actual 2010 population regarding each of those four methods; and (3) with respect to the degree of dissociation, we compare the effectiveness of those four methods. The degree is given by the sum of the absolute difference between the above two kinds of population enumerated by sex, age class and area. Table 22.2 shows the result of the simulation. This table enables us to realize that our new method is more effective than the other three methods, although there is little difference in the degree of dissociation between our new method and the uniform distribution based method. Meanwhile, the empirical Bayes method is not very effective, and this is possibly caused by the fact that the CCR and the CWR do not satisfy the above-mentioned hypothesis on the Poisson distribution.

Finally, this section adds a brief introduction on the original web system "The Web System of Small Area Population Projection for the Whole Japan," which the author developed and published in 2015, by applying our new method to the 2005 and 2010 censual small area population statistics of Japan.[10] This system provides the long-term small area population projected from 2015 to 2060 on a 5-year basis, by sex and 5-year age class. The territory of the whole of Japan consists of about 217 thousand small areas. All users can freely download such projected population data and can browse map pictures on proportion of elderly population, population density and population increase rate. URLs of this system are as follows:

http://arcg.is/1GkdZTX (English version)
http://arcg.is/1LqC6qN (Japanese version)

22.6 Discussion

This chapter considers a new method of estimating small area demographics, and explained how to apply the method to long-term small area population projection. This section discusses both the advantages and disadvantages of the method, focusing on its efficacy by comparison with the existing methods, especially with the empirical Bayes method.

[10] The author also described the brief manual for Japanese version of this system (Inoue 2016). This system is based on ArcGIS Online®, which is the Internet GIS system provided by Esri Corporation, so the interface and design of this website are changeable with updates of ArcGIS Online. In addition, please pay attention to possibility that the author may update its contents.

As regards the advantages, the following points are indicated: (1) the first point is that our new method consists of six formulas (the basic formula and applied formulas 1, 2a, 2b, 3a, and 3b), and that therefore the most appropriate formula is selectable according to the information amount of demographics. This is a unique advantage that existing methods do not have. The basic formula can utilize position or coordinate data of adjacent areas most satisfactorily. Even if such data cannot be obtained, the applied formulas 2b and 3b are available and estimators derived from these formulas can be easily calculated. The second point is that the applied formulas 2b and 3b can effectively smooth small area demographics, similarly to the empirical Bayes method, and that the smoothing by these formulas exceeds in strength that of the empirical Bayes method (Table 22.1). Therefore, our new method can adequately avoid yielding extraordinary values in projected population, especially in the case of long-term projection that requires iterative calculation. The third point is that our new method does not require the hypothesis that demographics follow the Poisson distribution and that nevertheless, the empirical Bayes method requires the hypothesis. This will be a crucial advantage, due to the fact that the smoothing by the empirical Bayes method did not lead to a desirable result in the simple simulation (Table 22.2), and this result is possibly caused by the fact that the CCR and CWR, indispensable to the cohort change ratio method, do not satisfy the hypothesis.

On the other hand, with regards to the disadvantages, the following points are indicated: (1) the first point is that the basic formula requires a large distance dataset between all small areas, and that the formula is not as practical, although this is the compensation for the above advantage; and (2) the second point is that the applied formulas 2b and 3b, especially the formula 3b, have a risk of smoothing demographics so strongly and shrinking their true variation excessively (Table 22.1). This probably resulted in the fact that there was little difference in the degree of dissociation between our new method and the uniform distribution based method (Table 22.2).

Our new method is formulated based on the following principle: the amount of population energy acts on an area from its neighborhood according to the observed demographics, and this enables us to estimate the true value of the demographics. On the other hand, the following principle is acceptable: the amount of population energy acts on an area from its neighborhood according to the true value of the demographics, and this leads to the observable demographics. Further mathematical discussion of the advantage (3), the disadvantage (2), and this point is beyond the scope of this chapter and requires future consideration.

Acknowledgements This chapter is a result of a study that received financial support from the Japan Society for the Promotion of Science by grant-in-aid (grant number: 25370919 and 16H03525). The author deeply appreciates the assistance of Mr. Atsushi Kajimoto (Esri Japan Corporation) for constructing the online system "The Web System of Small Area Population Projection for the Whole Japan."

References

Anselin, L., Lozano, N., & Koschinsky, J. (2006). *Rate transformations and smoothing*. Spatial Analysis Laboratory, Department of Geography, University of Illinois, Urbana-Champaign.

Clayton, D., & Kaldor, J. (1987). Empirical Bayes estimates of age-standardized relative risks for use in disease mapping. *Biometrics, 43*(3), 671–681.

Datta, G., & Ghosh, M. (2012). Small area shrinkage estimation. *Statistical Science, 27*, 95–114.

Efron, B., & Morris, C. (1973). Stein's estimation rule and its competitors: an empirical Bayes approach. *Journal of the American Statistical Association, 68*(341), 117–130.

Efron, B., & Morris, C. (1975). Data analysis using Stein's estimator and its generalizations. *Journal of the American Statistical Association, 70*(350), 311–319.

Farrell, P. J., MacGibbon, B., & Tomberlin, T. J. (1997). Empirical Bayes estimators of small area proportions in multistage designs. *Statistica Sinica, 7*, 1065–1083.

Fay, R. E., & Herriot, R. A. (1979). Estimates of income for small places: An application of James-Stein procedures to census data. *Journal of the American Statistical Association, 74*(366), 269–277.

Ghosh, M., Kim, M. J., & Kim, D. H. (2008). Constrained Bayes and empirical Bayes estimation under random effects normal ANOVA model with balanced loss function. *Journal of Statistical Planning and Inference, 138*(7), 2017–2028.

Gonzalez, M. E., & Hoza, C. (1978). Small-area estimation with application to unemployment and housing estimates. *Journal of the American Statistical Association, 73*(361), 7–15.

Griffith, D. A. (2003). *Spatial autocorrelation and spatial filtering: Gaining understanding through theory and scientific visualization*. Berlin: Springer.

Haining, R. P., Kerry, R., & Oliver, M. A. (2010). Geography, spatial data analysis, and geostatistics: An overview. *Geographical Analysis, 42*(1), 7–31.

Hamilton, C. H., & Perry, J. (1962). A short method for projecting population by age from one decennial census to another. *Social Forces, 41*(2), 163–170.

Holt, D., Smith, T. M. F., & Tomberlin, T. J. (1979). A model-based approach to estimation for small subgroups of a population. *Journal of the American Statistical Association, 74*(366), 405–410.

Inoue, T. (2007). The concept of population potential and small area population statistics. *Statistical Magazine*, *58*(12), 12–16. (in Japanese)

Inoue, T. (2014). A new method of estimating small area demographics using population potential. *Working Paper Series, Institute of Economic Research, Aoyama Gakuin University 2014*(3), 1–16.

Inoue, T. (2016). Development of the web system of small area population projection for whole Japan. *Expertise in Statistics and Research for Leaders and Administrators (ESTRELA) 262*, 2–9. (in Japanese)

Kubokawa, T. (2013a). Constrained empirical Bayes estimator and its uncertainty in normal linear mixed models. *Journal of Multivariate Analysis 122*, 377–392.

Kubokawa, T. (2013b). The benchmark problem on small area estimation and error estimate of constrained empirical Bayes estimation. *CIRJE Research Report Series, CIRJE-R-12*, 143–165. (in Japanese)

Laake, P. (1979). A predictive approach to subdomain estimation in finite populations. *Journal of the American Statistical Association, 74*(366), 355–358.

Louis, T. A. (1984). Estimating a population of parameter values using Bayes and empirical Bayes methods. *Journal of the American Statistical Association, 79*(386), 393–398.

MacGibbon, B., & Tomberlin, T. J. (1989). Small area estimates of proportions via empirical Bayes techniques. *Survey Methodology, 15*, 237–252.

Maritz, J. S. (1969). Empirical Bayes estimation for the Poisson distribution. *Biometrika, 56*(2), 349–359.

Marshall, R. J. (1991). Mapping disease and mortality rates using empirical Bayes estimators. *Applied Statistics, 40*(2), 283–294.

Mfungahema, R., & Kitamura, T. (1997). The significance of population potential at the local level and in rural areas: A case of Okayama and Tottori prefectures in Japan. *Journal of Rural Planning Association, 16*(3), 251–262.

Morris, C. N. (1983). Parametric empirical Bayes inference: Theory and applications. *Journal of the American Statistical Association, 78*(381), 47–55.

Mosteller, F., & Tukey, J. W. (1977). *Data analysis and regression: A second course in statistics*. Boston: Addison-Wesley.

Rich, D. C. (1980). *Potential models in human geography: Concepts and techniques in modern geography, volume 26 of concepts and techniques in modern geography* (Study Group in Quantitative Methods). Institute of British Geographers, University of East Anglia, Norwich: GeoBooks.

Shinozaki, N. (1991). Stein-type shrinkage estimators and their applications. *Japanese Journal of Applied Statistics, 20*(2), 59–76. (in Japanese)

Smith, S. K., Tayman, J., & Swanson, D. A. (2002). *State and local population projections: Methodology and analysis*. New York: Kluwer.

Smith, S. K., Tayman, J., & Swanson, D. A. (2013). *A practitioner's guide to state and local population projections*. Dordrecht: Springer.

Stewart, J. Q. (1947). Empirical mathematical rules concerning the distribution and equilibrium of population. *Geographical Review, 37*(3), 461–485.

Swanson, D. A., & Tayman, J. (2012). *Subnational population estimates*. Dordrecht: Springer.

Swanson, D. A., Schlottmann, A., & Schmidt, B. (2010). Forecasting the population of census tracts by age and sex: An example of the Hamilton-Perry method in action. *Population Research and Policy Review, 29*(1), 47–63.

Tango, T. (1988). Empirical Bayes estimation for mortality indices: Applications to disease mapping. *Japanese Journal of Applied Statistics, 17*(2), 81–96. (in Japanese)

Tocalis, T. R. (1978). Changing theoretical foundations of the gravity concept of human interactions. In B. J. L. Berry (Ed.), *Perspectives in geography 3: The nature of change in geographical ideas* (pp. 67–124). DeKalb: Northern Illinois University Press.

Tsutakawa, R. K., Shoop, G. L., & Marienfeld, C. J. (1985). Empirical Bayes estimation of cancer mortality rates. *Statistics in Medicine, 4*(2), 201–212.

Waller, L. A., & Gotway, C. A. (2004). *Applied spatial statistics for public health data*. New York: Wiley.

Warntz, W. (1964). A new map of the surface of population potentials for the United States, 1960. *Geographical Review, 54*(2), 170–184.

Chapter 23
A Long Term Test of the Accuracy of the Hamilton-Perry Method for Forecasting State Populations by Age

David A. Swanson and Jeff Tayman

Abstract The Hamilton-Perry method is a variant of the Cohort-Component population forecasting method that has minimal data input requirements. It only requires the age distributions for a population at two points in time, which generally are two successive census enumerations. Although the method has gained acceptance, tests of its accuracy are limited. In this chapter we evaluate the forecast accuracy of the Hamilton-Perry method both in terms of age and total population (obtained by summing up the forecasted age groups). This evaluation is based on a sample of four states (one from each of the four census regions) and decennial census data from 1900 to 2010, which yield 10 census test points (1920, 1930, 1940,. . ., 2010) that provide a wide range of characteristics in regard to population size, growth, and age-composition that affect forecast accuracy. We conclude that the results are encouraging and suggest that the Hamilton-Perry method be considered when either a 10-year forecast of state populations by age or a total population are desired and components of change are not required.

Keywords Cohort change ratio • MAPE • MALPE • Allocation error

23.1 Introduction

In a seminal paper, Hamilton and Perry (1962) proposed cohort-change ratios as a variant of the cohort-component method for purposes of short-term population forecasts.[1] The major advantage of the Hamilton-Perry method is it has much

[1] Although the name "Hamilton-Perry method" is virtually universal today, the first published instance of cohort change ratios for purposes of population forecasting is found in Hardy and

D.A. Swanson (✉)
University of California, Riverside, CA 92521, USA
e-mail: David.swanson@ucr.edu

J. Tayman
University of California, San Diego, CA 92093, USA
e-mail: jtayman@ucsd.edu

D.A. Swanson (ed.), *The Frontiers of Applied Demography*, Applied Demography Series 9, DOI 10.1007/978-3-319-43329-5_23

smaller data requirements than its more data-intensive cousin while still providing a forecast of population by age (as well as sex, race, ethnicity, if so desired), which is the hallmark of the full cohort-component method (George et al. 2004; Smith et al. 2013: 176; Swanson et al. 2010). Instead of mortality, fertility, migration, and population data by age and sex, the Hamilton-Perry method simply requires age (and sex, race, ethnicity, if these characteristics are desired in a forecast) data from the two most recent censuses. Consequently, it is much quicker, easier, and cheaper to implement than a full cohort-component model. Not surprisingly, it has mainly been used for small geographic areas in which mortality, fertility, and migration data are non-existent, unreliable, or very difficult to obtain (Baker et al. 2011; Smith and Tayman 2003; Swanson et al. 2010).

Although the Hamilton-Perry method has primarily been used for small geographic areas, its minimal data input requirements combined with its capability for producing age and other characteristics in a forecast make it attractive for use at high levels of geography such as states and counties. To our knowledge only one study by Smith and Tayman (2003) has evaluated the accuracy of the Hamilton-Perry method. They examined forecast errors by age group for all states and for counties in Florida using 1990 and 2000 as test point years and both 10– and 20-year forecast horizons. They found that its accuracy was equivalent to that of cohort-component method forecasts. This paper is intended to supplement their findings by using a sample of states in conjunction with tests covering a long period of time. Unlike Smith and Tayman (2003), we evaluate only a sample of four states (one randomly selected from each of the four census regions, respectively); but unlike them we use ten test point years, from 1920 to 2010, each representing a 10-year forecast horizon.

We begin by describing the Hamilton-Perry method and then move on to discuss three major dimensions of forecast accuracy and how they are measured. We then describe the data used in our empirical examination and our results, which include an evaluation of the accuracy of the forecasted age groups and the total populations. We conclude the paper with a discussion of our findings, their implications and limitations, which, in turn, lead to suggestions for future research in this area.

Wyatt (1911), who built cohort change ratios from the 1901 and 1906 census counts of England and applied them to the 1906 census to generate a forecast for 1911. Hamilton and Perry acknowledge that they learned about this method from a description of it found in Wolfenden (1954), but they were unable to secure a copy of the 1911 article and were, therefore, not exactly certain what was done by Hardy and Wyatt. In any event, Hamilton and Perry deserve credit for providing a clear and detailed description of this approach to population forecasting in a journal was read by many demographers in the United States and elsewhere prior to the founding of demographic journals such as *Canadian Studies in Population* (first published in 1973) *Demography* (first published in 1966) and *Population Research and Policy Review* (first published in 1982).

23.2 The Hamilton-Perry Method

Before describing the Hamilton-Perry method, it is useful to recall that any quantitative approach to forecasting is constrained to satisfy various mathematical identities (Land 1986). In regard to population forecasting, an approach should ideally satisfy demographic accounting identities, which is summarized in the identity known as the fundamental demographic equation:

$$P_t = P_0 + \text{Births} - \text{Deaths} + \text{In-migrants} - \text{Out-migrants}. \tag{23.1}$$

That is, the population at some time in the future, P_t, must be equal to the population at an earlier time, P_0, plus the births and in-migrants and less the deaths and out-migrants that occur between time *0* and time *t*. The most commonly used approach to population forecasting, the cohort-component method satisfies the fundamental equation, but it is data-intensive (George et al. 2004; Smith et al. 2013: 180; Murdock and Ellis 1991; Pittenger 1976). In Appendix A.1, we show that the Hamilton-Perry method satisfies the fundamental demographic equation. Moreover, it has far less intensive input data requirements than does the cohort-component method. Instead of mortality, fertility, migration, and population data by age and sex, which are required by the full-blown cohort-component method, the Hamilton-Perry method requires age data only from the two most recent censuses (Hamilton and Perry 1962; Smith et al. 2013: 176–179; Swanson et al. 2010; Swanson and Tedrow 2012).

The Hamilton-Perry method moves a population by age (and sex) from time *t* to time $t + k$ using cohort-change ratios (CCR) computed from data in the two most recent censuses. It consists of two steps. The first uses existing data to develop CCRs and the second applies the CCRs to the cohorts of the launch year population to move them into the future. As shown by Swanson et al. (2010), the formula for developing a CCR is:

$${}_{n}CCR_{x,i,t} = {}_{n}P_{x,i,t} / {}_{n}P_{x\text{-}k,i,t\text{-}k} \tag{23.2}$$

where,

${}_{n}P_{x,i,t}$ is the population aged x to $x + n$ in area i at the most recent census (t),

${}_{n}P_{x\text{-}k,i,t\text{-}k}$ is the population aged $x\text{-}k$ to $x\text{-}k + n$ in area i at the 2nd most recent census ($t\text{-}k$), and

k is the number of years between the most recent census at time t and the one preceding it at time $t\text{-}k$.

The formula for the second step, moving the cohorts of a population into the future, is:

$$_{n}P_{x+k,i,t+k} = {_{n}CCR_{x,i,t}} \times {_{n}P_{x,i,t}} \tag{23.3}$$

where,

${_nP_{x+k,i,t+k}}$ is the population aged $x + k$ to $x + k + n$ in area i at time $t + k$, and ${_nCCR_{x,i,t}}$ and ${_nP_{x,i,t}}$ are as defined in Eq. (23.2).

Given the nature of the CCRs, 10–14 is the youngest 5-year age group for which forecasts can be made if there are 10 years between censuses. To forecast the population aged 0–4 and 5–9 one can use the Child Woman Ratio (CWR) or more generally a "Child Adult Ratio" (CAR). These ratios do not require any data beyond what is available in the decennial census. In calculating a CAR, we believe that the age group most closely associated with the children in question should be used. For example, for children aged 0–4, we use adults aged 20–34. In a different country where fertility starts earlier, it may be appropriate to use adults aged 15–34. So, we forecast the population aged 0–4 using a CAR defined as the population aged 0–4 divided by the population aged 20–34. Similarly, for forecasting the population aged 5–9 the CAR is defined as the population aged 5–9 divided by the population aged 25–39. There are other "adult" age groups that could be used to define CAR (Smith et al. 2013: 278), but we prefer the ones we describe here. The equations for forecasting the population aged 0–4 and 5–9 are:

$$\text{Population } 0\text{–}4 : {_5P_{0,t+k}} = \left({_5P_{0,t}}/{_{15}P_{20,t}}\right) \times {_{15}P_{20,t+k}}, \text{ and} \tag{23.4a}$$

$$\text{Population } 5\text{–}9 : {_5P_{5,t+k}} = \left({_5P_{5,t}}/{_{15}P_{25,t}}\right) \times {_{15}P_{25,t+k}} \tag{23.4b}$$

where,

P is the population,
t is the year of the most recent census, and
$t + k$ is the forecast year.

Another way to forecast the youngest age groups is to take their ratios at two points in time and apply that ratio to the launch year age group (t). In the first step, the ratios are:

$$\text{Population } 0\text{–}4 : {_5R_{0,t}} = {_5P_{0,t}}/{_5P_{0,t-k}} \text{ and} \tag{23.5a}$$

$$\text{Population } 5\text{–}9 : {_5R_{5,t}} = {_5P_{5,t}}/{_5P_{5,t-k}}. \tag{23.5b}$$

In the second step, the forecasted population at $t + k$ is found by:

$$\text{Population } 0\text{–}4 : {_5P_{0,t+k}} = {_5P_{0,t}} \times {_5R_{0,t}}, \text{ and} \tag{23.6a}$$

$$\text{Population } 5\text{–}9 : {_5P_{5,t+k}} = {_5P_{5,t}} \times {_5R_{5,t}}. \tag{23.6b}$$

We prefer the "CAR" approach because it directly incorporates changes in the age groups associated with childbearing.

Forecasts of the oldest open-ended age group also differ slightly from the forecasts for the age groups beyond age 10 up to the oldest open-ended age group. If, for example, the final closed age group is 70–74 and 75+ is the terminal open-ended age group, calculations for the $CCR_{i,x+}$ require the summation of the three oldest age groups to get the population age 65+ at time t–k:

$$_{\infty}CCR_{75,i,t} = {}_{\infty}P_{75,i,t} / {}_{\infty}P_{65,i,t-k}. \tag{23.7a}$$

The formula for estimating the population 75+ of area i for the year $t + k$ is:

$$_{\infty}P_{75+,i,t+k} = {}_{\infty}C^{\infty}CR_{75,i,t} \times {}_{\infty}P_{65,i,t}. \tag{23.7b}$$

23.3 Measuring Forecast Error

There are three major dimensions to forecast (and estimation) error: (1) bias; (2) precision; and (3) allocation (Swanson 2015; Swanson et al. 2011). Three commonly used measures that correspond to these three dimensions are: (1) Mean Algebraic Percent Error (MALPE); (2) Mean Absolute Percent Error (MAPE); and (3) The Index of Allocation Error (IOAE). In moving toward definitions of these three measures, we begin by defining forecast error (E), which is the difference between a given forecast (F) for a particular population and an observed count such as the 2010 census (CEN): E = F – CEN. The error will be positive when the forecast is larger than the census count and it will be negative when it "underforecasts" the census. The definition of error given above can be broadened to include age, sex, race, and ethnicity. For example, the definition of error for a particular age group (a) would be $E_a = F_a - CEN_a$.

Errors are often expressed as percent differences rather than absolute differences. The use of percent errors is particularly helpful when making comparisons across geographic areas. A forecast error of 2000 has a very different meaning for a place with 20,000 residents than a place with 200,000 residents. Without adjustments for population size, errors for places with large populations (or age groups with large numbers of people) would dominate the effects of errors for places with small populations (or age groups with small numbers of people). Thus, the definition of error can be broadened to provide a "relative" perspective as follows:

$$\text{ALPE} = (\text{E} / \text{CEN}) \times 100, \text{ and} \tag{23.8}$$

$$\text{APE} = |(\text{E} / \text{CEN}) \times 100|. \tag{23.9}$$

The ALPE (algebraic percent error) preserves the sign of the percent error; it has a theoretical minimum of -100 % and no upper bound, while the APE (absolute percent error) has a minimum at zero and no upper bound. ALPE and APE represent individual forecast errors for the set of geographic areas under study and form a distribution of forecast errors. As before, the definition of relative error given above can be broadened to include age, sex, race, and ethnicity. For example, the definition of relative error for a particular age group *(a)* would be:

$$\mathrm{ALPE_a} = (\mathrm{E_a}/\mathrm{CEN_a}) \times 100, \text{and} \tag{23.10}$$

$$\mathrm{APE_a} = |(\mathrm{E_a}/\ \mathrm{CEN_a})\ \times\ 100|. \tag{23.11}$$

Two common summary measures of relative error are MALPE and MAPE, both of which are arithmetic means:

$$\mathrm{MALPE} = \sum \mathrm{ALPE}\ /\ \mathrm{n\ and} \tag{23.12}$$

$$\mathrm{MAPE} = \sum \mathrm{APE}\ /\ \mathrm{n}, \tag{23.13}$$

where,

n is the number of observations (usually geographic areas).

MALPE is a measure in which positive and negative values offset each other. Consequently, it is often used as an average measure of bias. A positive MALPE reflects the average tendency for forecasts to be too high and a negative MALPE reflects the average tendency for forecasts to be too low. A zero MALPE would indicate no bias in the set of forecasts because the sum of the positive percentage errors would equal the sum of the negative percentage errors.

MAPE is a measure in which positive and negative values do not offset each other; it measures the precision of the forecasts by showing the average percent difference between forecasts and observed populations regardless of whether the individual forecasts were too high or too low. MAPE has several desirable properties including reliability and ease of use and interpretation. It also incorporates all of the information in its calculation, but MAPE has a major drawback. Like any average MAPE is affected by extreme values which most often occur at the high end of the distribution. Thus, the error distribution of the APEs is often asymmetrical and right-skewed because it is bounded on the left by zero and unbounded on the right. Therefore, MAPE, and to a much lesser extent the MALPE, is susceptible to being pulled upward and to overstating the error represented by most of the observations.[2]

[2] Other measures of precision can be used when the distribution of APEs is highly asymmetrical (Swanson et al. 2011, 2012). Two such measures are the "MEDAPE" (Median Absolute Percent Error); and "MAPE-R" (MAPE–Rescaled). The MAPE can be supplemented with MEDAPE, which is the median value of the APEs, or MAPE-R, a measure that minimizes the effect of outliers while preserving more information about the original distribution than MEDAPE (Swanson et al. 2011). One drawback of MAPE-R is that it is much more complicated to calculate than MEDAPE. Since MAPE is so widely known and forecast errors are generally stable across a variety of error measures (Rayer 2007), we forego using MEDAPE and MAPE-R in this study.

Based on previous research and our experience (e.g., Smith and Tayman 2003; Smith et al. 2013: 364–366; Swanson and Beck 1994; Swanson et al. 2011), we believe that for a 10-year forecast of total populations for states, a MALPE less than +5 % and greater than -5 % indicates that a forecast is not substantially biased. A MALPE either greater than +5 % but less than +15 % or less than -5 % but not less than -15 % indicates considerable bias and one greater than +15 % indicates sub stantial upward bias (the forecast is way too high) and one less than -15 % indicates substantial downward bias (the forecast is way too low). A MAPE less than 5 % indicates that the forecast is highly precise, while a MAPE greater than 5 % but less than 10 % indicates a modest level of precision, and, finally, a MAPE greater than 10 % indicates a low level of precision.

The summary measures discussed above are based on the error for a particular geographic area. Another perspective views the misallocation of the forecast across geographic space or a given variable such as age. Misallocation focuses on how well the forecast distribution matches the observed distribution such as the total population distribution across states or the age distribution in a particular state. Our focus here is not on geographic misallocation, but on the accuracy of the age distribution forecast from the Hamilton-Perry method for each state. The Index of Allocation Error (IOAE), also known as the Index of Dissimilarity, can be used to measure the extent that the forecast misallocates across the age distribution:

$$\mathrm{IOAE} = 100 \times \left\{0.5 \times \sum \left|(\mathrm{F_a}/\sum \mathrm{F_a}) - (\mathrm{CEN_a}/\sum \mathrm{CEN_a})\right|\right\} \tag{23.14}$$

where,

$\sum$ = the summation across age groups (a).

IOAE compares the percent distributions of the forecast and observed shares across the categories of a given variable (e.g., age) and measures the percentage that one distribution (i.e. based on the forecasts) would have to be re-allocated to match the other (i.e. based on the census). The IOAE ranges from 0 to 100; a score of zero is means that there is no allocation error, and 100 means that the maximum allocation error exists. This can mean several things, but a common interpretation is that half of the forecast numbers would have to be re-allocated and half of the census counts would have to be re-allocated.

As a simple example, suppose we were forecasting a population divided into two age groups and that the forecast has all of the people in one age group while the census has them in the other. In this case, $\sum\left|(\mathrm{F_a}/\sum\mathrm{F_a}) - (\mathrm{CEN_a}/\sum\mathrm{CEN_a})\right|$ will equal 2 and the and IOAE will n equal 100 % (100 × (0.5 × 2). Again, based on previous research and our experience (e.g., Fonseca and Tayman 1989; Swanson and Beck 1994; Swanson et al. 2011), we believe that an IOAE of less than five percent indicates a very close match between the two distributions, while an IOAE between five and ten percent portrays a reasonable match between the two distributions, and an IOAE greater than ten percent suggests the distributions are quite distinct.

23.4 Empirical Data

To empirically examine the accuracy of the Hamilton-Perry method, we selected a sample made up of one state from each of the four census regions in the United States. Within each census region, a state was randomly selected. The states selected were Georgia (South Region), Minnesota (Midwest Region), New Jersey (Northeast Region), and Washington (West Region). We assembled census data for these four states for each decade from 1900 to 2010. We used "75 years and over" as the terminal age group because that age group was consistent throughout the sample time periods. As a result, 16 age groups (0–4, 5–9,. . ., 70–74, and 75+) were used in the empirical evaluation.

We constructed CCRs over two successive decennial periods (e.g., 1900 and 1910) over the entire sample time period and apply CCRs to the "launch year" (e.g. 1910) to develop a 10-year forecast for the 16 age groups for the simulated horizon year (e.g., 1920). Successively applying this approach provided ten simulated horizon years at which the forecast can be evaluated, 1920, 1930, . . ., 2000, and 2010. Some applications control the Hamilton-Perry forecast to an independent forecast of total population (Smith and Tayman 2003; Swanson et al. 2010).[3] We opted not to control the Hamilton-Perry forecasts by age because we wanted to evaluate the total population forecast error derived directly from the method itself. Total population forecasts are derived by summing the forecast over all age groups. These forecasts were compared to the corresponding census data in ex post facto tests used to construct the, MALPEs, MAPEs, and IOAEs described previously.

This sample provided a wide range of demographic characteristics in terms of variation in population size, age-composition, and rates of change. Table 23.1 provides an overview of this range by displaying the population of each of the four states in 1900 and in 2010 and decennial rates of annual population change from 1900 to 2010. Although we do not show a summary of the changes in age composition by state and census year, they are extensive as can be seen in Appendix A.2.

The population in Georgia increases by almost five-fold between 1900 and 2010. In 1900 it has the largest population of the four sample states and it retains that position in 2010. Its annual average growth rates (by decade) range from 0.05 % between 1920 and 1930 to 1.68 % between 2000 and 2010. Growth rates in Georgia, generally higher since the 1960s, are associated with in-migration to the Sunbelt states. The population of Minnesota triples from 1900 to 2010. Its average annual growth rates range from a low of 0.66 % between 1940 and 1950 to a high of 1.70 %

[3] In rapidly changing areas it is advisable to control Hamilton-Perry (HP) forecasts by age to an independent total population forecast (Smith et al. 2013:180–181). The application of constant CCRs can often lead to forecasts that are too high (low) in rapidly increasing (decreasing) areas. Smith and Tayman (2003) found that while uncontrolled HP forecasts generally had larger errors than the controlled forecasts, the patterns of errors by age groups very generally similar for both sets.

Table 23.1 Total population 1900 and 2010 and annual rate of change by decade, sample states

Census year	Georgia	Minnesota	New Jersey	Washington
1900	2,216,311	1,751,394	1,881,018	518,103
1900–1910	1.64 %	1.70 %	2.99 %	7.97 %
1910–1920	1.05 %	1.41 %	2.19 %	1.75 %
1920–1930	0.05 %	0.72 %	2.47 %	1.44 %
1930–1940	0.72 %	0.86 %	0.30 %	1.06 %
1940–1950	0.98 %	0.66 %	1.50 %	3.14 %
1950–1960	1.35 %	1.35 %	2.27 %	1.83 %
1960–1970	1.52 %	1.08 %	1.67 %	1.78 %
1970–1980	1.74 %	0.69 %	0.27 %	1.92 %
1980–1990	1.70 %	0.71 %	0.48 %	1.64 %
1990–2000	2.34 %	1.17 %	0.85 %	1.92 %
2000–2010	1.68 %	0.75 %	0.44 %	1.32 %
2010	9,687,653	5,303,925	8,791,894	6,724,540

between 1900 and 1910, a period when the state was still receiving a large number of immigrants from Europe. Minnesota's growth rates are the most temporally stable of any state in the sample. The population of New Jersey grows from 1,879,890 in 1900 to 8,791,894 in 2010. New Jersey has the second highest population in 1900 and again in 2010. Its average annual growth rates range from a low of 0.27 % between 1970 and 1980 to a high of 2.99 % between 1900 and 1910. New Jersey's growth rates have slowed considerably since 1970. Washington is largely a frontier state in 1900 and has the smallest population (511,844) of any of the four states in the sample in that year. However, by 2010 it has grown to 6,724,540 which surpass the population of Minnesota. Its annual rates of population change vary the most of any state during the first half of the twentieth century. Between 1900 and 1910 it posts an annual growth rate of 7.97 %, the highest of any of the decennial growth rates in the sample. Washington's growth rate drops to 1.06 % between 1930 and 1940, then increases substantially to 3.14 % during the subsequent decade. Since 1950, Washington's growth rates have been relatively stable.

23.5 Evaluation Results

23.5.1 Total Population

Forecasts of the total population of Georgia produce that algebraic percent errors (PE) that range from -8.64 in 2000 to 9.67 in 1930 year and absolute percent errors (APE) that range from 0.16 in 1990 to 9.67 in 1930 (See Table 23.2). The mean of the PEs is -0.96 while the SD is 6.19 and the coefficient of variation (CV) is -6.48. The mean of the APEs is 5.13 while the SD is 3.18 and the CV is 0.62.

Table 23.2 Forecast errors for total population by state and horizon year[a]

Target year	Georgia		Minnesota		New Jersey		Washington		All states	
	Pe[b]	APE[c]	Pe[b]	APE[c]	Pe[b]	APE[c]	Pe[b]	APE[c]	Pe[b]	APE[c]
1920	5.38	5.38	3.80	3.80	8.82	8.82	88.68	88.68	26.67	26.67
1930	9.67	9.67	4.37	4.37	−5.20	5.20	−0.65	0.65	2.05	4.07
1940	−6.77	6.77	−3.33	3.33	24.00	24.00	2.78	2.78	4.17	9.22
1950	−3.86	3.86	0.83	0.83	−13.07	13.07	−20.30	20.30	−9.10	9.52
1960	−7.45	7.45	−10.46	10.46	−11.84	11.84	11.45	11.45	−4.58	10.30
1970	−3.52	3.52	1.02	1.02	4.16	4.16	0.33	0.33	0.50	2.26
1980	0.55	0.55	9.54	9.54	20.26	20.26	3.89	3.89	8.56	8.56
1990	−0.16	0.16	0.81	0.81	−1.68	1.68	2.71	2.71	0.42	1.34
2000	−8.64	8.64	−7.99	7.99	−6.04	6.04	−6.62	6.62	−7.32	7.32
2010	5.25	5.25	3.34	3.34	1.74	1.74	5.40	5.40	3.93	3.93
Mean	−0.96	5.13	0.19	4.55	2.12	9.68	8.77	14.26	2.53	8.41
Std. Dev.	6.19	3.18	5.98	3.58	12.55	7.64	29.33	26.82	10.09	7.15
CV[d]	−6.48	0.62	30.99	0.79	5.93	0.79	3.35	1.88	3.99	0.85

[a]Sum of the age group forecasts
[b]Algebraic percent error
[c]Absolute percent error
[d]Coefficient of variation (Std. Deviation/Mean)

Forecasts of the total population of Minnesota produce PEs that range from -10.46 in 1960 to 9.54 in 1980 and APEs that range from 0.81 in 1990 to 10.46 in 1960. The mean of the PEs is 0.19 while the SD is 5.98 and the CV is 30.99. The mean of the APEs is 4.55 while the SD is 3.58 and the CV is 0.79.

Forecasts of the total population of New Jersey produce PEs that range from -13.07 in 1950 to 24.00 in 1940 and APEs that range from 1.68 in 1990 to 24.00 in 1940. The mean of the PEs is 2.12 while the SD is 12.55 and the CV is 5.93. The mean of the APEs is 9.68 while the SD is 7.64 and the CV is 0.79.

Forecasts of the total population of Washington produce PEs that range from -20.30 in 1950 to 88.68 in 1920 and APEs that range from 0.33 in 1970 to 88.68 in 1920. The mean of the PEs is 8.77 while the SD is 29.33 and the CV is 3.35. The mean of the APEs is 14.26 while the SD is 26.82 and the CV is 1.88. Washington's PE and APE means, SDs, and CV are heavily influenced by the outlying error in the 1920 horizon year. For example, the mean PE drops from 8.77 to -0.11 and the mean APE drops from 14.26 to 6.01 when the 1920 forecast error is excluded.

Forecasts of the total population for all states combined produce PEs that range from -9.10 in 1950 to 26.67 in 1920 and APEs that range from 1.34 in 1990 to 26.67 in 1920. The mean of the PEs is 2.53 while the SD is 10.09 and the CV is 3.99. The mean of the APEs is 8.41 while the SD is 7.15 and the CV is 0.85. Removing the large 1920 error in Washington would lower the PE and APE by around 2.0 %age points to 0.46 and 6.34, respectively.

23.5.2 Age Groups

Table 23.3 shows forecast errors that represent the average of the PE for each age group (MALPE), the average of the APE for each age group (MAPE), and the percentage misallocation across all age groups (IOAE) by horizon year and state. The MALPEs for Georgia range from -8.39 in 1940 to 8.12 in 1930. The mean of the MALPEs is -1.47 while the SD is 5.57 and the CV is -3.78. The MAPEs range from 5.04 in 1920 to 11.28 in 1940. The mean of the MAPEs is 7.38 while the SD is 2.49 and the CV is 0.34. The IOAEs range from 1.80 in 1990 to 6.30 in 1980. The mean of the IOAEs is 3.31 while the SD is 1.34 and the CV is 0.41.

The MALPEs for Minnesota range from −7.57 in 2000 to 2.98 in 2010. The mean of the MALPEs is 0.00 while the SD is 4.86 and the CV is undefined since the mean is zero. The MAPEs range from a low of 2.01 in 1990 to a high of 7.57 in 2000. The mean of the MAPEs is 5.95 while the SD is 2.34 and the CV is 0.39. The IOAEs range from 1.19 in 1990 to 8.57 in 1980. The mean of the IOAEs is 3.47 while the SD is 2.35 and the CV is 0.68.

The MALPEs for New Jersey range from −11.55 in 1950 to 21.37 in 1940. The mean of the MALPEs is 1.92 while the SD is 11.39 and the CV is 5.94. The MAPEs range from a low of 4.14 in 1990 to a high of 21.37 in 1940. The mean of the MAPEs is 9.62 while the SD is 6.10 and the CV is 0.63. The IOAEs range from 1.45

Table 23.3 Forecast errors by horizon year and state[a]

	Georgia			Minnesota			New Jersey			Washington		
Horizon year	MALPE[b]	MAPE[c]	IOAE[d]	MALPE[b]	MAPE[c]	IOAE[d]	MALPE[b]	MAPE[c]	IOAE[d]	MALPE[b]	MAPE[c]	IOAE[d]
1920	3.65	5.04	2.27	2.42	4.78	3.00	7.30	7.30	2.43	77.15	77.15	7.03
1930	8.12	8.22	3.11	2.74	4.46	2.46	−6.06	8.25	2.46	−0.76	4.96	2.77
1940	−8.39	11.28	4.29	−3.60	6.35	2.40	21.37	21.37	5.88	2.15	5.59	3.23
1950	−4.13	6.16	3.50	2.06	6.44	3.91	−11.55	11.55	4.04	−16.49	16.49	6.09
1960	−6.36	7.09	3.84	−7.42	7.80	6.13	−9.28	9.28	6.27	11.43	14.86	5.85
1970	−3.41	8.29	3.59	0.11	6.30	4.16	3.56	6.40	3.74	−0.19	7.39	4.37
1980	−0.75	10.86	6.30	7.78	10.20	8.57	19.62	18.97	8.47	2.50	11.38	7.19
1990	0.19	3.29	1.80	0.50	2.01	1.19	−1.36	4.14	1.82	1.93	3.96	1.61
2000	−7.94	7.94	2.39	−7.57	7.57	1.62	−6.01	6.01	2.63	−5.93	5.93	2.12
2010	4.29	5.64	2.01	2.98	3.58	1.28	1.59	2.95	1.45	4.76	5.65	1.63
Mean	−1.47	7.38	3.31	0.00	5.95	3.47	1.92	9.62	3.92	7.66	15.34	4.19
Std. Dev.	5.57	2.49	1.34	4.86	2.34	2.35	11.39	6.10	2.28	25.48	22.15	2.21
CV[e]	−3.78	0.34	0.41	N/A	0.39	0.68	5.94	0.63	0.58	3.33	1.44	0.53

[a]Summary of the percentage errors for each age group
[b]Mean algebraic percent error
[c]Mean absolute percent error
[d]Index of allocation error
[e]Coefficient of variation (Std. Deviation/Mean)

in 2010 to 8.47 in 1980. The mean of the IOAEs is 3.92 while the SD is 2.28 and the CV is 0.58.

The MALPEs for Washington range from −16.49 in 1950 to 77.15 in 1920. The mean of the MALPEs is 7.66 while the SD is 25.48 and the CV is 3.33. The MAPEs range from a low of 3.96 in 1990 to a high of 77.15 in 1920. The mean of the MAPEs is 15.34 while the SD is 22.15 and the CV is 1.44. The IOAEs range from 1.61 in 1990 to 7.19 in 1980. The mean of the IOAEs is 4.19 while the SD is 2.21 and the CV is 0.53. Washington's summary measures of error for the MALPE and MAPE are heavily impacted by the 1920 average errors across age groups, but the IOAE is much less affected. The average MALPE drops from 7.66 to −0.77 and the average MAPE drops from 15.34 to 8.47 when the 1920 error is excluded. However, the average IOAE only decreases to 3.78; a drop of around 7 %.

Table 23.4 shows summary measures of forecast error for the PEs and APEs for each age group and the misallocation across age groups for all states combined. Beginning with the PEs, we see the MALPE ranges from -7.53 in 1950 to 22.63 in 1920. The mean of the MALPEs is 2.03 and its SD and CV are 8.55 and 4.21, respectively. The SD ranges from 1.04 in 2000 to 36.41 in 1920. The mean of the SDs is 8.88 and its SD and CV are 10.51 and 1.18, respectively. The CVs range from −3.31 in 1960 to 162.79 in 1970. The mean of the CVs is 17.62 and its SD and CV are 51.8 and 2.90, respectively. Obviously, the summary measures for the CV are pulled upward by its large value for 1970. If this value is removed, the mean, SD and CV drop to 1.49, 2.96, and 1.99, respectively. Removing the large 1920 error in Washington would lower the 1920 MALPE from 22.63 to 4.46 and the mean of the MALPEs from 2.03 to 0.21.

Turning to the APEs, the MAPEs range from 3.35 in 1990 to 23.57 in 1920. The mean of the MAPEs is 9.32 and its SD and CV are 5.64 and 0.61, respectively. The SD ranges from 0.94 in 1970 to 35.74 in 1920. The mean of the SDs is 6.56 and its SD and CV are 10.57 and 1.61, respectively. The CV ranges from 0.13 in 1970 to 1.52 in 1920. The mean of the CVs is 0.50 and its SD and CV are 0.41 and 0.82, respectively. Removing the large 1920 error in Washington would lower the 1920 MAPE from 23.57 to 5.71 and the mean of the MAPEs from 9.32 to 7.53.

Finally turning to the IOAEs, the mean of the IOAEs ranges from 1.61 in 1990 to 7.63 1980. The mean of the means of the IOAE is 3.72 and its SD and CV are 1.87 and 0.50, respectively. The SD ranges from 0.29 in 1990 to 2.25 in 1920. The mean of the SDs of the IOAE is 0.88 and its SD and CV are 0.66 and 0.75, respectively. The CV ranges from 0.09 in 1970 to 0.61 in 1920. The mean of the CVs of the IOAE is 0.24 and its SD and CV are 0.15 and 0.63, respectively. The large 1920 error in Washington does not affect the IOAE nearly as much as the PEs and APEs. The mean IOAE in 1920 drops from 3.68 to 2.57 and the mean of the means of the IOAE drops from 3.72 to 3.61 when this observation is removed.

Table 23.5 shows MAPE and MALPE for each age group for all states and horizon years combined. These data indicate a positive bias for age groups younger than 55 a negative bias for age groups 55 and older. Precision tends to improve from the youngest to the oldest age groups. These generalizations are useful but there are more subtle trends, especially in regard to precision. For the two youngest age

Table 23.4 Forecast errors by horizon year, all states[a]

Target year	Algebraic percent error			Absolute percent error			Index of allocation error		
	MALPE	Std. Dev.	CV[b]	MAPE	Std. Dev.	CV[b]	Mean	Std. Dev.	CV[b]
1920	22.63	36.41	1.61	23.57	35.74	1.52	3.68	2.25	0.61
1930	1.01	5.96	5.90	6.47	2.05	0.32	2.70	0.31	0.11
1940	2.88	13.06	4.53	11.15	7.27	0.65	3.95	1.50	0.38
1950	−7.53	8.16	−1.08	10.16	4.89	0.48	4.39	1.16	0.26
1960	−2.91	9.63	−3.31	9.76	3.52	0.36	5.52	1.14	0.21
1970	0.02	2.85	162.79	7.10	0.94	0.13	3.97	0.36	0.09
1980	7.29	8.94	1.23	10.33	7.76	0.75	7.63	1.09	0.14
1990	0.32	1.35	4.28	3.35	0.97	0.29	1.61	0.29	0.18
2000	−6.86	1.04	−0.15	6.86	1.04	0.15	2.19	0.43	0.20
2010	3.41	1.43	0.42	4.46	1.40	0.31	1.59	0.31	0.20
Mean	2.03	8.88	17.62	9.32	6.56	0.50	3.72	0.88	0.24
Std. Dev.	8.55	10.51	51.08	5.64	10.57	0.41	1.87	0.66	0.15
CV[b]	4.21	1.18	2.90	0.61	1.61	0.82	0.50	0.75	0.63

[a]Summary of the percentage errors for each age group
[b]Coefficient of variation (Std. Dev./MALPE, MAPE or Mean)

groups, 0–4 and 5–9, the MAPEs are very high at 23.61 and 22.61, respectively. Precision increases for the next two age groups, 10–14 and 15–19, with MAPEs of 6.85 and 7.32, respectively. Precision then decreases for the next two age groups 20–24 and 25–29, which have MAPEs of 11.34 and 13.05, respectively. Precision then improves, going from a MAPE of 11.79 for age group 30–34 to a MAPE of 4.85 for age group 50–54. The MAPEs tend to stabilize around 5.00 % for all remaining age groups.

23.6 Discussion

Recalling the standards described earlier, the MALPEs for the age forecasts indicated low levels of bias in that they were less than +/-5 % across all states and horizon years. The MAPEs for the age forecasts indicated a moderate degree of precision of between around five percent and less than 10 percent across all states and horizon years. While there was a great deal of variation by state and horizon year in the MALPE, MAPE, and IOAE associated with the age group forecasts, we find that on the whole the Hamilton-Perry method produced forecasts with relatively low bias, moderate to high levels of precision, and low allocation errors for the 10-year forecast horizons. This analysis also shows the potential utility of using the Hamilton Perry method to forecast the total population for 10-year forecast horizons. The average levels of bias and precision for all states and horizon years were 2.53 % and 8.41 %, respectively, and were around 2 percentage points lower

Table 23.5 Forecast errors by age group, all states and horizon years

Age	MALPE[a]	MAPE[b]
0–4	6.35	23.61
5–9	5.66	22.61
10–14	1.09	6.85
15–19	1.67	7.32
20–24	4.73	11.34
25–29	4.64	13.05
30–34	2.25	11.79
35–39	1.92	8.97
40–44	1.19	7.23
45–49	0.15	6.15
50–54	0.81	4.85
55–59	−0.26	4.92
60–64	−0.74	4.80
65–69	−0.24	5.07
70–74	−1.84	5.39
75+	−1.86	5.06

[a]Mean algebraic percent error
[b]Mean absolute percent error

when the large 1920 error in Washington was omited.[5] Considering the variability in the bias and precision of the total population forecasts across states and horizon years, the Hamilton-Perry method also produced total population forecasts of with relatively low levels of bias and moderate precision.

In terms of the results by age group, it should not be surprising that the Hamilton-Perry method was better able to forecast older age groups, on average, than the very youngest. Births have changed dramatically, for example, at different points in time (e.g. the depression era "baby bust" and the subsequent "baby boom") and migration likely comes into play in that errors generally decreased when moving from age groups 0–4 and 5–9 to age groups 10–14 and 15–19, where migration rates tend to be lowest. Errors became worse moving from age groups 10–14 and 15–19 to age groups 20–24 and 25–29 and 30–34, where migration rates tend to be highest. To some extent, migration also affects the forecast accuracy of the youngest two age groups (0–4 and 5–9) since they would move with their parents who are likely to be in age groups with high migration. Similar results have been seen in other studies of population forecast errors by age (Smith and Tayman 2003). We suggest these effects are consistent with theory that postulates that migrants tend to be less socially integrated into communities than those who tend not to move and that community social integration tends to increase with age (Goldscheider 1978).

In considering these results, two horizon years (1940 and 1950) with generally larger errors than the other horizon years tested were directly related to major events that significantly affected demographic behaviors. The 1940 point encompassed the economic boom experienced in the 1920s and the economic depression during the 1930s and the large scale "baby bust" associated with it. The 1950 point encompassed the depression and baby bust period of the 1930s and the economic recovery stimulated by World War II and the initial part of the large scale "baby boom" from 1946 to 1950. These points, especially the latter one, are well-known in terms of being "unexpected" events. In terms of population forecasting, the task of capturing demographic change associated with them is very difficult.

At this stage in our understanding, we suggest caution in using the Hamilton-Perry method beyond a 10-year forecast horizon. This is consistent with observations about the use of the Hamilton-Perry method in general (Swanson and Tayman 2014; Swanson et al. 2010 and findings by Smith and Tayman (2003).[4] As such, this caution is not a major limitation. We also suggest that the Hamilton-Perry method be used with care when applied to small populations, such as those found at the county and sub-county levels. While our sample of states provided a wide range of demographic characteristics in terms of size, age composition, and population changes, it does not account for the high variability in demographic characteristics found at sub-state levels (Baker et al. 2011; Swanson et al. 2010). We further suggest that research using the Hamilton-Perry method should examine state level accuracy over forecast horizons longer than 10 years and the forecast accuracy of smaller populations over a variety of forecast horizons and geographic stratifications.

Appendices

Appendix A.1: Hamilton Perry Method and the Fundamental Demographic Equation

In this Appendix we demonstrate that the Hamilton-Perry method satisfies the fundamental demographic equation. We begin by restating the fundamental demographic equation as follows:

$$P_{i,t+k} = P_{i,t} + B_i - D_i + I_i - O_i \tag{A.1.1}$$

where,

$P_{i,t}$ = Population of area i at time t (the launch year),

[4] While we caution against using the Hamilton-Perry method for forecast horizons longer than 10-years, we note that Smith and Tayman (2003) found that the Hamilton-Perry method produces levels of accuracy comparable to the cohort-component method over a twenty year horizon and Swanson (2016) found it useful over a long-term "reverse" projection used to reconstruct the population of Native and Part-Hawaiians in Hawaii (see also Chapter 25 in this book). Thus, we find that the method can be used over longer horizons, given the context and the population in question.

$P_{i,t+k}$ = Population of area i at time $t+k$ (the forecast year),
B_i = Births in area i between time t and $t+k$,
D_i = Deaths in area i between time t and $t+k$,
I_i = In-migrants in area i between time t and $t+k$, and
O_i = Out-migrants in area i between time t and $t+$

then,

$$_{n}CCR_{x,i,t} = ({}_{n}P_{x-k,i,t-k} + B_i - D_i + I_i - O_i)/{}_{n}P_{x-k,i,t-k} \tag{A.1.2}$$

Expressing Eq. (A.1.2) in terms of Eq. (A.1.1):

$$_{n}P_{x+k,i,t+k} = \left(({}_{n}P_{x-k,i,t-k} + B_i - D_i + I_i - O_i)/({}_{n}P_{x-k,i,t-k})\right) \times ({}_{n}P_{x,i,t}) \tag{A1.3}$$

where $x+k>=10$, then,

$$_{n}CCR_{x,i,t} = ({}_{n}P_{x-k,i,t-k} - D_i + I_i - O_i)/{}_{n}P_{x-k,i,t-k}, \text{ and}$$

since $N_i = I_i - Oi$, where $x+k \geq 10$, we have

$$_{n}CCR_{x,i,t} = ({}_{n}P_{x-k,i,t-k} - D_i + N_i)/{}_{n}P_{x-k,i,t-k}. \tag{A.1.4}$$

Equations (A.1.2), (A1.3) and (A.1.4) show that the Hamilton-Perry method is not only consistent with the fundamental demographic equation, but also closely related to the cohort-component method. The Hamilton-Perry method simply expresses the individual components of change—births, deaths, and migration—in terms of CCRs (Cohort Change Ratios). An important reason for a demographic forecasting method to be consistent with the fundamental demographic equation is to minimize the potential errors associated with hidden heterogeneity (Vaupel and Yashin 1985).

Appendix A.2: Population by Age, 1900 to 2010 by Decade (Tables A.2.1, A.2.2, A.2.3, and A.2.4)

Table A.2.1 Population by age, 1900 to 2010, Georgia

Age	1900	1910	1920	1930	1940	1950	1960	1970	1980	1990	2000	2010
0–4	325,473	376,641	363,229	316,404	313,122	422,486	471,901	421,709	414,935	495,535	595,150	686,785
5–9	313,524	347,369	382,373	353,910	319,056	355,208	440,198	470,311	446,831	483,952	615,584	695,161
10–14	277,865	315,217	365,312	338,860	325,009	311,293	411,650	480,924	469,598	466,614	607,759	689,684
15–19	241,478	280,383	307,549	334,836	328,410	291,806	331,554	442,571	530,773	497,152	596,277	709,999
20–24	229,199	260,140	272,814	288,126	304,638	276,193	271,211	416,949	516,084	522,634	592,196	680,080
25–29	172,819	214,250	230,373	222,930	277,500	276,270	251,770	330,790	481,276	589,952	641,750	673,935
30–34	127,782	169,314	180,749	183,399	236,138	255,385	256,351	273,995	448,765	584,944	657,506	661,625
35–39	111,711	152,232	185,500	186,959	209,545	254,264	260,063	256,934	356,263	531,619	698,735	698,059
40–44	97,256	109,644	140,477	151,156	174,120	219,640	244,981	260,140	291,069	484,079	654,773	699,481
45–49	78,565	85,850	125,849	133,154	156,489	182,855	229,397	252,278	266,793	374,918	573,017	722,661
50–54	78,307	96,240	106,175	131,455	134,244	153,118	196,204	232,825	261,211	294,033	506,975	668,591
55–59	46,756	61,442	66,256	84,633	102,773	126,309	161,507	207,126	246,907	259,735	375,651	573,551
60–64	42,863	55,526	64,125	67,562	83,965	100,096	125,668	175,565	215,869	238,779	285,805	496,006
65–69	27,942	35,469	44,269	45,142	75,095	95,556	113,144	137,744	188,897	218,078	236,634	356,007
70–74	18,887	21,911	29,550	33,738	42,732	60,606	81,647	97,362	141,977	169,973	199,061	250,422
75+	19,547	23,349	28,292	34,398	40,887	63,493	95,870	132,352	185,857	266,219	349,580	425,606
Age not reported	6357	4144	2940	1844	0	0	0	0	0	0	0	0
Total	**2,216,331**	**2,609,121**	**2,895,832**	**2,908,506**	**3,123,723**	**3,444,578**	**3,943,116**	**4,589,575**	**5,463,105**	**6,478,216**	**8,186,453**	**9,687,653**

Sources: U.S. Census Bureau, Table QT-P1, 2000 and 2010 census
U.S. Census Bureau, Table 19, General Population Characteristics, 1990 census
U.S. Census Bureau, Table 19, General Population Characteristics, 1980 census
U.S. Census Bureau, Characteristics of the Population (Georgia, Vol. 1, Part 12), March 1973 (years 1900 through 1970)

Table A.2.2 Population by age, 1900 to 2010, Minnesota

Age	1900	1910	1920	1930	1940	1950	1960	1970	1980	1990	2000	2010
0–4	228,290	226,840	261,394	231,001	230,057	332,460	416,005	331,771	307,249	336,800	329,594	355,504
5–9	217,447	220,233	248,599	256,751	220,176	267,652	380,650	402,635	296,295	345,840	355,894	355,536
10–14	192,064	214,402	233,961	253,788	238,918	223,787	324,710	415,021	333,378	313,297	374,995	352,342
15–19	170,177	215,148	219,609	239,946	257,349	207,460	251,352	373,405	399,818	297,609	374,362	367,829
20–24	160,674	216,670	217,919	214,432	245,592	213,712	194,883	292,037	393,566	316,046	322,483	355,651
25–29	148,607	187,438	213,646	193,469	225,097	220,780	193,160	249,516	363,435	381,759	319,826	372,686
30–34	131,055	153,195	189,778	189,705	204,311	212,765	206,487	206,769	313,104	397,984	353,312	342,900
35–39	121,193	135,612	168,540	192,934	192,452	205,447	211,163	192,863	246,356	361,274	412,490	328,190
40–44	100,646	117,256	135,353	172,980	187,196	189,729	204,868	202,710	202,860	304,810	411,692	352,904
45–49	72,042	105,289	122,435	147,143	182,525	176,212	194,149	202,904	187,051	237,050	364,247	406,203
50–54	57,896	88,110	105,208	122,171	162,931	170,805	176,190	193,956	193,199	191,410	301,449	401,695
55–59	45,293	59,272	87,437	100,813	129,941	157,690	159,840	177,011	189,457	173,066	226,857	349,589
60–64	35,137	45,188	69,827	84,372	103,137	134,854	146,056	155,454	170,638	171,220	178,012	279,775
65–69	28,251	34,825	45,827	69,079	82,635	105,188	131,315	130,155	149,114	160,036	153,169	202,570
70–74	19,424	23,536	30,188	48,256	60,455	73,705	102,086	110,251	121,034	134,486	142,656	151,857
75+	19,096	27,696	34,751	46,145	69,528	90,237	120,950	168,513	209,416	252,412	298,441	328,694
Age not reported	4102	4998	2653	968	0	0	0	0	0	0	0	0
Total	**1,751,394**	**2,075,708**	**2,387,125**	**2,563,953**	**2,792,300**	**2,982,483**	**3,413,864**	**3,804,971**	**4,075,970**	**4,375,099**	**4,919,479**	**5,303,925**

Sources: U.S. Census Bureau, Table QT-P1, 2000 and 2010 census
U.S. Census Bureau, Table 19, General Population Characteristics, 1990 census
U.S. Census Bureau, Table 19, General Population Characteristics, 1980 census
U.S. Census Bureau, Characteristics of the Population (Minnesota, Vol. 1, Part 23), March 1973 (years 1900 through 1970)

Table A.2.3 Population by age, 1900 to 2010, New Jersey

Age	1900	1910	1920	1930	1940	1950	1960	1970	1980	1990	2000	2010
0–4	206,446	266,942	338,696	329,668	256,264	458,906	642,197	589,226	463,289	532,637	563,785	647,731
5–9	196,725	242,279	322,958	380,918	280,722	371,826	582,212	692,648	508,447	493,044	604,529	671,855
10–14	174,347	228,695	291,236	384,342	337,776	290,544	524,380	710,409	605,841	480,983	590,577	685,713
15–19	166,746	236,541	255,161	364,396	375,112	295,859	396,363	611,831	670,665	505,388	525,216	652,864
20–24	178,228	250,613	271,042	350,402	376,912	350,403	321,054	509,198	614,828	566,594	480,079	609,920
25–29	176,408	236,172	286,617	332,810	361,291	409,890	362,373	463,164	574,135	668,917	544,917	657,765
30–34	158,858	213,082	263,733	331,332	340,976	409,434	435,080	403,475	563,758	691,734	644,123	703,929
35–39	144,124	199,647	251,252	338,222	322,760	393,917	472,429	413,929	479,749	622,963	727,924	702,384
40–44	117,887	166,638	207,122	291,871	315,720	357,760	446,139	465,492	400,074	573,696	707,182	675,301
45–49	92,115	136,295	185,551	246,388	297,595	318,504	406,721	477,978	394,038	466,481	611,357	609,260
50–54	78,915	112,003	151,688	205,434	259,570	305,235	350,531	439,103	432,520	376,528	547,541	558,208
55–59	60,248	75,739	108,505	157,128	198,622	263,516	304,112	380,677	430,048	355,677	423,338	493,551
60–64	49,226	62,678	86,297	124,676	158,024	215,546	262,777	314,045	367,660	363,521	330,646	427,084
65–69	33,955	45,948	56,135	88,449	119,172	164,921	222,457	245,757	303,670	340,232	293,196	372,200
70–74	23,186	31,193	38,149	58,951	80,239	109,441	163,149	194,112	227,037	269,960	281,473	306,975
75+	22,476	29,946	39,197	53,643	79,410	119,627	174,808	257,120	329,064	421,833	538,467	488,842
Age not reported	1128	662	792	244	0	0	0	0	0	0	0	0
Total	**1,881,018**	**2,535,073**	**3,154,131**	**4,038,874**	**4,160,165**	**4,835,329**	**6,066,782**	**7,168,164**	**7,364,823**	**7,730,188**	**8,414,350**	**9,263,582**

Sources: U.S. Census Bureau, Table QT-P1, 2000 and 2010 census
U.S. Census Bureau, Table 19, General Population Characteristics, 1990 census
U.S. Census Bureau, Table 19, General Population Characteristics, 1980 census
U.S. Census Bureau, Characteristics of the Population (New Jersey, Vol. 1, Part 32), March 1973 (years 1900 through 1970)

Table A.2.4 Population by age, 1900 to 2010, Washington

Age	1900	1910	1920	1930	1940	1950	1960	1970	1980	1990	2000	2010
0–4	53,243	108,756	126,434	114,854	121,918	263,326	315,633	280,442	306,123	366,780	394,306	439,657
5–9	56,423	99,678	128,258	136,013	116,762	203,786	301,051	328,397	296,011	371,093	425,909	429,877
10–14	48,233	92,802	117,553	138,393	127,842	159,695	275,510	348,892	321,995	337,662	434,836	438,233
15–19	44,104	99,647	106,485	137,922	146,725	157,695	208,575	329,903	369,023	322,711	427,968	462,128
20–24	46,403	122,058	111,014	130,401	148,867	175,619	173,804	295,964	400,542	351,680	390,185	461,512
25–29	46,093	126,074	120,421	120,651	146,594	195,087	166,376	238,704	389,997	411,822	403,652	480,398
30–34	47,118	106,963	119,446	115,448	134,757	188,636	179,899	193,398	354,645	443,366	437,478	453,383
35–39	46,368	90,149	117,587	122,833	124,990	180,749	198,495	181,020	273,382	427,690	483,950	448,607
40–44	37,863	77,286	95,805	118,105	118,525	159,090	189,191	192,828	213,832	376,073	491,137	459,698
45–49	26,027	64,992	81,764	108,280	117,709	136,714	176,071	203,880	193,473	284,674	454,223	492,909
50–54	20,754	52,413	69,451	90,223	112,915	125,939	150,495	188,774	198,548	216,869	391,749	495,296
55–59	14,127	33,661	55,053	69,260	96,698	115,306	129,003	166,878	203,986	191,602	285,505	453,078
60–64	10,407	24,144	42,352	57,530	77,569	103,916	110,066	138,028	179,037	189,382	211,075	382,087
65–69	7195	16,585	27,298	44,440	57,963	86,551	98,659	107,008	151,324	186,679	176,225	270,474
70–74	4161	10,374	16,647	30,075	41,943	59,655	80,938	84,335	112,023	149,355	160,941	186,746
75+	3325	9614	16,266	26,988	44,414	65,199	99,448	130,718	168,215	239,254	324,982	370,457
Age Not Reported	6259	6794	4787	1980	0	0	0	0	0	0	0	0
Total	**518,103**	**1,141,990**	**1,356,621**	**1,563,396**	**1,736,191**	**2,376,963**	**2,853,214**	**3,409,169**	**4,132,156**	**4,866,692**	**5,894,121**	**6,724,540**

Sources: U.S. Census Bureau, Table QT-P1, 2000 and 2010 census
U.S. Census Bureau, Table 19, General Population Characteristics, 1990 census
U.S. Census Bureau, Table 19, General Population Characteristics, 1980 census
U.S. Census Bureau, Characteristics of the Population (Washington, Vol. 1, Part 49), March 1973 (years 1900 through 1970)

References

Baker, J., Ruan, X., & Alcantara, A. (2011). *Incorporating spatial effects into Hamilton-Perry based projections improves accuracy and precision at the census tract level*. Paper presented at the annual meeting of the Population Association of America, Washington, DC.

Fonseca, L., & Tayman, J. (1989). Postcensal estimates of household income distributions. *Demography, 26*(1), 149–159.

George, M. V., Swanson, D. A., & Tayman, J. (2004). Population projections. In J. Siegel & D. Swanson (Eds.), *The methods and materials of demography* (2nd ed., pp. 561–601). New York: Elsevier Academic Press.

Goldscheider, C. (1978). *Modernization, migration, and urbanization*. Paris: International Union for the Scientific Study of Population.

Hamilton, C. H., & Perry, J. (1962). A short method for projecting population by age from one decennial census to another. *Social Forces, 41*, 163–170.

Hardy, G. F., & Wyatt, F. B. (1911). Report of the actuaries in relation to the scheme of insurance against sickness, disablement, etc. *Journal of the Institute of Actuaries, XLV*, 406–443.

Land, K. (1986). Methods for national population forecasts: A review. *Journal of the American Statistical Association, 81*, 888–901.

Murdock, S., & Ellis, D. (1991). *Applied demography: An introduction to basic concepts, methods, and data*. Denver: Westview Press.

Pittenger, D. (1976). *Projecting state and local populations*. Cambridge, MA: Ballinger.

Rayer, S. (2007). Population forecast accuracy: Does the choice of summary measure of error matter? *Population Research and Policy Review, 26*, 163–184.

Smith, S. K., & Tayman, J. (2003). An evaluation of population projections by age. *Demography, 40*(4), 741–757.

Smith, S. K., Tayman, J., & Swanson, D. A. (2013). *A practitioner's guide to state and local population projections*. Dordrecht: Springer.

Swanson, D. A. (2015). On the relationship among values of the same summary measure of error when it is used across multiple characteristics at the same point in time: An examination of MALPE and MAPE. *Review of Economics and Finance, 5*(1), 1–14.

Swanson, D. A. (2016, June 1–3). *The number of native Hawaiians and part-Hawaiians in Hawai'i, 1778 to 1900: Demographic estimates by age, with discussion*. Paper presented at the annual conference of the Canadian Population Association, Calgary, Alberta.

Swanson, D. A., & Beck, D. (1994). A new short-term county population projection method. *Journal of Economic and Social Measurement, 20*, 1–26.

Swanson, D. A. & Tayman, J. (2014). Measuring uncertainty in population forecasts: A new approach. In M. Marsili, & G. Capacci (Eds.), *Proceedings of the 6th EUROSTAT/UNECE work session on demographic projections* (pp. 203–215). Rome: National Institute of Statistics.

Swanson, D. A., & Tedrow, L. (2012). Using cohort change ratios to estimate life expectancy in populations with negligible migration: A new approach. *Canadian Studies in Population, 39*, 83–90.

Swanson, D. A., Schlottmann, A., & Schmidt, R. (2010). Forecasting the population of census tracts by age and sex: An example of the Hamilton–Perry method in action. *Population Research and Policy Review, 29*, 47–63.

Swanson, D. A., Tayman, J., & Bryan, T. (2011). MAPE-R: A rescaled measure of accuracy for cross-sectional, sub-national forecasts. *Journal of Population Research, 28*, 225–243.

Swanson, D. A., Tayman, J., McKibben, J. & Cropper, M. (2012). *A "blind" ex post facto evaluation of total population and total household forecast for small areas made by five vendors for 2010: Results by geography and error criteria*. Presented at the 2012 conference of the Canadian Population Society, Waterloo, Ontario, Canada.

Vaupel, J., & Yashin, A. (1985, August). Heterogeneity's ruses: Some surprising effects of selection on population dynamics. *The American Statistician, 39*, 176–185.

Wolfenden, H. H. (1954). *Population statistics and their compilation* (Rev. ed.). Chicago: University of Chicago Press.

Index

D.A. Swanson (ed.), *The Frontiers of Applied Demography*, Applied Demography Series 9, DOI 10.1007/978-3-319-43329-5

Kolibri 360 GmbH
Ferdinand-Jühlke-Straße 7
99095 Erfurt, Deutschland
produktsicherheit@kolibri360.de